P9-DGH-794

							VIIIA
							2 **He** Helium 4.00260

		IIIA	IVA	VA	VIA	VIIA	
		5 **B** Boron 10.81	6 **C** Carbon 12.011	7 **N** Nitrogen 14.0067	8 **O** Oxygen 15.9994	9 **F** Fluorine 18.998403	10 **Ne** Neon 20.179

IB	IIB	13 **Al** Aluminum 26.98154	14 **Si** Silicon 28.0855	15 **P** Phosphorus 30.97376	16 **S** Sulfur 32.06	17 **Cl** Chlorine 35.453	18 **Ar** Argon 39.948
29 **Cu** Copper 63.546	30 **Zn** Zinc 65.38	31 **Ga** Gallium 69.72	32 **Ge** Germanium 72.59	33 **As** Arsenic 74.9216	34 **Se** Selenium 78.96	35 **Br** Bromine 79.904	36 **Kr** Krypton 83.80
47 **Ag** Silver 107.868	48 **Cd** Cadmium 112.41	49 **In** Indium 114.82	50 **Sn** Tin 118.69	51 **Sb** Antimony 121.75	52 **Te** Tellurium 127.60	53 **I** Iodine 126.9045	54 **Xe** Xenon 131.30
79 **Au** Gold 196.9665	80 **Hg** Mercury 200.59	81 **Tl** Thallium 204.37	82 **Pb** Lead 207.2	83 **Bi** Bismuth 208.9804	84 **Po** Polonium (209)	85 **At** Astatine (210)	86 **Rn** Radon (222)

Left-edge partial cells: **Ni** Nickel 8.70 (28); **Pd** Palladium 06.4 (46); **Pt** Platinum 95.09 (78)

65 **Tb** Terbium 158.9254	66 **Dy** Dysprosium 162.50	67 **Ho** Holmium 164.9304	68 **Er** Erbium 167.26	69 **Tm** Thulium 168.9342	70 **Yb** Ytterbium 173.04	71 **Lu** Lutetium 174.97
97 **Bk** Berkelium (247)	98 **Cf** Californium (251)	99 **Es** Einsteinium (254)	100 **Fm** Fermium (257)	101 **Md** Mendelevium (258)	102 **No** Nobelium (255)	103 **Lr** Lawrencium (260)

Left-edge partial cells: **Gd** Gadolinium 57.25 (64); **Cm** Curium 47 (96)

metals

nonmetals

metalloids

*Name not officially assigned

Values of atomic mass in parentheses are estimated and represent, in most cases, the mass number of the most stable isotope.

FUNDAMENTALS OF
ORGANIC AND
BIOLOGICAL CHEMISTRY

131

Fundamentals of Organic and Biological Chemistry

DENIS M. CALLEWAERT

JULIEN GENYEA

DEPARTMENT OF CHEMISTRY

OAKLAND UNIVERSITY

WORTH PUBLISHERS, INC.

Dedicated to Our Families

Worth Publishers, Inc.

444 Park Avenue South

New York, New York 10016

PREFACE

Fundamentals of Organic and Biological Chemistry and its companion volume, *Fundamentals of College Chemistry*, are a result of many years of experience teaching a two-semester sequence of courses in general, organic, and biological chemistry for students who have not necessarily had any prior exposure to chemistry. With these books, or with the combined volume that contains them both, students in nursing, allied health, agriculture, home economics, and similar programs — where the study of chemistry is limited to two courses — can acquire a good understanding of general chemical concepts and basic biochemistry. *Fundamentals of Organic and Biological Chemistry* is well suited for use in any short course in organic and biological chemistry for students who have completed a college course in general chemistry. *Fundamentals of College Chemistry* is ideally suited as a preceding text because of its integration of organic and biological material into the discussion of fundamental chemical concepts. However, any sound introductory course in basic chemistry should provide an adequate preparation for this textbook.

Anyone teaching an introductory science course must contend with a common dilemma — how to treat material at an elementary level without introducing drastic oversimplifications that might result in misinterpretations. We have attempted to explain concepts in sufficient qualitative detail to avoid such oversimplification. Based on our classroom experience, we have tried to provide especially careful explanations of difficult topics and to anticipate the most common student misconceptions. Students in our classes have found that the approach used here stimulates them intellectually without overtaxing their mathematical skills.

Both *Fundamentals of Organic and Biological Chemistry* and *Fundamentals of College Chemistry* contain the essential features of a student study guide as an integral part of the textbook. Chapters begin with an introduction that puts the material in the chapter into perspective, followed by a list of study objectives. These two brief sections are designed to help students focus on what they are expected to learn and why it is important. At the end of nearly every section in a chapter there are exercises designed to test whether the student has grasped the basic points covered in that section. Students are strongly encouraged to work these exercises before going on. Detailed solutions to these exercises are presented at the end of the chapter so that students can test their comprehension. Each chapter ends with a concise summary list of the important concepts and relationships discussed in the chapter, followed by a set of problems. Answers to all of these problems are provided at the back of the book.

Both texts also include self-contained units called Essential Skills. Wherever an essential skill is needed in a chapter, students are reminded to turn to the appropriate Essential Skills unit in order to increase their facility with that skill.

The indexes for each book use a boldface number to identify the page on which a term is defined. Thus, students can easily find the definitions of terms in their correct context.

For the sake of flexibility, a few sections in each text are clearly indicated as optional. These optional sections have been written so that they may be omitted without affecting the comprehension of subsequent topics.

Fundamentals of Organic and Biological Chemistry

The primary aim of this text is to help students who previously have completed at least one semester of introductory chemistry to acquire a basic understanding of fundamental biochemical processes. We have attempted to make the many illustrations simple enough to be readily understandable without erring in the direction of inaccurate oversimplification.

Throughout our discussions of organic molecules and biochemistry, we point out both the similarities and the differences between related reactions as they occur in the human body and as they are carried out under laboratory conditions. We have made every effort to keep our discussions as up-to-date as possible, emphasizing the dynamic nature of our knowledge of human biochemistry and the rapidly progressing medical applications of that knowledge.

Fundamentals of Organic and Biological Chemistry begins with a brief review of those basic and essential chemical concepts with which students should be familiar from their previous study of chemistry. The text is then divided into three parts: organic molecules, biomolecules, and metabolism. Each part begins with an overview that introduces the material covered in that group of chapters and places it in proper perspective.

ORGANIC MOLECULES

In these seven chapters, the physical and chemical properties of simple organic molecules are discussed, with special emphasis on those classes of compounds whose functional groups are important components of biomolecules. Our treatment of organic chemistry is carefully limited to those topics that are fundamental to the understanding of basic biochemistry. Biochemical examples are often used in these chapters to illustrate the various types of organic reactions.

BIOMOLECULES

These five chapters treat the structures, properties, and reactivities of the major classes of biomolecules (with the exception of lipids, which are described later). Carbohydrates are covered in Chapter 20, and then three chapters are devoted to the crucial topic of proteins. The first discusses amino acids and the general structural features of proteins. The second describes the many functional roles of proteins in the human body, with separate sections devoted to transport proteins and to antibody structure and function. The next chapter presents the structure-function relationship of enzymes. The topics of enzyme kinetics, inhibition, and allosteric regulation are developed in sufficient qualitative detail to give students an appreciation for the precision of metabolic regulation and the molecular basis of drug action. The functions of coenzymes and their relationship to vitamins are discussed, but detailed reaction mechanisms are not. The structural formulas for major coenzymes are presented at appropriate places in the text. These coenzymes, with their vitamin portions indicated, are also included in an appendix at the back of the book for reference. The final chapter in this unit treats nucleic acids and protein biosynthesis. This chapter has been written to give students a basic appreciation for human genetic processes, as opposed to the detailed molecular genetics of bacteria. Recent advances in our understanding of genetic organization, function, and regulation in higher organisms are included where appropriate.

METABOLISM

In our presentation of the major metabolic pathways, we first present an overview of the process. Only then do we take up a step-by-step description of the individual reactions involved. Finally, we provide a summary of these reactions. We also discuss the integration and regulation of metabolic pathways in some detail, concentrating on the underlying control mechanisms.

It is our experience that many students find the study of metabolic pathways a hopeless exercise in the memorization of complex charts. For this reason we believe it is essential to relate each step in a pathway to the reaction of one or a few functional groups, to point out similarities in reactions, and to keep a constant eye on the overall process and its utility to the cell. We feel that the unified discussion of the interrelated topics of carbohydrate metabolism and bioenergetics in Chapter 25 is preferable to fragmenting these topics into separate chapters.

The structure and function of lipids are discussed in Chapter 26. This topic is placed here in order to aid students in their subsequent study of lipid metabolism (Chapter 27), as well as to provide some relief at this point from the study of metabolic pathways. The interrelationships of the various metabolic pathways are discussed in the final chapter, together with human nutrition and body fluids, as these topics relate to metabolism. A discussion of the important topic of diabetes is included here in order to illustrate the necessity for proper metabolic regulation in the normal human body.

Acknowledgments

We have been fortunate to have the assistance of many capable people during the preparation of these books. We thank Robert Stern for his help in getting this project started and for his assistance with early drafts. Steven Miller has been a continual source of constructive criticism through various drafts, including the preliminary edition and galley proofs. We also appreciate the contributions of the following chemists who reviewed drafts of the manuscript and/or the preliminary edition; they provided us with a great many valuable suggestions.

Wasi Ahmed, State University of New York at Binghamton
P. Wayne Ayers, East Carolina University
David Bak, Hartwick College
Jack Dalton, Boise State University
Leslie N. Davis, Community College of the Finger Lakes
Mary Delton, Oakland University
H. Ed Fiehler, Miami University
David Johnson, Wayne State University
Paul Ketchum, Oakland University
Gordon Lillis, San Joaquin Delta College
Tom Mines, St. Louis Community College at Florissant Valley
Clarence R. Perisho, Mankato State University
Susan Poulter, University of Utah
Wilmer Reed, DeAnza College
Joan Reeder, Eastern Kentucky University

William Schulz, Eastern Kentucky University
Cynthia Sevilla, Oakland University
Michael Sevilla, Oakland University
Peter Sheridan, State University of New York at Binghamton
Harry M. Smiley, Eastern Kentucky University
Carol B. Swezey, St. Elizabeth Hospital Medical Center,
    Lafayette, Indiana
L. G. Wade, Jr., Colorado State University
David Warr, Bristol Community College
Albert Zabady, Montclair State College

We thank the many students at Oakland University who used the preliminary editions of these texts, and who provided us with vitally important student feedback. We are indebted also to Geraldine Felton for her helpful advice.

We thank Susan Forgette for her skillful and rapid typing of many drafts, the preliminary edition, and the final manuscript. We are grateful to our families; they have been a source of strength for us. They have also contributed in more direct ways: We would like to thank Karen Callewaert for her help in typing initial drafts, and Carol Genyea for illustrating the preliminary edition.

Finally, we express our gratitude to Worth Publishers for their patience, guidance, and insistence upon the highest standards. In particular we wish to thank Ken Ekkens, whose attention to design and layout contributed so much to the "look" of the book, and Gordon Beckhorn, our editor, who served as a constant source of help and encouragement.

Denis M. Callewaert
Julien Genyea

Rochester, Michigan
March 1980

TO THE STUDENT

Fundamentals of Organic and Biological Chemistry is intended to make your study of chemistry an interesting and enjoyable experience. Whether or not we are successful depends to a large extent on you, because learning chemistry will require your active intellectual participation. To learn chemistry you cannot merely memorize a collection of facts or a rote method for solving problems. You need to acquire sufficient understanding of general principles to be able to apply them. There are several features of this book that will assist you in this.

Each chapter begins with a brief **introduction,** which provides an overview of the topics to be discussed in the chapter, their relevance to our daily lives, and how they fit into the overall picture of chemistry.

Following the introduction there is a list of **study objectives** setting forth what you are expected to know and what types of problems you are expected to solve after you have learned the material in the chapter. These study objectives can be useful in reviewing for tests. For example, if one study objective is to be able to define a term, as part of your review you should write down an appropriate definition without looking at the chapter. Similarly, if a particular study objective is to be able to solve a certain type of problem, then as part of your review you should actually work out a few problems of that type.

There are **exercises** at the end of almost every section of every chapter; *these are your most important learning aid.* After you have studied the material in a particular section, it is imperative that you test your comprehension by working out the exercises before going on to the next section. At the end of each chapter, you will find worked-out solutions to all the exercises in that chapter. If you have any difficulty with an exercise, you should spend more time studying the material in that section.

Also, at the end of each chapter is a list of **summary** statements highlighting the major points in the chapter. These statements can be used as a brief outline for reviewing the chapter.

Finally, there is a set of **problems** at the end of every chapter, and it is important that you work all of them. The only way to gain an adequate understanding of a chemical principle is to work a number of problems that require use of that principle. As many students have learned the hard way, it is not sufficient merely to follow the worked-out examples presented in the text or in a lecture. Working problems yourself is the key to success in the study of chemistry.

The **answers** to all problems are given at the back of the book. A word of caution: Do not be satisfied with just getting the correct answer to a particular problem. If you had difficulty solving the problem, or if you are unclear about the method you used, then you should consider the problem further and perhaps ask your instructor for help.

There are two units called **Essential Skills** at the back of the book. One of them, entitled "Naming Organic Compounds," is a list of the rules and components used in constructing the names of the organic compounds discussed in Chapters 13 through 18. This unit should prove helpful when you need to review the fundamentals of organic nomenclature. The other Essential Skills unit, "Predicting the Products of Organic Reactions," outlines a systematic approach to solving problems in which you are asked to draw structural formulas for the products of an organic reaction, given the structural formulas for the reactants.

To some extent, the study of chemistry is like learning a new language, in that you will encounter many new terms whose precise meanings you will be required to know. Since you will want on occasion to refresh your memory, the **index** at the back of the book uses boldface type to indicate the page on which each term is defined. Thus you can easily find the definition for every scientific term used in the text and the context in which it is used.

Chemistry is a fascinating subject that affects every aspect of our lives. Learning chemistry is not easy—it is a challenge. But it need not be a very difficult task. What is required is a moderate amount of time spent studying the basic principles of chemistry and working all the problems you are assigned. We believe that with a reasonable effort you will find the study of chemistry both interesting and rewarding.

CONTENTS

Occasional references are made throughout the text to its companion volume, *Fundamentals of College Chemistry*. For that reason an abbreviated version of its table of contents is included here.

FUNDAMENTALS OF COLLEGE CHEMISTRY

General Chemistry

REVIEW OF SOME BASIC
CONCEPTS OF CHEMISTRY

REVIEW OF SOME BASIC CONCEPTS OF CHEMISTRY

From your previous study of chemistry, you are already familiar with a number of basic concepts and principles. These fundamentals will be useful to you throughout this textbook. The following section reviews succinctly those basic concepts of chemistry that you will encounter most often. For a more extensive discussion of any particular topic, you should consult the book used in your previous chemistry course.

Chemical Bonds

The **chemical formula** for a substance indicates the relative number of atoms of each element in the substance. For a molecular substance (such as water or carbon dioxide) the chemical formula is called a **molecular formula,** and it indicates the number of atoms of each element in any molecule of that substance.

The forces that hold atoms close together in a molecule are called **chemical bonds.** The energy required to break a chemical bond is called the **bond energy,** which is a measure of the strength of the bond. The distance between the centers of two bonded atoms is called the **bond length.** Bond energies and bond lengths are determined experimentally. In general, a strong bond has a larger bond energy and a shorter bond length than a weak bond.

The **electronegativity** of an atom is a measure of its relative ability to attract the shared electrons in a bond. Nonmetals (such as N, O, and F) have large electronegativities, whereas metals (such as Na, Mg, and K) have small electronegativities. Those electrons that are primarily involved when an atom bonds to other atoms are called **valence electrons.** For the representative elements (groups IA through VIIIA in the periodic table), the number of valence electrons for an atom of an element is equal to the group number of that element.

Chemists use a variety of models to describe chemical bonds and use these models to classify bonds as ionic, polar covalent, or nonpolar covalent. An **ionic bond** between two atoms is described by a model in which one or more valence electrons are considered to be transferred completely from the atom with the smaller electronegativity to the atom with the much larger electronegativity. The resulting positive and negative ions are held together by the attractive electrical forces between these oppositely charged ions.

Chemists describe a **covalent bond** by a model in which two atoms share a pair of valence electrons. When the pair of valence electrons is shared equally (or nearly equally), the bond is classified as a **nonpolar covalent bond.** When the valence electrons are shared unequally—but not significantly enough to be called an ionic bond—a bond is classified as a **polar covalent bond.** The bond between two atoms, A and B, is considered to be ionic if the difference in electronegativities between atoms A and B is 2.0 or larger; polar covalent if the difference is greater than 0.4 but less than 2.0; and nonpolar covalent if the difference is less than or equal to 0.4.

Representations for Molecules and Molecular Shape

In addition to its molecular formula, the most basic feature of a molecule is the way in which its component atoms are bonded together, that is, the bonding arrangement. **Structural formulas** are used to represent bonding arrangements in molecules. In a structural formula, a line between the symbols for two elements represents a covalent bond between atoms of those elements. For example, the structural formula for a molecule of hydrogen chloride, HCl, is

H—Cl

A **Lewis structure** is used to represent both the bonding arrangement and the approximate arrangement of valence electrons in a molecule. In a Lewis structure, a pair of dots next to the symbol for an element represents an unshared pair of valence electrons for that atom. The Lewis structure for HCl is

H—C̈l̈:

The noble gases have a particularly stable configuration of valence electrons. The bonding arrangement for many molecules can be considered to be a consequence of the tendency of its atoms to attain a noble gas arrangement of valence electrons by forming covalent bonds. Since all the noble gases except helium have eight valence electrons, this generalization is referred to as the **octet rule.** For most molecules involving the representative elements in groups IVA, VA, VIA, and VIIA, a Lewis structure that satisfies the octet rule provides a good qualitative picture of how the valence electrons are distributed in the molecule.

Structural formulas and Lewis structures for many simple covalent compounds can be constructed as follows:

1. Draw a structural formula for the molecule or ion using the following generalizations as a guide:
   (a) In covalent compounds, atoms of the representative elements in groups IVA, VA, VIA, and VIIA usually form four, three, two, and one covalent bond, respectively, and a hydrogen atom forms one covalent bond.
   (b) When a molecule or polyatomic ion contains two elements, A and B, with one atom of A and more than one atom of B, the A atom is usually at the center of the molecule or polyatomic ion, all the B atoms are bonded to the A atom, and all the A—B bonds have identical properties.
2. Determine the total number of valence electrons present in the molecule or the ion.
3. Draw a Lewis structure by distributing the valence electrons in a manner consistent with the octet rule.

For several molecules and polyatomic ions one cannot draw a single Lewis structure that is consistent with the octet rule and the known experimental facts. For these molecules and ions chemists use a composite picture involving two or more Lewis structures to represent the approximate arrangement of valence electrons. The term **resonance structures** is used to describe a picture constructed from more than one Lewis structure. For example, the resonance structure representation for benzene, C_6H_6, is

This representation agrees with the experimental facts that all of the carbon-carbon bonds in benzene are equivalent, and that each carbon-carbon bond is intermediate in strength between a carbon-carbon single bond and a carbon-carbon double bond.

Structural formulas and Lewis structures usually do not represent the shapes of molecules. The **shape of a molecule** refers to the relative position of the nuclei of all of its atoms or, equivalently, the orientation in space of all of the bonds in the molecule. For example, a molecule of methane has a tetrahedral shape in which the carbon atom can be considered to be at the center of a regular tetrahedron, with the four carbon-hydrogen bonds oriented toward the four corners. Similarly, for *any* carbon atom that forms four single bonds, the four bonds have a tetrahedral arrangement.

Intermolecular Attractions

The attractive forces between molecules are called **intermolecular attractions** and are divided into three types: dipolar forces, London forces, and hydrogen bonds. Diatomic molecules, such as HCl, in which there is a separation of positive and negative charge (due to an unequal sharing of the pair of bonding electrons) are said to have a **dipole moment.** For a molecule with more than two atoms, each bond in which there is unequal sharing of electrons is said to have a **bond dipole moment,** and the dipole moment of the whole molecule depends on all the bond dipole moments, as well as on the molecular shape. Molecules that have dipole moments are called **polar molecules,** and molecules with essentially zero dipole moments are called **nonpolar molecules.** The attractive forces between polar molecules are called **dipolar forces.**

When two nonpolar molecules are close to one another, they induce temporary charge separations (dipole moments) in each other, and the resulting attractive forces between the two molecules are called **London forces.** London forces are the only type of intermolecular attractions that exist between nonpolar molecules, whereas both London forces and dipolar forces contribute to the intermolecular attractions between polar molecules. For a group of similar molecules, the magnitude of the London forces tends to increase as the molecular weight increases.

Certain molecules, such as H_2O, NH_3, and HF, exhibit a particularly strong type of intermolecular attraction called a hydrogen bond, in which a hydrogen atom can be considered to be a bridge linking together the two interacting molecules. In general, a **hydrogen bond** is a particularly strong intermolecular attraction between a hydrogen atom bonded covalently to a small atom with a large electronegativity (principally, N, O, or F) in one molecule (the donor molecule) and a small atom with a large electronegativity (principally, N, O, or F) in another molecule (the acceptor molecule).

Molecules (or parts of large molecules or large ions) can be divided into two groups, depending on how they interact with water molecules. Molecules that form strong attractive interactions with water molecules are **hydrophilic** ("water-liking"). Molecules that form hydrogen bonds with water are especially hydrophilic. Molecules (or parts of large molecules or large ions) that do not form strong attractive interactions with water molecules are called **hydrophobic** ("water-fearing"). Nonpolar molecules are hydrophobic. Hydrophobic molecules do not repel water molecules; rather, the force of attraction between a hydrophobic molecule and a water molecule is much weaker than the force of attraction between water molecules.

Many substances composed of polar molecules that can form strong hydrogen bonds with water molecules (hydrophilic molecules) are quite soluble in water. Roughly speaking, if the concentration of a substance in a saturated aqueous solution is greater than $0.1 M$, we say that the substance is soluble in water; and if the concentration is much less than $0.1 M$, we say that the substance is insoluble in water. Nonpolar hydrophobic substances are generally insoluble in water, but they are often soluble in nonpolar solvents.

In general, for substances whose molecules contain a hydrophilic part and a hydrophobic part, the solubility in water decreases the larger the hydrophobic part. Hydrophobic molecules (or hydrophobic parts of molecules) cluster together in an aqueous environment; this clustering is called a **hydrophobic interaction.**

When a soluble ionic compound dissolves in water, separate cations and anions are present and there is a strong attractive interaction between water molecules and the cations and anions. The resulting close association of water molecules around an ion is called **hydration of the ion.**

Rates of Chemical Reactions

For a chemical reaction, we usually say that the **reaction rate** is the increase in *product* concentration per unit time. In general, the rate of a chemical reaction increases as the *reactant* concentrations increase, and for most chemical reactions, there is a large increase in the reaction rate for even a small increase in temperature.

A model describing the molecular changes that occur during a chemical reaction is called a **reaction mechanism.** For example, for the overall reaction

$$AB + CD \longrightarrow AC + BD$$

one possible reaction mechanism is as follows: When AB and CD collide with sufficient kinetic energy and with a favorable orientation, molecular changes occur (involving stretching and weakening of A—B and C—D bonds, followed by formation of new A—C and B—D bonds), which result in the formation of products AC and BD. When bonds are stretched, the potential energy increases, and the potential energy decreases when bonds are formed. A **potential energy-profile diagram** is used to show the relationship between the molecular changes that occur during a reaction and the changes in potential energy. As the reactants come together, the particular configuration with the greatest potential energy is called the **activated complex,** and the potential energy increase that must occur to form this configuration is called the **activation energy.** In order for products to form, the reactants must collide with kinetic energy equal to or greater than the activation energy. All other things being equal, the larger the activation energy, the fewer collisions with sufficient energy, and the slower the reaction rate.

Reaction rates can be drastically altered by the presence of a **catalyst**—a substance that speeds up a reaction by providing an alternative path (reaction mechanism) with a lower activation energy. A catalyst is involved in the reaction mechanism and is chemically altered during the course of the reaction, but it does not undergo any net overall change.

Chemical Equilibrium

Many reactions can proceed in either direction and are said to be **reversible.** When writing an equation for such a reaction, double arrows are used to indicate reversibility. For example, the reaction between nitrogen and hydrogen to form ammonia in the gas state is written as

(1) $$N_2 + 3H_2 \rightleftharpoons 2NH_3$$

In general, a state of **dynamic equilibrium,** characterized by equal rates in the forward and reverse directions, is reached by any reversible reaction after a sufficient period of time, regardless of the composition of the starting mixture.

For any reversible reaction, there is a relationship among the equilibrium concentrations of the reactants and products, which is called the **equilibrium constant expression** for that reaction. For the general reversible reaction

(2) $$aA + bB + cC \cdots \rightleftharpoons dD + eE + fF \cdots$$

in which the capital letters represent chemical substances and the lowercase letters represent stoichiometric coefficients, the equilibrium constant expression is

(3) $$K_{eq} = \frac{[D]^d[E]^e[F]^f}{[A]^a[B]^b[C]^c}$$

The term K_{eq} is called the **equilibrium constant** for the reaction, and [A], [B], and so on, represent the concentrations of A, B, and so on, in moles per liter. For example, the equilibrium constant expression for reaction (1) is

(4) $$K_{eq} = \frac{[NH_3]^2}{[N_2][H_2]^3}$$

The effect of an increase or a decrease in one or more of the substances involved in a reversible reaction upon the position of equilibrium can be predicted qualitatively by using Le Châtelier's principle. **Le Châtelier's principle** states that when a stress is applied to a system in dynamic equilibrium, the equilibrium readjusts so as to reduce the effect of the stress as much as possible. For example, if some additional H_2 is added to an equilibrium mixture of N_2, H_2, and NH_3, some of the additional H_2 will react with some of the N_2 so that more NH_3 will be formed.

Acids and Bases

The Brønsted-Lowry model is quite useful for acid-base reactions in aqueous solutions. According to this model, an **acid** is defined as a substance that donates a proton, a **base** is defined as a substance that accepts a proton, and an **acid-base reaction** is a proton-transfer reaction in which one substance donates a proton and another substance accepts it.

When a proton (an H^+ ion) interacts strongly with a water molecule, a hydrated species, H_3O^+, called a **hydronium ion,** is formed. In water, protons also form $H_5O_2^+$, $H_9O_4^+$, and other ions. For simplicity, we usually represent a hydrated proton by the symbol H_3O^+ or, less frequently, by the symbol $H^+(aq)$.

Two chemical species that differ by only a single H^+, such as HF and F^-, are called a **conjugate acid-base pair.** In the reaction of HF with water,

(5) $$HF + H_2O \rightleftharpoons F^- + H_3O^+$$

HF acts as an acid and its conjugate base, F^-, is formed, whereas H_2O acts as a base and its conjugate acid, H_3O^+, is formed.

Acids that dissociate completely in water to form H_3O^+ (such as HCl, HI, and HNO_3) are called **strong acids**. A **weak acid** is an acid for which a 1 M solution is significantly less than 100% dissociated. The general reversible reaction for any weak acid (represented as HA) with water is

(6) $$HA + H_2O \rightleftharpoons A^- + H_3O^+$$

and the associated equilibrium constant expression is

(7) $$K_a = \frac{[H_3O^+][A^-]}{[HA]}$$

The **strength of an acid** is defined as the percent dissociation of a 1 M solution of the acid or, equivalently, as the percentage of its conjugate base (A^-) in a 1 M solution. The larger the K_a, the stronger the acid.

A **strong base** is a substance that dissociates completely in water, or reacts completely with water to produce OH^- ions. The general reversible reaction of any weak base, A^-, with water is

(8) $$A^- + H_2O \rightleftharpoons HA + OH^-$$

and the associated equilibrium constant is

(9) $$K_b = \frac{[HA][OH^-]}{[A^-]}$$

The strength of a base is defined as the percentage of HA in a 1 M solution of A^-. The larger the K_b, the farther the position of equilibrium lies to the right in reaction (8) and the stronger the base.

Some substances, such as water, can act either as an acid or as a base, depending on the other chemical species that are present in a given solution. Such substances are called **amphoteric** substances. The acid-base reaction involving one molecule of water acting as an acid and another molecule of water acting as a base is

(10) $$H_2O + H_2O \rightleftharpoons H_3O^+ + OH^-$$

This reaction is called the **self-ionization of water.** The equilibrium constant expression for reaction (10) is $K_w = [H_3O^+][OH^-]$. The value of K_w at 25°C is 1.0×10^{-14} mole^2/liter^2. In pure water, $[H_3O^+] = [OH^-] = 10^{-7}$ mole/liter.

Any solution in which $[H_3O^+] = [OH^-]$ is called a **neutral solution.** A solution in which $[H_3O^+] > [OH^-]$ is called an **acidic solution.** A solution in which $[H_3O^+] < [OH^-]$ is called a **basic solution.**

The **pH** of a solution is an alternative way of expressing $[H_3O^+]$ for the solution. Mathematically, pH is defined by the equation $pH = -\log[H_3O^+]$. Because $\log 10^{-n} = -n$, the pH of a solution with $[H_3O^+] = 10^{-n}$ is n. The pH of an acidic solution is less than 7, the pH of a basic solution is greater than 7, and the pH of a neutral solution is 7.

A solution that can maintain the pH at a nearly constant level, even when some strong acid or strong base is added, is called a **buffer solution.** In general, a solution with a large concentration of both a weak acid and its conjugate base can act as a buffer solution. A buffer solution stabilizes the pH because there is a large amount of weak acid available to react with added OH^-, and a large amount of weak base present to react with added H_3O^+.

Oxidation-Reduction Reactions

Oxidation is defined as a loss of electrons or an increase in oxidation number. **Reduction** is defined as a gain of electrons or a decrease in oxidation number. The **oxidation number** of an atom is equal to the hypothetical "charge" on that atom when, for each covalent bond involving that atom, the shared electrons are assigned to the more electronegative element. Oxidation numbers do not represent actual electrical charges; they are assigned to atoms according to a set of rules.

In an oxidation-reduction reaction, an **oxidizing agent** oxidizes another substance while it is itself being reduced, and a **reducing agent** reduces another substance while it is itself being oxidized. We can determine if a reaction involving carbon-containing compounds is an oxidation-reduction reaction by assigning oxidation numbers to the carbon atom and the other atoms in the reactants and products. However, there is a simpler alternative procedure: We merely consider the bonds that carbon atoms form in the reactant molecules and what changes, if any, occur in these bonds as the result of the chemical reaction. For carbon-containing compounds, oxidation generally involves the loss of carbon-hydrogen bonds or the gain of carbon-oxygen bonds, whereas reduction involves the gain of carbon-hydrogen bonds or the loss of carbon-oxygen bonds. For example, in the reaction

(11)

$$HO-\overset{\overset{O}{\|}}{C}-\overset{\overset{H}{|}}{\underset{\underset{H}{|}}{C}}-\overset{\overset{H}{|}}{\underset{\underset{OH}{|}}{C^*}}-\overset{\overset{O}{\|}}{C}-OH + NAD^+ \longrightarrow HO-\overset{\overset{O}{\|}}{C}-\overset{\overset{H}{|}}{\underset{\underset{H}{|}}{C}}-\overset{*}{\underset{\underset{O}{\|}}{C}}-\overset{\overset{O}{\|}}{C}-OH + NADH + H^+$$

malic acid   oxaloacetic acid

the substance malic acid is oxidized since the labeled carbon (C*) loses one carbon-hydrogen bond and gains one carbon-oxygen bond. In cells in the human body, the oxidation of malic acid to oxaloacetic acid is catalyzed by an enzyme, and a substance associated with the enzyme, called a coenzyme, is reduced. Coenzymes have complex structures and long names, so we refer to them by acronyms, abbreviations made up of the initial letters of parts of their chemical name. The acronym for the coenzyme involved in reaction (11) is NAD^+, which is an ion with a single positive charge. When malic acid loses two hydrogen atoms and is oxidized, a hydrogen atom with an additional electron, called a hydride ion, $H:^-$, combines with NAD^+ to form a substance called NADH,

(12) $$NAD^+ + H:^- \longrightarrow NADH$$

The remaining hydrogen, which does not have an electron, is left as a separate proton, H^+. Note that two hydrogen atoms ($H\cdot + H\cdot$) contain two electrons, and so does the combination of one hydride ion and one proton ($H:^- + H^+$).

Energy Changes in Chemical Reactions: Spontaneous Reactions

Changes in energy usually accompany chemical reactions. **Potential energy** is stored energy that an object possesses because of its position in relation to another object or objects. Most chemical reactions involve a change in the relative positions of atoms because the atoms are bonded together differently in the reactants and products. Thus, a chemical reaction is generally accompanied by a change in potential energy.

In many instances, if a chemical reaction takes place in which the chemical system loses energy, an equivalent amount of heat is released to the environment. Likewise, if a chemical reaction takes place in which the chemical system gains energy, an equivalent amount of heat is absorbed from the surroundings. When a system at constant pressure absorbs heat energy, we say that the **enthalpy** of the system increases.

One natural tendency is for a system to undergo a spontaneous change to a state of lower enthalpy (lower potential energy). A **spontaneous change** is one that can take place by itself, without work having to be done on the system. Diffusion, evaporation, and a ball rolling down a hill are examples of spontaneous changes. Another natural tendency is for a system to undergo a spontaneous change to a state of greater randomness. Scientists use the term **entropy** to describe the amount of randomness in a system. The larger the entropy, the less order or more randomness a system has. The tendency of a system to go to both a state of lower enthalpy and a state of higher entropy can be taken into account with a quantity called **free energy.** The symbol ΔG represents the change in free energy for a chemical reaction.

Work can be obtained only from spontaneous reactions. Spontaneous chemical reactions have a negative ΔG and are called **exergonic.** Thus, work can be obtained only from exergonic reactions. The maximum amount of work (other than the small amount involved in any change in volume) that can possibly be obtained from an exergonic reaction is equal to the magnitude of the free-energy change, ΔG, for the reaction. For example, for the oxidation of glucose,

(13) $$C_6H_{12}O_6 + 6CO_2 \longrightarrow 6CO_2 + 6H_2O$$

ΔG is equal to -688 kcal/mole. Thus, the maximum amount of work that can possibly be obtained when 1 mole of glucose is oxidized is 688 kcal.

FUNDAMENTALS OF ORGANIC AND BIOLOGICAL CHEMISTRY

ORGANIC MOLECULES

Overview

An interesting episode of the science fiction television program, *Star Trek*, involved a visit to a planet where the only life form is a creature who eats rocks. This being obtains its energy from chemical reactions involving the element silicon, a major constituent of the rocks it munches. Thus the life of this fictional creature depends on silicon-containing compounds. Now, as far as we know, no such form of life actually exists, or even could exist. All the living systems we know about—human beings, plants, animals, even single-celled organisms—use compounds containing carbon, rather than silicon, as the source of both their energy and their substance. Carbon-containing compounds are the basis of life as we know it.

Carbon atoms are unique in their ability to bond with one another and with other elements. Each carbon atom in virtually every molecule forms a total of four bonds to other atoms. Because of carbon's versatile bonding properties, more than 3 million carbon-containing compounds are known to exist. Chemists are continually synthesizing new ones, and identifying previously unknown carbon-containing compounds derived from natural sources. Oil, natural gas, and coal are natural sources of many carbon-containing compounds, such as those used in the synthesis of nylon, plastics, and the majority of modern medicines.

As a consequence of the importance and the diversity of carbon-containing compounds, chemists have separated the study of these compounds into a special field, called organic chemistry. Chemists call most compounds that contain the element carbon organic compounds.

Proteins, carbohydrates, vitamins, DNA, RNA, and most other substances unique to living systems are very large, complicated organic molecules, often called bio-organic molecules, or simply biomolecules. In the next seven chapters, we shall study the chemistry of some of the simpler organic molecules. Once we have become familiar with some of their chemical properties, we can apply what we have learned to the study of biomolecules.

How can we possibly study the 3 million and more organic compounds? Obviously we cannot investigate them one by one. We must group together compounds with similar chemical properties and study the characteristics that are common to each group.

The classification of organic compounds begins with their separation into three broad groups, based on the elements they contain: (1) those that contain the elements carbon and hydrogen only; (2) those that contain the element oxygen in addition to carbon and hydrogen; and (3) those that contain nitrogen, sulfur, or other atoms in addition to carbon, hydrogen, and possibly oxygen.

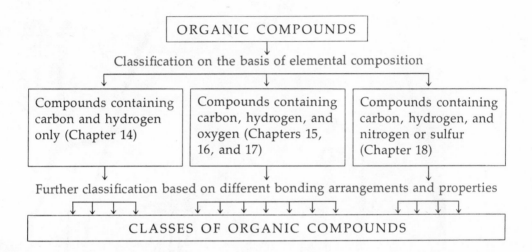

Further classification of organic compounds beyond these three broad groups is based on the different bonding arrangements that are found in compounds composed of the same elements, and on the properties associated with these bonding arrangements. For example, both ethyl alcohol (in alcoholic beverages) and acetic acid (in vinegar) contain the elements carbon, hydrogen, and oxygen. However, the bonding arrangements and the properties of ethyl alcohol and acetic acid are quite different. Ethyl alcohol and acetic acid are therefore placed in different classes of organic compounds. Note the arrows in the diagram, which indicate that we shall be concerned with four classes of organic compounds made up of only carbon and hydrogen, seven classes that have carbon, hydrogen, and oxygen as their constituent elements, and four classes of compounds containing nitrogen or sulfur, in addition to carbon and hydrogen.

In Chapter 13 we shall describe several classes of organic compounds and the basis for naming organic compounds. Then, in Chapters 14, 15, 16, 17, and 18 we shall study the structure and properties of several classes of organic compounds. In Chapter 19 we shall discuss stereoisomers—compounds with the same bonding arrangement but with different shapes, and therefore different properties.

After our study of relatively simple organic compounds, we shall be ready to study the structure and properties of more complex biomolecules.

CHAPTER 13

Organic Chemistry

13-1 INTRODUCTION

Several thousand of the organic compounds that occur in nature have been isolated and studied by chemists. An even larger number of organic compounds have been synthesized in the laboratory. Even a casual glance at the widespread use of synthetic chemicals in modern society attests to the success of the organic chemist in synthesizing new compounds that can benefit society.

Both the isolation of naturally occurring organic compounds and the synthesis of new organic compounds require a detailed understanding of the physical and chemical properties of a relatively small number of structural features that are common to large numbers of organic compounds.

For example, one structural feature found in many organic compounds consists of a nitrogen atom bonded to a carbon atom and to two hydrogen atoms,

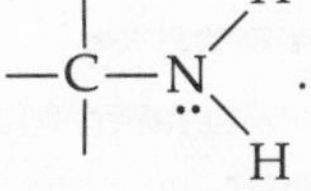

$$-\overset{\displaystyle |}{\underset{\displaystyle |}{C}}-N\overset{\displaystyle H}{\underset{\displaystyle \cdot\cdot H}{}}$$. This group of bonded atoms is called an amine group. The nitrogen

atom of an amine group can act as a Brønsted-Lowry base (Chapter 10) and both the nitrogen atom and the hydrogen atoms of the amine group can participate in hydrogen bonds with water molecules (Chapter 7). These properties of amine groups can be used to separate compounds containing amine groups from compounds with other structural features. Many synthetic drugs contain the amine group, and since the ammonium salt of an amine is usually more soluble in aqueous solutions than the corresponding amine, chemists frequently convert drugs containing amines to the corresponding ammonium salts in order to increase their solubility in body fluids.

In this chapter we shall discuss how chemists classify and name organic compounds. The vast number of organic compounds requires a systematic way of naming them. Learning these names is akin to learning a new language. It is very important that you learn this new language well because, as in any language, slight differences in a word can impart a very different meaning. Finally, we shall discuss the shapes of organic molecules. We will then be ready to study individual classes of organic compounds in subsequent chapters.

A very large number of naturally occurring organic compounds have been isolated from green plants. Petroleum is the starting material for many synthetic organic compounds.

13-2   STUDY OBJECTIVES

After studying the material in this chapter, you should be able to:

1. Define the terms functional group and class of organic compounds.

2. Define the terms isomer, structural formula, structural isomer, functional group isomer, and positional isomer.

3. Tell whether two compounds are, or are not, structural isomers, given structural formulas for the compounds.

4. Draw the structural formula for a simple alkane, given its common or IUPAC name, and for a complex alkane, given its IUPAC name.

5. Write the IUPAC name for an alkane, given its structural formula.

6. Define the term alkyl group and know the structural formulas and names for all of the alkyl groups with one to four carbon atoms.

7. Explain the meaning of the phrase, shape of a molecule.

8. Describe how the bonds of a carbon atom are oriented in space for a molecule in which the carbon atom forms four single bonds.

13-3   FUNCTIONAL GROUPS: CLASSES OF ORGANIC COMPOUNDS

The chemical properties of a compound depend on the elements it contains, on which atoms in a molecule are bonded together, and on whether these bonds are single, double, or triple bonds. Recall from Chapter 4 that:

1. The representation of a molecule that indicates its bonding arrangement is called a structural formula.

2. In structural formulas we use a line between two atoms to represent a bond that consists of a shared electron pair. A double bond, represented by two parallel lines, consists of two shared electron pairs.

3. Each carbon atom in virtually every molecule forms a total of four bonds, whereas every hydrogen atom forms only one bond.

4. A nitrogen atom tends to form a total of three bonds, and an oxygen atom tends to form a total of two bonds.

Consider the structural formulas for the compounds ethene and 2-butene:

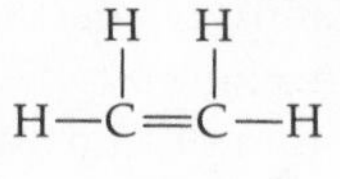

ethene 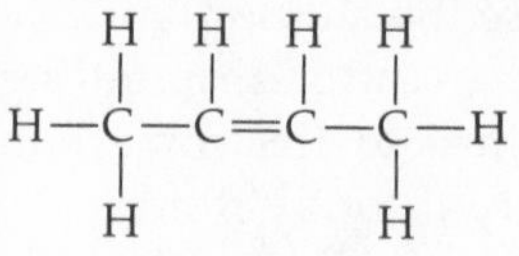   2-butene

Notice that a carbon-carbon double bond is a structural feature that both ethene and 2-butene have in common. When molecules of two different compounds have structural formulas with a common feature, the two compounds usually have a number of similar chemical properties that are characteristic of this common structural feature. For example, both ethene and 2-butene react with water, in the presence of a suitable catalyst, to give product compounds in which an —OH group is bonded to a carbon atom:

(13-1)

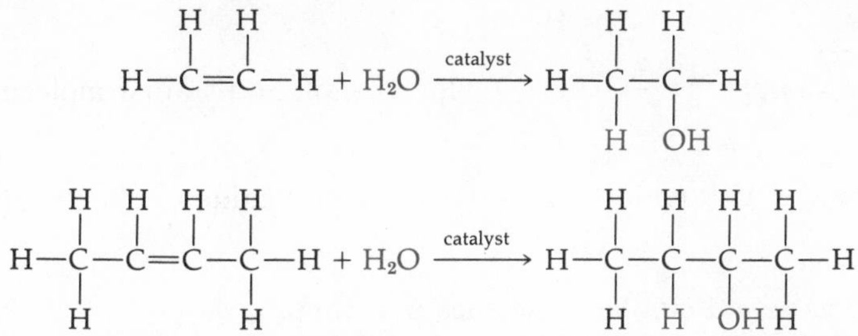

$$H-\overset{\overset{\displaystyle H}{|}}{C}=\overset{\overset{\displaystyle H}{|}}{C}-H + H_2O \xrightarrow{\text{catalyst}} H-\overset{\overset{\displaystyle H}{|}}{\underset{\underset{\displaystyle H}{|}}{C}}-\overset{\overset{\displaystyle H}{|}}{\underset{\underset{\displaystyle OH}{|}}{C}}-H$$

$$H-\overset{\overset{\displaystyle H}{|}}{\underset{\underset{\displaystyle H}{|}}{C}}-\overset{\overset{\displaystyle H}{|}}{C}=\overset{\overset{\displaystyle H}{|}}{C}-\overset{\overset{\displaystyle H}{|}}{\underset{\underset{\displaystyle H}{|}}{C}}-H + H_2O \xrightarrow{\text{catalyst}} H-\overset{\overset{\displaystyle H}{|}}{\underset{\underset{\displaystyle H}{|}}{C}}-\overset{\overset{\displaystyle H}{|}}{\underset{\underset{\displaystyle H}{|}}{C}}-\overset{\overset{\displaystyle H}{|}}{\underset{\underset{\displaystyle OH}{|}}{C}}-\overset{\overset{\displaystyle H}{|}}{\underset{\underset{\displaystyle H}{|}}{C}}-H$$

Experiments with many other compounds show that, in general, compounds with a carbon-carbon double bond react with water in a manner similar to reactions 13-1. We can say that the addition of water is a characteristic chemical property of molecules with a carbon-carbon double bond.

Another characteristic chemical property of molecules with a carbon-carbon double bond is their reaction with hydrogen in the presence of a suitable catalyst. The reaction of 2-butene with hydrogen is an example:

(13-2)

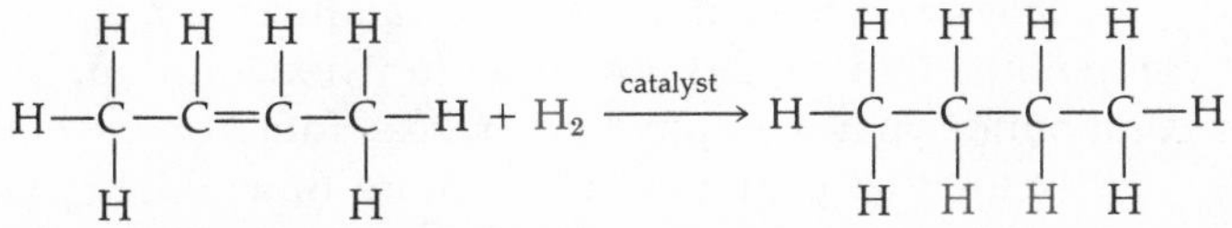

$$H-\overset{\overset{\displaystyle H}{|}}{\underset{\underset{\displaystyle H}{|}}{C}}-\overset{\overset{\displaystyle H}{|}}{C}=\overset{\overset{\displaystyle H}{|}}{C}-\overset{\overset{\displaystyle H}{|}}{\underset{\underset{\displaystyle H}{|}}{C}}-H + H_2 \xrightarrow{\text{catalyst}} H-\overset{\overset{\displaystyle H}{|}}{\underset{\underset{\displaystyle H}{|}}{C}}-\overset{\overset{\displaystyle H}{|}}{\underset{\underset{\displaystyle H}{|}}{C}}-\overset{\overset{\displaystyle H}{|}}{\underset{\underset{\displaystyle H}{|}}{C}}-\overset{\overset{\displaystyle H}{|}}{\underset{\underset{\displaystyle H}{|}}{C}}-H$$

Chemists call a group of atoms and bonds that behave in a similar manner in many different compounds a **functional group.** Functional groups are used to classify organic compounds. A carbon-carbon double bond is one example of a functional group. Organic compounds that contain a carbon-carbon double bond and no other functional group are called **alkenes.** Examples of some of the classes of organic compounds and their corresponding functional groups are given in Table 13-1.

Table 13-1 Some Classes of Organic Compounds and Their Functional Groups

Class	Functional Group	Example Formula	Name
Alkene	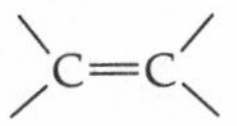 $\overset{\diagdown}{\diagup}C=C\overset{\diagup}{\diagdown}$	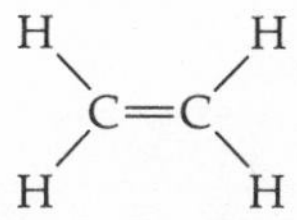 $\overset{H}{\diagdown}\overset{}{\underset{H}{\diagup}}C=C\overset{\diagup}{\underset{H}{\diagdown}}\overset{H}{}$	Ethylene or ethene
Alcohol	$-\overset{|}{\underset{|}{C}}-O-H$	$H-\overset{\overset{H}{|}}{\underset{\underset{H}{|}}{C}}-\overset{\overset{H}{|}}{\underset{\underset{H}{|}}{C}}-O-H$	Ethyl alcohol or ethanol
Carboxylic acid	$-\overset{\overset{\displaystyle O}{\|}}{C}-O-H$	$H-\overset{\overset{H}{|}}{\underset{\underset{H}{|}}{C}}-\overset{\overset{\displaystyle O}{\|}}{C}-O-H$	Acetic acid or ethanoic acid
Ester	$-\overset{\overset{\displaystyle O}{\|}}{C}-O-\overset{|}{\underset{|}{C}}-$	$H-\overset{\overset{\displaystyle O}{\|}}{C}-O-\overset{\overset{H}{|}}{\underset{\underset{H}{|}}{C}}-H$	Methyl formate
Amine	$-\overset{|}{\underset{|}{C}}-N\overset{\diagup}{\diagdown}$	$H-\overset{\overset{H}{|}}{\underset{\underset{H}{|}}{C}}-\overset{\overset{H}{|}}{\underset{\underset{H}{|}}{C}}-N\overset{H}{\underset{H}{\diagdown}}\overset{\diagup}{}$	Ethylamine

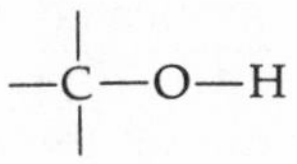

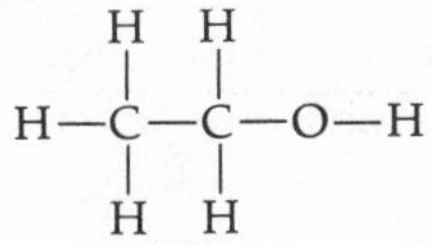

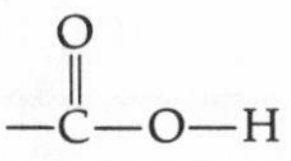

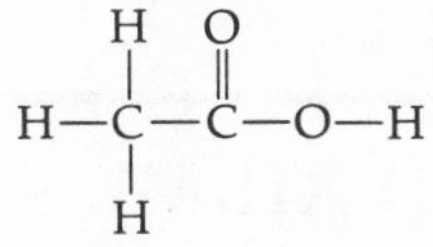

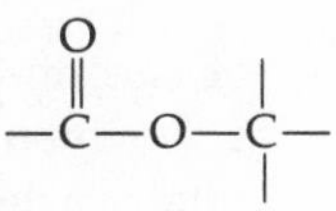

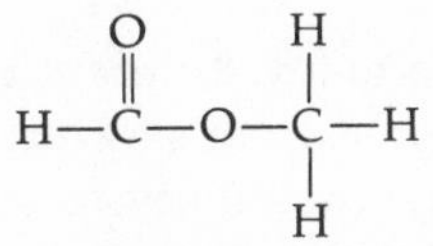

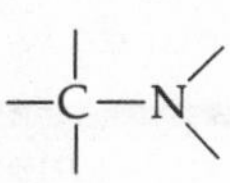

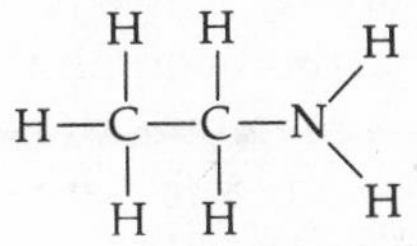

For example, $H-\overset{\overset{\displaystyle H}{|}}{\underset{\underset{\displaystyle H}{|}}{C}}-\overset{\overset{\displaystyle H}{|}}{\underset{\underset{\displaystyle H}{|}}{C}}-OH$, ethanol, and similar molecules that contain

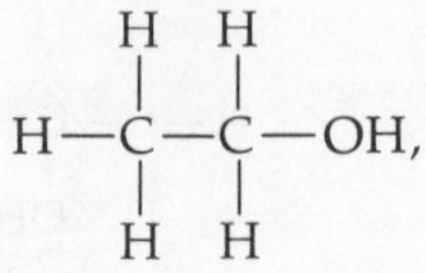

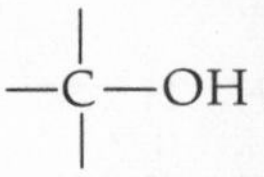

the $-\overset{|}{\underset{|}{C}}-OH$ functional group are called **alcohols,** whereas $H-\overset{\overset{\displaystyle H}{|}}{\underset{\underset{\displaystyle H}{||}}{C}}-\overset{\overset{\displaystyle O}{||}}{C}-OH$,

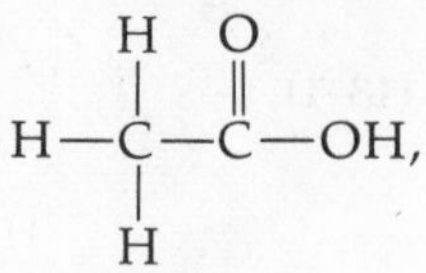

acetic acid, and similar molecules that contain the $-\overset{\overset{\displaystyle O}{||}}{C}-OH$ functional group are called **carboxylic acids.**

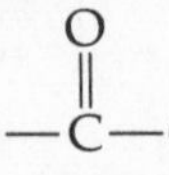

The simplest class of organic molecules, the alkanes, contain only carbon-carbon and carbon-hydrogen single bonds. Alkanes are reactants in only a few types of reactions (see Chapter 14), so we do not consider alkanes to contain a functional group. Most organic compounds in other classes contain one or more portions that are alkanelike in that they contain only C—C and C—H single bonds. The alkanelike portions of organic compounds are also relatively nonreactive.

In the next few chapters we shall discuss the characteristic physical and chemical properties and the rules used for naming the major classes of simple organic compounds that contain only one functional group. More complicated organic compounds may contain two, three, four, or even more functional groups. A molecule that contains two or more functional groups can usually, but not always, undergo all of the characteristic reactions of each of the functional groups. For example, the compound β-hydroxybutyric acid, which is present in excessive amounts in the bloodstream and urine of diabetic individuals, has both an alcohol functional group and a carboxylic acid functional group (see Figure 13-1). The compound β-hydroxybutyric acid can undergo the chemical reactions characteristic of *both* alcohols and carboxylic acids.

Figure 13-1   The compound β-hydroxybutyric acid has both an alcohol and a carboxylic acid functional group.

carboxylic acid group

alcohol group

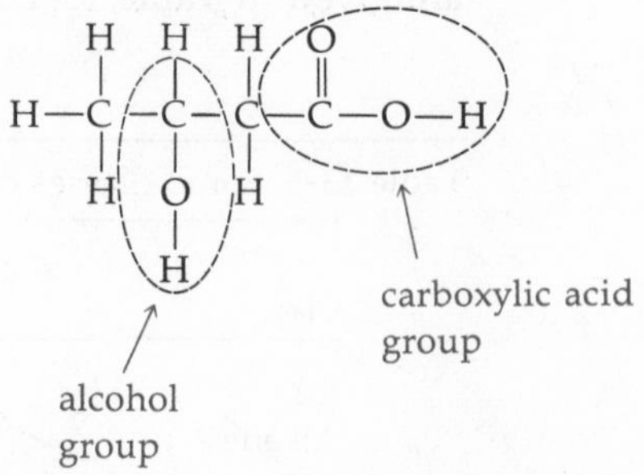

In general, we can think of a complicated organic molecule or biomolecule as being composed of several functional groups and capable of reacting in a variety of ways depending on the nature of the functional groups in the molecule.

13-4   ISOMERISM: STRUCTURAL ISOMERS

There are many different classes of carbon-containing compounds. The basic reason for carbon's versatility, as we mentioned before, lies in the carbon atom's unique ability to bond to other carbon atoms and to a variety of other atoms. Another consequence of carbon's bonding ability is the existence of different carbon-containing compounds with the same molecular formula. Different compounds that have the same molecular formula are called **isomers.** For compounds to be different, (1) there must be some difference in the chemical and physical properties of the compounds, and (2) it must be possible to separate one compound from another. The compounds ethyl alcohol and di-methyl ether, shown in Figure 13-2, are examples of isomers. Ethyl alcohol and dimethyl ether are compounds with very different chemical and physical prop-erties, but they both have the same molecular formula, C_2H_6O.

Figure 13-2   Ethyl alcohol and dimethyl ether are structural isomers.

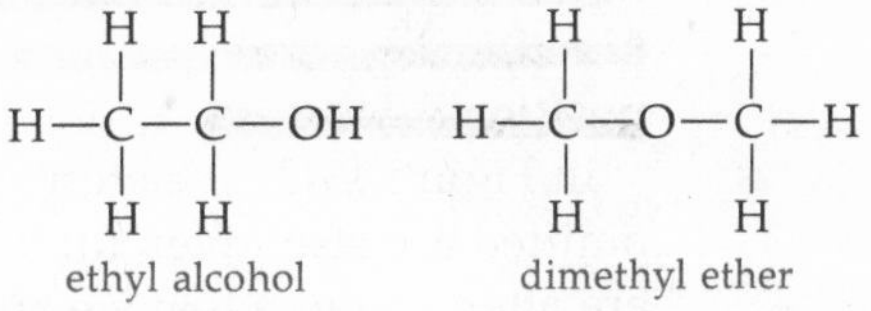

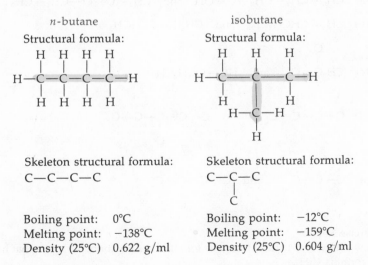

There are two major types of isomers: **structural isomers** and **stereoisomers.** We shall discuss structural isomers in this section. Stereoisomers will be dis-cussed later, in Chapters 14 and 19.

Recall that a structural formula for a molecule indicates which atoms in the molecule are bonded together. We can say that a structural formula is a picture representing the bonding arrangement in the molecule. We can see from the structural formulas given in Figure 13-2 that ethyl alcohol and dimethyl ether have different bonding arrangements. For example, the oxygen atom in ethyl al-cohol is bonded to a carbon atom and a hydrogen atom, whereas in dimethyl ether the oxygen atom forms bonds with two different carbon atoms. Ethyl al-cohol and dimethyl ether are examples of structural isomers. **Structural isomers are isomers that have different structural formulas; in other words, they are dif-ferent compounds that have the same molecular formula but differ in the way the atoms are bonded together.** Ethyl alcohol and dimethyl ether are structural isomers that belong to different classes of organic compounds. Structural is-omers that have different functional groups, and thus belong to different classes of organic compounds, are referred to as **functional-group isomers.**

There are also many examples of structural isomers that belong to the same class of organic compounds. For example, there are two different alkane com-pounds with the molecular formula C_4H_{10}. One compound is called normal bu-tane or n-butane, and the other is isobutane. (How alkanes are named is dis-cussed in the next section.) The physical properties of n-butane and isobutane are different (see Figure 13-3), and there are some ways in which the chemical properties of these compounds differ as well.

Figure 13-3   The structural isomers n-butane and isobutane have different physi-cal properties, and some different chemical properties as well.

Notice that the bonding arrangement in *n*-butane is different than the bonding arrangement in isobutane. Each carbon atom in *n*-butane is bonded to either one or two other carbon atoms. In isobutane, on the other hand, each carbon atom is bonded to either one or three other carbon atoms. Thus *n*-butane and isobutane have different structural formulas and are examples of structural isomers. Structural isomers that contain *identical* functional groups (for the alkanes there is no functional group) but have *different* bonding arrangements are referred to as **positional isomers.** Thus, *n*-butane and isobutane are positional isomers.

In Figure 13-3, skeleton structural formulas in which the hydrogen atoms are omitted are also given for *n*-butane and isobutane. Skeleton structural formulas are often useful when we are concerned primarily with carbon-carbon bonding. When looking at a skeleton structural formula, we just have to keep in mind that each carbon atom in a molecule forms four bonds and that the bonds that are omitted in a skeleton structural formula are carbon-hydrogen bonds.

We can draw pictures that look different but have the same bonding arrangement. Notice that there is the same bonding arrangement in each of the representations in Figure 13-4, which are all equivalent skeleton structural formulas for isobutane.

Figure 13-4   All of these skeleton structural formulas represent isobutane.

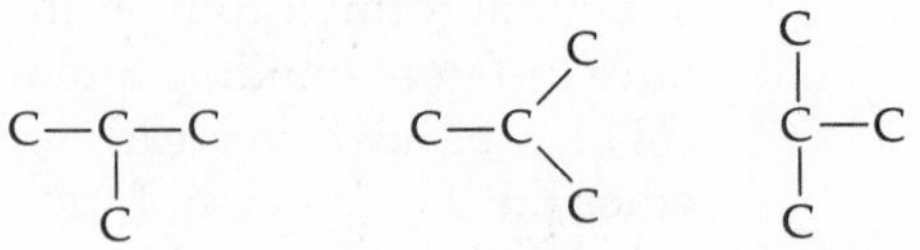

Another abbreviated way of representing the compounds *n*-butane and isobutane is: $CH_3-CH_2-CH_2-CH_3$ (*n*-butane) and $CH_3-CH-CH_3$ (isobutane).

$$\overset{|}{C}H_3$$

In these representations, called **condensed structural formulas,** subscripts are used to indicate the number of hydrogen atoms bonded to each carbon atom. This is often more convenient than drawing all of the C—H bonds.

Exercise 13-1

Which of the following pairs are structural formulas for structural isomers? Indicate which structural isomers are functional-group isomers and which are positional isomers.

(a) 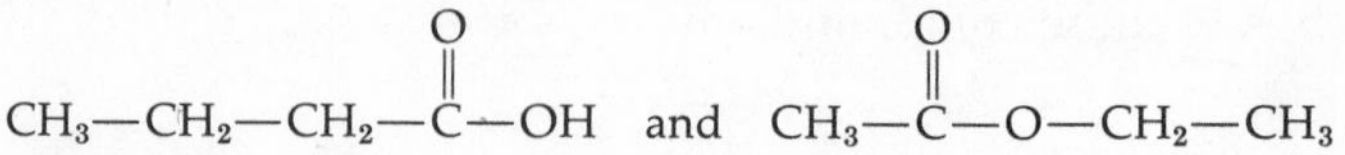 $CH_3-CH_2-CH_2-\overset{\overset{O}{\|}}{C}-OH$ and $CH_3-\overset{\overset{O}{\|}}{C}-O-CH_2-CH_3$

(b) $CH_2{=}CHCH_2CH_3$ and $CH_3CH{=}CHCH_3$

(c) 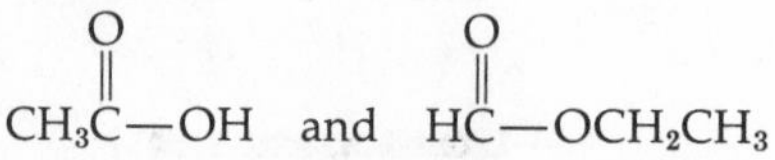 $CH_3\overset{\overset{O}{\|}}{C}-OH$ and $H\overset{\overset{O}{\|}}{C}-OCH_2CH_3$

(d) 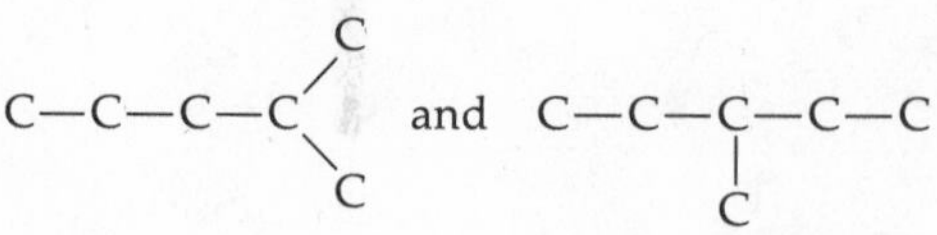   and

(e) 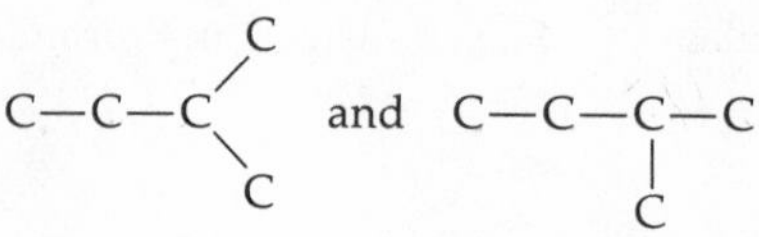   and

Exercise 13-2

Draw skeleton structural formulas for all of the positional isomers that have the molecular formula C_5H_{12}.

13-5 NAMING ORGANIC MOLECULES: THE ALKANES

The first organic compounds to be isolated and studied were named in a non-systematic fashion. For example, the compound

$$
\begin{array}{c}
\quad\ \ \text{H}\quad\ \text{H}\quad\ \text{H}\qquad\quad \text{O} \\
\quad\ \ |\qquad |\qquad |\qquad\quad \| \\
\text{H}-\text{C}-\text{C}-\text{C}-\text{C} \\
\quad\ \ |\qquad |\qquad |\qquad\quad \diagdown \\
\quad\ \ \text{H}\quad\ \text{H}\quad\ \text{H}\qquad\quad \text{OH}
\end{array}
$$

is commonly called butyric acid because it was originally found in rancid butter (the Latin word for butter is *butyrum*). As more and more organic compounds were discovered, attempts were made to develop a systematic method for naming them. At the present time the system devised by the International Union of Pure and Applied Chemistry (IUPAC) is universally accepted by chemists. The systematic IUPAC names for many organic compounds are, how-ever, very long and cumbersome. For this reason we often use the less formal common names for many compounds.

Naming Alkanes

The alkanes comprise the simplest class of organic compounds; they contain only C—C and C—H single bonds. The condensed structural formulas and common names for the unbranched alkanes containing one to 10 carbon atoms are given in Table 13-2. An unbranched alkane is an alkane whose structural formula can be written as a single linear chain of carbon atoms. The common names (and also the IUPAC names) for all alkanes end in *-ane*. Notice that the names for the unbranched alkanes with five or more carbon atoms begin with the Greek word for the number of carbon atoms. Also notice that alkanes with four or more carbon atoms have positional isomers. In order to distinguish one positional isomer from another, we can attach prefixes to their names. The prefix *n-* (for normal) is used for an unbranched alkane.

Table 13-2 Unbranched Alkanes

Common Name	Structural Formula	No. of Carbon Atoms	No. of Positional Isomers
Methane	CH_4	1	1
Ethane	CH_3—CH_3	2	1
Propane	CH_3—CH_2—CH_3	3	1
n-Butane	CH_3—CH_2—CH_2—CH_3	4	2
n-Pentane	CH_3—CH_2—CH_2—CH_2—CH_3	5	3
n-Hexane	CH_3—CH_2—CH_2—CH_2—CH_2—CH_3	6	5
n-Heptane	CH_3—CH_2—CH_2—CH_2—CH_2—CH_2—CH_3	7	9
n-Octane	CH_3—CH_2—CH_2—CH_2—CH_2—CH_2—CH_2—CH_3	8	18
n-Nonane	CH_3—CH_2—CH_2—CH_2—CH_2—CH_2—CH_2—CH_2—CH_3	9	35
n-Decane	CH_3—CH_2—CH_2—CH_2—CH_2—CH_2—CH_2—CH_2—CH_2—CH_3	10	75

As we saw in the previous section, *n*-butane has one positional isomer, which is called isobutane (Figure 13-3). There are three alkanes with five carbon atoms. The names of these positional isomers are *n*-pentane, isopentane, and neopentane (see Figure 13-5). Alkanes with more carbon atoms have larger and larger numbers of positional isomers.

Figure 13-5   Positional isomers of pentane.

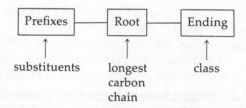

n-Pentane $CH_3—CH_2—CH_2—CH_2—CH_3$

Isopentane

Neopentane

IUPAC Nomenclature

To identify each of the 75 alkanes with 10 carbon atoms by a different prefix would be ridiculous. The use of IUPAC nomenclature makes this unnecessary. The IUPAC name for an organic compound consists of three component parts:

```
┌──────────┐   ┌──────┐   ┌────────┐
│ Prefixes │───│ Root │───│ Ending │
└──────────┘   └──────┘   └────────┘
     ↑            ↑            ↑
 substituents  longest       class
               carbon
               chain
```

1. The **root** of the IUPAC name. <u>The root specifies the longest continuous chain of carbon atoms in a molecule of the compound</u>. It is derived from the name for the unbranched alkane with that number of carbon atoms. For this reason the names for the unbranched alkanes in Table 13-2 should be committed to memory.

2. The **ending** of an IUPAC name specifies either the class of compounds to which the compound belongs or the major functional group in a molecule of the compound. The ending *-ane* specifies an alkane. (Some other IUPAC endings include *-ene* for alkenes, with a $\diagdown C{=}C \diagup$ functional group, and *-ol* alcohols, with a $—\overset{|}{\underset{|}{C}}—OH$ functional group.)

3. **Prefixes** are used to specify the identity and location of atoms or groups of atoms (other than hydrogen), called **substituents,** which are attached to the longest carbon chain. The order, numbering, and punctuation used for prefixes follow a set of rules that we shall discuss shortly.

Let us consider a few examples of how to draw a structural formula for a compound given its IUPAC name.

EXAMPLE 1 Draw a structural formula for pentane.

Begin by identifying the root and ending of this IUPAC name:

pentane

root   ending

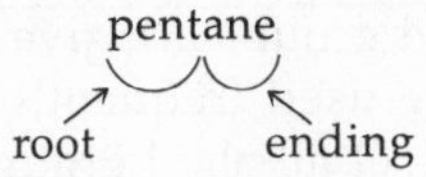

The root of this IUPAC name specifies that the longest continuous chain of carbon atoms is five carbons long. The ending -ane specifies that the compound is an alkane. The structural formula for this compound is therefore drawn by first constructing a chain of five carbon atoms,

C—C—C—C—C

then adding hydrogen atoms for a total of four bonds for each carbon atom:

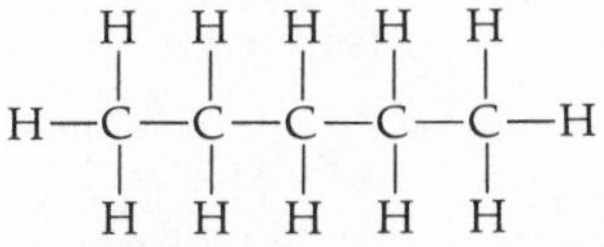

$$
\begin{array}{ccccc}
H & H & H & H & H \\
| & | & | & | & | \\
H-C- & C- & C- & C- & C-H \\
| & | & | & | & | \\
H & H & H & H & H
\end{array}
$$

The condensed structural formula for a molecule of pentane is

$H_3C—CH_2—CH_2—CH_2—CH_3$

The prefix n-, which is used in the common name for this compound, is not needed, because the root of the IUPAC name specifies a single continuous carbon chain.

EXAMPLE 2 Draw a structural formula for methylbutane.

First, identify the components of this IUPAC name:

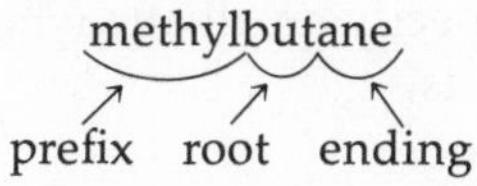

methylbutane

prefix   root   ending

The root and ending of this IUPAC name specify a continuous chain or back-bone of four carbon atoms,

C—C—C—C

and that the compound is an alkane. The prefix, methyl-, specifies that a methyl substituent, —CH_3, is bonded to this chain. The condensed structural formula for a molecule of methylbutane is therefore

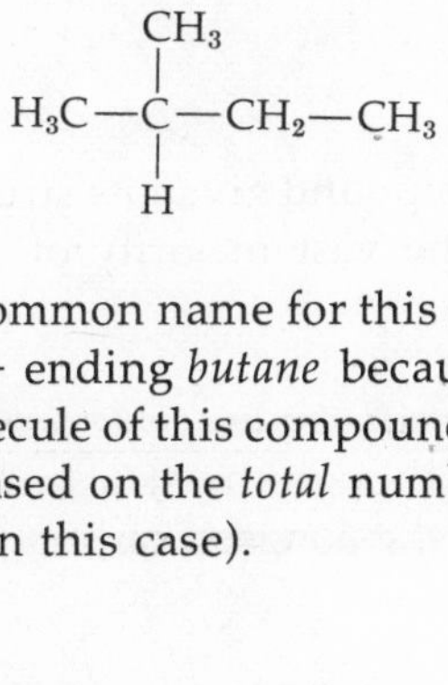

$$
\begin{array}{c}
CH_3 \\
| \\
H_3C-C-CH_2-CH_3 \\
| \\
H
\end{array}
$$

The common name for this compound is isopentane. The IUPAC name uses the root + ending butane because the longest continuous chain of carbon atoms in a molecule of this compound is 4. On the other hand, common names for alkanes are based on the total number of carbon atoms in a molecule of the compound (five in this case).

The methyl group refers to the group of atoms that would result if one hydrogen were removed from methane. In general, an **alkyl group** refers to a group of atoms that would result if one hydrogen atom were removed from an alkane. The names for the most common alkyl groups are given in Table 13-3. The names of these common alkyl groups are used frequently and should be committed to memory. Notice that the names of all alkyl groups end in -*yl*.

Table 13-3 Common Alkyl Groups

Structural Formula	Name
CH_3—	Methyl
CH_3—CH_2—	Ethyl
CH_3—CH_2—CH_2—	*n*-Propyl
CH_3—CH—CH_3 |	Isopropyl
CH_3—CH_2—CH_2—CH_2—	*n*-Butyl
CH_3—CH_2—CH—CH_3 |	*sec*-Butyl (secondary butyl)
CH_3 | CH_3—CH—CH_2—	Isobutyl

CH_3
    |
CH_3—C—CH_3
    |
	tert-Butyl or *t*-butyl (tertiary butyl)

EXAMPLE 3 Draw the structural formula for 3,5-diethyl-4-methyloctane.

The root + ending of this IUPAC name specify an alkane whose longest continuous chain has eight carbon atoms, that is,

$$C_1—C_2—C_3—C_4—C_5—C_6—C_7—C_8$$

The prefixes specify ethyl substituents bonded to the third and fifth carbon atoms, and a methyl group bonded to the fourth carbon atom. The structural formula for a molecule of 3,5-diethyl-4-methyloctane is thus

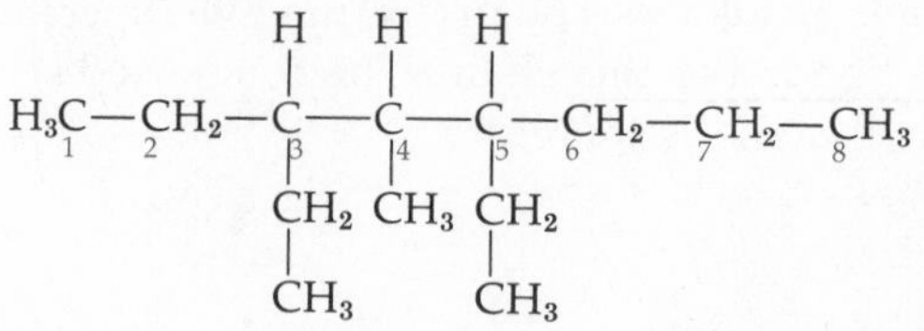

In order to write the IUPAC name of a compound given its structural formula, you must learn a few additional rules. The vast majority of alkanes can be named using the following rules:

1. The root of the IUPAC name must specify the longest continuous chain of carbon atoms in the molecule, even though the structural formula for a compound may be drawn so that the longest carbon chain is not a straight line.

For example, in the skeleton structural formula

$$
\begin{array}{c}
\text{C} \\
| \\
\text{C}-\!\!\left[\begin{array}{c}\text{C}-\text{C}-\text{C}\\|\\\text{C}\\|\\\text{C}\end{array}\right]
\end{array}
\longleftarrow \text{longest carbon chain}
$$

the longest continuous chain consists of five carbon atoms.

2. Substituents are identified by name and by a number that indicates the carbon atom of the longest chain to which they are attached. The longest continuous chain must be numbered so that the positions of the substituents will have the lowest possible numbers. The prefixes *di-*, *tri-*, and *tetra-* before the name of a substituent indicate two, three, or four of that substituent in the molecule. For example, *diethyl* indicates two ethyl groups in a molecule.

3. When a molecule contains more than one substituent, the substituents are arranged alphabetically. Notice that di*ethyl* comes before *methyl* in Example 3; the prefixes *di-*, *tri-*, and so on, are not counted when arranging substituents alphabetically.

4. IUPAC names are written as a single word with numbers separated from one another by commas and numbers separated from letters by hyphens (see Example 3). No punctuation or space is used between the name of a substituent and the root name.

The following examples illustrate how these rules are used.

EXAMPLE 4 Write the IUPAC name for the compound with the condensed structural formula

$$
\begin{array}{c}
\text{H} \\
| \\
\text{H}_3\text{C}-\text{C}-\text{CH}_3 \\
| \\
\text{H}-\text{C}-\text{CH}_3 \\
| \\
\text{CH}_2 \\
| \\
\text{CH}_3
\end{array}
$$

Proceed as follows: (a) Identify the longest continuous chain of carbon atoms and thus determine the root of the name. (b) Identify the class of compound and thereby obtain the ending for the IUPAC name. (c) Identify the substituents and number the longest carbon chain so that the substituents have the lowest possible numbers. (d) Name the substituents as prefixes of the IUPAC name. Thus:

$$
\begin{array}{c}
\text{H} \\
| \\
\text{H}_3\underset{1}{\text{C}}-\underset{2}{\text{C}}-\text{CH}_3 \\
| \\
\text{H}-\underset{3}{\text{C}}-\text{CH}_3 \\
| \\
\underset{4}{\text{CH}_2} \\
| \\
\underset{5}{\text{CH}_3}
\end{array}
$$

(c) Methyl substituents on carbons 2 and 3

(d) Prefix = 2,3-dimethyl

(a) Longest carbon chain (root = pent)

(b) An alkane (ending = -ane)

The IUPAC name for this compound is therefore 2,3-dimethylpentane.

EXAMPLE 5 Write the IUPAC name for the compound with the condensed structural formula

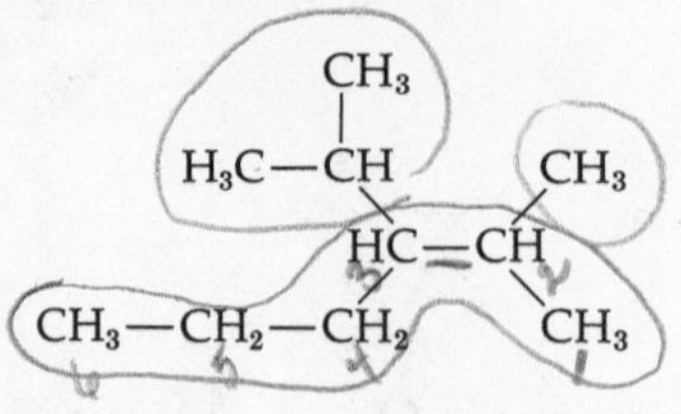

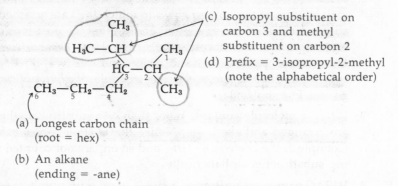

Proceeding as we did in Example 4, we determine the following:

(c) Isopropyl substituent on carbon 3 and methyl substituent on carbon 2

(d) Prefix = 3-isopropyl-2-methyl (note the alphabetical order)

(a) Longest carbon chain (root = hex)

(b) An alkane (ending = -ane)

Thus, the IUPAC name for this compound is 3-isopropyl-2-methylhexane. (Notice that in this example the carbon chain could be numbered differently, but the name of this compound would be the same.)

Exercise 13-3
Draw condensed structural formulas for the following compounds:
(a) 3-Ethylpentane
(b) 3-Ethyl-2-methyloctane
(c) 2,2,4-Trimethylpentane

Exercise 13-4
Write the IUPAC name for each of the following compounds:

(a) H_3C\ /H
        C
   H_3C/ \ H

(b) CH_3
         |
   $H_3C-C-CH_3$
         |
         H

(c) H_2C-CH_3
        |
  H_3C-CH
        |
      H_2C-CH_3

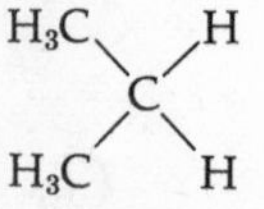

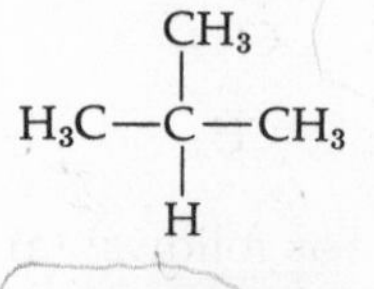

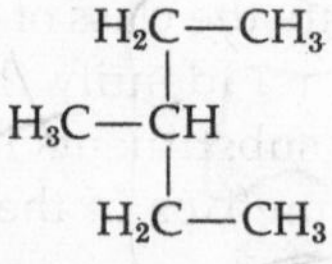

(d) $H_3C-CH-CH_3$
              |
             CH_2
              |
           $CH-CH_3$
              |
             CH_3

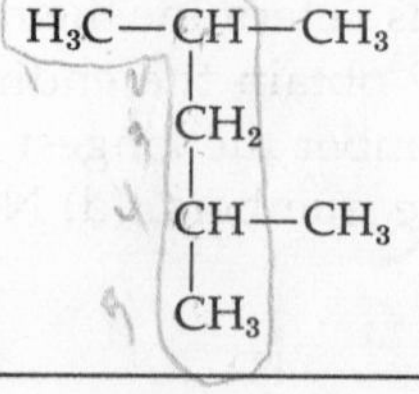

13-6   SHAPES OF ORGANIC MOLECULES

Recall from Chapter 4 that a structural formula depicts the bonding arrangement in a molecule, but it is *not* a picture of the shape of a molecule. The **shape** of a molecule is specified by the positions of the centers of all atoms in the molecule, or the orientation in space of the bonds in the molecule. It is important to consider this orientation, since the chemical and physical properties of a compound depend critically on the shape of the molecules of the compound.

The molecular shape is a particularly crucial factor determining the chemical properties of large biomolecules. We shall use the phrase "structure of a molecule" to refer to a combination of both the bonding arrangement in the molecule and the shape of the molecule. Thus a structural formula for a molecule gives only a partial description of the structure of that molecule. Other pictures must be used to indicate the shape of the molecule.

Molecules of different compounds sometimes have the same structural formula but different shapes. This leads to a more subtle, but extremely important, kind of isomerism called stereoisomerism. Two different compounds that have the same molecular formula and the same structural formula but different shapes are called **stereoisomers.** One type of stereoisomerism is discussed in Chapter 14 and another type is discussed in Chapter 19. We shall consider the shape of a few simple molecules here. We shall discuss the shape of more complicated molecules and see numerous examples of the intimate relationship between molecular shape and chemical behavior throughout the remainder of this textbook.

The simplest organic compound is methane, which has the molecular formula CH_4. Recall that experimental evidence shows that the carbon atom in methane is at the center of a regular tetrahedron with a hydrogen atom at each of the four corners (Section 4-13). We can think of the tetrahedral shape of a methane molecule as being determined by four equivalent single bonds on the carbon atom oriented toward the corners of a regular tetrahedron. Chemists use various types of perspective drawings, such as that in Figure 13-6, to picture the shape of methane and other molecules. Most compounds have molecules whose atoms do not all lie in the same plane. Therefore, the best way to visualize the shape of a molecule, and to make correct inferences about those properties of the molecule that depend on its shape, is to construct a three-dimensional model of the molecule. Chemists often use wooden balls and sticks, or plastic tetrahedra and tubing, for this purpose.

Figure 13-6   Methane molecules are tetrahedral in shape.

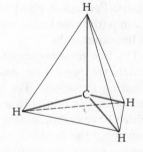

Each bond angle in a molecule of methane with its tetrahedral shape—that is, the angle between any two carbon-hydrogen bonds—is 109.5°. This tetrahedral orientation of the four single bonds about the carbon atom is the typical orientation of bonds about any carbon atom that forms four single bonds. This is true, for example, for each carbon atom in ethane and ethyl alcohol (Figure 13-7). We can think of the shape of ethane or ethyl alcohol as consisting of two tetrahedra joined together by a carbon-carbon single bond. In order to have a full understanding of the shape of ethane or ethyl alcohol, however, we have to know how the two tetrahedra in these molecules are oriented with respect to one another. Is one orientation of the two tetrahedra preferred over all other orientations?

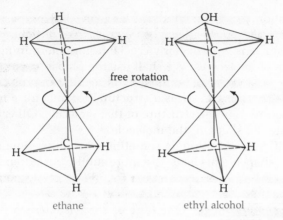

ethane   ethyl alcohol

Figure 13-7   In ethane and ethyl alcohol   rotation about the carbon-carbon
molecules, there is essentially free   single bond.

On the basis of experimental evidence, it has been found that there is very little energy difference between different orientations of the two tetrahedra in ethane or ethyl alcohol. Thus, for example, ethane is a gas at room temperature, and we can think of the ethane molecules as moving about with the two tetrahedra essentially rotating freely about the carbon-carbon single bond. Because of this unrestricted rotation of the two tetrahedra about the carbon-carbon single bond, it is not possible to separate two (or more) kinds of ethane—one form corresponding to one orientation of the two tetrahedra with respect to one another, and the other corresponding to a different relative orientation of the two tetrahedra. Thus ethane does *not* have any stereoisomers.

What we have said about the carbon-carbon bond in ethane is generally true of carbon-carbon single bonds in simple molecules. At room temperature, there is essentially free rotation about any carbon-carbon single bond in a simple molecule. Thus we should not think of molecules containing carbon-carbon single bonds as rigid entities. Rather, we must consider them as having a large degree of flexibility. For example, when we see the skeleton structural formula for butane, C—C—C—C, we should not think of a rigid linear arrangement of atoms. Perhaps a good way to visualize the shape of butane is to think of a chain with three links, with each link capable of moving randomly with respect to the other links.

The free rotation about the carbon-carbon single bond in simple molecules, which we have just described, is not always found in some complicated bonding situations. For example, there is no free rotation about the carbon-carbon single bonds in a class of compounds called cycloalkanes (Chapter 14), which have carbon atoms joined in a ring. There is also an absence of free rotation about the double bond in alkene molecules. In addition, as we shall discuss later, various nonbonding interactions between parts of large biomolecules, such as proteins and nucleic acids, cause these molecules to adopt very definite shapes. The shape of a biomolecule is a primary factor in determining its chemical behavior.

13-7 SYMBOLS, PICTURES, AND REPRESENTATIONS OF MOLECULES

As you know, we use various symbols and pictures to represent molecules. It is essential that you understand what information is contained in a given representation of a molecule. Moreover, it is equally important that you understand what information is *not* conveyed by a given representation. This is particularly important to keep in mind when you consider isomers. Remember that a molecular formula for a compound, such as C_4H_{10}, does not imply a particular bonding arrangement for molecules of that compound. In fact, we have seen that there are two different compounds, the isomers *n*-butane and isobutane, which have the same molecular formula, C_4H_{10}.

Likewise, a structural formula is not a picture of the shape of a molecule. Thus, for example, we must not draw the erroneous conclusion that the carbon atom in a methane molecule is at the center of a square by merely looking at the structural formula

$$\begin{array}{ccc} H & & H \\ & \diagdown \diagup & \\ & C & \\ & \diagup \diagdown & \\ H & & H \end{array}$$

Once we have learned something about the shapes of molecules, however, we should be able to use this information to infer the correct shape of a molecule from its structural formula. For example, when we see the condensed structural formula CH_3CH_3 for ethane, we should be able to infer that the shape of ethane can be described as two tetrahedra joined together by a carbon-carbon single bond (Figure 13-7). In this sense, a structural formula for a molecule can be considered an abbreviation for the structure of the molecule.

13-8 SUMMARY

1. A group of atoms and bonds that behaves in a similar manner in many different compounds is called a functional group. Organic compounds are classified on the basis of their functional groups.

2. The simplest class of organic molecules, the alkanes, contain only C—C and C—H single bonds, are generally not reactive, and are therefore not considered to possess a functional group.

3. A molecule that possesses two or more functional groups can usually undergo all of the reactions characteristic of each of its component functional groups.

4. Isomers are different compounds with the same molecular formula; structural isomers have different structural formulas, whereas stereoisomers have the same structural formulas but different shapes. There are two types of structural isomers: functional-group isomers and positional isomers.

5. The IUPAC method for naming organic compounds is universally accepted by chemists. It consists of a set of rules whereby names are constructed of a root to which an ending and prefixes are attached. The root specifies the longest carbon chain in molecules of that compound. The ending (-*ane* for alkanes) specifies the classification of the compound or the major functional group in the molecule, and the prefixes specify the identity and location of substituents bonded to the longest carbon chain.

6. If one hydrogen is removed from an alkane, the remaining group of atoms is called an alkyl group.

7. The shape of a molecule is specified by the positions of the centers of all of the atoms in the molecule. The typical arrangement of atoms about a carbon atom with four single bonds is tetrahedral.

8. There is essentially free rotation about C—C single bonds, except those C—C single bonds that are part of a ring structure.

PROBLEMS

1. Draw structural formulas for and name all of the alkyl groups with the formula C_4H_9-.

2. Draw structural formulas for each of the following alkanes:
   (a) Isobutane   (b) *n*-Octane   (c) Octane
   (d) 2,2-Dimethylnonane   (e) 3-Ethylpentane

3. Determine which pairs of the following compounds are structural isomers:
   (a) 2,2-Dimethylbutane and 2-methylpentane
   (b) *n*-Octane and 4-ethyloctane

   (c) 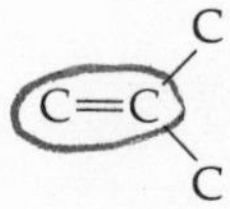
   $$C=C \quad \text{and} \quad C=C$$

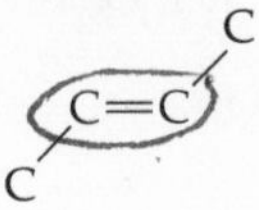

   (d) 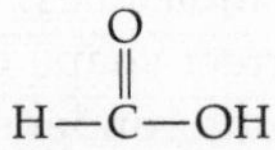 $H-\overset{\overset{\displaystyle O}{\|}}{C}-OH$ and $H-\overset{\overset{\displaystyle O}{\|}}{C}-H$

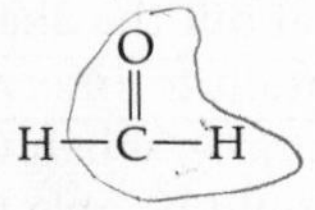

   For those pairs that are structural isomers, indicate whether they are functional-group or positional isomers.

4. Write the IUPAC names for the following alkanes:

   (a) 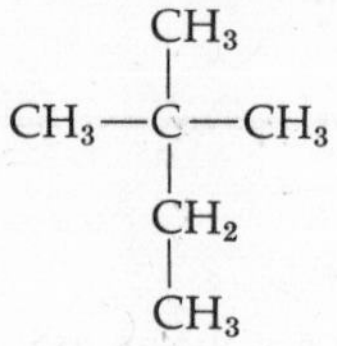 $CH_3-\overset{\overset{\displaystyle CH_3}{|}}{\underset{\underset{\displaystyle CH_3}{|}}{\underset{\overset{\displaystyle CH_2}{|}}{C}}}-CH_3$

   (b) 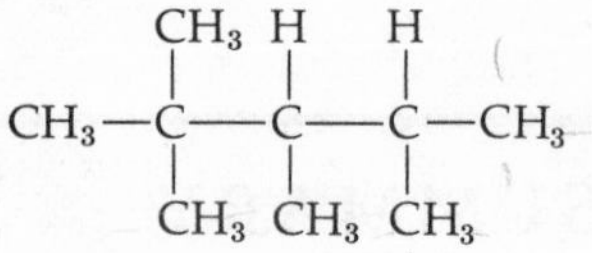 $CH_3-\overset{\overset{\displaystyle CH_3}{|}}{\underset{\underset{\displaystyle CH_3}{|}}{C}}-\overset{\overset{\displaystyle H}{|}}{\underset{\underset{\displaystyle CH_3}{|}}{C}}-\overset{\overset{\displaystyle H}{|}}{\underset{\underset{\displaystyle CH_3}{|}}{C}}-CH_3$

   (c) $CH_3-CH_2-CH_2-CH_3$

   (d) 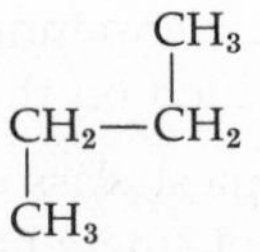 $CH_2-\overset{\overset{\displaystyle CH_3}{|}}{CH_2}$ $\underset{\displaystyle CH_3}{|}$

   (e) 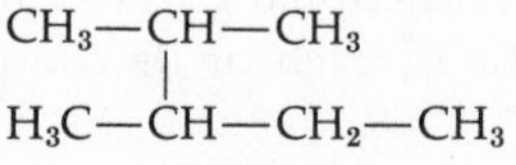 $\overset{\displaystyle CH_3-CH-CH_3}{\underset{\displaystyle H_3C-CH-CH_2-CH_3}{|}}$

   (f) 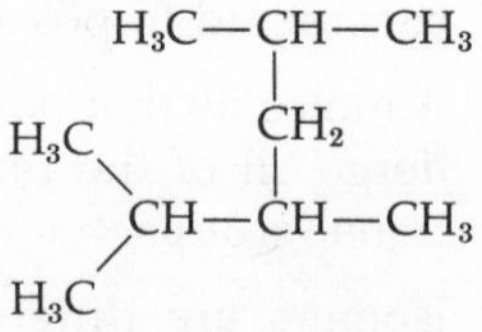 $H_3C \diagdown \atop H_3C \diagup CH - \overset{\overset{\displaystyle H_3C-CH-CH_3}{|}}{\underset{\underset{\displaystyle CH-CH-CH_3}{|}}{CH_2}}$

SOLUTIONS TO EXERCISES

13-1 (a) Both of these molecules have the same molecular formula, $C_4H_8O_2$. However, they have different functional groups and are thus functional-group isomers.
(b) These are positional isomers, because the $C=C$ double bond is located in different positions in these two molecules.
(c) These have different molecular formulas and are therefore not isomers.
(d) These are skeleton structural formulas for two positional isomers with the molecular formula C_6H_{14}.
(e) These two skeleton structural formulas both represent the same compound.

13-2 (a) C—C—C—C—C

(b) 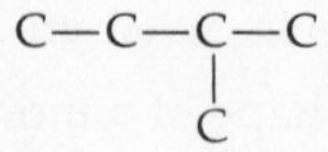 $C-C-\overset{\overset{\displaystyle C}{|}}{C}-C$

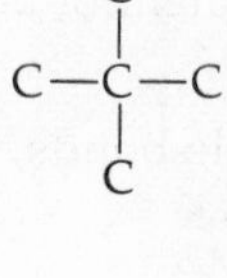

(c) $C-\overset{\overset{\displaystyle C}{|}}{\underset{\underset{\displaystyle C}{|}}{C}}-C$

13-3 (a)

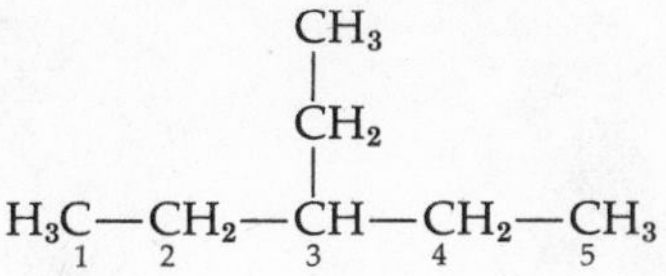

$$CH_3$$
$$|$$
$$CH_2$$
$$|$$
$$\underset{1}{H_3C}-\underset{2}{CH_2}-\underset{3}{CH}-\underset{4}{CH_2}-\underset{5}{CH_3}$$

Root = pent, so the longest chain has five carbon atoms

Ending = -ane, an alkane

Prefix = 3-ethyl; an ethyl substituent is located on carbon 3 of the longest chain

(b)

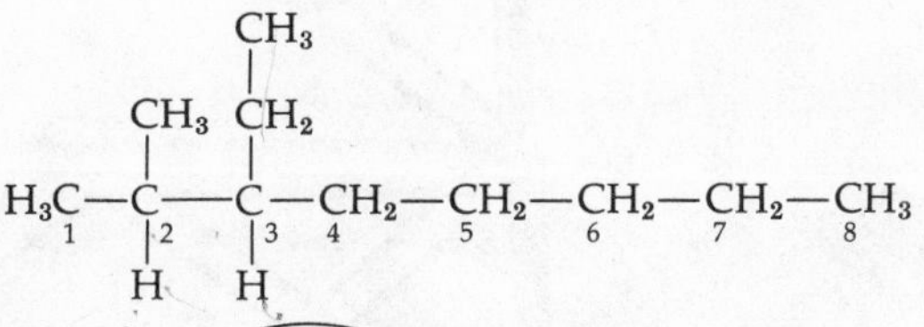

$$CH_3$$
$$|$$
$$CH_3 \ CH_2$$
$$| \quad |$$
$$\underset{1}{H_3C}-\underset{2}{C}-\underset{3}{C}-\underset{4}{CH_2}-\underset{5}{CH_2}-\underset{6}{CH_2}-\underset{7}{CH_2}-\underset{8}{CH_3}$$
$$| \quad |$$
$$H \quad H$$

(c)

$$CH_3 \qquad CH_3$$
$$| \qquad |$$
$$\underset{1}{H_3C}-\underset{2}{C}-\underset{3}{CH_2}-\underset{4}{CH}-\underset{5}{CH_3}$$
$$|$$
$$CH_3$$

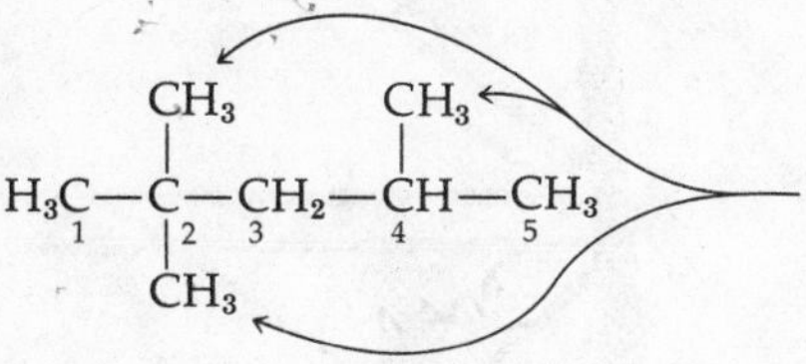

Prefix = 2,2,4-trimethyl

13-4 (a) Longest chain = three carbons; root = prop.
An alkane; ending = -ane.
No substituents.
Thus the IUPAC name is propane.
(b) Root + ending = propane.
A methyl substituent is located on the second carbon but need not be numbered, because if it were on one of the other carbon atoms the longest chain would have four carbons.
Thus the IUPAC name is methylpropane.
(c) Longest carbon chain = 5; root = pent.
An alkane; ending = -ane.
A methyl substituent is located on carbon 3.
The IUPAC name is thus 3-methylpentane.
(d) Root + ending = pentane.
Two methyl substituents located on carbons 2 and 4; prefix = 2,4-dimethyl.
Thus the IUPAC name is 2,4-dimethylpentane.

CHAPTER 14

Hydrocarbons

14-1 INTRODUCTION

The simplest broad group of organic compounds are those that contain only the elements carbon and hydrogen—the hydrocarbons. Hydrocarbons have a number of common physical and chemical properties. For example, they are all nonpolar and thus quite insoluble in water, and they can all burn, a process in which they react vigorously with oxygen.

None of the organic compounds found in the human body are hydrocarbons; these compounds all contain other elements, especially oxygen and nitrogen. But molecules of most of the chemical compounds in human cells do contain hydrocarbonlike parts consisting solely of carbon and hydrogen. Thus, in order to understand the chemical properties of more complex biomolecules, we must have some understanding of the structure, physical properties, and chemical reactivity of hydrocarbons.

There is another important and timely reason to become familiar with the properties and reactivity of hydrocarbons. The hydrocarbons found in natural gas and petroleum play a crucial role in the workings of modern industrial society. We depend on naturally occurring hydrocarbons for fuel and as a source of raw materials for the manufacture of plastics, synthetic rubber, fertilizers, and hundreds of compounds that we use in our daily lives. A large number of synthetic drugs are also manufactured from the hydrocarbons found in petroleum. Since the world's supply of naturally occurring petroleum is limited, a great deal of chemical research is currently aimed at discovering economical alternative sources of hydrocarbons.

Coal, shale oil, and even garbage are potential sources of vast amounts of hydrocarbons. There are major problems, however, with each of these potential sources. For example, although there is a huge amount of coal in the world, with our existing technology most of it is very costly to mine. In addition, much of the coal in nature contains some sulfur. When coal with a high sulfur content is burned, large amounts of atmospheric pollutants are produced. These pollutants are a major source of environmental damage.

The large number of derricks that pump oil from beneath the ocean floor off the coast of Venezuela are a graphic example of our dependence on limited reserves of naturally occurring petroleum.

14-2   STUDY OBJECTIVES

After careful study of the material in this chapter, you should be able to:

1. Classify a hydrocarbon as an alkane, alkene, alkyne, aromatic hydrocarbon, cycloalkane, or cycloalkene, given its structural formula.

2. Explain the use of the symbols R and R'.

3. Draw the structural formula for a hydrocarbon molecule, given its IUPAC name, and vice versa.

4. Draw structural formulas for structural and/or geometric isomers of a hydrocarbon molecule, given its molecular formula.

5. Describe the polarity and solubility of each class of hydrocarbons.

6. Identify the hydrocarbonlike portions of a molecule, given its structural formula.

7. Explain why cyclohexane does not preferentially exist in a planar shape, and why some disubstituted cycloalkanes possess stereoisomers.

8. Describe the shape of an ethene molecule, and explain why some disubstituted alkenes possess geometric isomers.

9. Draw the structural formulas and give the names of the products of reactions in which alkenes are (a) reduced, (b) hydrated, or (c) halogenated.

10. Draw the structural formulas for the products obtained in reactions that form carbon-carbon double bonds.

11. Describe the chemical reactivity of aromatic molecules.

12. Define the term polymer, and describe the structure of polyolefins.

14-3   CLASSES OF HYDROCARBONS

Hydrocarbons are subdivided into four classes that have markedly different properties. Recall that alkanes which contain only C—C and C—H single bonds participate in relatively few chemical reactions and do not have a characteristic functional group. Alkanes are combustible, however, and are used extensively as a primary source of energy. For example, methane is the major component of natural gas. Although there are no alkanes in human cells, most biomolecules do contain alkanelike portions.

Hydrocarbons that contain one or more C=C double bond functional groups belong to the class of molecules called **alkenes.** In living cells there are many molecules that contain C=C double bonds. Thus, after studying the properties of alkenes, we shall be able to predict those properties of more complex molecules that depend on the presence of C=C double bonds. **Alkynes** are a class of hydrocarbons that contain C≡C triple bonds. There are only a handful of compounds in nature that contain the C≡C functional group.

Some hydrocarbons contain rings of carbon atoms. **Cycloalkanes** are a subclass of alkanes that contain rings of carbon atoms and C—C and C—H single bonds only. Cycloalkanes are interesting because a number of biomolecules contain rings of carbon atoms joined by single bonds. Alkenes containing rings of carbon atoms with one or more C=C functional groups also exist; they are called **cycloalkenes.** Compounds that contain a benzene ring belong to a class of hydrocarbons called **aromatic hydrocarbons.**

Exercise 14-1

For each class of hydrocarbons, name the characteristic functional group.

14-4 ALKANES

In the last chapter we discussed the structure and nomenclature of alkanes. Recall that in an alkane each carbon atom forms four covalent bonds either with hydrogen atoms or with other carbon atoms. Since alkanes contain only C—C and C—H single bonds, and each carbon atom in the skeleton formula forms the maximum number of C—H bonds, we say they are **saturated.** You should also recall that the IUPAC name for an alkane consists of: (1) a root derived from the longest continuous chain of carbon atoms, (2) the ending -*ane*, and (3) prefixes that indicate the nature and position of substituent alkyl groups that are bonded to the longest continuous chain of carbon atoms. The rules for naming all classes of organic compounds are reviewed in Essential Skills 8.

Properties of Alkanes

Alkanes are relatively unreactive. The most important exception is that alkanes and all other classes of hydrocarbons can react with oxygen to produce CO_2 and H_2O. The process is known as **combustion** (burning). For example, propane reacts with oxygen as follows:

(14-1) $$CH_3—CH_2—CH_3 + 5O_2 \longrightarrow 3CO_2 + 4H_2O$$

Reaction 14-1 is highly exothermic (heat releasing), $\Delta H = -531$ kcal/mole, and this is one reason for the use of bottled propane gas as a fuel for heating homes and campers. Larger alkanes, such as octane and its isomers, are also used as fuels (for example, in gasoline). The human body cannot use the combustion of alkanes as a source of energy. In fact, most alkanes are poisonous.

Alkyl groups and alkanelike portions of large molecules are also relatively nonreactive. There are, however, a few important enzyme-catalyzed reactions in the human body in which C—C single bonds are oxidized to produce a C=C double bond. In the majority of these reactions, however, the C—C single bond is very close to a functional group that strongly enhances its reactivity. A typical example is the oxidation of the —CH_2—CH_2— portion of succinic acid:

(14-2)

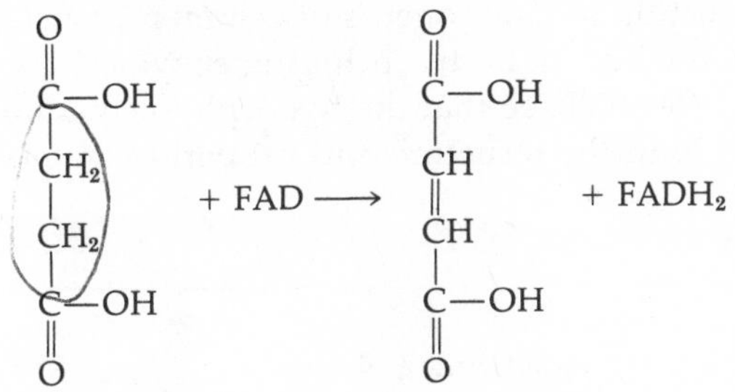

succinic acid   fumaric acid

Reaction 14-2 is necessary for energy production in the human body (Chapter 25). In this reaction, succinic acid is oxidized (loses hydrogen) and the complex molecule represented by the symbol FAD (Chapter 23) is reduced (gains hydrogen).

Natural gas, a mixture of simple gaseous alkanes, is burned off at this well in a Libyan oil field.

Transport of natural gas to countries in need of fuel is often not economical.

Not only are alkanes generally unreactive, but they are also nonpolar. Recall that C—C and C—H are nonpolar covalent bonds (Chapter 4). Thus, alkanes cannot form hydrogen bonds with water and are insoluble in it. Modern society is very aware of this fact, as a result of the unfortunate incidences of oil spills in the oceans. Recall that we use the word **hydrophobic** to refer to molecules or parts of molecules that tend to be insoluble in water (Chapter 7). The alkanelike portions of biomolecules are also nonpolar and hydrophobic, and thus biomolecules with large alkanelike portions are not soluble in water.

For example, fatty acids with a large alkanelike portion, such as palmitic acid, which has the structural formula

$$
H-\overset{\overset{\displaystyle H}{|}}{\underset{\underset{\displaystyle H}{|}}{C}}-\overset{\overset{\displaystyle H}{|}}{\underset{\underset{\displaystyle H}{|}}{C}}-\overset{\overset{\displaystyle H}{|}}{\underset{\underset{\displaystyle H}{|}}{C}}-\overset{\overset{\displaystyle H}{|}}{\underset{\underset{\displaystyle H}{|}}{C}}-\overset{\overset{\displaystyle H}{|}}{\underset{\underset{\displaystyle H}{|}}{C}}-\overset{\overset{\displaystyle H}{|}}{\underset{\underset{\displaystyle H}{|}}{C}}-\overset{\overset{\displaystyle H}{|}}{\underset{\underset{\displaystyle H}{|}}{C}}-\overset{\overset{\displaystyle H}{|}}{\underset{\underset{\displaystyle H}{|}}{C}}-\overset{\overset{\displaystyle H}{|}}{\underset{\underset{\displaystyle H}{|}}{C}}-\overset{\overset{\displaystyle H}{|}}{\underset{\underset{\displaystyle H}{|}}{C}}-\overset{\overset{\displaystyle H}{|}}{\underset{\underset{\displaystyle H}{|}}{C}}-\overset{\overset{\displaystyle H}{|}}{\underset{\underset{\displaystyle H}{|}}{C}}-\overset{\overset{\displaystyle H}{|}}{\underset{\underset{\displaystyle H}{|}}{C}}-\overset{\overset{\displaystyle H}{|}}{\underset{\underset{\displaystyle H}{|}}{C}}-\overset{\overset{\displaystyle H}{|}}{\underset{\underset{\displaystyle H}{|}}{C}}-\overset{\overset{\displaystyle O}{\|}}{C}-OH
$$

are not water soluble, despite the presence of the polar carboxylic acid functional group on these molecules. We shall see that this feature of fatty acids is important in their function as components of cell membranes (Chapter 26).

Other molecules in the human body, including enzymes, also contain hydrophobic alkyl groups. We shall see that the hydrophobic nature of alkyl groups plays an important role in the structure and properties of enzymes and other proteins (Chapter 21).

Exercise 14-2
Draw structural formulas for the following molecules:

    (a) 3-Methylheptane
    (b) 3-Ethyl-4-*tert*-butyloctane
    (c) 3-Isopropylpentane

Exercise 14-3
Draw the structural formulas and write the IUPAC names for all of the positional isomers of hexane.

Exercise 14-4
Identify the alkanelike portions of the following molecules:

(a) 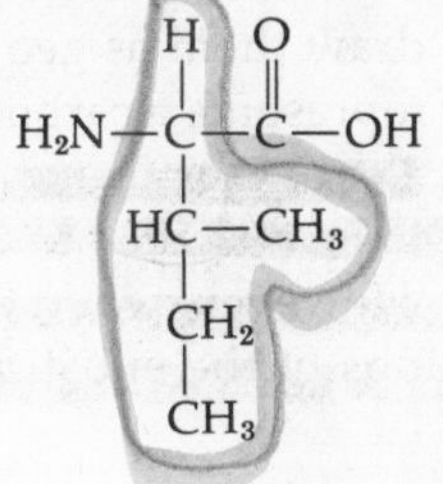 $H_2N-\underset{\underset{\underset{\underset{CH_3}{|}}{CH_2}}{\underset{|}{HC-CH_3}}}{\overset{\overset{H}{|}}{C}}-\overset{\overset{O}{\|}}{C}-OH$ isoleucine (an amino acid)

(b) 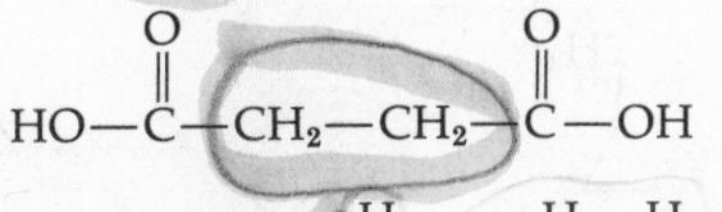 $HO-\overset{\overset{O}{\|}}{C}-CH_2-CH_2-\overset{\overset{O}{\|}}{C}-OH$ succinic acid

(c) 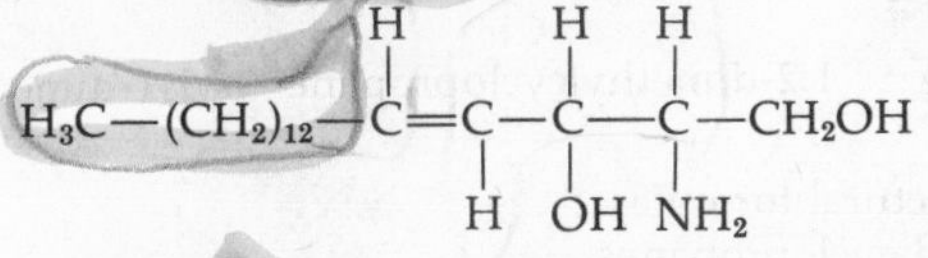 $H_3C-(CH_2)_{12}-\overset{\overset{H}{|}}{C}=C-\overset{\overset{H}{|}}{\underset{\underset{OH}{|}}{C}}-\overset{\overset{H}{|}}{\underset{\underset{NH_2}{|}}{C}}-CH_2OH$ sphingosine (a component of cell membranes)

(d) 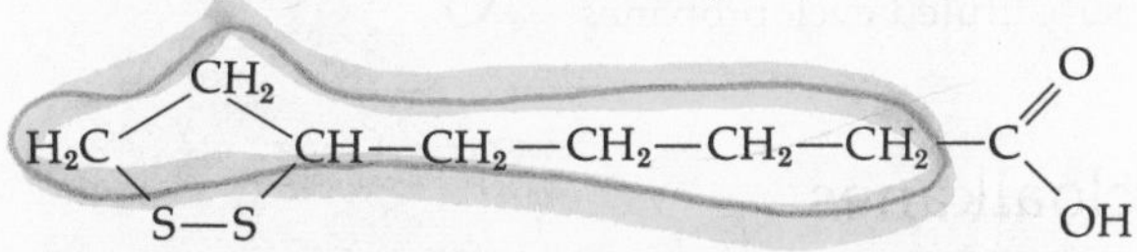 $\underset{S-S}{H_2C}\overset{CH_2}{\diagdown}CH-CH_2-CH_2-CH_2-CH_2-\overset{\overset{O}{\|}}{C}\diagdown_{OH}$ lipoic acid (a coenzyme)

14-5 CYCLOALKANES

We have just discussed alkanes, which consist of linear chains of carbon atoms. Alkanes with rings of carbon atoms also exist. They are called **cycloalkanes.** The C—C and C—H bonds in cycloalkanes are generally similar to those in alkanes, and like the alkanes, cycloalkanes are not components of human cells. However, several important molecules in human cells contain rings of five or six atoms. The study of cycloalkanes will aid our understanding of these complex molecules.

Structure and Nomenclature

Cycloalkanes are named by placing the prefix *cyclo-* before the name of the corresponding alkane. The structures and names of some simple cycloalkanes are given in Table 14-1. Since an additional C—C single bond is used to close the

Table 14-1 Some Cycloalkanes

Name	Structural Formula	Simplified Representation
Cyclopropane	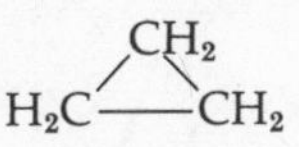	
Cyclobutane	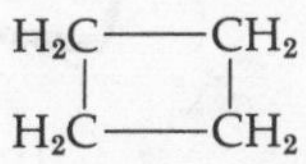	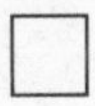
Cyclopentane	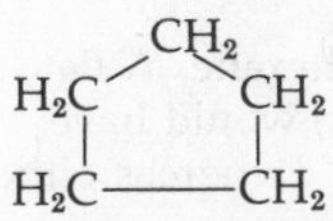	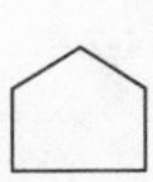
Cyclohexane	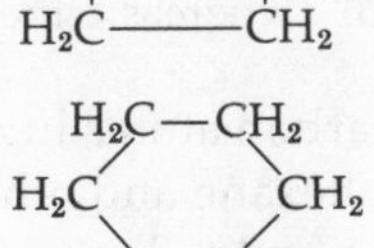	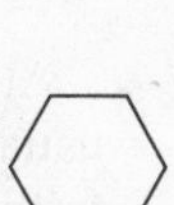

ring, a cycloalkane contains two fewer hydrogen atoms than the open-chain alkane with the same number of carbon atoms. Chemists often abbreviate the
structural formulas for cycloalkanes, and just draw them as geometrical figures
(triangles, squares, etc.) in which each corner represents a carbon atom, and the
hydrogen atoms are omitted (see Table 14-1). For a substituted cycloalkane the
position of a single substituent does not need to be specified in the name, since
all of the positions in the ring are equivalent. However, when there are two or
more substituents in a cycloalkane, the positions of the substituents are numbered as for alkanes (see Figure 14-1).

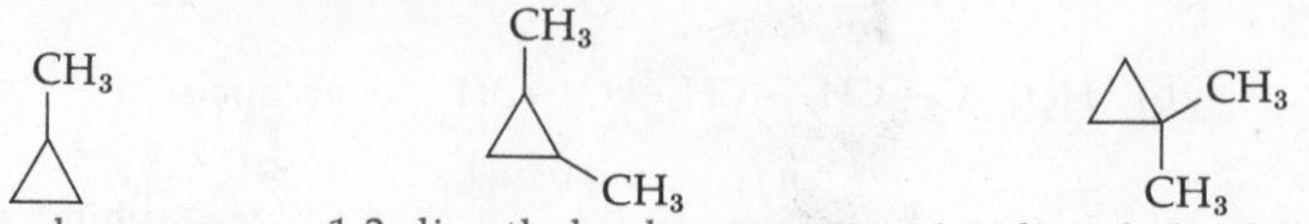

| methylcyclopropane | 1,2-dimethylcyclopropane | 1,1-dimethylcyclopropane |

Figure 14-1   The names and structural formulas
for some substituted cyclopropanes.

The Shape of Cycloalkanes

As we saw in Section 13-6, a tetrahedral orientation of bonds with bond angles
of 109.5° is typical for carbon atoms that form four single bonds. The tetrahedral
arrangement of four bonds around a carbon atom is preferred whenever possible because this arrangement is the one with the lowest energy. For some
cycloalkanes, however, a tetrahedral arrangement for all carbon atoms is not
possible. The $C\!\!\!\curvearrowright\!\!\!C$ bond angles in cyclopropane must be 60°, and they are
close to 90° in cyclobutane.

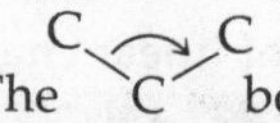

All of the carbon atoms in larger cycloalkanes such as cyclohexane can possess
a tetrahedral arrangement of bonds when all of the carbon atoms do not lie in
the same plane. Cyclohexane with the carbon atoms at the corners of a flat hexagon would have bond angles of 120° (see Figure 14-2a). However, if the carbon
atoms in a molecule of cyclohexane are not all in the same plane, the bond
angles can be the 109.5° required for tetrahedral bonding. Cyclohexane preferentially exists in a nonplanar shape, called the **chair conformation,** where all
of the bonds angles are 109.5° (see Figure 14-2b). The chair conformation of
cyclohexane has the lowest energy. However, cyclohexane is not rigid and can
assume other shapes. Another shape where all of the bond angles are 109.5° is
called the **boat conformation** (see Figure 14-2c). Since there is very little energy
difference between these and other conformations for cyclohexane, it is not
possible to isolate cyclohexane molecules with different shapes. Hence, these
conformations of cyclohexane are *not* stereoisomers.

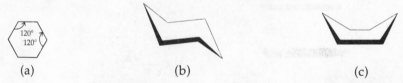

(a)   (b)   (c)

Figure 14-2   The shape of cyclohexane. A flat   the chair conformation (b) or the
hexagonal shape (a) would have   boat conformation (c) has bond
bond angles of 120°, whereas either   angles of 109.5°.

Although all of the carbon atoms in cyclohexane are not in the same plane, we
usually represent cyclohexane and other molecules containing a six-carbon ring
with a planar hexagon, and talk about bonds above and below this plane.

Geometric Isomers of Cycloalkanes

Although there is essentially free rotation around C—C single bonds in alkanes (Section 13-6), this is not possible for cycloalkanes. The ring structure prevents free rotation (see Figure 14-3). For example, rotation of one carbon atom 180° around another carbon atom would require a C—C single bond to be broken, which requires a large amount of energy.

Figure 14-3   Free rotation occurs around C—C single bonds in open-chain alkanes (left) but is not possible around the C—C single bonds in a cycloalkane (right).

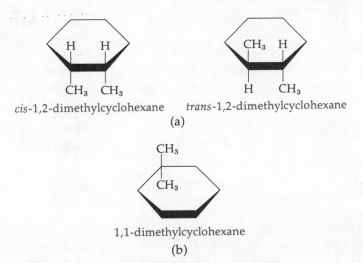

free rotation   free rotation not possible

Because of the lack of free rotation around C—C bonds in cycloalkanes, disubstituted cycloalkanes with the same structural formula may have different shapes; that is, they may be stereoisomers. For example, consider a molecule of 1,2-dimethylcyclohexane (Figure 14-4a). The two methyl groups in this molecule can be either on the same side of the ring or on opposite sides. Since the methyl groups cannot rotate from one side to the other, molecules of the two compounds depicted in Figure 14-4a are distinct. These two compounds have different properties and can thus be separated from each other. Since these compounds have the same structural formula but different shapes, they are stereoisomers (Section 13-6). Stereoisomers that differ from each other in the spatial arrangement (the geometry) of substituents are called **geometric** or **cis-trans isomers.** We use the prefix *cis-* to denote the isomer in which both substituents are on the same side of the ring and *trans-* to denote the isomer in which they are on opposite sides. Not all disubstituted cycloalkanes, however, possess cis-trans isomers. For example, there are no geometric isomers of 1,1-dimethylcyclohexane (Figure 14-4b).

cis-1,2-dimethylcyclohexane   *trans*-1,2-dimethylcyclohexane

(a)

1,1-dimethylcyclohexane

(b)

Figure 14-4   There are two geometric (cis-trans) isomers of 1,2-dimethylcyclohexane (a), but there are no geometric isomers of 1,1-dimethylcyclohexane (b).

Exercise 14-5

The molecular formula for cyclohexane is C_6H_{12}. Why are there two fewer hydrogen atoms in cyclohexane than there are in hexane?

Exercise 14-6

Why is the shape of cyclohexane molecules not simply a planar hexagon?

Exercise 14-7

Draw the structural formula of *trans*-1-ethyl-2-isopropylcyclohexane.

14-6 ALKENES

Hydrocarbons that contain the C=C double bond functional group are called alkenes. Molecules containing C=C double bonds abound in nature and are very important to the human body.

Structure and Nomenclature

In the IUPAC system of nomenclature, an alkene is named after the alkane with the same number of carbon atoms by changing the ending from *-ane* to *-ene*. A numerical prefix is used to indicate the position of the C=C double bond. Hence, the compound $H_2C=CH_2$ is named ethene. The common name for ethene is ethylene. The names and structural formulas of a few alkenes are given in Table 14-2. Notice that there are three different positional isomers that contain four carbon atoms and one C=C double bond. According to IUPAC rules, the positional isomer with the C=C double bond between the first and second carbon atoms is called 1-butene. The other two are 2-butene and 2-methyl-propene. Note that the IUPAC name for $CH_3—CH_2—CH=CH_2$ is 1-butene, and not 3-butene. The IUPAC rules require that the lowest possible numbers be used to specify the location or locations of all double bonds or substituents.

Table 14-2 Some Alkenes

No. of C Atoms	Structural Formula	Name
2	$H_2C=CH_2$	Ethene or ethylene (used to make polyethylene)
3	$CH_3—CH=CH_2$	Propene
4	$CH_2=CH—CH_2—CH_3$	1-Butene
4	$CH_3—CH=CH—CH_3$	2-Butene
4	$CH_3—\overset{\displaystyle CH_3}{\overset{\displaystyle \vert}{C}}=CH_2$	2-Methylpropene
4	$CH_2=CH—CH=CH_2$	1,3-Butadiene
5	$H_2C=\overset{\displaystyle CH_3}{\overset{\displaystyle \vert}{C}}—CH=CH_2$	2-Methyl-1,3-butadiene or isoprene (used to make rubber)

Alkenes that contain two or more double bonds also exist. For these alkenes the ending *-ene* is changed to *-diene* (or *-adiene*) for alkenes with two double bonds, *-triene* (or *-atriene*) for alkenes with three double bonds, and so on. A numerical prefix is also used to specify the position of each double bond. For example, the IUPAC name for $H_2C=CH—CH=CH_2$ is 1,3-butadiene. The compound 2-methyl-1,3-butadiene is commonly called isoprene (Table 14-2). An isoprene-type unit is a common structural feature of some biomolecules, including vitamins A and K (Chapter 26), as well as natural and synthetic rubber (Section 14-9).

Shapes of Alkenes

It has been determined experimentally that all six atoms in the ethene molecule lie in the same plane, with angles of 120° between the three bonds formed by each carbon atom (see Figure 14-5). We can view the shape of ethene as two CH_2

Figure 14-5   The shape of ethene molecules. All six atoms in ethene lie in the same plane, with bond angles of 120°.

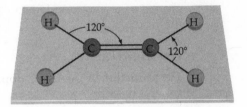

triangles joined together by the C=C double bond. It has also been determined experimentally that a great deal of energy is required to rotate one CH_2 triangle with respect to the other. Thus, the CH_2 triangles are kept in a rigid coplanar arrangement and we say that there is not free rotation about the C=C double bond. As a consequence of this fact, substituted alkenes may have geometric isomers. The simplest example is 2-butene:

cis-2-butene $trans$-2-butene

Two geometric isomers of 2-butene exist: cis-2-butene, in which both methyl groups are located on the same side of the C=C double bond; and $trans$-2-butene, in which the methyl groups are on opposite sides. These geometric isomers have somewhat different physical properties and chemical reactivity (see Table 14-3), so they can be separated from each other. Thus there are a total of four isomers (positional and geometric) of butene.

Table 14-3 Properties of Butene Isomers

Name	Structural Formula	Boiling Point (°C)
1-Butene	$H_2C{=}CH{-}CH_2{-}CH_3$	−6
cis-2-Butene	$\begin{array}{c} H_3C \quad\quad CH_3 \\ \diagdown C{=}C \diagup \\ \diagup \quad\quad \diagdown \\ H \quad\quad\quad H \end{array}$	+1
$trans$-2-Butene	$\begin{array}{c} H_3C \quad\quad H \\ \diagdown C{=}C \diagup \\ \diagup \quad\quad \diagdown \\ H \quad\quad\quad CH_3 \end{array}$	+4
2-Methylpropene	$H_2C{=}C\diagup^{CH_3}_{\diagdown CH_3}$	−7

Properties of Alkenes

The physical properties of alkenes are generally similar to those of the corresponding alkanes. They are nonpolar, insoluble in water, and less dense than water. However, the chemical reactivity of alkenes is quite different than that of alkanes. The C=C double bond functional group can participate in several reactions. Three important reactions of alkenes are reduction, halogenation, and hydration. All of these reactions involve breaking one of the two bonds between the doubly bonded carbon atoms, leaving a C—C single bond, and the formation of an additional single bond between each carbon atom and a hydrogen, oxygen, or halogen atom. These three types of reactions are referred to as **addition reactions,** since the product molecule has an additional atom (or groups of atoms) bonded to each of the carbon atoms of the double bond in the reactant.

CHEMICAL REACTIVITY

The **reduction** of the C=C double bond in an alkene produces an alkane in the general reaction:

(14-3)
$$H_2 + R_1'—CH=CH—R_2' \xrightarrow{catalyst} R_1'—CH_2—CH_2—R_2'$$

Chemists frequently use the symbol R for an alkyl group. Thus, R can represent —CH_3 (methyl), —CH_2—CH_3 (ethyl), and so on. The symbol R′ denotes an alkyl group or a hydrogen atom. The subscripts 1 and 2 indicate that the R groups may not be identical. Thus, if R_1' is the ethyl alkyl group and R_2' is a hydrogen atom, reaction 14-3 becomes

(14-3a)
$$H_2 + CH_3—CH_2—CH=CH_2 \xrightarrow{catalyst} CH_3—CH_2—CH_2—CH_3$$

The reduction of a C=C double bond is also referred to as the **hydrogenation** of a C=C double bond. Hydrogenation is commercially important in the production of gasoline and other fuels, shortenings, and some detergents. Hydrogenation reactions in industry and chemical laboratories require high pressures, high temperatures, and a catalyst such as platinum. On the other hand, the metabolism of fats and carbohydrates in the human body requires the enzyme-catalyzed reduction of C=C double bonds under much milder conditions. For example, one step in the biosynthesis of a fatty acid molecule is the reduction of a double bond in the reaction

(14-4)
$$R—CH_2—CH_2—CH=CH—\overset{\displaystyle O}{\overset{\displaystyle \|}{C}}—S—CoA + NADPH + H^+ \xrightarrow{enzyme}$$

$$R—CH_2—CH_2—\overset{\displaystyle H}{\underset{\displaystyle H}{C}}—\overset{\displaystyle H}{\underset{\displaystyle H}{C}}—\overset{\displaystyle O}{\overset{\displaystyle \|}{C}}—S—CoA + NADP^+$$

In reaction 14-4, NADPH represents a coenzyme that donates hydrogen and is oxidized to NADP$^+$. A coenzyme portion of the other reactant is represented as CoA. Another coenzyme, FAD, is frequently used as a hydrogen acceptor in reactions in which CH_2—CH_2 single bonds are oxidized to HC=CH double bonds in the human body. See reaction 14-2 for an example.

Another addition reaction of C=C double bonds is **halogenation.** The reaction between an alkene and Br_2 or Cl_2 results in the addition of two halogen atoms, forming a dibromo- or a dichloro- derivative of the alkene. For example, 2-methylpropene will react with Br_2 to produce 1,2-dibromo-2-methylpropane:

(14-5)
$$H_2C{=}\underset{\underset{\displaystyle CH_3}{|}}{C}{-}CH_3 + Br_2 \longrightarrow H{-}\underset{\underset{\displaystyle H}{|}}{\overset{\overset{\displaystyle Br}{|}}{C}}{-}\underset{\underset{\displaystyle CH_3}{|}}{\overset{\overset{\displaystyle Br}{|}}{C}}{-}CH_3$$

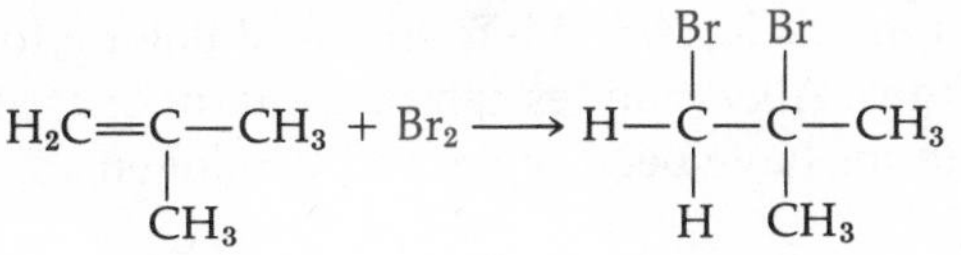

Notice in reaction 14-5 that one bromine atom adds to each of the two carbon atoms involved in the C=C double bond. When Br_2 or Cl_2 reacts with any C=C double bond, one halogen atom will bond to each of the two C atoms that formed the double bond.

Alkenes will also participate in addition reactions with hydrogen bromide or hydrogen chloride to produce derivatives that contain a single halogen atom. For example,

(14-6)

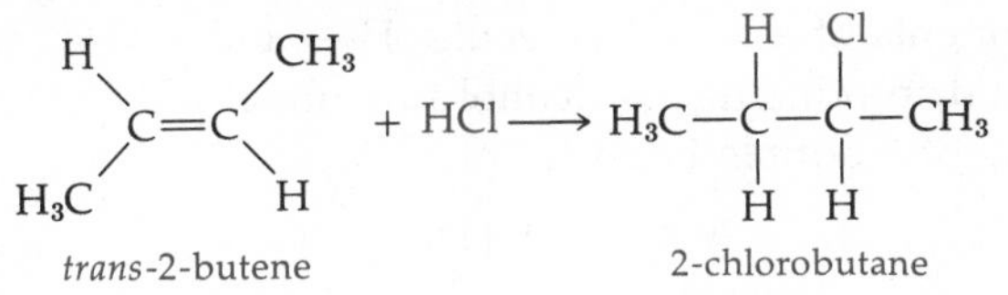

        trans-2-butene   2-chlorobutane

and

(14-7)

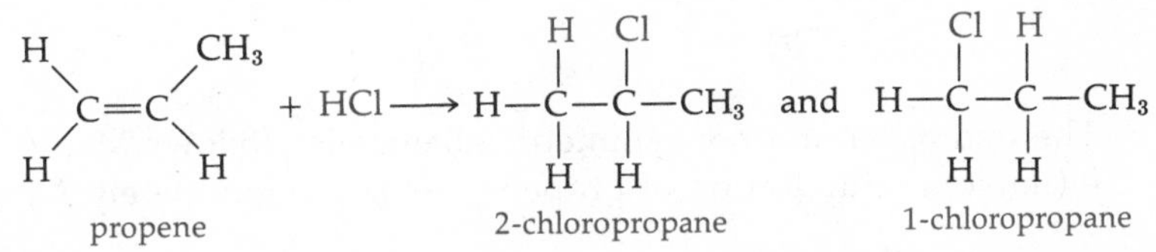

        propene   2-chloropropane   1-chloropropane

In reaction 14-6 the only product formed is 2-chlorobutane, since the reactant, *trans*-2-butene, is a symmetric molecule. Propene, on the other hand, is not a symmetric molecule, and the products of reaction 14-7 are a mixture of the positional isomers 2-chloropropane and 1-chloropropane.

It has been determined experimentally that the product of reaction 14-7 is almost entirely 2-chloropropane. Similarly, when any nonsymmetric alkene reacts with HCl, HBr, or certain other compounds such as H_2O, the hydrogen atom usually adds to the carbon atom of the C=C double bond that *already* has the most hydrogen atoms. This generalization is known as **Markovnikov's rule,** after the Russian chemist Vladimir Markovnikov, who studied the addition reactions of alkenes.

For some nonsymmetric alkenes the products of an addition reaction are roughly a 50–50 mixture of positional isomers. Consider the reaction of HBr with *trans*-2-pentene:

(14-8)

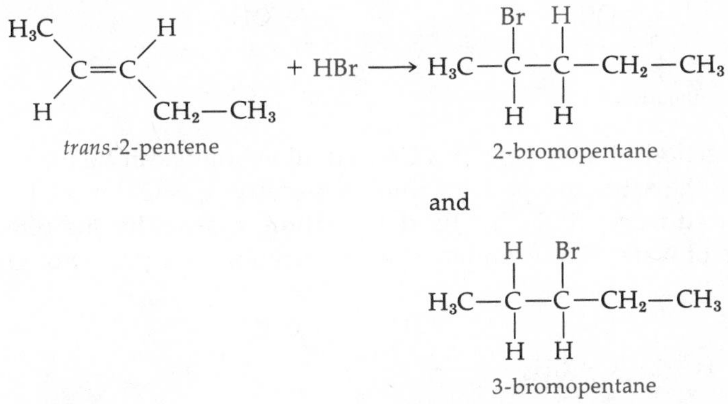

        trans-2-pentene   2-bromopentane

and

                    3-bromopentane

Although *trans*-2-pentene is nonsymmetric, each of the carbon atoms in the C=C double bond has the *same* number of hydrogen atoms (one) bonded to it. Therefore, both 2-bromopentane and 3-bromopentane are formed in roughly equal amounts.

The products of reactions 14-5, 14-6, 14-7, and 14-8 belong to a class of compounds called **alkyl halides.** Alkyl halides can be thought of as alkanes in which one or more hydrogen atoms have been replaced by halogen atoms (Table 14-4).

Table 14-4 Some Important Alkyl Halides

Formula	Common Name	Use
CCl_4	Carbon tetrachloride	Solvent; dry cleaning
$CHCl_3$	Chloroform	Solvent for fats, oils, and rubber
CH_3-CH_2Cl	Ethyl chloride	Local anesthetic

A very important addition reaction of C=C double bonds is **hydration,** where the components of a water molecule, H— and —OH, are added to the C=C double bond, producing an alcohol functional group. The general reaction for hydration of a double bond is

(14-9)

$$\underset{H}{\overset{R_1'}{\diagdown}}C=C\underset{R_2'}{\overset{H}{\diagup}} + H_2O \xrightarrow{\text{catalyst}} R_1'-\overset{\overset{\displaystyle H}{|}}{\underset{\underset{\displaystyle H}{|}}{C}}-\overset{\overset{\displaystyle OH}{|}}{\underset{\underset{\displaystyle H}{|}}{C}}-R_2'$$

The hydration of a nonsymmetric alkene also follows Markovnikov's rule. For example, the hydration of propene yields predominately 2-propanol:

(14-10)

$$\underset{H}{\overset{H}{\diagdown}}C=C\underset{CH_3}{\overset{H}{\diagup}} + H_2O \xrightarrow{\text{catalyst}} H-\overset{\overset{\displaystyle H}{|}}{\underset{\underset{\displaystyle H}{|}}{C}}-\overset{\overset{\displaystyle OH}{|}}{\underset{\underset{\displaystyle H}{|}}{C}}-CH_3$$

propene   2-propanol

The naming of alcohols is discussed in Chapter 15. Several enzyme-catalyzed reactions in the human body involve hydration of C=C double bonds. One example is the hydration of fumaric acid to produce malic acid:

(14-11)

$$H-\overset{\overset{\displaystyle O}{\parallel} \;\; \overset{\displaystyle C}{-}OH}{\underset{\underset{\displaystyle C-OH}{\parallel}}{\underset{\underset{\displaystyle O}{}}{C}}} + H_2O \xrightarrow{\text{an enzyme}} H-\overset{\overset{\displaystyle O}{\parallel}\;\;\overset{\displaystyle C}{-}OH}{\underset{\underset{\displaystyle C-OH}{\parallel}}{\underset{\underset{\displaystyle O}{}}{C}}}$$

fumaric acid   malic acid

In Section 14-4 we saw that C=C double bonds can be formed by the oxidation of the saturated parts of some molecules. C=C double bonds can also be produced from an alcohol by **dehydration,** that is, by the removal of the elements of water. For example, the dehydration of 2-propanol yields propene:

(14-12)

$$H-\overset{\overset{\displaystyle H}{|}}{\underset{\underset{\displaystyle H}{|}}{C}}-\overset{\overset{\displaystyle OH}{|}}{\underset{\underset{\displaystyle H}{|}}{C}}-CH_3 \xrightarrow{\text{catalyst}} \underset{H}{\overset{H}{\diagdown}}C=C\underset{H}{\overset{CH_3}{\diagup}} + H_2O$$

2-propanol   propene

We shall see several more examples of reactions involving C=C double bonds when we study human metabolism.

Exercise 14-8

Draw the structural formulas for the major products of the following reactions:

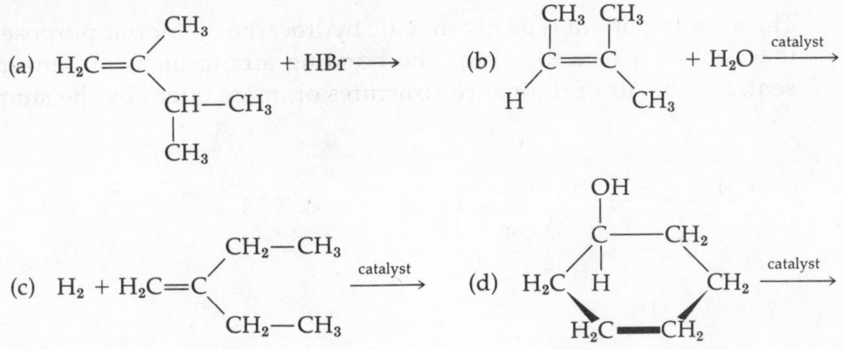

(a) $H_2C{=}C\substack{\text{---}CH_3 \\ \\ \text{---}CH{-}CH_3 \\ \\ \text{---}CH_3}$ + HBr $\longrightarrow$

(b) $\substack{CH_3 \;\; CH_3 \\ \\ C{=}C \\ \\ H \quad\quad CH_3}$ + H_2O $\xrightarrow{\text{catalyst}}$

(c) $H_2 + H_2C{=}C\substack{\text{---}CH_2{-}CH_3 \\ \\ \\ \text{---}CH_2{-}CH_3}$ $\xrightarrow{\text{catalyst}}$

(d) $\substack{\quad OH \\ \\ C{-}CH_2 \\ H_2C \;\; H \quad CH_2 \\ \\ H_2C{=\!=}CH_2}$ $\xrightarrow{\text{catalyst}}$

If you have difficulty with this type of problem, refer to Essential Skills 9 for some helpful hints.

Exercise 14-9

Draw structural formulas for:

(a) *trans*-4-Ethyl-3-methyl-3-heptene
(b) *cis*-3,4-Dimethyl-3-hexene
(c) *trans*-2-Pentene

14-7 ALKYNES

Hydrocarbons that contain one or more C≡C triple bonds are called **alkynes.** The IUPAC ending for an alkyne is *-yne*. Very few alkynes are found in nature. The simplest alkyne is ethyne, H—C≡C—H, commonly called acetylene, which is used as a fuel for welding torches and other processes that require very high temperatures.

Acetylene-fueled torches are used in a variety of industrial welding operations, but in this instance an artist uses an acetylene torch to construct a metal sculpture.

14-8   AROMATIC HYDROCARBONS

The most important type of aromatic hydrocarbons, for our purposes, are those that contain a benzene ring. The bonding arrangement in benzene is represented by a pair of resonance structures or, more often, by the simple symbols

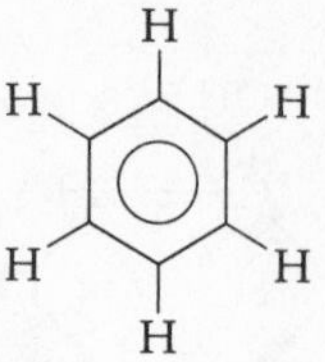

Nomenclature and Structure

Most molecules that contain benzene rings are referred to by their common names, rather than by their IUPAC names. For example, toluene is the name given to the aromatic hydrocarbon that results when a methyl group is substituted for a hydrogen atom in benzene. The names and structures of some substituted aromatic hydrocarbons are given in Table 14-5. Notice that there are only three positional isomers of xylene or dimethyl benzene. Unlike cyclohexane, aromatic molecules are planar. All of the six carbon and six hydrogen atoms in benzene lie in the same plane (see Figure 14-6). Therefore, substituted aromatic compounds such as xylene do not exhibit cis-trans isomerism. When there are two substituent groups on a benzene ring, their positions are

Table 14-5 Some Aromatic Hydrocarbons

Structural Formula	Name
	Benzene
	Toluene (methyl benzene)
	Dimethyl benzenes: o-Xylene (1,2-dimethylbenzene)
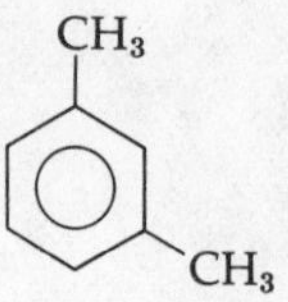	m-Xylene (1,3-dimethylbenzene)
	p-Xylene (1,4-dimethylbenzene)

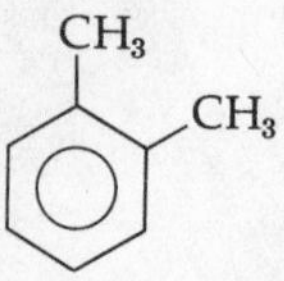

Positional isomers

ortho - 1,2
meta - 1,3
para - 1,4

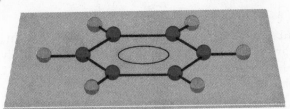

ortho 1,2
para 1,4
meta 1,3

H   C

Figure 14-6   The shape of benzene molecules. All six carbon atoms and all six   hydrogen atoms in benzene lie in the same plane.

indicated by the prefixes *o-* (for ortho), *m-* (for meta), and *p-* (for para) or by numbers (see Table 14-5). For a benzene ring with three or more substituents, numbers are used to indicate the positions of the substituents.

The group of atoms obtained by removing one hydrogen from benzene is called a **phenyl** group. One compound that contains a phenyl group is the amino acid phenylalanine (see Figure 14-7).

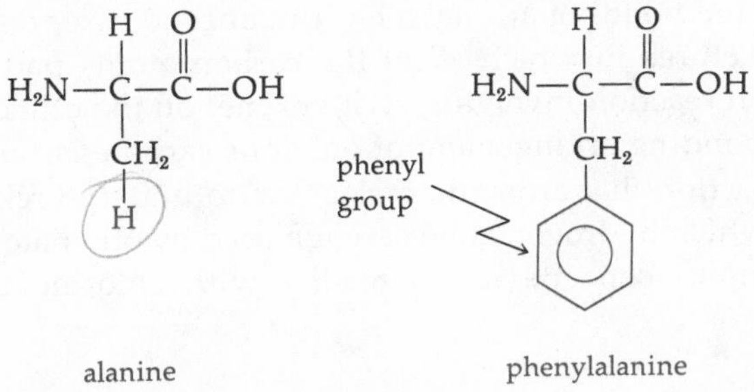

alanine   phenylalanine

Figure 14-7   The difference between the amino acids alanine (on the left) and phenylalanine (on the right) is in   the presence of a phenyl group in phenylalanine instead of one of the hydrogens in alanine.

Properties of Aromatic Compounds

PHYSICAL PROPERTIES

The physical properties of benzene and other aromatic hydrocarbons are similar to those of alkanes and alkenes. They are nonpolar and thus insoluble in water. The components of proteins and nucleic acids contain aromatic portions that, because of their hydrophobic nature, are excluded from the aqueous environment of cells. For example, the amino acid phenylalanine contains a benzene ring that associates with other hydrophobic amino acids in proteins to form a hydrophobic interior, which is partially responsible for the unique structure of these large molecules (see Figure 14-8).

Figure 14-8   Hydrophobic groups, such as aromatic rings, form the interior or hydrophobic core in protein molecules, whereas the hydrophilic portions of the protein are found on the exterior, exposed to the aqueous environment.

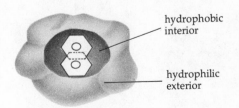

hydrophobic interior

hydrophilic exterior

CHEMICAL REACTIVITY

Benzene and substituted benzenes do not readily participate in the same type of addition reactions as alkenes. For example, cyclohexene rapidly reacts with hydrogen gas in the presence of nickel to give cyclohexane at 25°C and a pressure of about 1.5 atm:

(14-13)

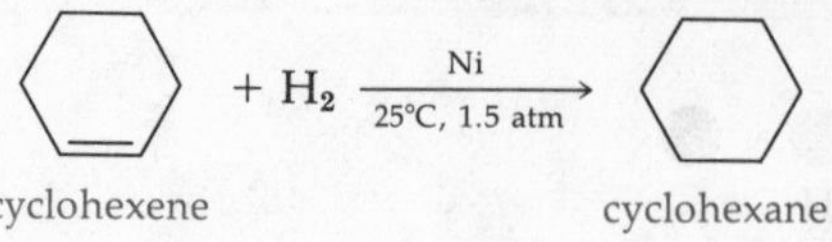

$$\text{cyclohexene} + H_2 \xrightarrow[25°C,\ 1.5\ atm]{Ni} \text{cyclohexane}$$

Benzene also reacts with hydrogen in the presence of nickel to give cyclohexane, but even at 200°C and a pressure of 100 atm, the reaction proceeds slowly:

(14-14)

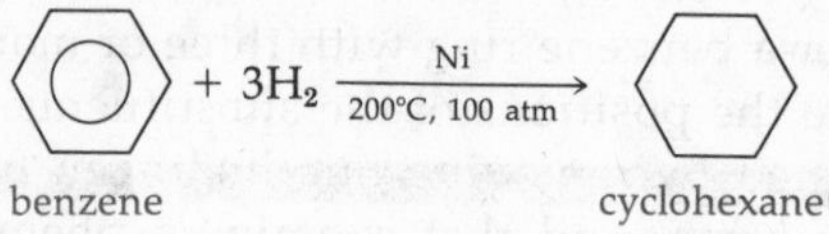

$$\text{benzene} + 3H_2 \xrightarrow[200°C,\ 100\ atm]{Ni} \text{cyclohexane}$$

An explanation for this difference in reactivity is that an addition reaction involves a change in the bonding arrangement of all of the carbon atoms in benzene. Recall that the bonding arrangement in benzene is represented by a pair of resonance structures in which *all* of the carbon atoms participate (Section 4-12). An addition reaction involving cyclohexene, on the other hand, requires a change in the bonding arrangement of only one carbon-carbon double bond.

One type of reaction that aromatic molecules do undergo relatively easily is **substitution,** in which hydrogen atoms are replaced by other atoms or groups of atoms. For example, benzene reacts readily with chlorine to form chlorobenzene:

(14-15)

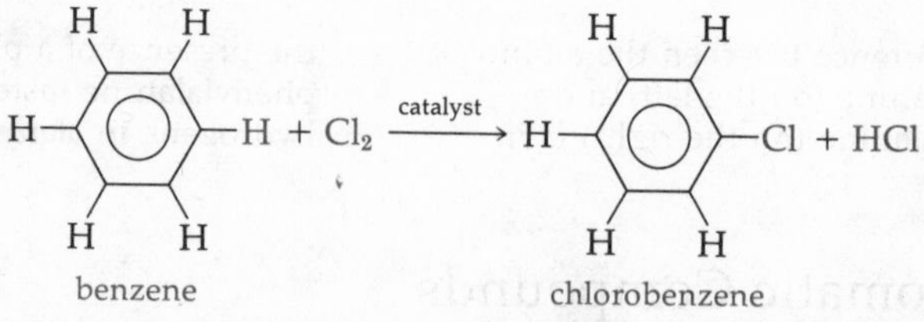

$$\text{benzene} + Cl_2 \xrightarrow{catalyst} \text{chlorobenzene} + HCl$$

A number of products that can be obtained in such substitution reactions are listed in Table 14-6. One aromatic substitution reaction that occurs in the human body is the substitution of a hydroxyl group for one of the hydrogen atoms in the amino acid phenylalanine:

(14-16)

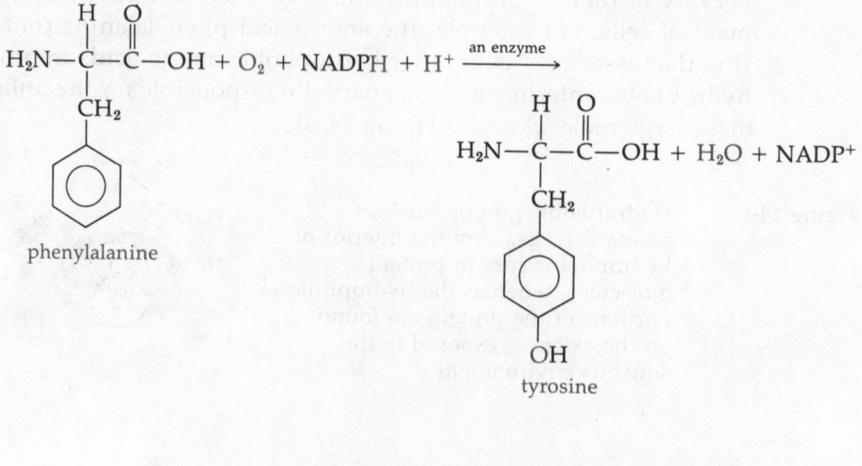

$$\text{phenylalanine} + O_2 + NADPH + H^+ \xrightarrow{\text{an enzyme}} \text{tyrosine} + H_2O + NADP^+$$

Table 14-6 Some Aromatic Hydrocarbons Used Commercially

Structural Formula	Name	Some Uses
 (benzene ring)	Benzene	Manufacture of leather, linoleum, varnishes, etc. (benzene is a carcinogen)
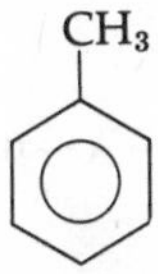 CH_3 (benzene ring)	Toluene	Manufacture of explosives, dyes, etc., and to extract chemicals from plants
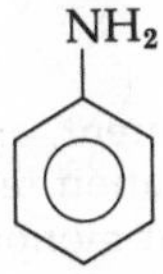 NH_2 (benzene ring)	Aniline	Manufacture of dyes, medicines, perfume, and as a solvent
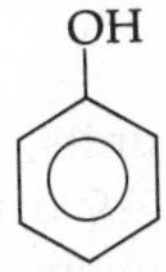 OH (benzene ring)	Phenol	Used as a disinfectant and for manufacture of a variety of materials
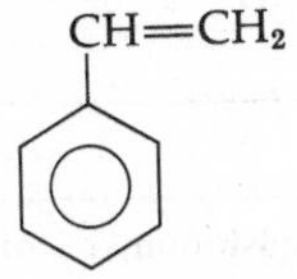 $CH=CH_2$ (benzene ring)	Styrene	Used to make polystyrene and other plastics and synthetic rubbers

Tyrosine is also an amino acid required for protein synthesis. Interestingly, some tyrosine is also converted to hormones and neurotransmitters, which carry chemical messages from one cell to another. These conversions involve additional substitution reactions in which other hydroxyl groups (or in one case, iodine atoms) replace hydrogens on the benzene ring (Chapter 28).

Exercise 14-10
Draw structural formulas for:
  (a) *m*-Ethylphenol
  (b) 2,4-Dichlorotoluene
  (c) 1,2,4-Trimethylbenzene

Exercise 14-11
Describe the shape of a toluene molecule.

14-9 HYDROCARBON POLYMERS

A **polymer** is a very large molecule composed of simpler units or building blocks, called **monomers,** linked together in a long chain. Many naturally occurring compounds (e.g., starch and proteins) are polymers, whereas a large number of other polymers (e.g., nylon and polystyrene) are synthesized commercially. In order for a simple molecule to be a monomer unit, it must be capable of reacting with at least two other monomers. The reaction that links monomers together is called **polymerization** (see Figure 14-9).

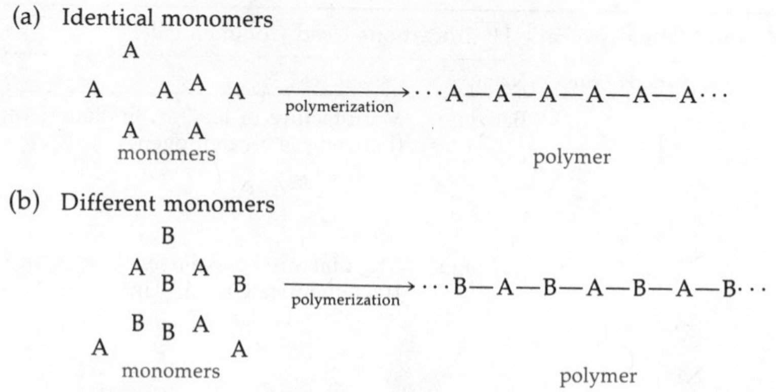

(a) Identical monomers

(b) Different monomers

Figure 14-9   Thousands of monomers can join together to form long polymer chains. The monomers may all be identical in a polymer (a), or they may be different, as in (b), where a polymerization reaction involving two different monomers is shown.

Some polymers are composed of identical monomer units, whereas others are composed of at least two different monomers. Table 14-7 lists the repeating monomer units found in some natural and synthetic polymers. Notice in Table

Table 14-7 Some Important Polymers

Name and Source	Monomer(s)	Repeating Monomer Unit (n Very Large)
Polyethylene (synthetic)	$CH_2{=}CH_2$ ethylene	$-(CH_2{-}CH_2)_n-$
Rubber (natural and synthetic)	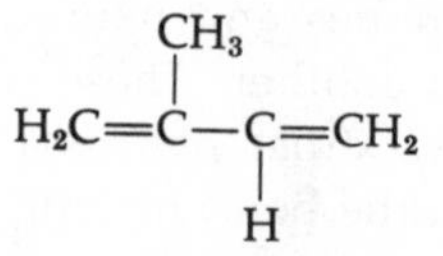 isoprene	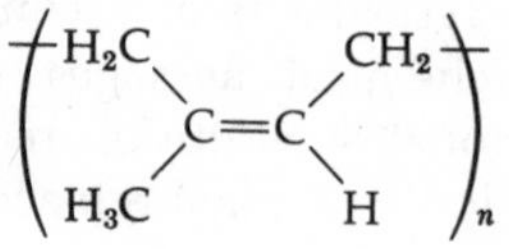
Nylon (synthetic)	$H_2N-(CH_2)_x-NH_2$ a diamine and 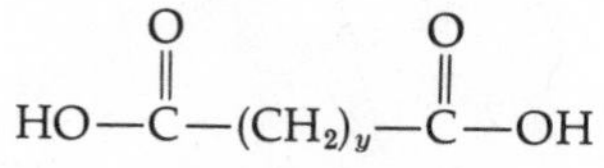 a dicarboxylic acid	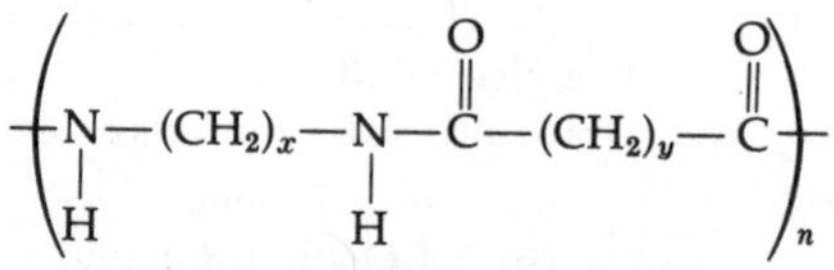 where x and y refer to the lengths of alkanelike carbon chains
Proteins (naturally occurring)	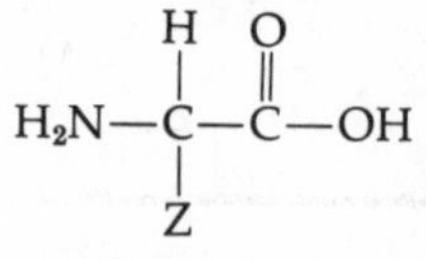 amino acids	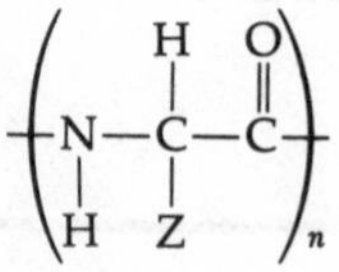 where Z is a group of atoms specific for each amino acid monomer
Starch (naturally occurring)	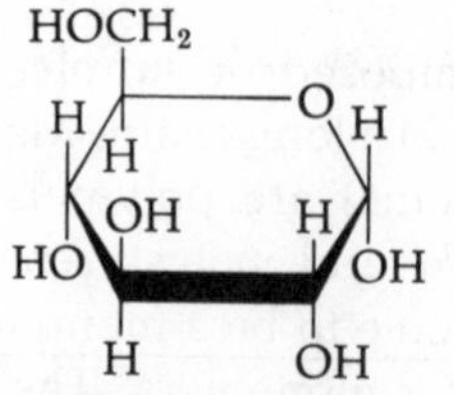 glucose	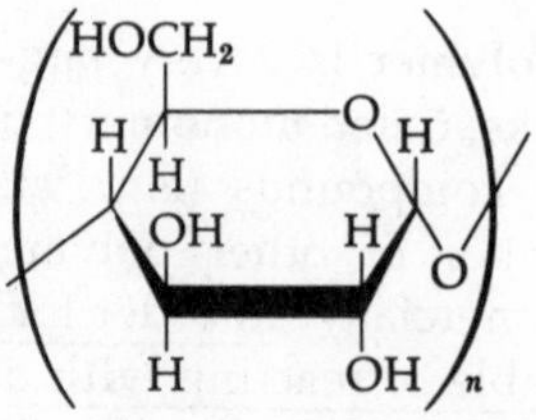

14-7 that the repeating monomer unit of a polymer has a bonding arrangement that is different from that in the individual monomers, since some bonds are broken and other bonds are formed during polymerization. Also notice that polymerization reactions can involve a variety of functional groups, including C=C double bonds. Compounds containing C=C double bonds (i.e., alkenes) were originally called olefins. Hence, polymers formed from alkene monomers are called **polyolefins.** Polymers formed from monomers containing other types of functional groups will be discussed later in this text.

The simplest polyolefin is polyethylene, which is formed by the polymerization of the monomer ethylene:

(14-17) $n(CH_2=CH_2) \xrightarrow[\text{polymerization}]{\text{a catalyst}} \cdots CH_2-CH_2-CH_2-CH_2-CH_2 \cdots$

            ethylene   polyethylene

It is important to note that polymers such as polyethylene do not form automatically when their monomers are mixed together. Ethylene will react with itself to form polyethylene only under certain special conditions and in the presence of a catalyst. The particular polyethylene product obtained is somewhat dependent on the specific set of reaction conditions and the specific catalyst used. Many commonly used items are manufactured from polyethylene, including packaging materials, buckets, bottles, and syringes. Finished products made from synthetic polymers are called **plastics.**

A number of other commercial hydrocarbon polymers are similar to polyethylene in that their monomer unit is a substituted ethylene molecule. Compare the monomer units in polypropylene, polystyrene, polyvinyl chloride, and in Teflon with the monomer unit of polyethylene (see Table 14-8). Polypropylene, for example, has a methyl group attached to every other carbon atom, which gives it somewhat different properties from polyethylene.

Table 14-8 Polyolefins Related to Polyethylene

Name	Structure	Monomer	Uses
Polypropylene	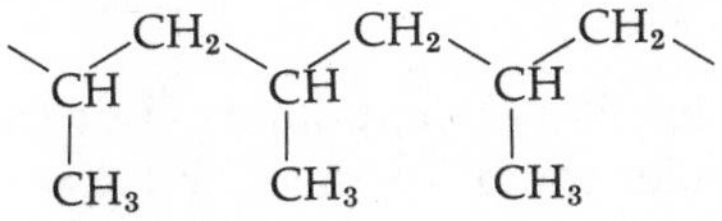 (repeating $-CH_2-CH(CH_3)-$ units)	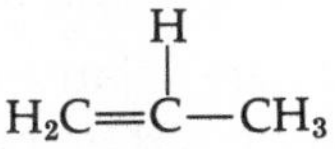 propylene $H_2C=C(H)-CH_3$	Many—as for polyethylene
Polystyrene	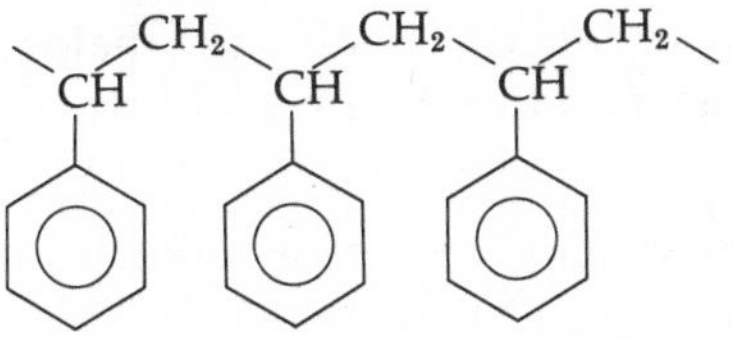 (repeating $-CH_2-CH(C_6H_5)-$ units)	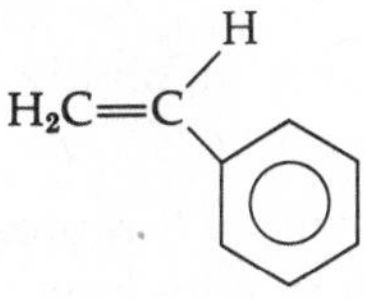 styrene $H_2C=C(H)(C_6H_5)$	Polystyrene foam used for insulation
Polyvinyl chloride (PVC)	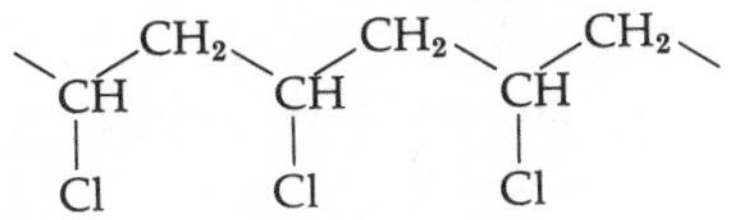 (repeating $-CH_2-CH(Cl)-$ units)	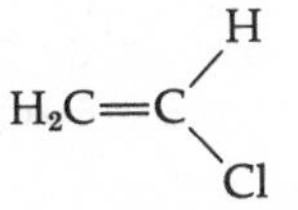 vinyl chloride $H_2C=C(H)(Cl)$	Clear plastic bottles and tubing (vinyl chloride is a carcinogen)
Teflon	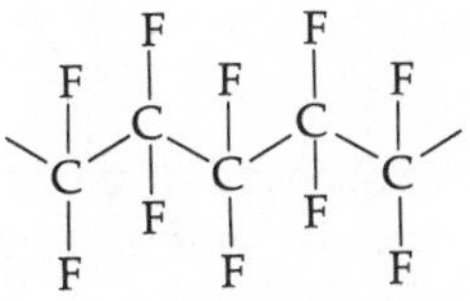 (repeating $-CF_2-CF_2-$ units)	tetrafluoroethylene $F_2C=CF_2$	Chemically inert slippery coatings

Synthetic rubber is formed by the polymerization of isoprene monomer units (Table 14-7), which contain two C=C double bonds. One of these bonds is broken during polymerization, while the other is shifted to another carbon atom (see Figure 14-10). Natural rubber is identical in structure to one of the synthetic rubber polymers but is polymerized in a different manner.

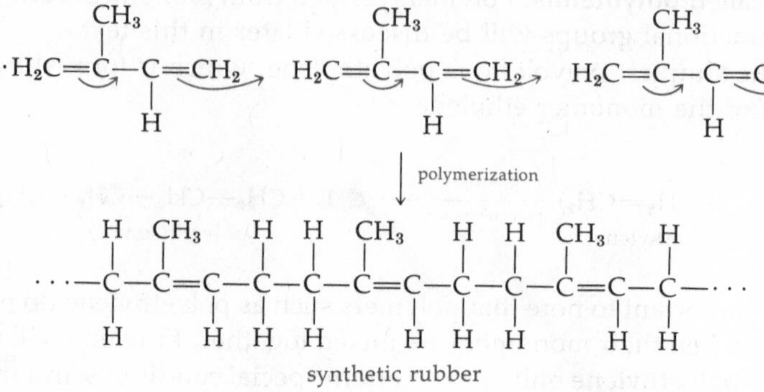

Figure 14-10   The polymerization of isoprene monomers to form synthetic rubber can be represented by curved arrows to indicate the changes in the bonding arrangement and the linking together of monomer units.

14-10 SUMMARY

1. Hydrocarbons are a group of organic molecules that contain only carbon and hydrogen. Hydrocarbons that contain only C—C and C—H single bonds are called alkanes. The classes of alkene and alkyne compounds contain C=C double bond and C≡C triple bond functional groups, respectively. Hydrocarbons that contain a benzene ring are called aromatic hydrocarbons.

2. Hydrocarbons are nonpolar, and hydrocarbonlike portions of larger molecules are also nonpolar.

3. Alkanes are relatively unreactive compounds, but the —H$_2$C—CH$_2$— portions of certain molecules can be oxidized to HC=CH double bonds.

4. Alkenes readily participate in addition reactions, including reduction to form alkanes, hydration to form alcohols, and halogenation. Addition reactions involving nonsymmetric compounds usually follow Markovnikov's rule.

5. The symbol R represents an alkyl group, whereas R' represents an alkyl group or a hydrogen atom.

6. Aromatic hydrocarbons can undergo substitution reactions, in which hydrogen atoms are replaced by substituents. The H group is

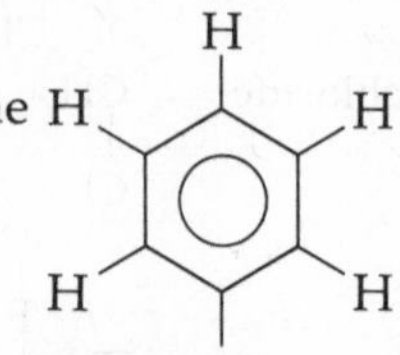

called a phenyl group.

7. A polymer is a very large molecule composed of monomer units that are polymerized to form a long chain. Polymers formed from alkene monomers are called polyolefins.

PROBLEMS

1. To which class of compounds do molecules with the following structural formulas belong?

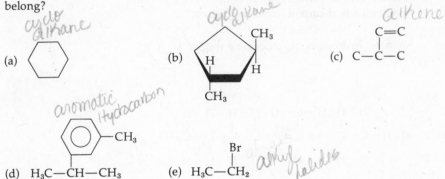

(a) *cyclo alkane*

(b) *cyclo alkane*   CH₃

(c) *alkene*   C=C / C—C—C

(d) *aromatic hydrocarbon*   —CH₃ H₃C—CH—CH₃

(e) *alkyl halides*   H₃C—CH₂ (Br)

2. Draw the structural formulas for each of the following:
   (a) *cis*-2-Pentene
   (b) *trans*-1,3-Dimethylcyclobutane
   (c) 2,3-Dimethyl-2-butene
   (d) 1,4-Diethylbenzene
   (e) *trans*-1,2-Diethylcyclopentane

3. Write IUPAC names for each of the following:

   (a) 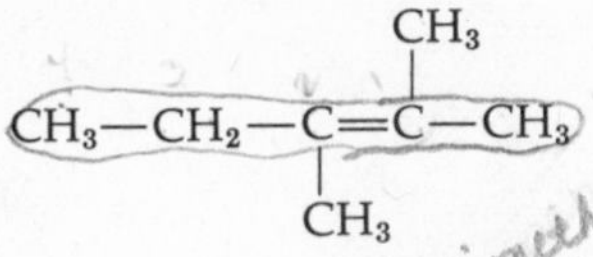 CH₃—CH₂—C=C—CH₃ with CH₃ groups
      2,3-dimethyl-2-pentene

   (b) H₃C   CH₂—CH₃
        C=C
       H   H   *cis-2-pentene*

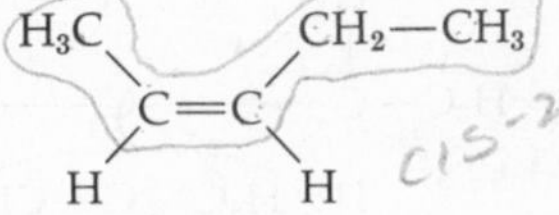

   (c) ⬡—CH₃   *methylcyclohexane*

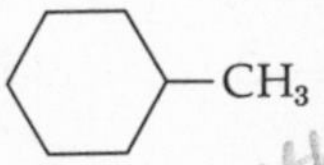

   (d) 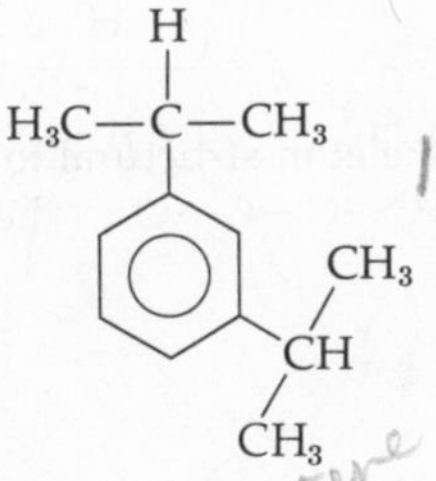 H₃C—C(H)—CH₃ on benzene ring with CH—(CH₃)₂   *1,3-diisopropylbenzene*

   (e) 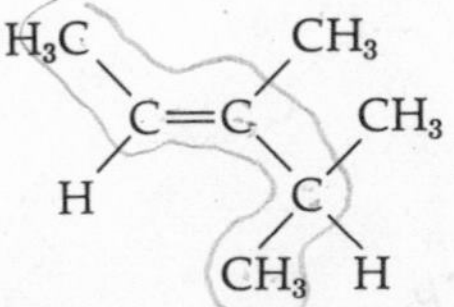 H₃C   CH₃
        C=C   CH₃
       H   C
            CH₃ H   *cis-3,4-dimethyl-2-pentene*

4. Draw structural formulas for all of the hydrocarbons with the formula C₄H₈. Which of these are geometric isomers, and which are structural isomers?

5. Draw structural formulas for the products of the following reactions:

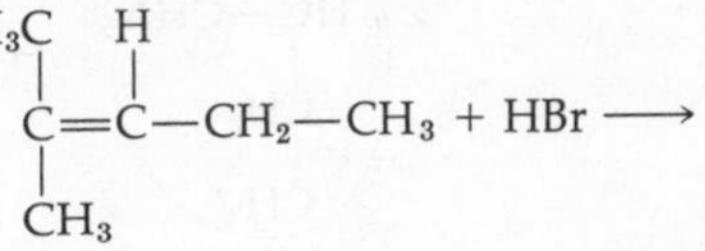

   (a) 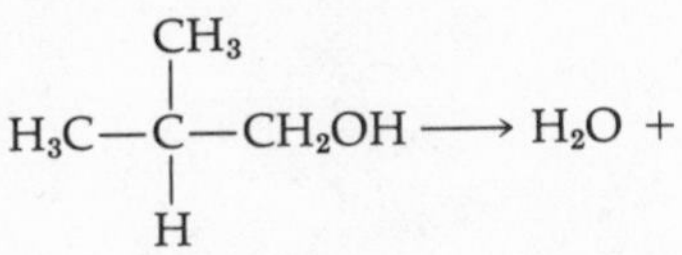 H₃C—C(CH₃)(H)—CH₂OH ⟶ H₂O +

   (b) H₃C   H
        C=C—CH₂—CH₃ + HBr ⟶
       CH₃

   (c) ⬡ with H, CH₃ (cyclohexene) + H₂O ⟶

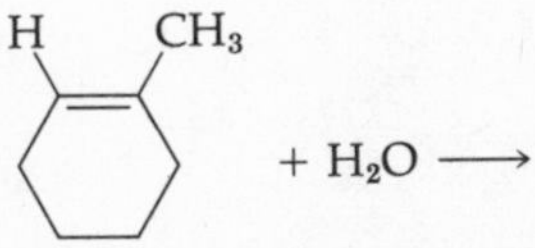

   (d) Propene + H₂ ⟶

6. The polyolefin Saran, which is used in packaging, is polymerized from the monomer vinylidene chloride, CH₂=CCl₂. Draw a structural formula for a short segment of Saran.

SOLUTIONS TO EXERCISES

14-1 Alkanes; no characteristic functional group.
Alkenes; carbon-carbon double bond, $C{=}C$.
Alkynes; carbon-carbon triple bond, $C{\equiv}C$.

Aromatic hydrocarbons; a benzene ring, ⬡ .

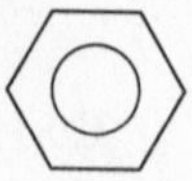

14-2 (a)

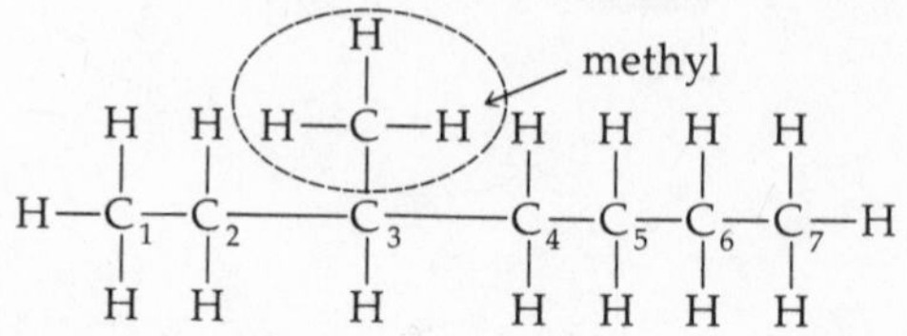

(b)

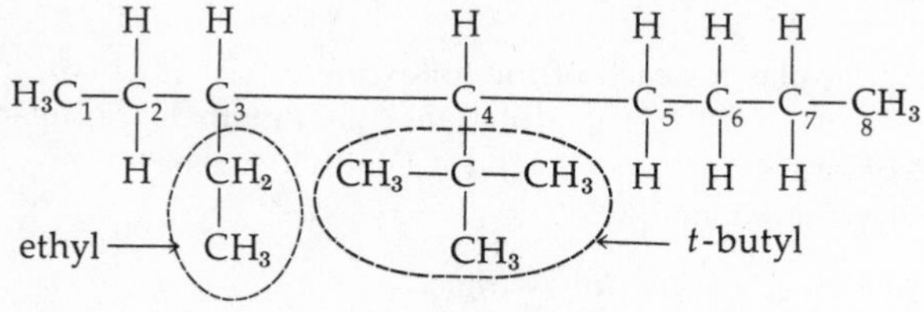

(c)

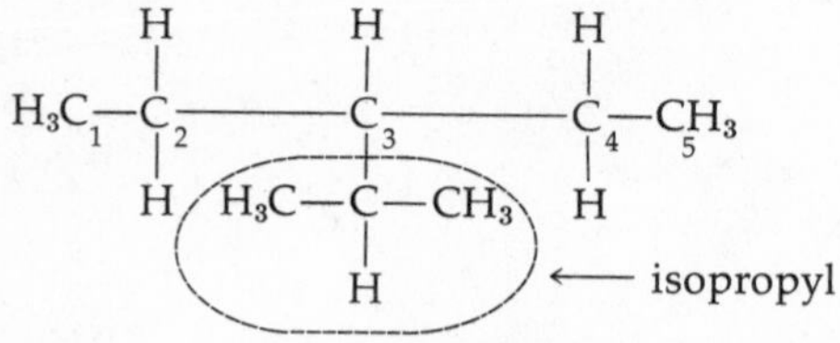

14-3 Using skeleton structural formulas:

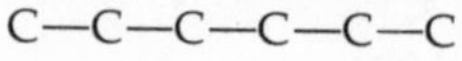

C—C—C—C—C—C   hexane

C—C—C—C—C   3-methylpentane
       |
       C

C—C—C—C   2,2-dimethylbutane

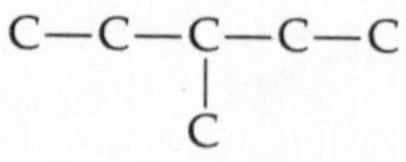

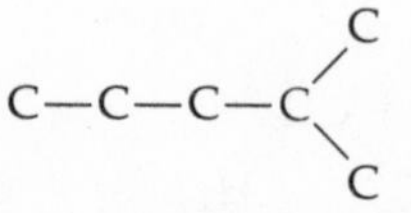

C—C—C—C   2-methylpentane

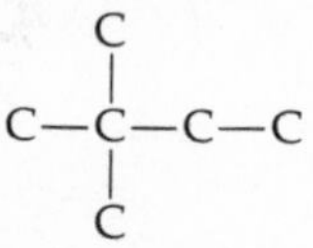

C—C—C—C   2,3-dimethylbutane
| |
C C

14-4 (a)

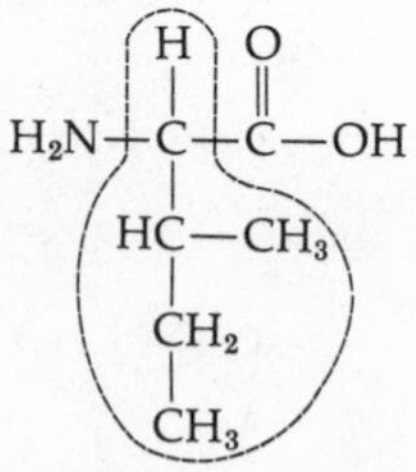

(b) 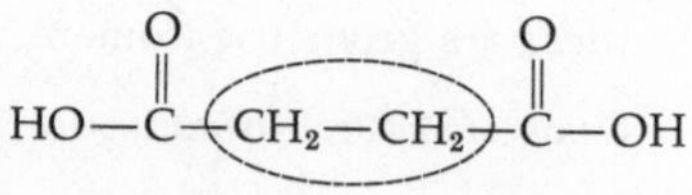 $HO-C-(CH_2-CH_2)-C-OH$ (with O's above the C's)

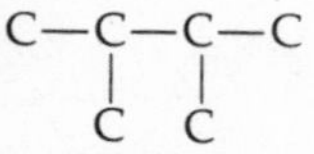

(c)

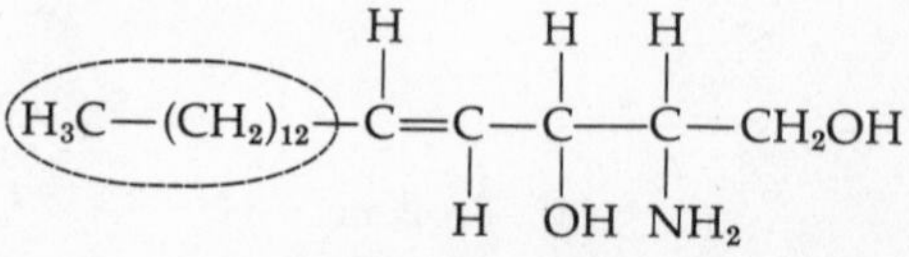

(d)

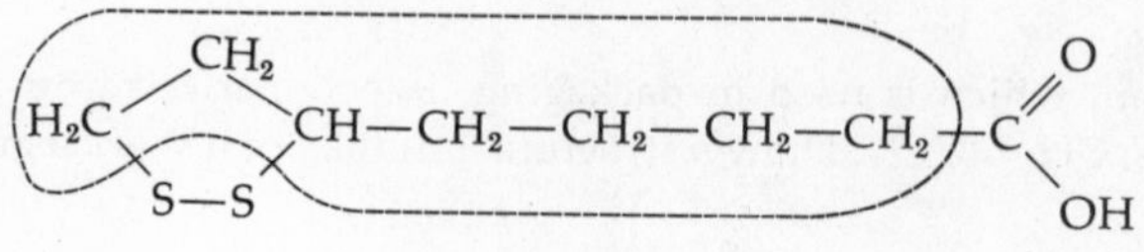

14-5 Cycloalkane rings, such as cyclohexane, involve one more carbon-carbon single bond than alkanes with the same number of carbon atoms. The carbon atoms involved in the additional C—C single bond can each bond to one fewer H atom than they could if this C—C bond were not present.

14-6 A planar hexagon would have bond angles of 120°, as opposed to the tetrahedral (109.5°) bond angles preferred by carbon atoms involved in four single bonds.

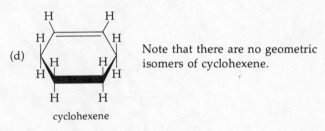

14-7

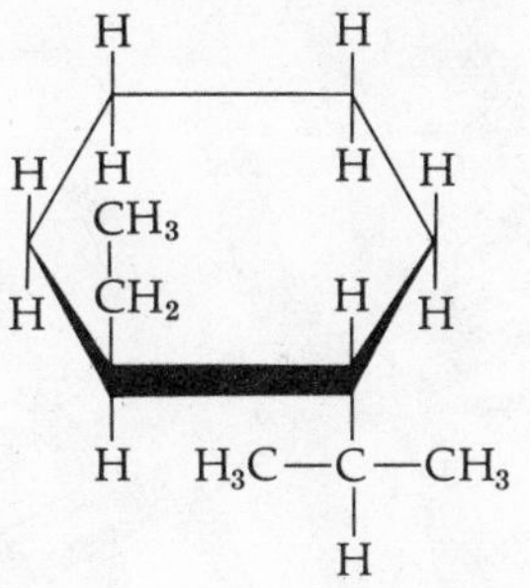

14-8 (a)   H₃C—C—CH with CH₃ CH₃ Br CH₃
2-bromo-2,3-dimethylbutane

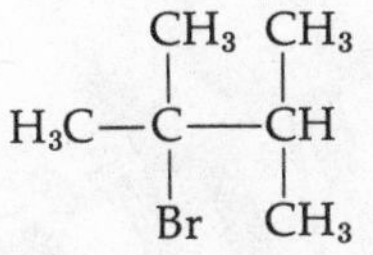

(b) H₃C—CH₂—C—OH with CH₃ CH₃
2-methyl-2-butanol

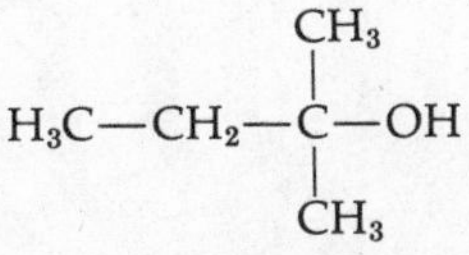

(c) H₃C—CH with CH₂—CH₃ CH₂—CH₃
3-methylpentane

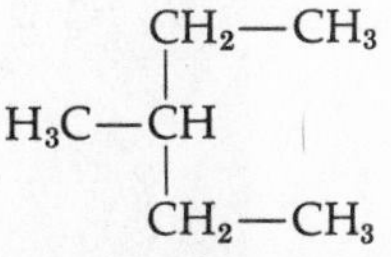

(d) cyclohexene

Note that there are no geometric isomers of cyclohexene.

14-9 (a) H₃C—CH₂—C=C—CH₂—CH₂—CH₃ with CH₃ CH₂ CH₃

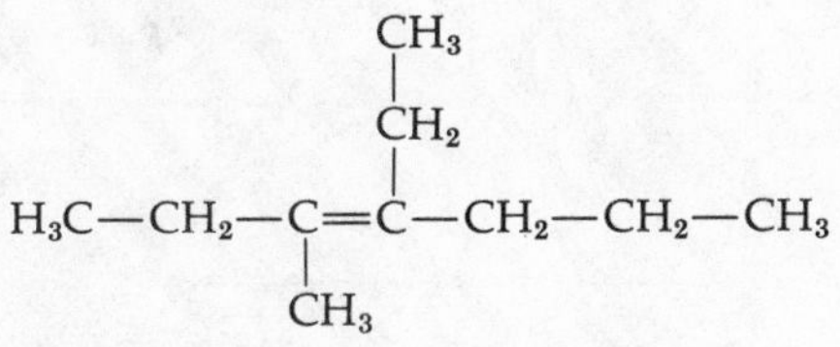

(b) H₃C—CH₂—C=C—CH₂—CH₃ with CH₃ CH₃

(c) H₃C—C=C—CH₂—CH₃ with H H

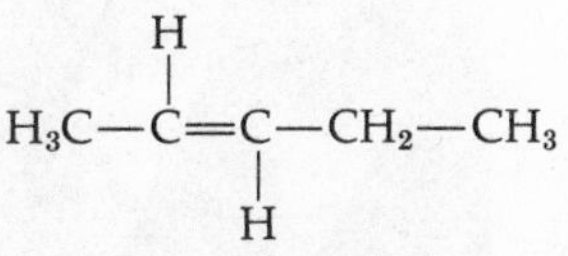

14-10 (a) OH, CH₂—CH₃   (b) CH₃, Cl, Cl   (c) CH₃, CH₃, CH₃

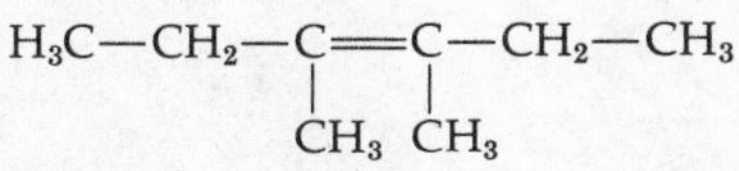

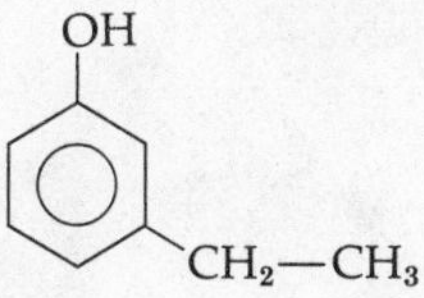

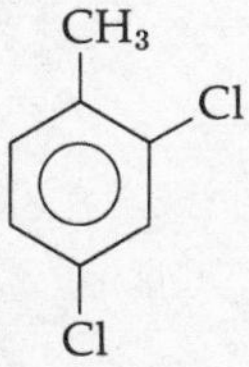

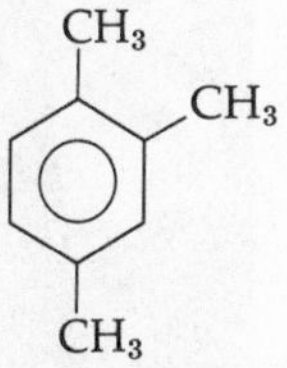

14-11 The methyl carbon atom is located at the center of a tetrahedron. At three corners of this tetrahedron are hydrogen atoms. At the fourth corner of this tetrahedron is one carbon atom of the phenyl group. The entire phenyl group is located in one plane. Thus, the shape of toluene can be represented as

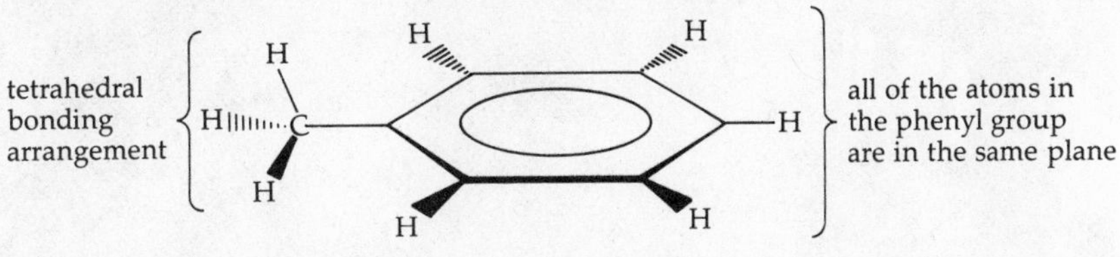

tetrahedral bonding arrangement { ... } all of the atoms in the phenyl group are in the same plane

CHAPTER 15

Alcohols, Phenols, and Ethers

15-1 INTRODUCTION

Several billion years ago, certain early inhabitants of our planet developed the ability to make O_2 from H_2O by photosynthesis. Their closest relatives among organisms existing today are blue-green algae, and we humans are literally indebted to their ancestors for the air we breathe! Today there is a considerable supply of O_2 in the Earth's atmosphere, and the O_2 that is used by humans and other living organisms is continually replenished by the photosynthetic activity of plants and algae.

Now, in addition to requiring molecular oxygen for survival, humans make and degrade a large number of oxygen-containing organic molecules. In fact, just about every molecule in the human body contains the element oxygen.

The large group of oxygen-containing organic compounds is subdivided into several classes on the basis of their component functional groups. Three classes of organic compounds—alcohols, ethers, and phenols—all contain a C—O single bond. We shall discuss the structure, nomenclature, physical properties, and chemical reactivity of simple alcohols, phenols, and ethers in this chapter. In the next two chapters, we shall discuss those classes of oxygen-containing organic compounds that contain a C=O double bond: aldehydes and ketones (Chapter 16), and carboxylic acids, esters, and anhydrides (Chapter 17).

After studying Chapters 15, 16, and 17, you will not only know much about the chemistry of oxygen-containing compounds with one functional group, but you will also know much of the chemistry of compounds containing two or more functional groups. For example, the compound β-hydroxybutyric acid,

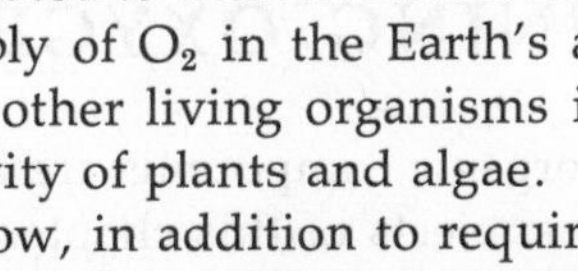

$$H_3C-\overset{\overset{\displaystyle OH}{|}}{\underset{\underset{\displaystyle H}{|}}{C}}-CH_2-\overset{\overset{\displaystyle O}{\|}}{C}-OH$$

contains both an alcohol and a carboxylic acid functional group, and it undergoes chemical reactions characteristic of both of these functional groups.

Wine is stored in huge vats for aging. The anaerobic (absence of free O_2) aging process involves several chemical reactions catalyzed by enzymes present in yeast cells and culminates in the reduction of acetaldehyde to ethanol. When the concentration of alcohol in wine reaches about 12%, the yeast cells die and further production of ethanol ceases.

15-2   STUDY OBJECTIVES

After careful study of this chapter, you should be able to:

1. Identify alcohol, ether, and phenol functional groups, given the structural formula for a compound, and draw the general structural formula for each of these classes of compounds.

2. Determine whether a specific alcohol, ether, or phenol can form hydrogen bonds in the pure state and/or with water.

3. Describe the acid-base properties of a specific alcohol, phenol, or ether.

4. Explain how hydrogen bonding affects the boiling point and solubility of alcohols, ethers, or phenols.

5. Draw the structural formulas for simple alcohols, phenols, and ethers, given their names, and vice versa.

6. Draw structural formulas for the products of reactions in which alcohols are oxidized or dehydrated.

7. Describe the major differences between an enzyme-catalyzed reaction in a living cell and the same reaction performed in the laboratory using nonenzyme catalysts.

15-3   GENERAL PROPERTIES OF ORGANIC COMPOUNDS CONTAINING OXYGEN

The major classes of oxygen-containing organic compounds are listed in Table 15-1. Alcohols, ethers, and phenols, the subjects of this chapter, all contain a C—O single bond. Aldehydes and ketones (Chapter 16) contain the carbonyl group, whereas carboxylic acids, esters, and anhydrides (Chapter 17) contain the carboxyl group.

The physical properties and the chemical reactivity of all of these oxygen-containing compounds are quite different from those of hydrocarbons. Recall that an oxygen atom has six valence electrons and a large electronegativity, and that in covalent compounds an oxygen atom usually forms two covalent bonds. In a carbon-oxygen bond or in an oxygen-hydrogen bond, the bonding electrons are pulled closer to the more electronegative oxygen atom, making these bonds polar, with a partial negative charge on the oxygen atom. Thus, organic functional groups that contain oxygen are polar groups and can participate in hydrogen bonding with water (Chapter 7). Therefore, oxygen-containing functional groups are hydrophilic, and the presence of these functional groups in a molecule *tends* to make it water soluble. Of course, the solubility of a molecule is determined by *all* of the atoms in the molecule and how they are bonded together.

Oxygen-containing organic molecules can undergo many types of reactions. We shall discuss those reactions that are important in the human body, and some others that are important for the commercial synthesis of drugs and plastics.

The kinds of reactions that occur inside living cells can also be carried out in the laboratory, but there are several important differences. Generally speaking, organic compounds that are capable of reacting do so only very slowly in the laboratory. A catalyst must usually be added in order for reactions to take place

Table 15-1 Classes of Oxygen-Containing Organic Compounds

Characteristic Functional Group		Class
$C-O$ Single bond	$-\overset{\vert}{\underset{\vert}{C}}-OH$	Alcohols
	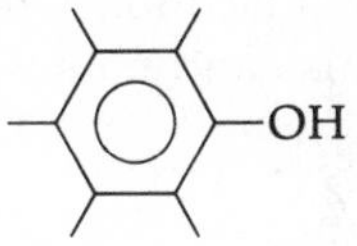 (aromatic ring)$-OH$	Phenols
	$-\overset{\vert}{\underset{\vert}{C}}-O-\overset{\vert}{\underset{\vert}{C}}-$ or 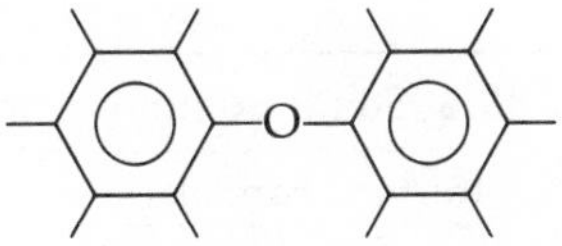 (ring)$-O-$(ring)	Ethers
$-\overset{O}{\overset{\|}{C}}-$ Carbonyl group	$-\overset{O}{\overset{\|}{C}}-H$	Aldehydes
	$-\overset{\vert}{\underset{\vert}{C}}-\overset{O}{\overset{\|}{C}}-\overset{\vert}{\underset{\vert}{C}}-$ or (ring)$-\overset{O}{\overset{\|}{C}}-$(ring)	Ketones
$-\overset{O}{\overset{\|}{C}}-O-$ Carboxyl group	$-\overset{O}{\overset{\|}{C}}-OH$	Carboxylic acids
	$-\overset{O}{\overset{\|}{C}}-O-\overset{\vert}{\underset{\vert}{C}}-$ or $-\overset{O}{\overset{\|}{C}}-O-$(ring)	Esters
	$-\overset{O}{\overset{\|}{C}}-O-\overset{O}{\overset{\|}{C}}-$	Anhydrides

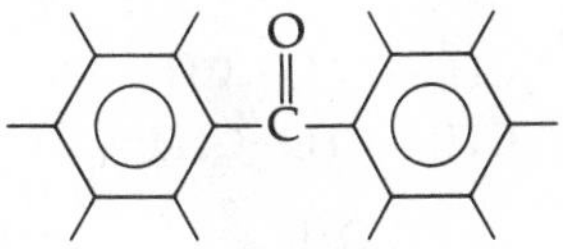

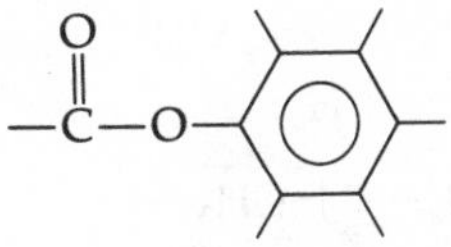

in a reasonable period of time. In living cells, highly efficient catalysts (called enzymes) are naturally present. Enzymes enable reactions to proceed at rates that are much faster than those that can be achieved using nonenzyme catalysts. In addition, enzymes are highly *specific* catalysts, assisting in the formation of only a single set of products. Nonenzyme catalysts, on the other hand, catalyze other reactions, in addition to the one desired, and so a mixture of products is often formed. Thus in the laboratory or industrial situation one has the additional task of separating the desired product from undesired ones. These differences between enzyme- and nonenzyme-catalyzed reactions should be kept in mind when studying specific reactions of organic compounds.

15-4 ALCOHOLS   R—OH

Alcohols are organic derivatives of water, in which one of the hydrogen atoms of water is replaced by an alkyl group. Thus, an alcohol can be viewed as an alkyl group, R, bonded to a **hydroxyl** group, —OH, so that an alcohol has the

general formula R—OH. Note that the alcohol functional group is $-\overset{\vert}{\underset{\vert}{C}}-OH$,

whereas the group of two atoms, —OH, is called the hydroxyl group.

Structure and Nomenclature

The common names of alcohols are formed by adding the word "alcohol" to the name of the alkyl group to which the hydroxyl group is attached. Thus,

$$
\begin{array}{c}
\text{OH} \\
| \\
\text{H}_3\text{C}\text{—CH}\text{—CH}_3
\end{array}
$$

is called isopropyl alcohol, which is frequently used as rubbing alcohol. The structural formulas and names of some common alcohols are given in Table 15-2.

Table 15-2 Some Common Alcohols: General formula = R—OH

Structural Formula	Common Name	IUPAC Name
CH_3—OH	Methyl alcohol (wood alcohol)	Methanol
CH_3—CH_2—OH	Ethyl alcohol (grain alcohol)	Ethanol
CH_3—CH_2—CH_2—OH	n-Propyl alcohol	1-Propanol

CH_3—CH—CH_3
      |
     OH
	Isopropyl alcohol	2-Propanol

CH_3—CH_2—CH_2—CH_2—OH	n-Butyl alcohol	1-Butanol

CH_3—CH_2—CH—CH_3
           |
         OH
	sec-Butyl alcohol	2-Butanol

CH_3
      |
 CH_3—C—CH_3
     |
     OH
	t-Butyl alcohol	2-Methyl-2-propanol

H_2C——CH_2 | | OH OH	Ethylene glycol	1,2-Ethanediol
H_2C——CH—CH_2 | | | OH OH OH	Glycerol (glycerine)	1,2,3-Propanetriol

The IUPAC names for simple alcohols are formed by adding the ending *-ol* to the name of the parent alkane (the *e* is dropped). Thus, the IUPAC name for methyl alcohol is methanol. The position of the hydroxyl group on the carbon chain of propanol and larger alcohols is indicated by a numerical prefix. Thus

$$
\begin{array}{c}
\text{OH} \\
| \\
\text{H}_3\text{C}\text{—CH}\text{—CH}_3
\end{array}
$$

the IUPAC name for H$_3$C—CH—CH$_3$ is 2-propanol. When a compound contains two, three, or more alcohol functional groups, the IUPAC ending is *-diol, -triol*, and so on (the *e* is not dropped in this case).

Ethanol, also called grain alcohol, is the only simple alcohol that is tolerated by the human body. In small doses it often produces pleasant sensations. However, ethanol acts as a physiological depressant. Many drivers who drink have become painfully aware of ethanol's ability to slow down their response time and decrease their muscle coordination. In very large doses, ethanol is lethal.

Methanol, also called wood alcohol, is frequently used as a solvent and as a fuel for cooking. Ethylene glycol is the major component of the antifreeze used in car radiators, and glycerol is a component of many oils and creams. We shall see in Chapter 26 that glycerol, also called glycerine, is a major component of the fats stored in the human body.

Hydrogen Bonding

The large electronegativity of the oxygen atom makes the hydroxyl group quite polar. Thus, an organic molecule that contains a hydroxyl group can act as either a hydrogen bond donor or a hydrogen bond acceptor. For example, alcohol molecules can form hydrogen bonds with water and other alcohol molecules (see Figure 15-1).

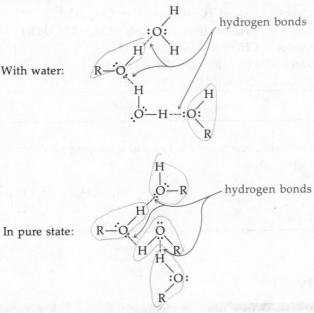

Figure 15-1   Alcohols form hydrogen bonds with water molecules and, in the pure state, with other alcohol molecules.

Because of their ability to form hydrogen bonds between identical molecules, pure alcohols have higher boiling points than molecules of similar molecular weight that do not form hydrogen bonds in the pure state, such as the hydrocarbons (see Table 15-3).

Table 15-3 Boiling Points of Some Hydrocarbons and Alcohols with Similar Molecular Weights

Hydrocarbons			Alcohols		
Name and Structural Formula	MW	Boiling Point (°C)	Name and Structural Formula	MW	Boiling Point (°C)
Ethane $H_3C—CH_3$	30	−89	Methanol $H_3C—OH$	32	+65
Propane $H_3C—CH_2—CH_3$	44	−42	Ethanol $H_3C—CH_2—OH$	46	+78
Butane $H_3C—CH_2—CH_2—CH_3$	58	0	1-Propanol $H_3C—CH_2—CH_2—OH$	60	+97

The formation of hydrogen bonds between the hydroxyl group of alcohol and water molecules also tends to make alcohols soluble in water, whereas hydrocarbon molecules are hydrophobic. However, alcohols that contain long alkyl chains are relatively insoluble in water because of the nonpolar hydrophobic character of the alkyl group. Notice that, in the series of alcohols in Table 15-4, the larger the nonpolar portion, the lower the solubility.

Table 15-4 Solubility of Alcohols

Name	Structural Formula	Solubility (g/100 ml of water)
Methanol	CH_3-OH	
Ethanol	CH_3-CH_2-OH	Completely miscible
1-Propanol	$CH_3-CH_2-CH_2-OH$	
1-Butanol	$CH_3-CH_2-CH_2-CH_2-OH$	7.9
1-Pentanol	$CH_3-CH_2-CH_2-CH_2-CH_2-OH$	2.3
1-Hexanol	$CH_3-CH_2-CH_2-CH_2-CH_2-CH_2-OH$	0.6
1-Heptanol	$CH_3-(CH_2)_5-CH_2-OH$	0.2
1-Octanol	$CH_3-(CH_2)_6-CH_2-OH$	0.05
1-Decanol	$CH_3-(CH_2)_8-CH_2-OH$	~0

Exercise 15-1

Arrange the following compounds in order of increasing solubility in water:

(a) $CH_3-CH_2-CH_2-OH$ (b) $CH_3-CH_2-CH_3$ (c) 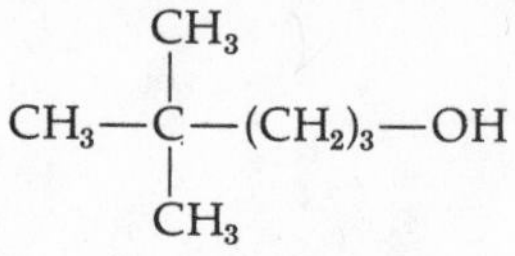 $CH_3-\overset{\displaystyle CH_3}{\underset{\displaystyle CH_3}{C}}-(CH_2)_3-OH$

Acidity-Basicity

Alcohols do not act as acids in aqueous solutions. Let us consider the alcohol functional group, $-\overset{|}{\underset{|}{C}}-O-H$. The hydroxyl proton found in this group would be acidic if electrons were "pulled" away from the O—H bond, facilitating the loss of a proton (H^+). This is not the case, however. The electronegativity of the carbon atom in the alcohol functional group is not very large, and the carbon atom is not bonded to other atoms that strongly attract electrons. On the other hand, phenols are weak acids, as we shall see shortly.

Alcohols (and phenols and ethers) are extremely weak (weaker than water) Brønsted-Lowry bases. The oxygen atom in an alcohol, phenol, or ether has two unshared pairs of electrons, but the ability of these unshared electrons to accept a proton is quite small.

Chemical Reactivity

DEHYDRATION

In Section 14-6 we saw that an alkene could be formed by the dehydration of an alcohol in the reaction

(15-1)

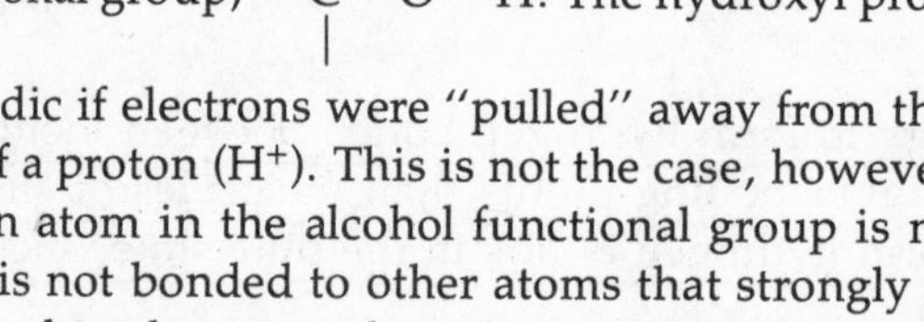

$$R_1-\overset{\displaystyle OH}{\underset{\displaystyle H}{C}}-\overset{\displaystyle H}{\underset{\displaystyle H}{C}}-R_2 \longrightarrow R_1-\overset{\displaystyle H}{C}=\overset{\displaystyle H}{C}-R_2 + H_2O$$

alcohol   alkene

Consider the following specific example. When a chemist dehydrates 2-butanol,

$CH_3—CH_2—\underset{\underset{OH}{|}}{CH}—CH_3$, a mixture of three products is obtained—the alkenes 1-butene, $CH_3—CH_2—CH{=}CH_2$, and *cis*-2-butene and *trans*-2-butene, $CH_3—CH{=}CH—CH_3$ (*trans*-2-butene is the major product). In later chapters we shall see several important reactions in human cells where the dehydration of the alcohol functional group to form a carbon-carbon double bond takes place. In the human body, however, a mixture of products is usually *not* obtained, since enzyme-catalyzed reactions yield only a single set of products.

Ethers can also be formed by the dehydration of alcohols under appropriate conditions. In this dehydration reaction, the elements of water are split off from two reacting alcohol molecules:

(15-2)
$$R—OH + R—OH \longrightarrow R—O—R + H_2O$$
two alcohol molecules　　　　an ether

OXIDATION

Another type of reaction that alcohols can undergo is oxidation. Recall from Chapter 12 that for organic molecules we can say that oxidation involves the loss of hydrogen or the gain of oxygen, whereas reduction involves the gain of hydrogen or the loss of oxygen. Hydration and dehydration reactions, on the other hand, involve neither oxidation nor reduction.

An alcohol functional group, $—\overset{\overset{|}{}}{\underset{\underset{|}{}}{C}}—OH$, can lose hydrogen and be oxidized to a carbonyl group, $—\overset{O}{\overset{\|}{C}}—$. The ability of a given alcohol to be oxidized (and the product obtained) depends on the nature of the alkyl portion of the alcohol molecule. In an alcohol the carbon atom to which the hydroxyl group is attached is called the hydroxylic carbon atom (see Figure 15-2). If the hydroxylic carbon atom is bonded to none or one other carbon atom, it is called a *primary* carbon atom and the alcohol is called a **primary alcohol.** If the hydroxylic carbon atom is bonded to two or three other carbon atoms, then this carbon atom is called a *secondary* or *tertiary* carbon atom, respectively, and the alcohol is called a **secondary** or **tertiary alcohol.**

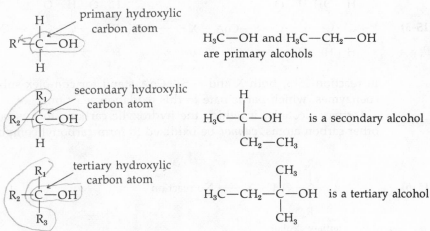

Figure 15-2　Alcohols are classified as primary, secondary, or tertiary depending on the bonding arrangement of the hydroxylic carbon atom.

The oxidation of a primary alcohol produces an **aldehyde:**

(15-3)
$$R'—\overset{\overset{\displaystyle H}{|}}{\underset{\underset{\displaystyle H}{|}}{C}}—OH + X \longrightarrow R'—\overset{\overset{\displaystyle O}{\|}}{C}—H + XH_2$$

primary   aldehyde
alcohol

Reaction 15-3 represents the general reaction for the oxidation of a primary alcohol. In this representation, X symbolizes the oxidizing agent. Thus, for example, the primary alcohols methyl alcohol and ethyl alcohol can be oxidized to the aldehydes $H_2C{=}O$ (formaldehyde) and $CH_3—\overset{\overset{\displaystyle O}{\|}}{C}—H$ (acetaldehyde).

The laboratory synthesis of aldehydes from primary alcohols is quite difficult, since the aldehyde product is generally further oxidized to a carboxylic acid. The enzyme-catalyzed oxidation of a primary alcohol to an aldehyde, on the other hand, readily produces only the aldehyde. We shall discuss how aldehydes are named in the next chapter.

The oxidation of a secondary alcohol produces a **ketone:**

(15-4)
$$R_1—\overset{\overset{\displaystyle OH}{|}}{\underset{\underset{\displaystyle H}{|}}{C}}—R_2 + X \longrightarrow R_1—\overset{\overset{\displaystyle O}{\|}}{C}—R_2 + XH_2$$

secondary   ketone
alcohol

Reaction 15-4 represents the general reaction for the oxidation of any secondary alcohol. Again, X symbolizes the oxidizing agent. For example, the secondary alcohol isopropyl alcohol can be oxidized to the ketone $H_3C—\overset{\overset{\displaystyle O}{\|}}{C}—CH_3$ (acetone).

Another example of the oxidation of a secondary alcohol to a ketone takes place in human cells as part of the oxidation of fatty acid molecules (Chapter 27):

(15-5)
$$R—\overset{\overset{\displaystyle H}{|}}{\underset{\underset{\displaystyle H}{|}}{C}}—\overset{\overset{\displaystyle OH}{|}}{\underset{\underset{\displaystyle H}{|}}{C}}—\overset{\overset{\displaystyle H}{|}}{\underset{\underset{\displaystyle H}{|}}{C}}—\overset{\overset{\displaystyle O}{\|}}{C}—S—CoA + X \xrightarrow{\text{enzyme}} R—\overset{\overset{\displaystyle H}{|}}{\underset{\underset{\displaystyle H}{|}}{C}}—\overset{\overset{\displaystyle O}{\|}}{C}—\overset{\overset{\displaystyle H}{|}}{\underset{\underset{\displaystyle H}{|}}{C}}—\overset{\overset{\displaystyle O}{\|}}{C}—S—CoA + XH_2$$

In reaction 15-5, both X and —S—CoA stand for complex substances, called coenzymes, which participate in this oxidation reaction.

Tertiary alcohols, in which the hydroxylic carbon atom is bonded to three other carbon atoms, *cannot* be oxidized to form carbonyl compounds:

(15-6)
$$R_1—\overset{\overset{\displaystyle R_2}{|}}{\underset{\underset{\displaystyle R_3}{|}}{C}}—OH + X \longrightarrow \text{No reaction}$$

tertiary alcohol

As we shall see in the following chapters, alcohols can be formed by reducing aldehydes and ketones in reactions that are similar to the reverse of reactions 15-3 and 15-4. Alcohols will also react with aldehydes, ketones, and carboxylic acids. The reactions of alcohols are summarized in Essential Skills 9.

Exercise 15-2

Why does pure methanol, with a molecular weight of 32, exist as a liquid at room temperature, whereas methane (MW = 16), ethane (MW = 30), and propane (MW = 44) are gases at room temperature?

Exercise 15-3

Classify as primary, secondary, or tertiary the hydroxylic carbon atoms in glycerol,

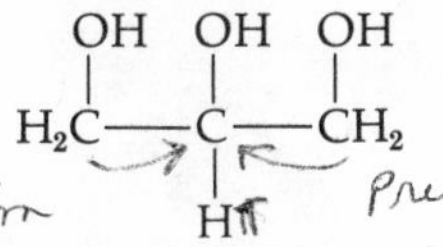

$$OH \quad OH \quad OH$$
$$H_2C \text{—} C \text{—} CH_2$$

Prim   *HⁿT Primary*
              Secondary

Exercise 15-4

Write the IUPAC names for each of the following alcohols, and draw structural formulas for the products of the given reactions (X represents an oxidizing agent):

(a) $2(CH_3\text{—}CH_2\text{—}OH) \rightarrow H_2O +$ (b) $H_3C\text{—}CH_2\text{—}OH \rightarrow H_2O +$

(c) 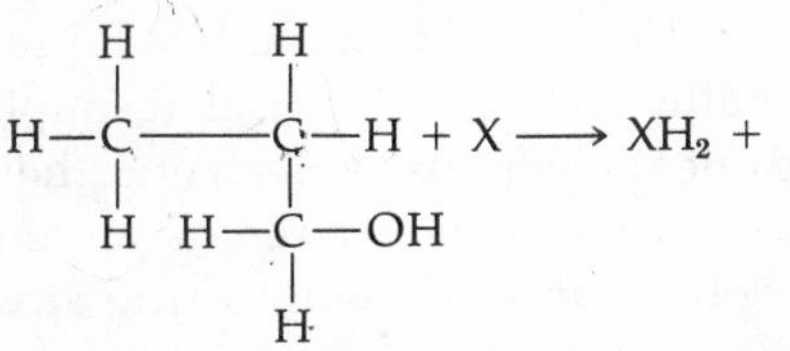
$$H\text{—}\underset{H}{\overset{H}{C}}\text{——}\underset{\underset{\underset{H}{|}}{\overset{|}{H\text{—}C\text{—}OH}}}{\overset{H}{C}}\text{—}H + X \longrightarrow XH_2 +$$

(d) 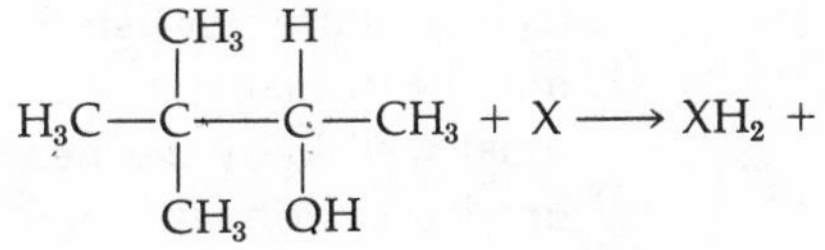
$$H_3C\text{—}\underset{CH_3}{\overset{CH_3}{C}}\text{——}\underset{QH}{\overset{H}{C}}\text{—}CH_3 + X \longrightarrow XH_2 +$$

If you have difficulty working this type of problem, consult Essential Skills 9.

$$CH_3\text{—}CH$$
$$CH_2\text{—}OH$$

15-5   PHENOLS

Phenols are compounds in which a hydroxyl group is bonded to a benzene ring or to a substituted benzene ring. The simplest member of this class of compounds is phenol,

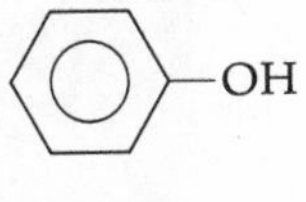

which is sometimes called carbolic acid. Note that phenol (feenol) is pronounced differently than phenyl (fenil). Phenol was the first antiseptic used in modern medicine. Phenols are similar to alcohols in many ways. They can form hydrogen bonds with phenols, with alcohols, and with water.

As we mentioned previously, phenols, in contrast to alcohols, are weak acids in aqueous solution. The phenyl group tends to pull electrons away from the atoms to which it is bonded. Thus, in a phenol, the phenyl group pulls electrons away from the —OH group. As a consequence, the hydrogen present in this —OH group is weakly acidic (Figure 15-3). We shall not discuss the complex nomenclature and reactions of phenols. Some phenols are used in photographic developing processes, whereas others are used in the manufacture of drugs, plastics, and other complex compounds.

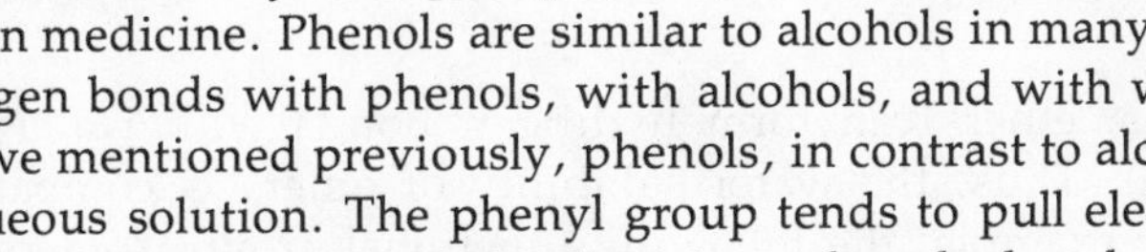

Figure 15-3   Phenol is a weak acid. The phenyl group pulls electrons away from the —OH group, thus facilitating the transfer of a proton to a water molecule

15-6   ETHERS

Ethers can also be considered organic derivatives of water in which *both* of the hydrogen atoms in a water molecule, H—O—H, are replaced by alkyl or phenyl groups, that is, R_1—O—R_2, or

Structure and Nomenclature

The common names of ethers are formed by adding the word "ether" to the names of the R groups. The structural formulas and names of some common ethers are shown in Table 15-5. The IUPAC names for simple ethers are rarely used and are not listed. Notice in Table 15-5 that the two R groups on an ether may be the same or different. Diethyl ether, usually called simply "ether," is the most well-known ether. It was once widely used as an anesthetic and is still so used occasionally. Divinyl ether has also been used as an anesthetic. Notice that ethylene oxide is an example of an ether with a ring structure. Ethylene oxide is a gas that is used to sterilize plastic syringes and other materials that cannot be sterilized by high temperatures.

Table 15-5 Some Common Ethers

Structural Formula	Common Name
CH_3—O—CH_3	Dimethyl ether
CH_3—CH_2—O—CH_3	Methyl ethyl ether
CH_3—CH_2—O—CH_2—CH_3	Diethyl ether
CH_2=CH—O—CH=CH_2	Divinyl ether
—O—CH_3	Methyl phenyl ether (anisole)
H_2C———CH_2 \\ / O	Ethylene oxide

Properties

Like alcohols, ethers can form hydrogen bonds with water molecules (see Figure 15-4a). Thus, the water solubility of ethers is comparable to that of alcohols. For example, the following functional group isomers, diethyl ether, CH_3—CH_2—O—CH_2—CH_3, and 1-butanol, CH_3—CH_2—CH_2—CH_2—OH, have similar solubilities, namely, 7.5 g/100 g of H_2O and 9 g/100 g of H_2O. Unlike alcohols, however, ethers *cannot* form hydrogen bonds in the pure state because they do not contain any hydrogen atoms bonded to highly electronegative atoms (Figure 15-4b).

As a result of their inability to form hydrogen bonds in the pure state, ethers have much lower boiling points than alcohols with similar molecular weights. For example, the boiling point of diethyl ether (MW = 74) is 35°C, which is much closer to that of the alkane pentane (MW = 72; b.p. = 36°C) than it is to 1-butanol (MW = 74; b.p. = 117°C). Because of its low boiling point and high vapor pressure, diethyl ether is easily inhaled, which is helpful when it is used as an anesthetic. However, ether vapor is extremely flammable. Flames and even static electrical sparks must be avoided when it is used. Operating room

With water: In the pure state:

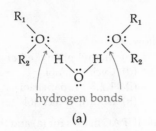

Do not form
hydrogen bonds

(a)   (b)

Figure 15-4   Ethers form hydrogen bonds with   hydrogen bonds with other ether
                   water molecules (a) but cannot form   molecules in the pure state (b).

personnel wear conductive shoes and take other special precautions to prevent static electrical sparks, which could have disastrous effects if ether or other highly flammable materials are present.

The only reaction of ethers that we need to consider is their formation from alcohols; see reaction 15-2: $R—OH + R—OH → R—O—R + H_2O$.

Exercise 15-5
Draw structural formulas for each of the following compounds:

(a) 3-Methyl-2-pentanol   (b) 2,3-Hexanediol   (c) Diisopropyl ether

Exercise 15-6
Identify the alcohol, ether, or phenol functional groups in the following structural formulas:

(a) ⬡–CH_3 –OH

(b) ⬡–CH_2—OH

(c) ⬡–CH_2—O—CH_2–⬡

(d) HO–⬡–O–⬡–OH

15-7 SUMMARY

1. The oxygen-containing portions of organic compounds are polar and hydrophilic.

2. The $—\overset{|}{\underset{|}{C}}—OH$ group is the alcohol functional group, and the —OH group of alcohols and phenols is called the hydroxyl group.

3. In aqueous solution, alcohols and ethers do not act as acids, and act as only extremely weak bases. Phenols act as weak acids and extremely weak bases.

4. Alcohols and ethers may be considered as organic derivatives of water. The general formula for an alcohol is $R—OH$; the general formula for an ether is $R_1—O—R_2$. The IUPAC ending for alcohols is -ol.

5. Alcohols can be dehydrated to form alkenes or ethers. Primary alcohols can be oxidized to form aldehydes. Secondary alcohols can be oxidized to ketones. Tertiary alcohols cannot be oxidized under similar conditions.

PROBLEMS

1. Draw structural formulas for each of the following:
   - (a) 2,2-Dimethyl-1-pentanol
   - (b) Methyl phenyl ether
   - (c) Methyl ethyl ether
   - (d) Cyclobutanol
   - (e) 2,3-Heptanediol
   - (f) 1,2,3-Propanetriol
   - (g) Phenol
   - (h) Isopropyl alcohol

2. Given the following structural formulas, write IUPAC names for (a) and (b), and write the common name for (c).

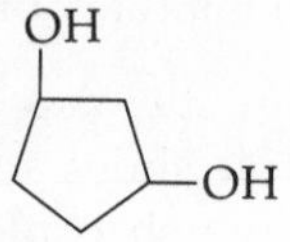

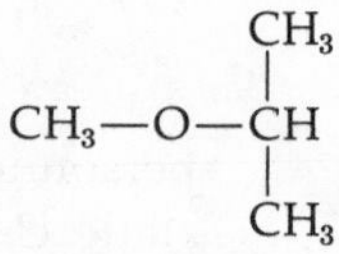

3. Arrange the compounds in Problem 2 in order of increasing solubility in water and predict which one has the lowest boiling point.

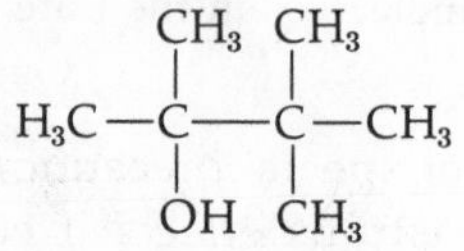

4. Draw the general structural formulas for
   - (a) A secondary alcohol
   - (b) A phenol
   - (c) A nonaromatic ether

5. Draw structural formulas for the products of the following reactions (X symbolizes an oxidizing agent, and XH_2 symbolizes a reducing agent):

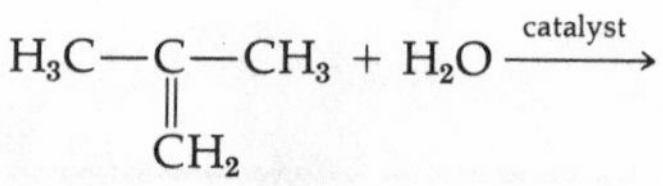

6. Draw structural formulas for all of the structural isomers with the following molecular formula: C_3H_8O.

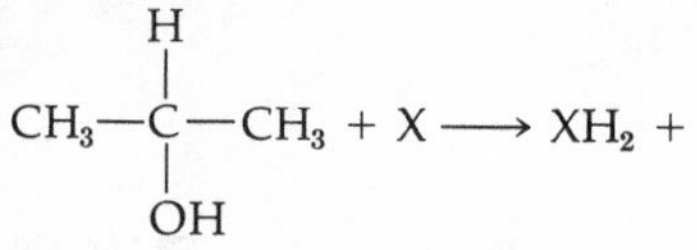

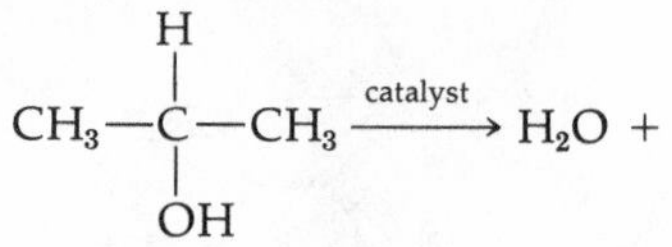

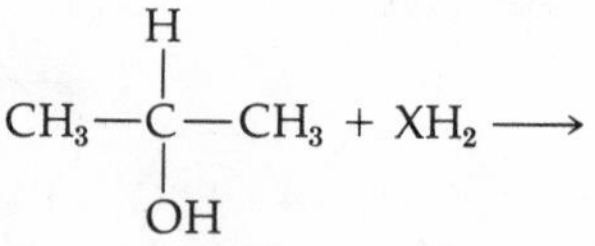

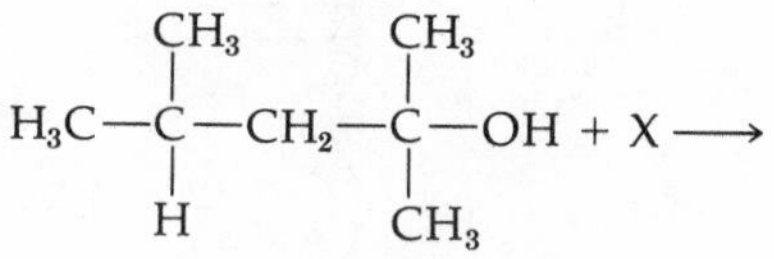

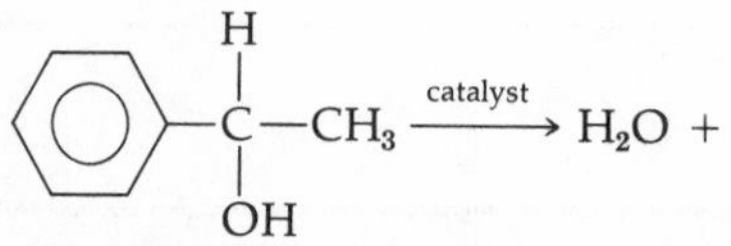

SOLUTIONS TO EXERCISES

15-1 $CH_2-CH_2-CH_3$ < $H_3C-\underset{\underset{CH_3}{|}}{\overset{\overset{CH_3}{|}}{C}}-(CH_2)_3-OH$ < $CH_3-CH_2-CH_2OH$

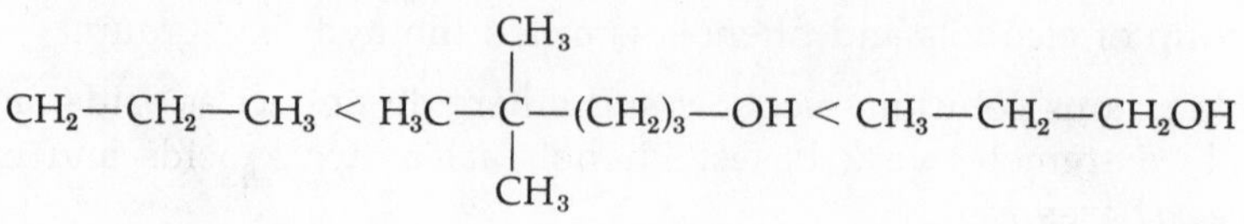

        insoluble alkane   soluble alcohols

Because of its smaller nonpolar portion, 1-propanol is most soluble.

15-2 Methanol is an alcohol and can act as both a hydrogen bond acceptor and donor, forming hydrogen bonds in the pure state. The other compounds are nonpolar alkanes, which cannot form hydrogen bonds.

15-3

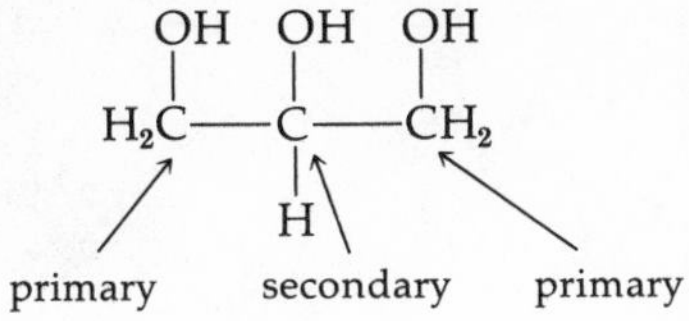

$$\underset{\text{primary}}{H_2C}\text{—}\underset{\substack{| \\ H \\ \text{secondary}}}{C}\text{—}\underset{\text{primary}}{CH_2}$$

OH OH OH

15-4 Name   Products

(a) Ethanol $CH_3\text{—}CH_2\text{—}O\text{—}CH_2\text{—}CH_3$

(b) Ethanol $H_2C\text{=}CH_2$

aldehyde

(c) 1-Propanol $CH_3\text{—}CH_2\text{—}\overset{\overset{\displaystyle O}{\|}}{C}\text{—}H$

Ketone

(d) 3,3-Dimethyl-2-butanol $CH_3\text{—}\underset{\underset{CH_3}{|}}{\overset{\overset{CH_3}{|}}{C}}\text{—}\overset{\overset{\displaystyle O}{\|}}{C}\text{—}CH_3$

15-5 (a) 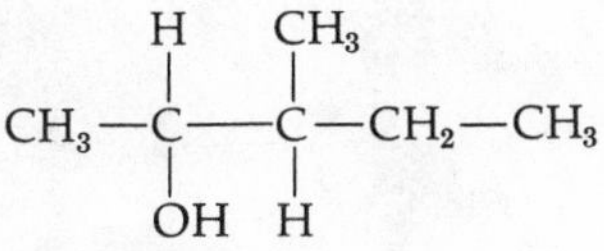 $CH_3\text{—}\underset{\underset{OH}{|}}{\overset{\overset{H}{|}}{C}}\text{—}\underset{\underset{H}{|}}{\overset{\overset{CH_3}{|}}{C}}\text{—}CH_2\text{—}CH_3$

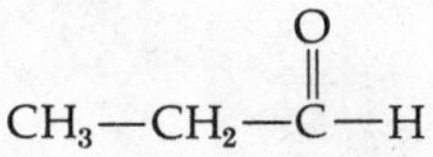

(b) $CH_3\text{—}\overset{\overset{OH}{|}}{CH}\text{—}\overset{\overset{OH}{|}}{CH}\text{—}CH_2\text{—}CH_2\text{—}CH_3$

(c) $H\overset{\overset{CH_3}{|}}{\underset{\underset{CH_3}{|}}{C}}\text{—}O\text{—}\overset{\overset{CH_3}{|}}{\underset{\underset{CH_3}{|}}{C}}H$

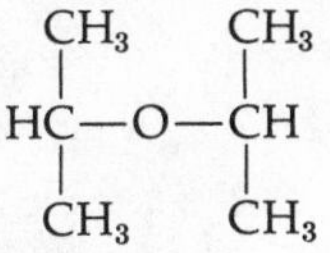

15-6 (a)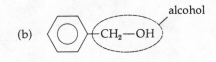
phenol

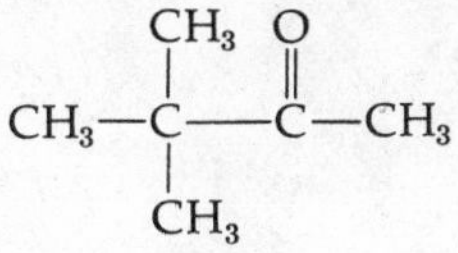

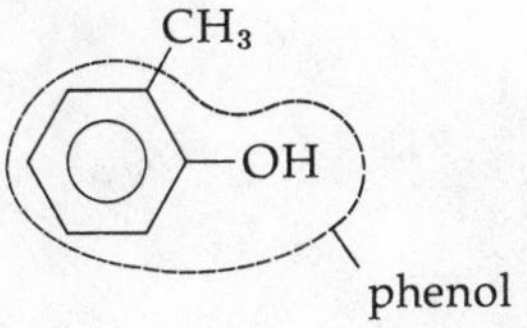

(b)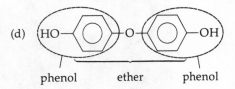
alcohol

(c)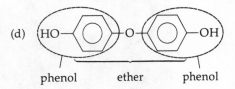
ether

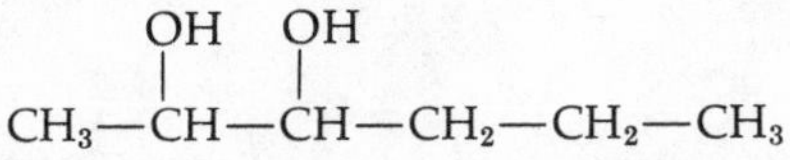

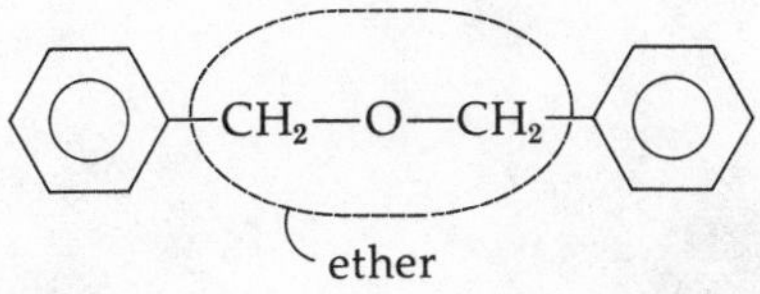

(d)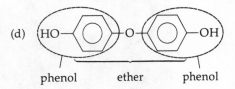
phenol   ether   phenol

CHAPTER 16

Aldehydes and Ketones

16-1 INTRODUCTION

The compounds we studied in Chapter 15 contained an oxygen atom bonded to a carbon atom with a single bond. All of the other classes of oxygen-containing organic compounds that we shall discuss contain a carbon-oxygen double

bond, $-\overset{\overset{\text{O}}{\|}}{\text{C}}-$, which is called a **carbonyl group.** Most of the organic compounds found in nature, including DNA, proteins, and simple carbohydrates, contain carbonyl groups. Many synthetic organic compounds, including a number of polymers, also contain carbonyl groups.

The simplest organic compound that contains a carbonyl group is formalde-

hyde, $\text{H}-\overset{\overset{\text{O}}{\|}}{\text{C}}-\text{H}$, the only compound with a carbonyl group that is a gas at room temperature. It is a widespread atmospheric pollutant, formed in the incomplete combustion of such fuels as coal and wood, and it tends to accumulate over large urban areas. Formaldehyde is quite soluble in water; formalin, a concentrated solution of formaldehyde in water (usually 40 grams of formaldehyde per 100 ml of solution), is widely used in modern society as a disinfectant and germicide, in photographic processing, and as one of the starting materials in the manufacture of polymers, dyes, explosives, and other complex synthetic organic chemicals. Formaldehyde solutions are also used as embalming fluids.

Aldehydes and ketones, the subject of this chapter, have only one oxygen atom bonded to the carbon atom of the carbonyl group. The properties of both of these classes of compounds are due primarily to the carbonyl group. On the other hand, carboxylic acids and their derivatives have an additional oxygen atom bonded to the carbonyl carbon atom, giving the carboxyl group,

$-\overset{\overset{\text{O}}{\|}}{\text{C}}-\text{O}-$. Many of the properties of carboxylic acids, which will be discussed in the next chapter, are quite different from those of aldehydes and ketones.

Formaldehyde has several uses. It is used as the starting material for the industrial synthesis of a number of complex chemicals. It is also the active ingredient in laboratory preservatives and embalming fluids.

16-2   STUDY OBJECTIVES

After careful study of this chapter, you should be able to:

1. Identify carbonyl groups and distinguish an aldehyde or a ketone from other compounds that contain a carbonyl group, given the structural formula for a compound.

2. Explain the role of hydrogen bonding in the water solubility of aldehydes and ketones.

3. Draw the structural formula for a simple aldehyde or ketone, given its name, or vice versa.

4. Draw structural formulas for the products of reactions in which: (a) aldehydes or ketones are oxidized or reduced; (b) aldehydes undergo an aldol condensation; (c) hemiacetals, acetals, hemiketals, or ketals are formed or are hydrolyzed, given the structural formulas of the reactants.

16-3   ALDEHYDES

Aldehydes have the general formula $R'-\overset{\displaystyle O}{\overset{\|}{C}}-H$, where R′ can be a hydrogen atom, an alkyl group, an aromatic group, or some other group of atoms.

Structure and Nomenclature

The IUPAC names for aldehydes are derived from the name of the parent alkane, with the ending *-al* to indicate the presence of the aldehyde functional group. Thus, the IUPAC names of the aldehydes $H-\overset{\displaystyle O}{\overset{\|}{C}}-H$ and $CH_3-\overset{\displaystyle O}{\overset{\|}{C}}-H$ are methanal and ethanal, respectively. Note that the final *-e* of the alkane name is dropped when forming the name of the aldehyde. The rules for naming aldehydes and other organic compounds are summarized in Essential Skills 8.

Most aldehydes are usually referred to by their common names. The common name of an aldehyde is derived from the common name of the carboxylic acid that is produced when the aldehyde is oxidized. For example, the common name for methanal is formaldehyde, since it can be oxidized to produce formic acid, $H-\overset{\displaystyle O}{\overset{\|}{C}}-OH$. The structural formulas and names of some simple aldehydes are given in Table 16-1.

The aromatic aldehyde benzaldehyde is used in the manufacture of dyes and for almond flavoring. In high concentrations benzaldehyde is a narcotic, and in very high concentrations it is lethal.

When common names of aldehydes are used, the Greek letters α, β, γ, δ, and so on, are used to indicate the positions of substituents on the hydrocarbon chain. For example, the structural formula of β-hydroxybutyraldehyde is

$$\underset{\gamma}{CH_3}-\underset{\beta}{\overset{\displaystyle OH}{\overset{|}{CH}}}-\underset{\alpha}{CH_2}-\overset{\displaystyle O}{\overset{\|}{C}}-H$$

In the IUPAC name for this compound, the position of the hydroxyl group is indicated by a numerical prefix. Its IUPAC name is 3-hydroxybutanal. Note that in the common names for aldehydes the α position refers to the carbon atom *next* to the carbonyl group carbon atom, whereas in the IUPAC name this is carbon atom 2 (the carbonyl group carbon atom is carbon atom 1). The components of IUPAC and common names must never be mixed. For example, β-hydroxybutanal and 3-hydroxybutyraldehyde are *not* acceptable names for β-hydroxybutyraldehyde (3-hydroxybutanal).

Table 16-1 Some Simple Aldehydes

IUPAC Name	Common Name	Structural Formula
Methanal	Formaldehyde	$\underset{\text{H}-\overset{\displaystyle\|}{\underset{}{\text{C}}}-\text{H}}{\overset{\text{O}}{}}$
Ethanal	Acetaldehyde	$\text{CH}_3-\overset{\overset{\text{O}}{\|\|}}{\text{C}}-\text{H}$
Propanal	Propionaldehyde	$\text{CH}_3-\text{CH}_2-\overset{\overset{\text{O}}{\|\|}}{\text{C}}-\text{H}$
	Benzaldehyde	$\bigcirc\!\!\!\!\bigcirc-\overset{\overset{\text{O}}{\|\|}}{\text{C}}-\text{H}$

Properties of Aldehydes

As we discussed in Chapter 15, the carbonyl group is polar as a result of the large electronegativity of the oxygen atom. Thus, low-molecular-weight aldehydes are soluble in water because hydrogen bonds form between the aldehyde molecules and water molecules (see Figure 16-1a). Aldehydes cannot, however, form hydrogen bonds in the pure state (see Figure 16-1b). The hydrogen atom bonded to the carbonyl carbon atom, $-\overset{\overset{\text{O}}{\|\|}}{\text{C}}-\text{H}$, does not participate in hydrogen bonds. Thus, the boiling points of aldehydes are lower than those of hydrogen bond-forming alcohols with about the same molecular weight (see Table 16-2).

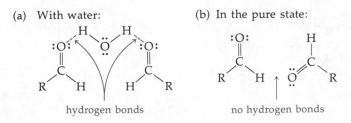

Figure 16-1   Aldehydes form hydrogen bonds with water (a), but they cannot form hydrogen bonds with each other in the pure state (b).

Note, however, that because of the polar nature of the carbonyl group, the intermolecular attractions are larger and thus the boiling points of aldehydes are higher than those of nonpolar alkanes of similar molecular weight.

In aqueous solution, aldehydes do not act as acids and only as extremely weak Brønsted-Lowry bases.

Table 16-2 Boiling Points of Aldehydes and Other Compounds with Similar Molecular Weights

Name	Structural Formula	MW	Boiling Point (°C)
Propane	$CH_3-CH_2-CH_3$	44	−42
Acetaldehyde	$CH_3-\overset{\displaystyle O}{\overset{\|}{C}}-H$	44	20
Ethanol	CH_3-CH_2-OH	46	78
Butane	$CH_3-CH_2-CH_2-CH_3$	58	0
Propionaldehyde	$CH_3-CH_2-\overset{\displaystyle O}{\overset{\|}{C}}-H$	58	49
1-Propanol	$CH_3-CH_2-CH_2-OH$	60	97

Reactions of Aldehydes

OXIDATION-REDUCTION

In the last chapter we saw that an aldehyde can be produced by the oxidation of a primary alcohol. The reverse reaction, formation of a primary alcohol by the reduction of an aldehyde, can also occur under appropriate conditions:

(16-1)
$$R'-\overset{\displaystyle O}{\overset{\|}{C}}-H + XH_2 \xrightarrow{\text{catalyst}} R'-\overset{\displaystyle OH}{\overset{\|}{\underset{\underset{\displaystyle H}{\|}}{C}}}-H + X$$

aldehyde   primary
                             alcohol

Reaction 16-1 represents the general reaction for the reduction of an aldehyde. In this representation, the reducing agent is symbolized by XH_2. For example, yeasts produce the primary alcohol ethanol in the enzyme-catalyzed reduction of acetaldehyde:

(16-2)
$$H_3C-\overset{\displaystyle O}{\overset{\|}{C}}-H + NADH + H^+ \xrightarrow{\text{a specific enzyme}} CH_3-CH_2-OH + NAD^+$$

acetaldehyde   ethanol

In this reaction, the symbol NAD^+ represents a complex substance called a coenzyme (Chapter 23), which is produced by the oxidation of the reducing agent NADH.

Aldehydes can also be oxidized to form carboxylic acids:

(16-3)
$$R'-\overset{\displaystyle O}{\overset{\|}{C}}-H + X + H_2O \longrightarrow R'-\overset{\displaystyle O}{\overset{\|}{C}}-OH + XH_2$$

aldehyde   carboxylic acid

Reaction 16-3 represents the general reaction for the oxidation of an aldehyde. In this representation, X symbolizes the oxidizing agent, and the additional oxygen atom in the carboxylic acid product is supplied by a water molecule. The reactant aldehyde loses a C—H bond and gains a C—O bond, so reaction 16-3 does indeed represent the oxidation of the aldehyde. The oxidation of ethyl alcohol to acetaldehyde and then to acetic acid occurs when wine is exposed to air. Vinegar (an aqueous solution of acetic acid) is the final product.

FORMATION OF HEMIACETALS AND ACETALS

An aldehyde can also react with an alcohol to form another compound, called a **hemiacetal,** in the general reaction

(16-4)
$$
\underset{\text{aldehyde}}{R'-\overset{\overset{\textstyle O}{\|}}{C}-H} + \underset{\text{alcohol}}{H-OR} \rightleftharpoons \underset{\text{hemiacetal}}{R'-\overset{\overset{\textstyle OH}{|}}{\underset{\underset{\textstyle H}{|}}{C}}-OR}
$$

Hemiacetal formation is an addition reaction in which the components of an alcohol, —OR and —H, add to the C=O bond. The addition of an alcohol to a C=O bond is similar to the hydration reaction in which the components of water are added to a C=C double bond (see Figure 16-2).

Figure 16-2 The addition of an alcohol to the C=O bond of an aldehyde forms a hemiacetal. This is similar to the addition of water to a C=C bond in which an alcohol is formed.

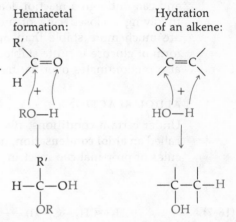

Reaction 16-4 is carried out in the laboratory by dissolving an aldehyde in an alcohol. In this alcoholic solution the equilibrium (reaction 16-4) is established, but most hemiacetals that do not involve a closed ring of atoms are not very stable and cannot be isolated from the alcoholic solution.

If an acid, such as HCl, is added to the equilibrium mixture in reaction 16-4, the hemiacetal can react further with another molecule of the alcohol to produce a compound called an **acetal:**

(16-5)
$$
\underset{\text{hemiacetal}}{R'-\overset{\overset{\textstyle OH}{|}}{\underset{\underset{\textstyle H}{|}}{C}}-OR} + \underset{\text{alcohol}}{ROH} \xrightarrow{H^+} \underset{\text{acetal}}{R'-\overset{\overset{\textstyle OR}{|}}{\underset{\underset{\textstyle H}{|}}{C}}-OR} + H_2O
$$

The H⁺ in reaction 16-5 acts as a catalyst. Acetals, in contrast to most open-chain hemiacetals, can be isolated by evaporating off the excess alcohol from the reaction mixture. On the other hand, if an acetal is dissolved in water and acid is added, a hydrolysis reaction equivalent to the reverse of reactions 16-4 and 16-5 takes place. The general reaction for the hydrolysis of an acetal is

(16-6)
$$
\underset{\text{acetal}}{R'-\overset{\overset{\textstyle OR}{|}}{\underset{\underset{\textstyle OR}{|}}{C}}-H} + H_2O \xrightarrow{H^+} \underset{\text{aldehyde}}{R'-\overset{\overset{\textstyle O}{\|}}{C}-H} + 2ROH
$$

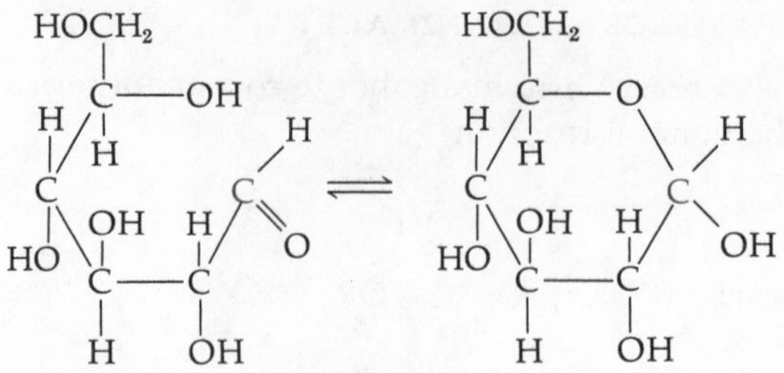

Figure 16-3   Internal hemiacetal formation by glucose. The free aldehyde form of glucose (left) reacts to form the hemiacetal (right). More than 99% of the dissolved glucose found in the human body is in the hemiacetal form, and solid glucose exists only as the hemiacetal.

Sugars, such as glucose, contain both aldehyde and alcohol functional groups and can undergo a reaction leading to the formation of an **internal hemiacetal** involving a closed ring of atoms (see Figure 16-3). Closed-ring hemiacetals are much more stable than open-chain hemiacetals. The internal hemiacetal form of glucose is quite stable and can be isolated from solution. In solution it also predominates over the free aldehyde form.

ALDOL REACTIONS

Under certain conditions, two aldehyde molecules can combine in a reaction called an **aldol condensation,** or aldol addition reaction. For example, two molecules of propanal can react in an aqueous solution of sodium hydroxide:

$$\text{(16-7)} \qquad 2(\text{CH}_3\!-\!\text{CH}_2\!-\!\overset{\overset{\text{O}}{\|}}{\text{C}}\!-\!\text{H}) \xrightarrow{\ \text{OH}^-\ } \text{CH}_3\!-\!\text{CH}_2\!-\!\underset{\underset{\text{CH}_3}{|}}{\overset{\overset{\text{OH}}{|}}{\text{CH}}}\!-\!\text{CH}\!-\!\overset{\overset{\text{O}}{\|}}{\text{C}}\!-\!\text{H}$$

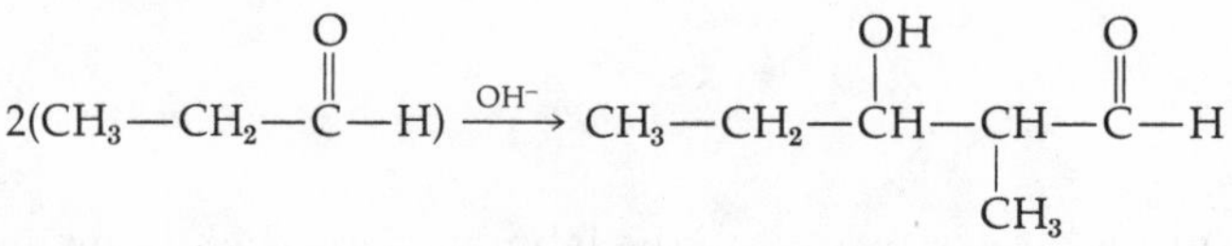

The product of reaction 16-7, 3-hydroxy-2-methylpentanal, has both **ald**ehyde and alc**ol** functional groups. This is where the name **aldol** comes from.

Aldol condensations, such as reaction 16-7, take place by means of a mechanism in which the base, OH$^-$, pulls off one of the weakly acidic α-hydrogens (Figure 16-4a). The resulting negatively charged ion combines with a second propanal molecule (Figure 16-4b). The oxygen atom of this second propanal then pulls a H$^+$ from a water molecule (Figure 16-4c). Note that the overall reaction for the mechanism shown in Figure 16-4 is reaction 16-7, and that the catalyst OH$^-$ is regenerated.

Reaction 16-7 can be viewed as the α-carbon and an α-hydrogen of one aldehyde adding to the carbonyl group of a second aldehyde to give a β-hydroxy-aldehyde product. The general aldol reaction for aldehydes that have an α-hydrogen can be represented as

$$\text{(16-8)} \qquad 2\left(\text{R}\!-\!\underset{\underset{}{|}}{\overset{\overset{\text{R}'}{|}}{\text{CH}}}\!-\!\overset{\overset{\text{O}}{\|}}{\text{C}}\!-\!\text{H}\right) \xrightarrow{\ \text{OH}^-\ } \text{R}\!-\!\underset{\underset{\text{H}}{|}}{\overset{\overset{\text{R}'}{|}}{\text{C}}}\!-\!\underset{\underset{\text{H}}{|}}{\overset{\overset{\text{OH}}{|}}{\text{C}}}\!-\!\underset{\underset{\text{R}}{|}}{\overset{\overset{\text{R}'}{|}}{\text{C}}}\!-\!\overset{\overset{\text{O}}{\|}}{\text{C}}\!-\!\text{H}$$

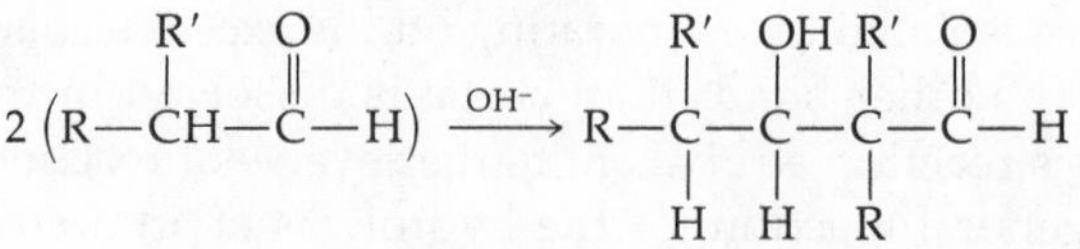

Recall that R' represents R or H. In order for this general reaction to take place, the reactant aldehyde must have an α-hydrogen (see Figure 16-4a). Thus, the α-carbon must be a primary or secondary carbon atom; aldehydes with a tertiary α-carbon atom do not undergo an aldol condensation.

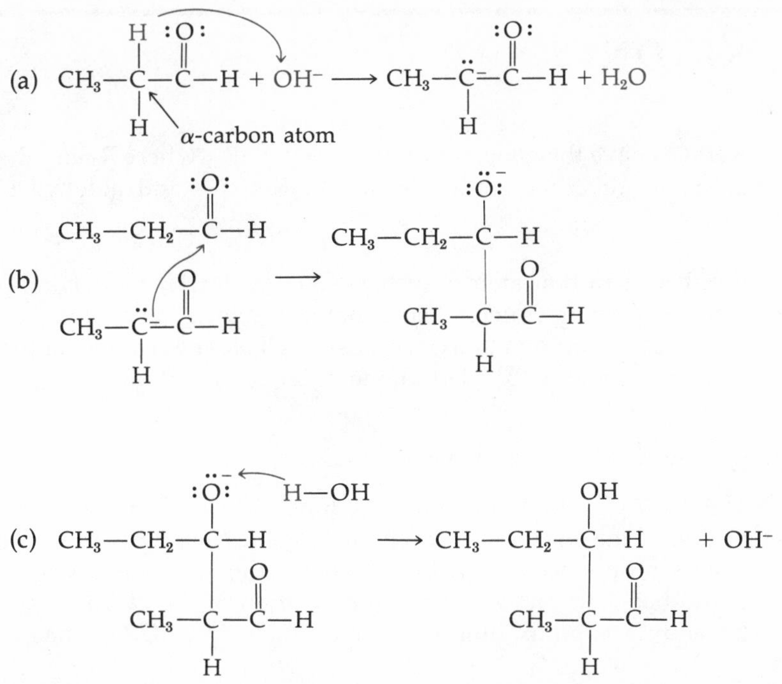

(a) $CH_3-\overset{\overset{\displaystyle H}{|}}{\underset{\underset{\displaystyle H}{|}}{C}}-\overset{\overset{\displaystyle :O:}{||}}{C}-H + OH^- \longrightarrow CH_3-\overset{\overset{\displaystyle :O:}{||}}{\underset{\underset{\displaystyle H}{|}}{C}}=C-H + H_2O$

α-carbon atom

(b) $CH_3-CH_2-\overset{\overset{\displaystyle :O:}{||}}{C}-H$ $CH_3-CH_2-\overset{\overset{\displaystyle :\overset{..}{O}:^-}{|}}{C}-H$

$CH_3-\overset{..}{C}=\overset{\underset{\underset{\displaystyle H}{|}}{}}{C}-H$ $\longrightarrow$ $CH_3-\overset{\overset{\displaystyle O}{||}}{C}-\overset{\underset{\underset{\displaystyle H}{|}}{}}{C}-H$

(c) $CH_3-CH_2-\overset{\overset{\displaystyle :\overset{..}{O}:^-}{|}}{C}-H$ H—OH $\longrightarrow$ $CH_3-CH_2-\overset{\overset{\displaystyle OH}{|}}{C}-H$ $+ OH^-$

$CH_3-\overset{\overset{\displaystyle O}{||}}{C}-\overset{\underset{\underset{\displaystyle H}{|}}{}}{C}-H$ $CH_3-\overset{\overset{\displaystyle O}{||}}{C}-\overset{\underset{\underset{\displaystyle H}{|}}{}}{C}-H$

Figure 16-4   Aldol condensation reactions take place by a three-step mechanism. (a) The α-hydrogen atoms in aldehydes are weakly acidic because of the pull on their electrons by the neighboring oxygen atom. A base, OH⁻ (or some other catalyst), first removes one of the α-hydrogens as a proton, leaving an unshared pair of electrons and a negative charge on the α-carbon atom. (b) The resulting negatively charged α-carbon atom attacks the carbonyl carbon atom of another aldehyde molecule, which has a partial positive charge. (c) The negatively charged oxygen atom of the product in (b) extracts a proton from a water molecule.

Aldol reactions are important in the commercial synthesis of many chemicals. Aldol-type reactions are also vital to the metabolism of carbohydrates in human cells (Chapter 25). In human cells, however, the catalysts for aldol reactions are specific enzymes.

Exercise 16-1

Draw structural formulas for the products of each of the following reactions (X symbolizes an oxidizing agent and XH₂ a reducing agent):

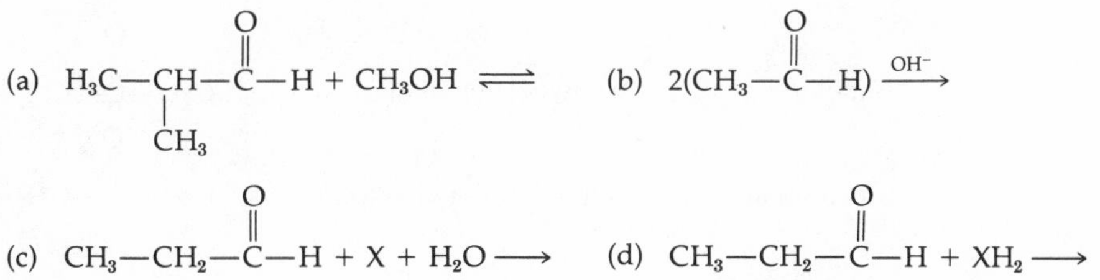

(a) $H_3C-\overset{\underset{\underset{\displaystyle CH_3}{|}}{}}{CH}-\overset{\overset{\displaystyle O}{||}}{C}-H + CH_3OH \rightleftharpoons$ (b) $2(CH_3-\overset{\overset{\displaystyle O}{||}}{C}-H) \xrightarrow{OH^-}$

(c) $CH_3-CH_2-\overset{\overset{\displaystyle O}{||}}{C}-H + X + H_2O \longrightarrow$ (d) $CH_3-CH_2-\overset{\overset{\displaystyle O}{||}}{C}-H + XH_2 \longrightarrow$

If you have difficulty with this type of problem, review the material in Essential Skills 9.

Exercise 16-2

Write the IUPAC names for the reactants in (a) and (b) of Exercise 16-1.

16-4 KETONES

Ketones have the general formula R_1—$\overset{\overset{\displaystyle O}{\|}}{C}$—$R_2$, where R_1 and R_2 represent alkyl, aromatic, or other groups of atoms. Ketones and aldehydes with the *same*

number of carbon atoms, such as CH_3—$\overset{\overset{\displaystyle O}{\|}}{C}$—$CH_3$ and CH_3—CH_2—$\overset{\overset{\displaystyle O}{\|}}{C}$—$H$, are functional group isomers. Note that in a ketone the carbonyl carbon is bonded to two other carbon atoms, whereas in all aldehydes, except formaldehyde, the carbonyl carbon is bonded to one other carbon atom and a hydrogen atom.

Structure and Nomenclature

The IUPAC names for simple ketones are derived from the parent alkane by use of the ending -*one* to indicate the ketone functional group. The ending -*one* is pronounced as in bone without the *b*. When a ketone has a chain of more than four carbon atoms, the position of the carbonyl group is indicated by adding a prefix number. For example, the IUPAC name for the ketone

CH_3—CH_2—$\overset{\overset{\displaystyle O}{\|}}{C}$—$CH_2$—$CH_3$ is 3-pentanone. Common, rather than IUPAC, names are often used for simple ketones. The common names for many ketones are formed by adding the name *ketone* to the names for the alkyl groups, R_1 and R_2, bonded to the carbonyl group. Thus, the common name for 3-pentanone is diethyl ketone. The structural formulas and names for some simple ketones are given in Table 16-3. Of the ketones listed in Table 16-3, acetone is the best known. It is an excellent solvent for many organic compounds. Acetone is also normally produced in small quantities in the human body. In diabetics, large quantities of acetone and other ketones may be produced. These "ketone bodies" are responsible for many of the serious effects of diabetes (Chapter 27). Acetophenone has a pleasant odor, like orange blossoms, and because of this property, it is used in some perfumes. This ketone is also used for the industrial synthesis of a number of compounds.

Table 16-3 Some Simple Ketones

IUPAC Name	Common Name	Structural Formula
Propanone	Acetone (dimethyl ketone)	CH_3—$\overset{\overset{\displaystyle O}{\|}}{C}$—$CH_3$
Butanone	Methyl ethyl ketone	CH_3—$\overset{\overset{\displaystyle O}{\|}}{C}$—$CH_2$—$CH_3$
2-Pentanone	Methyl *n*-propyl ketone	CH_3—$\overset{\overset{\displaystyle O}{\|}}{C}$—$CH_2$—$CH_2$—$CH_3$
3-Pentanone	Diethyl ketone	CH_3—CH_2—$\overset{\overset{\displaystyle O}{\|}}{C}$—$CH_2$—$CH_3$
	Acetophenone	CH_3—$\overset{\overset{\displaystyle O}{\|}}{C}$⟨⟩

Properties of Ketones

Low-molecular-weight ketones, like low-molecular-weight aldehydes, are water soluble because hydrogen bonds form between water molecules and the carbonyl oxygen atom (see Figure 16-5a). Ketones, like aldehydes, do not form hydrogen bonds in the pure state (see Figure 16-5b) and hence have boiling points comparable to their aldehyde isomers. For example, the boiling point of butanal is 76°C, whereas that of butanone is 80°C.

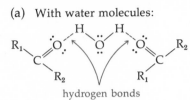

(a) With water molecules:

hydrogen bonds

(b) In the pure state:

no hydrogen bonds

Figure 16-5 Ketones form hydrogen bonds with water (a), but they cannot form hydrogen bonds with each other in the pure state (b).

Reactions of Ketones

OXIDATION-REDUCTION

Recall that the oxidation of a secondary alcohol produces a ketone (Chapter 15). In the reverse reaction a secondary alcohol can be formed by the reduction of a ketone:

$$(16\text{-}9) \qquad \underset{\text{ketone}}{R_1\overset{\displaystyle O}{\overset{\|}{-}C-R_2}} + XH_2 \longrightarrow \underset{\substack{\text{secondary}\\\text{alcohol}}}{R_1\overset{\displaystyle OH}{\underset{\displaystyle H}{-C-R_2}}} + X$$

Reaction 16-9 represents the general reaction for the reduction of a ketone, where the reducing agent is symbolized by XH_2.

A biologically important example is the reduction of the ketone functional group in the compound pyruvic acid to the hydroxyl group in the compound lactic acid:

$$(16\text{-}10) \qquad \underset{\text{pyruvic acid}}{HO-\overset{\displaystyle O}{\overset{\|}{C}}-\overset{\displaystyle O}{\overset{\|}{C}}-CH_3} + NADH + H^+ \xrightarrow{\substack{\text{a specific} \\ \text{enzyme}}} \underset{\text{lactic acid}}{HO-\overset{\displaystyle O}{\overset{\|}{C}}-\overset{\displaystyle OH}{\underset{\displaystyle H}{C}}-CH_3} + NAD^+$$

Reaction 16-10 occurs in active muscle cells. In this reaction the reducing agent is the complex coenzyme symbolized by NADH.

Unlike aldehydes, ketones *cannot* be oxidized to carboxylic acids.

FORMATION OF HEMIKETALS AND KETALS

Certain ketones, in reactions similar to those of aldehydes, can also react with some alcohols to produce compounds called **hemiketals** and **ketals.** Ketals can be hydrolyzed to yield an alcohol and a ketone in a reaction similar to the hydrolysis of an acetal. In Chapter 20 we shall see that sugars, such as fructose, that possess both alcohol and ketone functional groups can undergo a reaction leading to the formation of an **internal hemiketal** involving a closed ring of atoms. The formation of an internal hemiketal by fructose (see Figure 16-6) is similar to the formation of an internal hemiacetal by the aldehyde sugar glucose (Figure 16-3).

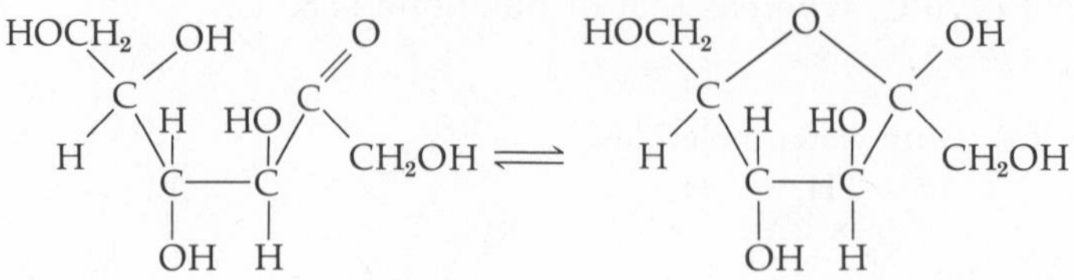

Figure 16-6   The free ketone form of fructose (left) reacts to form the internal hemiketal (right).

ALDOL CONDENSATION

Ketones that contain at least one primary or secondary α-carbon atom can undergo an aldol condensation reaction to give a β-hydroxyketone. For example,

(16-11)

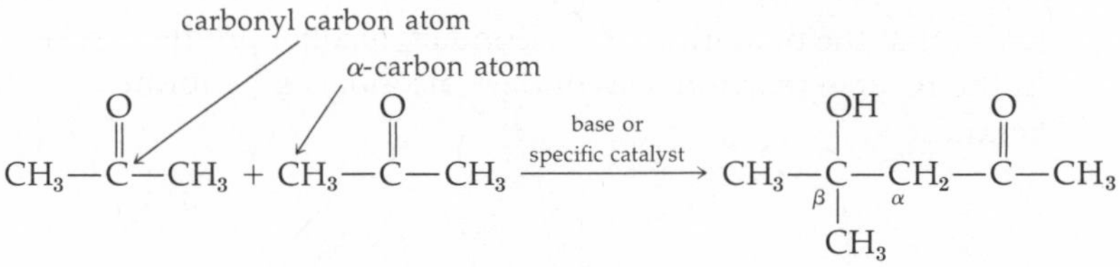

Note in the product of reaction 16-11 that the carbonyl carbon atom of one reactant ketone is bonded to an α-carbon atom of the other reactant ketone.

Exercise 16-3
Write the IUPAC name of the product of reaction 16-11.

Exercise 16-4
Draw structural formulas for the products of the following reactions (if no reaction occurs, write N.R.). Note that X is the symbol for an oxidizing agent and XH_2 represents a reducing agent.

(a) 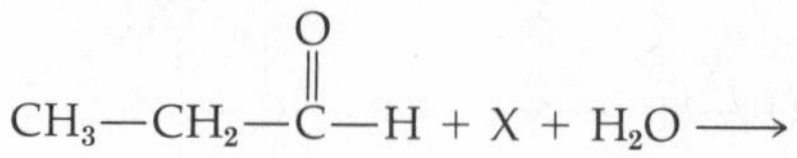 $CH_3-CH_2-\overset{\overset{\displaystyle O}{\|}}{C}-H + X + H_2O \longrightarrow$

(b) $H-\overset{\overset{\displaystyle CH_3}{|}}{\underset{\underset{\displaystyle CH_3}{|}}{C}}-\overset{\overset{\displaystyle O}{\|}}{C}-CH_3 + XH_2 \longrightarrow$

(c) $H_3C-\overset{\overset{\displaystyle CH_3}{|}}{\underset{\underset{\displaystyle CH_2-CH_3}{|}}{C}}-OH + X \longrightarrow$

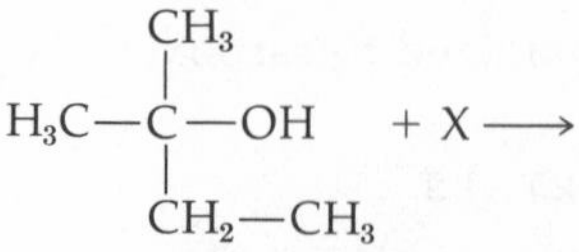

(d) 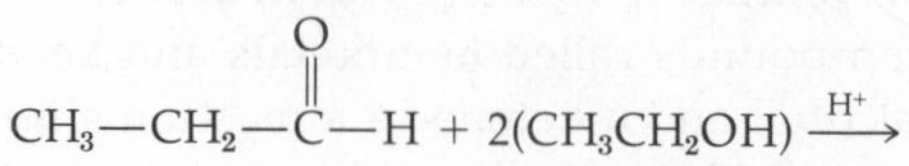 $CH_3-CH_2-\overset{\overset{\displaystyle O}{\|}}{C}-H + 2(CH_3CH_2OH) \overset{H^+}{\longrightarrow}$

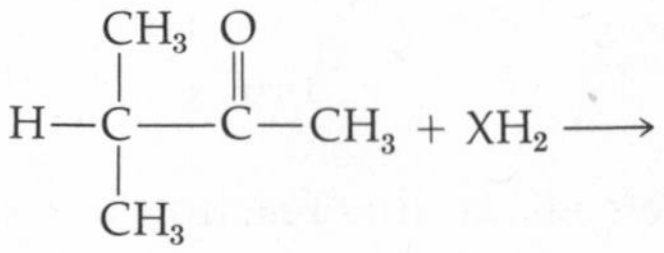

16-5 SUMMARY

1. The $-\overset{O}{\overset{\|}{C}}-$ group is called the carbonyl group.
2. Aldehydes have the general structural formula $R'-\overset{O}{\overset{\|}{C}}-H$, and their IUPAC names end in *-al*.
3. Ketones have the general structural formula $R_1-\overset{O}{\overset{\|}{C}}-R_2$, and their IUPAC names end in *-one*.
4. Aldehydes and ketones are polar and form hydrogen bonds with water. They do not form hydrogen bonds in the pure state.
5. Aldehydes and ketones can be formed by the oxidation of primary and secondary alcohols, respectively. In reverse reactions, primary and secondary alcohols can be formed by the reduction of aldehydes and ketones, respectively.
6. Aldehydes can be oxidized to form carboxylic acids; ketones cannot.
7. Aldehydes can react with alcohols to form hemiacetals and acetals. Some ketones can similarly form hemiketals and ketals. Acetals and ketals can be hydrolyzed to produce an aldehyde or ketone plus an alcohol.
8. Aldehydes and ketones that possess primary or secondary α-carbon atoms can undergo a type of reaction called aldol condensation to form β-hydroxyaldehydes or β-hydroxyketones, respectively.

PROBLEMS

1. Draw structural formulas for each of the following:
   (a) 2,2-Dimethylpentanal
   (b) 2,2-Dimethyl-3-pentanone
   (c) 4-Hydroxybutanal
   (d) α-Hydroxypropionaldehyde
2. Write the IUPAC names for the following compounds:
   (a) $CH_3-\underset{CH_3}{\underset{|}{CH}}-\overset{O}{\overset{\|}{C}}-H$
   (b) $\underset{CH_3}{\underset{|}{\overset{CH_3}{\overset{|}{HC}}}}-\overset{O}{\overset{\|}{C}}-\underset{CH_3}{\underset{|}{\overset{CH_3}{\overset{|}{CH}}}}$
   (c) $CH_3-\underset{H}{\underset{|}{\overset{OH}{\overset{|}{C}}}}-\underset{O}{\underset{\|}{C}}-CH_3$
   (d) 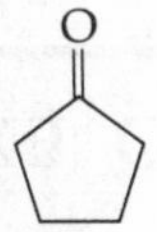
   (e) $CH_3-\overset{O}{\overset{\|}{C}}-CH_2-CH_3$
3. Which compounds in Problem 2 are isomers, and what kind of isomers are they?
4. Is compound (a) or (c) in Problem 2 more soluble in water? Which has the higher boiling point?

5. Draw the structural formulas for the products of the following reactions. If no reaction occurs, write N.R. (Note: X is the symbol for an oxidizing agent and XH_2 represents a reducing agent.)

(a) 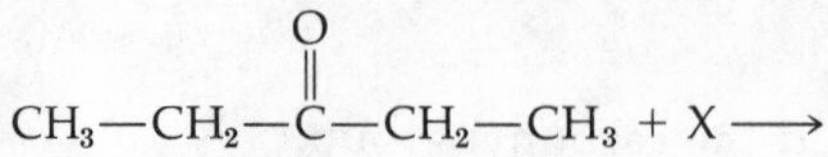 $CH_3-CH_2-\overset{\displaystyle O}{\overset{\|}{C}}-CH_2-CH_3 + X \longrightarrow$

(b) $CH_3-CH_2-\overset{\displaystyle O}{\overset{\|}{C}}-CH_2-CH_3 + XH_2 \longrightarrow$

(c) $2(H_3C-\overset{\displaystyle O}{\overset{\|}{C}}-H) \xrightarrow{OH^-}$

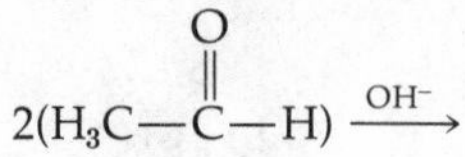

(d) 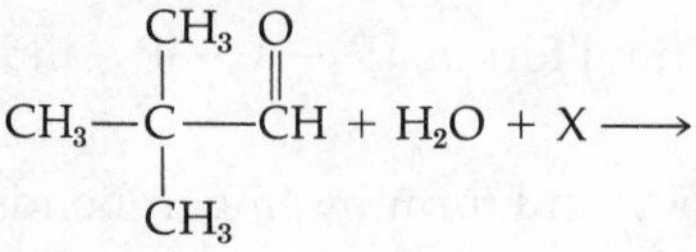 $CH_3-\overset{\displaystyle CH_3}{\underset{\displaystyle CH_3}{\overset{\displaystyle |}{\underset{|}{C}}}}-\overset{\displaystyle O}{\overset{\|}{C}}H + H_2O + X \longrightarrow$

(e) 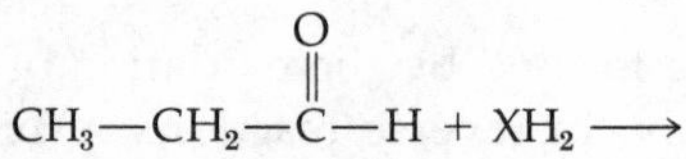 $CH_3-CH_2-\overset{\displaystyle O}{\overset{\|}{C}}-H + XH_2 \longrightarrow$

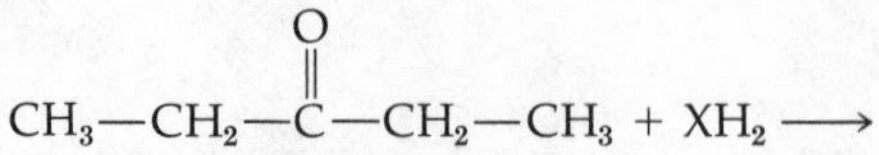

(f)  $CH_3-\overset{\displaystyle O}{\overset{\|}{C}}-H + 2(CH_3OH) \xrightarrow{H^+} H_2O +$

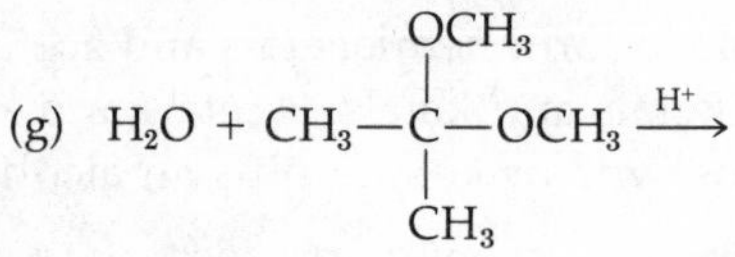

(g) $H_2O + CH_3-\overset{\displaystyle OCH_3}{\underset{\displaystyle CH_3}{\overset{\displaystyle |}{\underset{|}{C}}}}-OCH_3 \xrightarrow{H^+}$

(h) $CH_3-CH_2-CH_2-CH_2-\overset{\displaystyle O}{\overset{\|}{C}}-CH_2-CH_3 + X \longrightarrow$

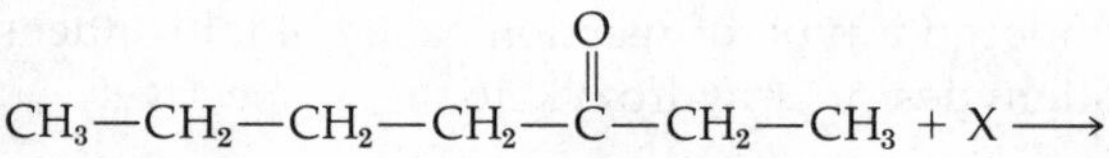

(i) 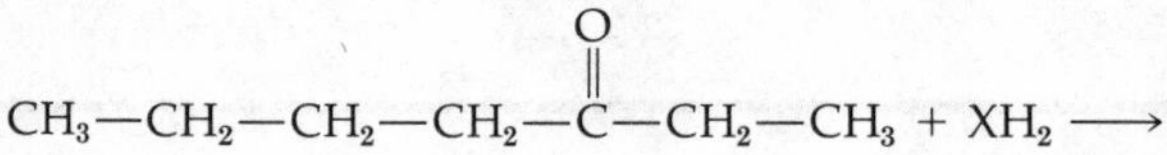 $CH_3-CH_2-CH_2-CH_2-\overset{\displaystyle O}{\overset{\|}{C}}-CH_2-CH_3 + XH_2 \longrightarrow$

6. Draw the structural formula for each of the following compounds:
   (a) 2-Pentanol
   (b) Propionaldehyde
   (c) Acetone
   (d) Dimethyl ketone
   (e) 3-Methyl-2-butanone
   (f) Methyl isopropyl ketone

7. Draw structural formulas for all of the structural isomers with the molecular formula C_3H_6O.

SOLUTIONS TO EXERCISES

16-1 (a) 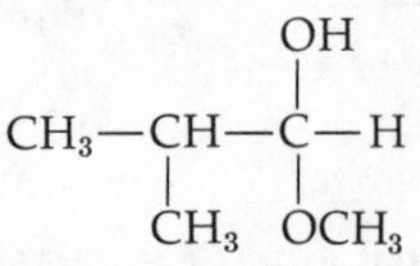 $CH_3-\overset{\displaystyle}{\underset{\displaystyle CH_3}{\overset{|}{\underset{|}{CH}}}}-\overset{\displaystyle OH}{\underset{\displaystyle OCH_3}{\overset{|}{\underset{|}{C}}}}-H$

(b) 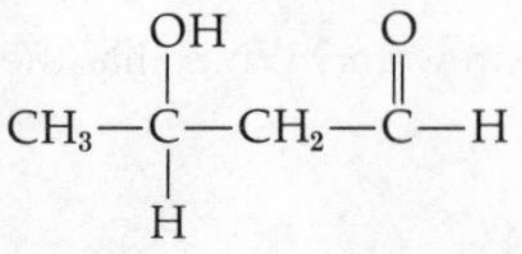 $CH_3-\overset{\displaystyle OH}{\underset{\displaystyle H}{\overset{|}{\underset{|}{C}}}}-CH_2-\overset{\displaystyle O}{\overset{\|}{C}}-H$

$$\text{(c)} \quad CH_3-CH_2-\overset{\overset{\displaystyle O}{\|}}{C}-OH + XH_2$$

(d) $CH_3-CH_2-CH_2OH + X$

16-2 (a) 2-Methylpropanal and methanol
(b) Ethanal

16-3 4-Hydroxy-4-methyl-2-pentanone

$$\text{16-4 (a)} \quad CH_3-CH_2-\overset{\overset{\displaystyle O}{\|}}{C}-OH + XH_2$$

$$\text{(b)} \quad \overset{\overset{\displaystyle CH_3}{|}}{\underset{\underset{\displaystyle CH_3}{|}}{HC}}-\overset{\overset{\displaystyle OH}{|}}{\underset{\underset{\displaystyle H}{|}}{C}}-CH_3 + X$$

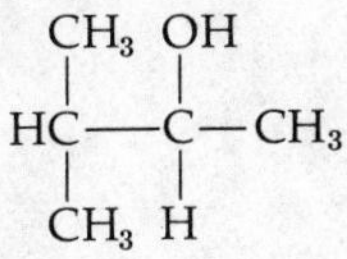

(c) N.R.

$$\text{(d)} \quad CH_3-CH_2-\overset{\overset{\displaystyle OCH_2CH_3}{|}}{\underset{\underset{\displaystyle OCH_2CH_3}{|}}{CH}} \quad + H_2O$$

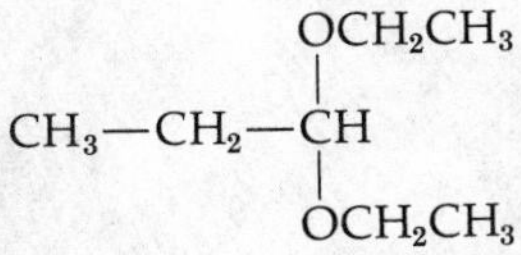

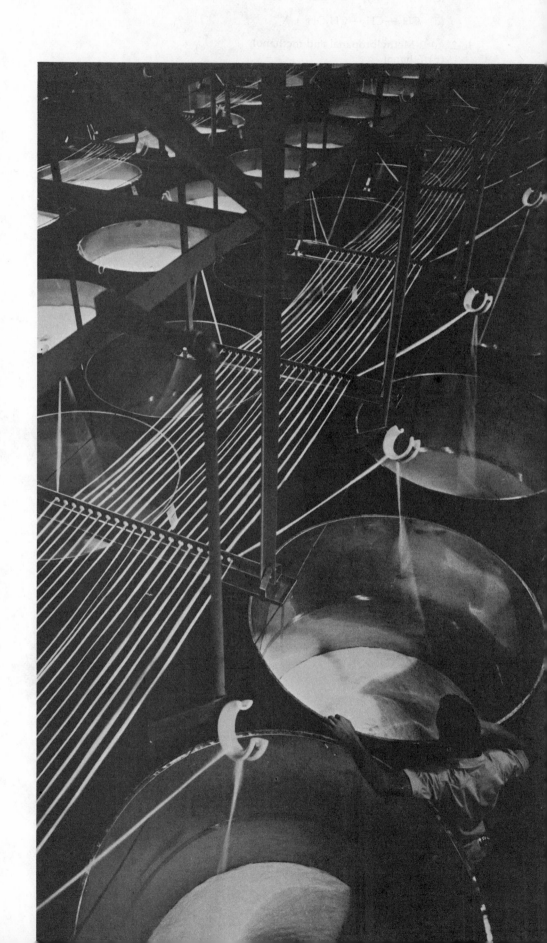

CHAPTER 17

Carboxylic Acids and Their Derivatives

17-1 INTRODUCTION

The carboxyl functional group, $-\overset{\overset{\displaystyle O}{\|}}{C}-OH$, contains two highly electronegative oxygen atoms. Molecules that contain this functional group, that is, carboxylic acids, behave quite differently from other oxygen-containing organic molecules.

In this chapter we shall discuss the unique properties of carboxylic acids that enable them to participate in large numbers of chemical reactions. Carboxylic acids found throughout nature include the amino acids used to make proteins (Chapter 21); lactic acid, which is produced by active muscle cells (Chapter 25); and citric acid, which is an important intermediate in cellular energy production (Chapter 25). Other naturally occurring carboxylic acids are found in bee venom, vinegar, soap, butter, cheese, and many other substances. Carboxylic acids are also extremely important commercially. Some are used to manufacture the synthetic fabrics Dacron and nylon. In the United States alone, more than 2000 tons of one carboxylic acid, acetylsalicylic acid (aspirin), are made yearly.

When a carboxylic acid reacts with an alcohol, a compound called an ester is formed. Similar reactions between two carboxylic acid molecules produce compounds called anhydrides. In this chapter we shall also study the structure, nomenclature, properties, and reactivity of these classes of oxygen-containing organic compounds. In later chapters we shall study several esters that are vital to the workings of cells in the human body. Simple anhydrides formed from carboxylic acids are not widespread in nature, although anhydrides are extremely important reactants in a number of industrial processes, including the production of some polymers.

After we have studied carboxylic acids, esters, and anhydrides, we shall briefly discuss the structure and properties of phosphoric acid and its derivatives. We shall see that phosphate esters and anhydrides are in many ways similar to the esters and anhydrides formed from carboxylic acids, and are involved in some of the most important chemical reactions that occur within human cells.

The manufacture of the synthetic fiber Dacron is an important application of the chemistry of carboxylic acids. Here continuous strands of Dacron polyester are being drawn from large vats into a stretching machine.

17-2   STUDY OBJECTIVES

After careful study of this chapter, you should be able to:

1. Draw the general structural formula for a carboxylic acid, an ester, or an anhydride.

2. Draw structural formulas for simple carboxylic acids, esters, and anhydrides, given their names, and vice versa.

3. Explain how hydrogen bonding affects the boiling point and solubility of carboxylic acids, esters, and anhydrides.

4. Explain why carboxylic acids are almost completely dissociated in the aqueous environment of human cells.

5. Draw structural formulas for the products of reactions in which carboxylic acids (a) react with bases, (b) are reduced, or (c) are decarboxylated, and for reactions in which anhydrides or esters are formed or hydrolyzed, given the structural formulas of the reactants.

6. Relate the properties of phosphoesters and phosphoanhydrides to esters and anhydrides of carboxylic acids.

17-3   CARBOXYLIC ACIDS

Structure and Nomenclature

The structural formulas of several carboxylic acids, together with their common and IUPAC names, are given in Table 17-1. The common names of some carboxylic acids are quite old and indicate some of the places where they are found in nature. The name formic acid, for instance, comes from the Latin word for ant (L. *formica* = ant). Acetic acid is named after vinegar (L. *acetum* = *vinegar*), of which it is the active ingredient. Butyric acid is named after butter (L. *butyrum* = *butter*) and is found in rancid butter. Common names of carboxylic acids end in -*ic* followed by the word "acid." The IUPAC names of carboxylic acids are formed from the parent alkane by using the suffix -*oic* to indicate the presence of the carboxylic acid functional group. The common names of carboxylic acids are frequently used. When common names of carboxylic acids are used, the Greek letters α, β, γ, δ, and so on, are used to indicate the positions of other substituents on the hydrocarbon chain in the same manner as for aldehydes (Chapter 16). For example, the formula of β-hydroxybutyric acid is

$$\underset{\gamma}{CH_3}-\underset{\beta}{\overset{\overset{\displaystyle OH}{|}}{CH}}-\underset{\alpha}{CH_2}-\overset{\overset{\displaystyle O}{\|}}{C}-OH$$

Carboxylic acids with long hydrocarbon chains, generally 12 to 20 carbon atoms, such as palmitic acid and stearic acid (Table 17-1) are called **fatty acids** and are used in several ways by the human body (Chapters 26 and 27). Fatty acids are not water soluble because of their long hydrophobic hydrocarbonlike chains. Some **dicarboxylic acids,** with two carboxyl functional groups (for example, succinic acid), and the **tricarboxylic acid** citric acid are also important in human biochemistry (Chapter 25).

Table 17-1 Some Carboxylic Acids

Structural Formula	Common Name	IUPAC Name
$\underset{\text{H—C—OH}}{\overset{\overset{\text{O}}{\|}}{}}$	Formic acid	Methanoic acid
$\underset{\text{CH}_3\text{—C—OH}}{\overset{\overset{\text{O}}{\|}}{}}$	Acetic acid	Ethanoic acid
$\underset{\text{CH}_3\text{—CH}_2\text{—C—OH}}{\overset{\overset{\quad\quad\quad\text{O}}{\|}}{}}$	Propionic acid	Propanoic acid
$\text{CH}_3\text{—CH}_2\text{—CH}_2\text{—C—OH}$ (C=O)	Butyric acid	Butanoic acid
$\text{CH}_3\text{—CH—C—OH}$ (CH₃, C=O)	α-Methylpropionic acid	2-Methylpropanoic acid
⬡—C(=O)—OH	Benzoic acid	
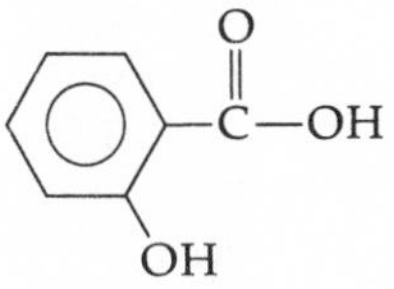 ⬡(OH)—C(=O)—OH	Salicylic acid	
$\text{CH}_3\text{—(CH}_2)_{14}\text{—C—OH}$ (C=O)	Palmitic acid, a fatty acid	
$\text{CH}_3\text{—(CH}_2)_{16}\text{—C—OH}$ (C=O)	Stearic acid, a fatty acid	
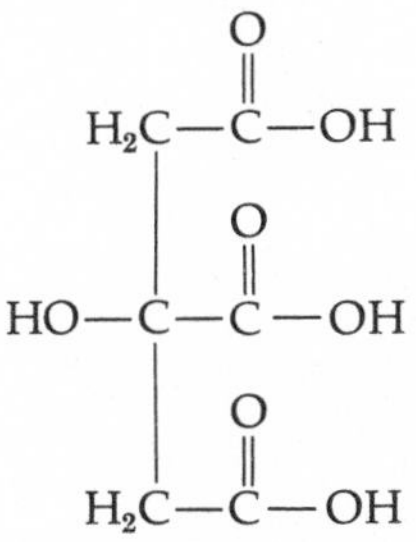 $\text{H}_2\text{C—C—OH}$ HO—C—C—OH $\text{H}_2\text{C—C—OH}$	Citric acid, a tricarboxylic acid	

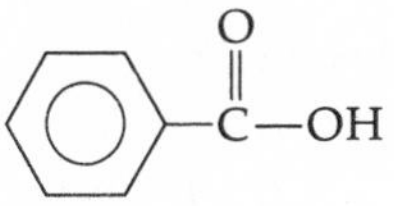

Properties of Carboxylic Acids

The carboxyl functional group, $-\overset{\overset{\text{O}}{\|}}{\text{C}}-\text{OH}$, contains the $-\overset{\overset{\text{O}}{\|}}{\text{C}}-$ group present in aldehydes and ketones and the $-\overset{}{\underset{|}{\text{C}}}-\text{OH}$ group present in alcohols. The properties of carboxylic acids, however, are enormously different from the properties of aldehydes, ketones, or alcohols, because in carboxylic acids both oxygen atoms are bonded to the *same* carbon atom. There is no carbon-carbon single bond "insulation" between the $-\overset{\overset{\textstyle O}{\|}}{C}-$ and $-\overset{|}{\underset{|}{C}}-OH$ groups. Therefore, the carboxyl group functions as a single functional group.

A molecule such as β-hydroxybutyraldehyde, $H_3C-\overset{\overset{\textstyle OH}{|}}{\underset{\underset{\textstyle H}{|}}{C}}-CH_2-\overset{\overset{\textstyle O}{\|}}{C}-H$, on the other hand, has enough carbon-carbon single bond "insulation" between the $-\overset{\overset{\textstyle O}{\|}}{C}-$ and $-\overset{|}{\underset{|}{C}}-OH$ groups, and it exhibits the separate characteristic properties of an aldehyde and an alcohol, and not those of a carboxylic acid.

HYDROGEN BONDING

The presence of two very electronegative oxygen atoms in carboxylic acids makes these molecules very polar. Carboxylic acids can therefore form hydrogen bonds with water molecules (see Figure 17-1). The formation of hydrogen bonds with water molecules makes low-molecular-weight carboxylic acids soluble in aqueous solution. Fatty acids (Table 17-1), however, are not water soluble because of the hydrophobic nature of their long alkyl groups. Carboxylic acids can also form hydrogen bonds with other carboxylic acid molecules (see Figure 17-1), which results in pure carboxylic acids having much higher boiling points than some other molecules of comparable molecular weight (see Table 17-2).

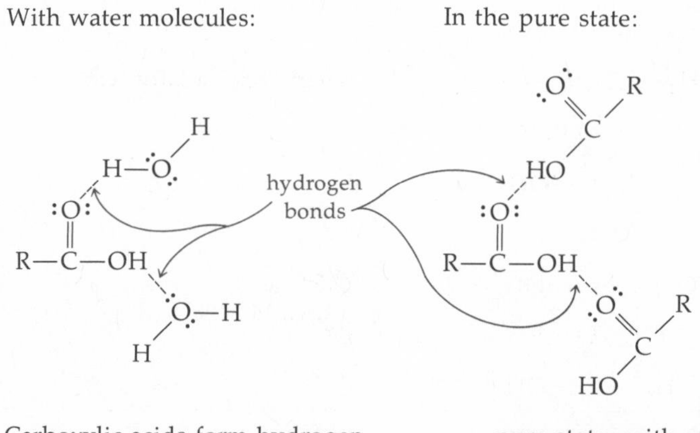

Figure 17-1   Carboxylic acids form hydrogen bonds with water (left), and in the pure state, with other carboxylic acid molecules (right).

ACIDIC PROPERTIES

We have already seen that the carboxylic acid functional group can serve as a proton donor, and thus that molecules containing this group are weak acids (Section 10-9). The doubly bonded oxygen in the carboxyl group pulls electrons toward itself and away from the O—H bond, thus facilitating the loss of this hydrogen as a proton (H$^+$). When a carboxylic acid is added to pure water, only a very small fraction of the carboxylic acid molecules lose a proton. Consider the following example: When 1 mole of acetic acid is dissolved in 1 liter of pure water (see reaction 17-1), only about 0.004 mole of acetate ions and 0.004 mole of hydronium ions are formed. Most of the acetic acid, 0.996 mole, is present

Table 17-2 Boiling Points of Compounds with Comparable Molecular Weights

Name	Structural Formula	MW	Boiling Point (°C)
Formic acid	$H-\overset{\displaystyle O}{\overset{\displaystyle \|}{C}}-OH$	46	101
Ethanol	CH_3-CH_2-OH	46	78
Acetaldehyde	$CH_3-\overset{\displaystyle O}{\overset{\displaystyle \|}{C}}-H$	44	20
Propane	$CH_3-CH_2-CH_3$	44	−42

as undissociated acetic acid molecules. The pH of this solution is about 2.4.

$$\text{(17-1)} \qquad CH_3-\overset{\displaystyle O}{\overset{\displaystyle \|}{C}}-OH + H_2O \rightleftharpoons CH_3-\overset{\displaystyle O}{\overset{\displaystyle \|}{C}}-O^- + H_3O^+$$

acetic acid   acetate ion

What happens, however, when some acetic acid is dissolved in a buffer that maintains the pH at about 7.4? Recall that pH 7.4 is the pH of normal body fluids. The pH does not drop significantly upon addition of the acetic acid, because most of the added acetic acid reacts with the base present in the buffer and is converted to acetate ions. We can determine the ratio of acetate ions to undissociated acetic acid molecules by using the equilibrium constant expression for reaction 17-1:

$$\text{(17-2)} \qquad K_a = \frac{[H_3O^+][A^-]}{[HA]} = \frac{[H_3O^+][\text{acetate ion}]}{[\text{acetic acid molecules}]}$$

Equation 17-2 can be rearranged algebraically to give Eq. 17-3:

$$\text{(17-3)} \qquad \frac{K_a}{[H_3O^+]} = \frac{[\text{acetate ion}]}{[\text{acetic acid molecules}]}$$

K_a is the dissociation constant for acetic acid and is about 1.8×10^{-5} mole/liter, and a pH of 7.4 corresponds to a $[H_3O^+]$ of about 4×10^{-8} M. Using these values in Eq. 17-3, the ratio of acetate ions to undissociated acetic acid molecules in a solution at pH 7.4 is

$$\text{(17-4)} \qquad \frac{[\text{acetate ion}]}{[\text{acetic acid molecules}]} = \frac{1.8 \times 10^{-5}}{4 \times 10^{-8}} \approx \frac{400}{1}$$

In other words, in an aqueous solution with the pH maintained at 7.4, virtually all of the acetic acid is dissociated and exists in the form of acetate ions. Similar considerations apply to all other carboxylic acids. Thus, in the human body, where buffers maintain the pH at about 7.4, virtually all carboxylic acid functional groups are ionized. The ionized form of a carboxyl group is called a

carboxylate ion $(R-\overset{\displaystyle O}{\overset{\displaystyle \|}{C}}-O^-)$. The name for a carboxylate ion is derived from the name of the carboxylic acid from which it is derived, by changing the -ic ending to -ate. Carboxylate ions are negatively charged and thus are attracted to positively charged parts of molecules. These ionic interactions play an important role in our body cells, as we shall see later.

Reactions of Carboxylic Acids

Some important reactions of carboxylic acids are summarized in Table 17-3. Reaction of a carboxylic acid in water with an equivalent amount of a strong base, such as NaOH, leads to the formation of a salt when water is evaporated:

(17-5)
$$R'-\overset{\overset{\textstyle O}{\|}}{C}-OH + NaOH \longrightarrow R'-\overset{\overset{\textstyle O}{\|}}{C}-O^-Na^+ + H_2O$$

The sodium salt of stearic acid, for example, is sodium stearate, which is commonly used as a soap.

$$CH_3-(CH_2)_{16}-\overset{\overset{\textstyle O}{\|}}{C}-O^-Na^+$$
sodium stearate

We have already seen that carboxylic acids can be formed by the oxidation of aldehydes (reaction 16-3). Alternatively, the reduction of a carboxylic acid produces an aldehyde:

(17-6)
$$R'-\overset{\overset{\textstyle O}{\|}}{C}-OH + XH_2 \longrightarrow R'-\overset{\overset{\textstyle O}{\|}}{C}-H + X + H_2O$$

where XH_2 represents the reducing agent. Under laboratory conditions, the aldehyde produced in reaction 17-6 will also react with the reducing agent, XH_2, to give the corresponding primary alcohol. It is therefore difficult to obtain significant amounts of an aldehyde using this reaction. In the human body, however, the reduction of a carboxylic acid to an aldehyde is catalyzed by a specific enzyme which produces only the aldehyde and not the corresponding alcohol.

Table 17-3 Reactions of Carboxylic Acids

Ionization
$$R'-\overset{\overset{\textstyle O}{\|}}{C}-OH + H_2O \rightleftharpoons R'-\overset{\overset{\textstyle O}{\|}}{C}-O^- + H_3O^+$$
a carboxylate ion

Salt formation
$$R'-\overset{\overset{\textstyle O}{\|}}{C}-OH + NaOH \rightleftharpoons R'-\overset{\overset{\textstyle O}{\|}}{C}-O^-Na^+ + H_2O$$
a sodium salt

Reduction
$$R'-\overset{\overset{\textstyle O}{\|}}{C}-OH + XH_2 \rightleftharpoons R'-\overset{\overset{\textstyle O}{\|}}{C}-H + X + H_2O$$
an aldehyde

Decarboxylation
$$R-\overset{\overset{\textstyle O}{\|}}{C}-OH \longrightarrow RH + CO_2$$

Anhydride formation
$$R_1-\overset{\overset{\textstyle O}{\|}}{C}-OH + HO-\overset{\overset{\textstyle O}{\|}}{C}-R_2 \rightleftharpoons R_1-\overset{\overset{\textstyle O}{\|}}{C}-O-\overset{\overset{\textstyle O}{\|}}{C}-R_2 + H_2O$$
an anhydride

Ester formation
$$R'-\overset{\overset{\textstyle O}{\|}}{C}-OH + HO-R \rightleftharpoons R'-\overset{\overset{\textstyle O}{\|}}{C}-O-R + H_2O$$
an ester

Another reaction that molecules containing carboxyl functional groups can undergo is **decarboxylation.** In a decarboxylation reaction, the carboxyl functional group is lost and a molecule of CO_2 is formed. Most simple carboxylic acids do not readily undergo decarboxylation reactions. However, carboxylic acids that have a ketone functional group located β to the carboxyl group (see reaction 17-7) are readily decarboxylated, both in the laboratory or by specific enzymes in the human body. Decarboxylation reactions reduce the length of a carbon chain by one carbon atom and are important in several normal biological processes. The decarboxylation of acetoacetic acid produces acetone in persons suffering from severe diabetes:

(17-7)

$$\underset{\substack{\beta\text{-ketone} \\ \\ \text{acetoacetic acid}}}{CH_3-\overset{\overset{\displaystyle O}{\|}}{C}-CH_2-\overset{\overset{\displaystyle O}{\|}}{C}-OH} \xrightarrow[\substack{\text{a specific} \\ \text{enzyme}}]{} \underset{\text{acetone}}{CH_3-\overset{\overset{\displaystyle O}{\|}}{C}-CH_3} + CO_2$$

Carboxylic acids also participate in a number of condensation reactions. A **condensation reaction** is a reaction in which two molecules condense or join together. We have already seen examples of condensation reactions, such as hemiacetal formation and aldol condensations, in Chapter 16. Carboxylic acids can participate in condensation reactions (1) with other carboxylic acid molecules to form anhydrides; (2) with alcohol molecules to form esters; (3) with phosphoric acid (see Section 17-5) to form phosphoesters; and (4) with other compounds, including some with nitrogen-containing or sulfur-containing functional groups (see Chapter 18).

When two carboxylic acid molecules react, an **anhydride** is formed in a condensation reaction accompanied by the elimination of a molecule of water:

(17-8)

$$\underset{\text{two carboxylic acids}}{R_1-\overset{\overset{\displaystyle O}{\|}}{C}-OH + HO-\overset{\overset{\displaystyle O}{\|}}{C}-R_2} \longrightarrow \underset{\text{an anhydride}}{R_1-\overset{\overset{\displaystyle O}{\|}}{C}-O-\overset{\overset{\displaystyle O}{\|}}{C}-R_2} + H_2O$$

Some anhydrides are formed just by heating the reactants, whereas the formation of other anhydrides requires the presence of a catalyst. Anhydrides can be hydrolyzed to carboxylic acids in the reverse of reaction 17-8.

When a carboxylic acid reacts with an alcohol, an **ester** is formed and a molecule of water is eliminated. The general condensation reaction for the formation of an ester is

(17-9)

$$\underset{\substack{\text{carboxylic acid} \qquad \text{alcohol}}}{R'-\overset{\overset{\displaystyle O}{\|}}{C}-OH + HO-R} \longrightarrow \underset{\text{an ester}}{R'-\overset{\overset{\displaystyle O}{\|}}{C}-O-R} + H_2O$$

Esters can also be hydrolyzed to produce a carboxylic acid and an alcohol:

(17-10)

$$R'-\overset{\overset{\displaystyle O}{\|}}{C}-O-R + H_2O \longrightarrow R'-\overset{\overset{\displaystyle O}{\|}}{C}-OH + HO-R$$

In the laboratory, reactions 17-9 and 17-10 are catalyzed by H^+ and usually result in a mixture of reactants and products in equilibrium. The desired product must then be separated from the other compounds in the mixture. The synthetic fabric Dacron is a polymer that is formed from dialcohol monomers and dicarboxylic acid monomers, which are joined together with ester bonds in an alternating arrangement (see Figure 17-2).

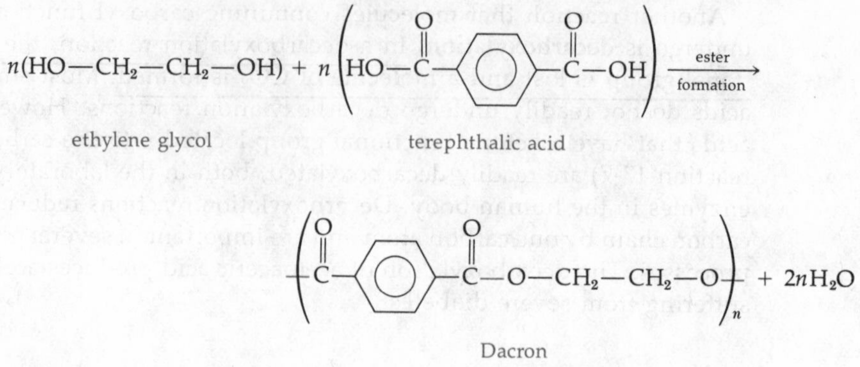

$$n(HO-CH_2-CH_2-OH) + n\left(HO-\overset{O}{\underset{\|}{C}}-\bigcirc-\overset{O}{\underset{\|}{C}}-OH\right) \xrightarrow[\text{formation}]{\text{ester}}$$

ethylene glycol   terephthalic acid

$$\left(\overset{O}{\underset{\|}{-C}}-\bigcirc-\overset{O}{\underset{\|}{C}}-O-CH_2-CH_2-O\right)_n + 2n\,H_2O$$

Dacron

Figure 17-2   The polymerization of Dacron from   monomers by the formation of ester
                   ethylene glycol and terephthalic acid   bonds (shown in color).

The hydrolysis of esters can also be catalyzed by OH^-. Commercially, esters of fatty acids are hydrolyzed by addition of a strong base, such as NaOH. This alkaline hydrolysis of fatty acid esters is called **saponification,** since the sodium salt of the acid that is formed is a soap. Hydrolysis and formation of esters are catalyzed by specific enzymes in the human body.

Exercise 17-1

Draw structural formulas for the products of the following reactions. (X represents an oxidizing agent, and XH_2 a reducing agent. If no reaction occurs, write N.R.)

(a) $CH_3-\overset{O}{\underset{\|}{C}}-OH + X \longrightarrow$ (b) $CH_3-\overset{O}{\underset{\|}{C}}-OH + XH_2 \longrightarrow$

(c) $2(CH_3-\overset{O}{\underset{\|}{C}}-OH) \xrightarrow[\text{catalyst}]{\text{heat}} H_2O +$

(d) $CH_3-CH_2-\overset{O}{\underset{\|}{C}}-OH + CH_3-CH_2OH \xrightarrow{\text{catalyst}} H_2O +$

If you have difficulty with this exercise, review Essential Skills 9.

17-4 ESTERS AND ANHYDRIDES

Structure and Nomenclature

Esters and anhydrides can be considered as derivatives of carboxylic acids. The structural formulas and common names of some esters are given in Table 17-4. Esters are named in a manner similar to ionic salts. For example, the ester

$CH_3-\overset{O}{\underset{\|}{C}}-O-CH_3$ is named methyl acetate, whereas $CH_3-\overset{O}{\underset{\|}{C}}-O^-Na^+$ is called sodium acetate. The *acetate* portion of these names comes from the

carboxylate ion of acetic acid, that is, $CH_3-\overset{O}{\underset{\|}{C}}-O^-$, the acetate ion. Recall that the names of carboxylate ions are formed by changing the suffix of the name for the carboxylic acid to *-ate*. The *methyl* part of the name methyl acetate comes from the name of the alkyl group that is bonded to the carboxylate part of the ester.

Table 17-4 Some Common Esters

Structural Formula	Common Name
$CH_3-\overset{\overset{O}{\|\|}}{C}-O-CH_3$	Methyl acetate
$CH_3-\overset{\overset{O}{\|\|}}{C}-O-CH_2-CH_3$	Ethyl acetate
$CH_3-CH_2-\overset{\overset{O}{\|\|}}{C}-O-CH_2-CH_3$	Ethyl propionate
benzene ring $-\overset{\overset{O}{\|\|}}{C}-O-CH_3$	Methyl benzoate
benzene ring $-\overset{\overset{O}{\|\|}}{C}-O-CH_3$, with OH	Methyl salicylate
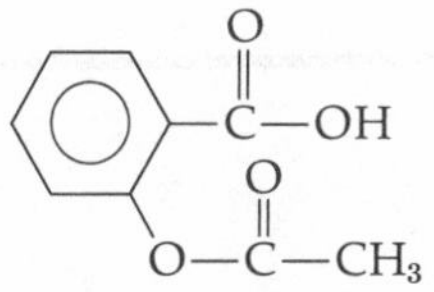 benzene ring $-\overset{\overset{O}{\|\|}}{C}-OH$, with $O-\overset{\overset{O}{\|\|}}{C}-CH_3$	Aspirin

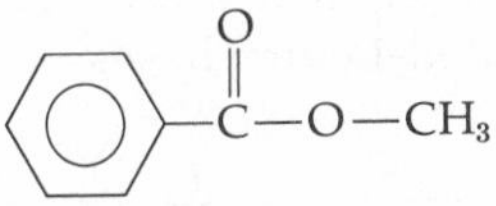

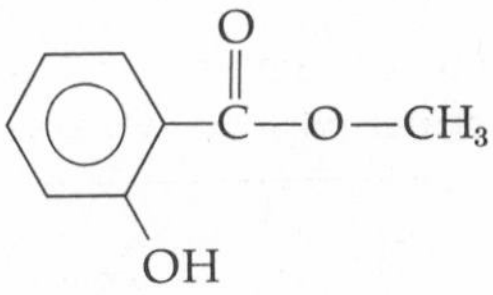

Many esters have pleasant odors. Methyl butyrate is the ester responsible for the smell of pineapples. Esters of the triol glycerol, $HOCH_2-\underset{\underset{OH}{|}}{CH}-CH_2OH$,

with three fatty acids, are called triglycerides. They are used for food storage in humans and are often referred to as just plain FAT!

Anhydrides are derivatives of two carboxylic acid molecules. The structural formulas and names of some simple anhydrides are given in Table 17-5. For symmetric anhydrides, that is, those derived from two molecules of the same carboxylic acid, the word *anhydride* is added to the name of the parent

carboxylic acid, after dropping the word *acid*. Thus, $CH_3-\overset{\overset{O}{\|\|}}{C}-O-\overset{\overset{O}{\|\|}}{C}-CH_3$ is called acetic anhydride.

Table 17-5 Some Simple Anhydrides

Structural Formula	Common Name
$CH_3-\overset{\overset{O}{\|\|}}{C}-O-\overset{\overset{O}{\|\|}}{C}-CH_3$	Acetic anhydride
$CH_3-CH_2-\overset{\overset{O}{\|\|}}{C}-O-\overset{\overset{O}{\|\|}}{C}-CH_2-CH_3$	Propionic anhydride
$CH_3-CH_2-CH_2-\overset{\overset{O}{\|\|}}{C}-O-\overset{\overset{O}{\|\|}}{C}-CH_2-CH_2-CH_3$	Butyric anhydride

Properties of Esters and Anhydrides

The ester $-\overset{\overset{\displaystyle O}{\|}}{C}-O-\overset{|}{\underset{|}{C}}-$ and anhydride $-\overset{\overset{\displaystyle O}{\|}}{C}-O-\overset{\overset{\displaystyle O}{\|}}{C}-$ functional groups are

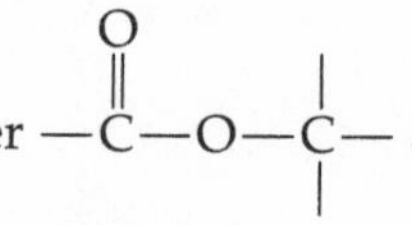

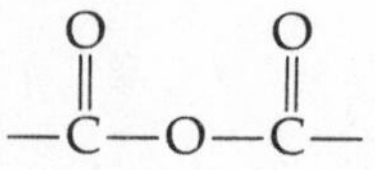

polar and can form hydrogen bonds with water. Thus, molecules containing these functional groups tend to be water soluble. Note, however, that <u>esters and anhydrides do not contain hydrogen-oxygen bonds, so they do not form hydrogen bonds in the pure state.</u> Recall that ethers, aldehydes, and ketones also do not form hydrogen bonds in the pure state.

   The formation and hydrolysis reactions of esters and anhydrides have already been discussed (Section 17-3). Esters and anhydrides are not acidic, and are extremely weak (weaker than water) Brønsted-Lowry bases.

Exercise 17-2

Would you expect acetic anhydride to have a lower, higher, or similar boiling point when compared to (a) hexane, (b) octanoic acid, (c) methyl acetate? Explain your answers.

17-5   PHOSPHOESTERS AND PHOSPHOANHYDRIDES

The structure of phosphoric acid is somewhat similar to that of a carboxylic acid (see Figure 17-3).

Figure 17-3   A carboxylic acid and phosphoric acid are somewhat similar in structure, but phosphoric acid has two additional acidic hydrogens.

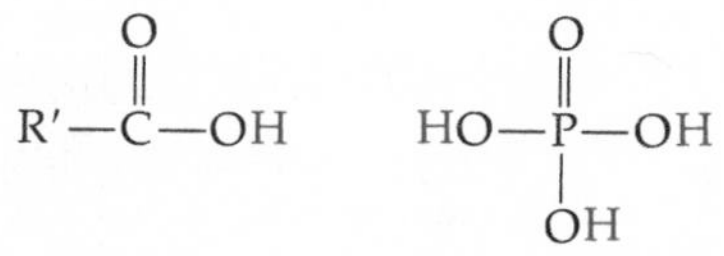

$$R'-\overset{\overset{\displaystyle O}{\|}}{C}-OH \qquad HO-\overset{\overset{\displaystyle O}{\|}}{\underset{\underset{\displaystyle OH}{|}}{P}}-OH$$

a carboxylic acid   phosphoric acid

Like carboxylic acids, phosphoric acid can form esters with alcohols:

(17-11) $R-OH + HO-\overset{\overset{\displaystyle O}{\|}}{\underset{\underset{\displaystyle OH}{|}}{P}}-OH \longrightarrow R-O-\overset{\overset{\displaystyle O}{\|}}{\underset{\underset{\displaystyle OH}{|}}{P}}-OH + H_2O$

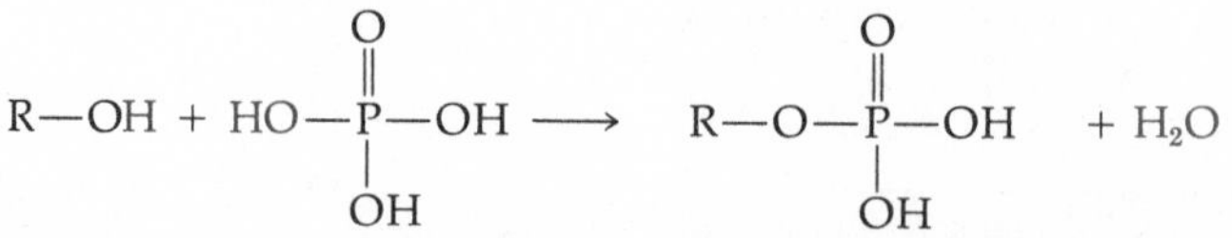

alcohol   phosphoric acid   organic phosphate ester

The resulting organic phosphate esters, or **phosphoesters,** are extremely important to living organisms, as we shall see in Chapter 25.

   Phosphoric acid can also form anhydrides with itself,

(17-12) $HO-\overset{\overset{\displaystyle O}{\|}}{\underset{\underset{\displaystyle OH}{|}}{P}}-OH + HO-\overset{\overset{\displaystyle O}{\|}}{\underset{\underset{\displaystyle OH}{|}}{P}}-OH \longrightarrow HO-\overset{\overset{\displaystyle O}{\|}}{\underset{\underset{\displaystyle OH}{|}}{P}}-O-\overset{\overset{\displaystyle O}{\|}}{\underset{\underset{\displaystyle OH}{|}}{P}}-OH + H_2O$

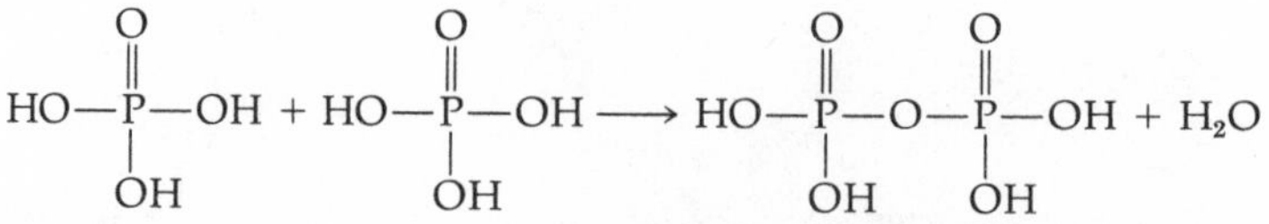

pyrophosphoric acid

as well as with a carboxylic acid,

(17-13)
$$R'-\overset{\overset{\displaystyle O}{\|}}{C}-OH + HO-\overset{\overset{\displaystyle O}{\|}}{\underset{\underset{\displaystyle OH}{|}}{P}}-OH \longrightarrow R'-\overset{\overset{\displaystyle O}{\|}}{C}-O-\overset{\overset{\displaystyle O}{\|}}{\underset{\underset{\displaystyle OH}{|}}{P}}-OH + H_2O$$

a carboxylic   phosphoric
acid   acid

Since phosphoric acid has *three* acidic hydrogens, a phosphoester or a phospho-anhydride can form further anhydrides and esters. For example, the acid part of a phosphoester molecule can react with a molecule of phosphoric acid and form a molecule that is both an ester and an anhydride:

(17-14)
$$R-O-\overset{\overset{\displaystyle O}{\|}}{\underset{\underset{\displaystyle OH}{|}}{P}}-OH + HO-\overset{\overset{\displaystyle O}{\|}}{\underset{\underset{\displaystyle OH}{|}}{P}}-OH \longrightarrow R-O-\overset{\overset{\displaystyle O}{\|}}{\underset{\underset{\displaystyle OH}{|}}{P}}-O-\overset{\overset{\displaystyle O}{\|}}{\underset{\underset{\displaystyle OH}{|}}{P}}-OH + H_2O$$

a diphosphate ester

Similarly,

(17-15)
$$R-O-\overset{\overset{\displaystyle O}{\|}}{\underset{\underset{\displaystyle OH}{|}}{P}}-O-\overset{\overset{\displaystyle O}{\|}}{\underset{\underset{\displaystyle OH}{|}}{P}}-OH + HO-\overset{\overset{\displaystyle O}{\|}}{\underset{\underset{\displaystyle OH}{|}}{P}}-OH \longrightarrow$$

$$R-O-\overset{\overset{\displaystyle O}{\|}}{\underset{\underset{\displaystyle OH}{|}}{P}}-O-\overset{\overset{\displaystyle O}{\|}}{\underset{\underset{\displaystyle OH}{|}}{P}}-O-\overset{\overset{\displaystyle O}{\|}}{\underset{\underset{\displaystyle OH}{|}}{P}}-OH + H_2O$$

a triphosphate ester

A very important property of phosphoanhydrides is the large amount of energy released upon hydrolysis. Consider the reaction

(17-16)
$$H_2O + R-O-\overset{\overset{\displaystyle O}{\|}}{\underset{\underset{\displaystyle OH}{|}}{P}}-O-\overset{\overset{\displaystyle O}{\|}}{\underset{\underset{\displaystyle OH}{|}}{P}}-O-\overset{\overset{\displaystyle O}{\|}}{\underset{\underset{\displaystyle OH}{|}}{P}}-OH \longrightarrow$$

a triphosphate ester

$$R-O-\overset{\overset{\displaystyle O}{\|}}{\underset{\underset{\displaystyle OH}{|}}{P}}-O-\overset{\overset{\displaystyle O}{\|}}{\underset{\underset{\displaystyle OH}{|}}{P}}-OH + HO-\overset{\overset{\displaystyle O}{\|}}{\underset{\underset{\displaystyle OH}{|}}{P}}-OH$$

a diphosphate ester   phosphoric acid

The hydrolysis reaction 17-16 is quite exergonic ($\Delta G \approx -8$ kcal/mole). Therefore, this reaction is a potential source of a large amount of work (Section 6-7). In the human body, hydrolysis reactions in which a phosphoanhydride bond in a triphosphate or a diphosphate is broken are an important source of energy (Chapter 25).

Exercise 17-3
Draw structural formulas for (a) ethyl phosphate and (b) glycerol 3-phosphate.

17-6 SUMMARY

1. Carboxylic acids, esters, and anhydrides are polar molecules and can form hydrogen bonds with water molecules. Carboxylic acids also form hydrogen bonds in the pure state; esters and anhydrides of carboxylic acids do not.

2. Carboxylic acids are weak acids with the general formula $R'-\overset{\displaystyle O}{\overset{\|}{C}}-OH$. The IUPAC suffix for carboxylic acids is -oic. In the human body, where the pH is normally 7.4, virtually all carboxylic acid molecules are ionized to carboxylate ions.

3. The general formulas for esters and anhydrides are $R'-\overset{\displaystyle O}{\overset{\|}{C}}-O-R$ and $R_1-\overset{\displaystyle O}{\overset{\|}{C}}-O-\overset{\displaystyle O}{\overset{\|}{C}}-R_2$, respectively.

4. Carboxylic acids can undergo several reactions, including: (a) ionization, (b) decarboxylation, (c) reduction to form an aldehyde, (d) condensation with another carboxylic acid molecule to form an anhydride, and (e) condensation with an alcohol to form an ester.

5. Anhydrides can be hydrolyzed to form carboxylic acids, and esters can be hydrolyzed to form carboxylic acids and alcohols. The alkaline hydrolysis of a fatty acid ester is called saponification.

6. Phosphoric acid behaves somewhat like a carboxylic acid. Phosphoesters and phosphoanhydrides are important components of human cells.

PROBLEMS

1. Which of the following has the highest boiling point? Which has the highest solubility in water? Explain.

   (a) 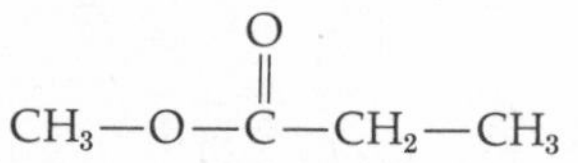 $CH_3-O-\overset{\displaystyle O}{\overset{\|}{C}}-CH_2-CH_3$ (b) 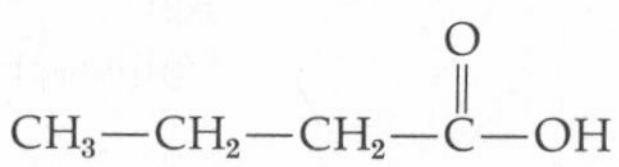 $CH_3-CH_2-CH_2-\overset{\displaystyle O}{\overset{\|}{C}}-OH$

2. Are the compounds in Problem 1 isomers? If so, what kind of isomers are they? Identify their functional groups.

3. Draw structural formulas for each of the following:
   (a) 2-Methylpropanoic acid
   (b) Methyl propionate
   (c) α-Methylpropionic acid
   (d) 3-Hydroxy-3-methylbutanoic acid
   (e) Butyric anhydride

4. Give the common names for each of the following:

   (a) 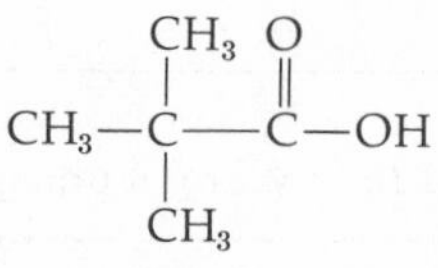 $CH_3-\overset{\displaystyle CH_3}{\underset{\displaystyle CH_3}{\overset{|}{\underset{|}{C}}}}-\overset{\displaystyle O}{\overset{\|}{C}}-OH$ (b) 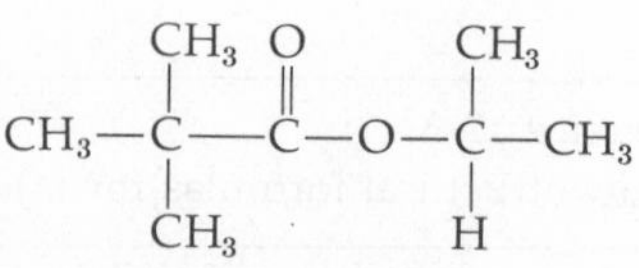 $CH_3-\overset{\displaystyle CH_3}{\underset{\displaystyle CH_3}{\overset{|}{\underset{|}{C}}}}-\overset{\displaystyle O}{\overset{\|}{C}}-O-\overset{\displaystyle CH_3}{\underset{\displaystyle H}{\overset{|}{\underset{|}{C}}}}-CH_3$

(c) $CH_3-CH_2-\overset{\displaystyle O}{\overset{\|}{C}}-O-CH_2-CH_3$

(d) 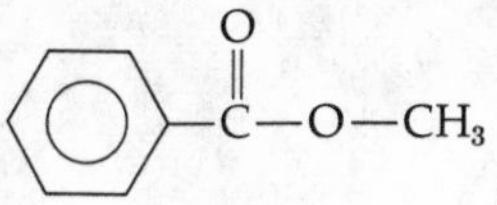 [benzene ring]$-\overset{\displaystyle O}{\overset{\|}{C}}-O-CH_3$

(e) $CH_3-CH_2-O-\overset{\displaystyle O}{\underset{\underset{\displaystyle OH}{|}}{\overset{\|}{P}}}-OH$

5. Draw structural formulas for the products of the following reactions. (X represents an oxidizing agent, and XH_2 a reducing agent. If no reaction occurs, write N.R.)

(a) $H_2O + CH_3-\overset{\displaystyle O}{\overset{\|}{C}}-O-CH_3 \xrightarrow{OH^-}$

(b) $OH^- + CH_3-CH_2-\overset{\displaystyle O}{\overset{\|}{C}}-OH \longrightarrow$

(c) $H_2O + CH_3-CH_2-\overset{\displaystyle O}{\overset{\|}{C}}-OH \xrightarrow{H^+}$

(d) $CH_3-\overset{\displaystyle O}{\overset{\|}{C}}-CH_3 + X + H_2O \longrightarrow$

(e) $CH_3-CH_2-\overset{\displaystyle O}{\overset{\|}{C}}-OH \longrightarrow CO_2 +$

(f) $CH_3-\overset{\displaystyle O}{\overset{\|}{C}}-O-\overset{\displaystyle O}{\overset{\|}{C}}-CH_3 + H_2O \xrightarrow{H^+}$

6. Are any of the reactants in Problem 5 functional group isomers? If so, which ones?

7. If you mix vodka (H_2O + ethanol) and vinegar (H_2O + acetic acid) over ice, what happens? What if you add a strong acid to this mixture and let it sit on the table for a day or so?

SOLUTIONS TO EXERCISES

17-1 (a) N.R.

(b) $CH_3-\overset{\displaystyle O}{\overset{\|}{C}}-H + X + H_2O$

(c) $CH_3-\overset{\displaystyle O}{\overset{\|}{C}}-O-\overset{\displaystyle O}{\overset{\|}{C}}-CH_3$

(d) $CH_3-CH_2-\overset{\displaystyle O}{\overset{\|}{C}}-O-CH_2-CH_3$

17-2 (a) Higher, since hexane is a nonpolar alkane with a lower molecular weight.
(b) Lower, since octanoic acid has a higher molecular weight and can form hydrogen bonds in the pure state.
(c) Higher, since methyl acetate has a lower molecular weight.

17-3 (a) $CH_3-CH_2-O-\overset{\overset{\displaystyle OH}{|}}{\underset{\underset{\displaystyle OH}{|}}{P}}=O$

(b) 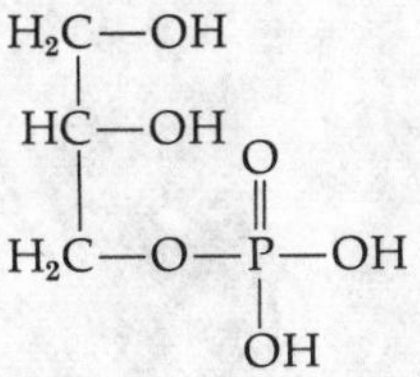
H_2C-OH
$|$
$HC-OH$
$|$
$H_2C-O-\overset{\displaystyle O}{\underset{\underset{\displaystyle OH}{|}}{\overset{\|}{P}}}-OH$

Note that according to our rules, the name for this compound should be glycerol 1-phosphate. The name glycerol 3-phosphate is used by biochemists and is derived from its reactions in body cells.

CHAPTER 18

Organic Compounds Containing Nitrogen or Sulfur

18-1 INTRODUCTION

The elements nitrogen and sulfur are important constituents of the human body. For example, 16% of the weight of a typical protein is nitrogen and 0.4% is sulfur. Nitrogen is also an important component of DNA.

How does the human body obtain the nitrogen and sulfur that it requires? Although about 80% of the air is molecular nitrogen, N_2, the human body cannot use nitrogen in the form of N_2. Our primary source of nitrogen is dietary protein. Large protein molecules in our diet are broken down during digestion into small organic molecules with nitrogen-containing functional groups. These small molecules, called amino acids, are then used in the synthesis of new proteins or to make other nitrogen-containing molecules. Dietary protein is also our primary source for the sulfur-containing functional groups needed to make proteins and other molecules. In some parts of the world there is tragic evidence of the lack of sufficient dietary protein. Newborn children in these regions thrive when they ingest their mother's milk, which usually contains an ample supply of proteins and other nutrients. However, these children often develop a disease called kwashiorkor when they stop breast-feeding. This disease results from a diet that is low in protein content, often one consisting primarily of starch. Early symptoms of kwashiorkor are diarrhea and loss of appetite; left untreated, the disease often leads to death. Kwashiorkor can be treated and prevented by providing children with ample amounts of dietary protein.

In preparation for our study of proteins, DNA, and other molecules with functional groups containing nitrogen or sulfur, we shall examine the structures and properties of simple organic molecules that contain these elements. As in the past few chapters, we shall look first at those classes of organic compounds that are determined by the presence of functional groups containing nitrogen or sulfur atoms. Then we shall see how these compounds react with some of the compounds we have studied previously.

The nitrogen and sulfur requirements of the human body are satisfied by dietary protein, which in many cultures is derived from meat. Before the advent of refrigeration, Native Americans preserved meat by drying it in the sun.

18-2 STUDY OBJECTIVES

After careful study of this chapter, you should be able to:

1. Identify amine, amide, sulfhydryl, thioester, and disulfide functional groups, given the structural formula of a compound.

2. Distinguish among primary, secondary, and tertiary amines and quaternary ammonium ions.

3. Write the structural formula for an amine, amide, thiol, or thioester, given its name, and vice versa.

4. Explain the solubility of amines and amides in water in terms of their ability to form hydrogen bonds.

5. Describe the acid-base properties of amines and amides, and how the predominant form of an amine in solution depends on the pH.

6. Draw the structural formula for an organic ammonium salt, given its name, and describe its acid-base properties and solubility.

7. Write structural formulas for the products of reactions in which (a) amines react with acids to form ammonium salts, (b) amines react with carboxylic acids to form amides, and (c) amides are hydrolyzed, given the structural formulas of the reactants.

8. Explain the difference between the boiling points of a thiol and an alcohol of similar molecular weight.

9. Draw structural formulas for the products of reactions in which (a) a thioester is formed, (b) a thioester is hydrolyzed, (c) a disulfide is formed, or (d) a disulfide is reduced, given the structural formulas of the reactants.

18-3 CLASSES OF COMPOUNDS CONTAINING NITROGEN OR SULFUR ATOMS

The names of the major classes of organic compounds that contain nitrogen or sulfur atoms are shown in Table 18-1. Notice that **amines** can be considered as organic derivatives of ammonia, NH_3. Primary, secondary, and tertiary amines can be considered as ammonia derivatives in which one, two, or three of the hydrogen atoms, respectively, have been replaced by alkyl groups. Like the base ammonia, which can accept an additional proton to form the positively charged ammonion ion, NH_4^+, primary, secondary, and tertiary amines are bases and can accept an additional proton. In addition, ions exist that contain four alkyl groups and have a charge of +1. These ions are called **quaternary ammonium ions.** Some of these ions are, in fact, components of cell membranes, as we shall see in Chapter 26.

An amine, like an alcohol, can react with a carboxylic acid in a dehydration reaction. The resulting compound is called an amide (see Table 18-1). **Amides** contain the functional group
$$-\overset{\displaystyle O}{\overset{\displaystyle \|}{C}}-\underset{\underset{\displaystyle H}{|}}{N}-.$$
The amide functional group is responsible for joining amino acids together in proteins (Chapter 21). Monomer units in the synthetic polymer nylon are also bonded together by amide bonds.

Table 18-1 Classes of Organic Compounds Containing Nitrogen or Sulfur

Name	Structure	Example
Primary amine	$R-NH_2$	CH_3-NH_2 methylamine
Secondary amine	$R_1-\underset{\underset{H}{\|}}{N}-R_2$	$CH_3-\underset{\underset{H}{\|}}{N}-CH_3$ dimethylamine
Tertiary amine	$R_1-\underset{\underset{R_2}{\|}}{N}-R_3$	$CH_3-CH_2-\underset{\underset{CH_3}{\|}}{N}-CH_2-CH_3$ diethylmethylamine
Quaternary ammonium salt	$\left[R_1-\overset{\overset{R_3}{\|}}{\underset{\underset{R_4}{\|}}{N^+}}-R_2\right]Cl^-$	$[(CH_3)_4N^+]Cl^-$ tetramethylammonium chloride
Simple amide	$R'-\overset{\overset{O}{\|\|}}{C}-NH_2$	$CH_3-\overset{\overset{O}{\|\|}}{C}-NH_2$ acetamide
N-substituted amide	$R'-\overset{\overset{O}{\|\|}}{C}-\underset{\underset{H}{\|}}{N}-R$ or $R'-\overset{\overset{O}{\|\|}}{C}-\underset{\underset{R_1}{\|}}{N}-R_2$	$H-\overset{\overset{O}{\|\|}}{C}-\underset{\underset{H}{\|}}{N}-CH_3$ N-methylformamide
Thiol (mercaptan)	$R-SH$	CH_3-CH_2-SH ethanethiol
Thioether	R_1-S-R_2	$CH_3-CH_2-S-CH_3$ methyl ethyl thioether
Thioester	$R'-\overset{\overset{O}{\|\|}}{C}-S-R$	$CH_3-\overset{\overset{O}{\|\|}}{C}-S-CH_2-CH_3$ ethyl thioacetate
Disulfide	$R_1-S-S-R_2$	$CH_3-S-S-CH_3$ methyl disulfide

Since sulfur and oxygen are both group VI elements, many sulfur-containing compounds have a functional group that can be considered as the sulfur analog of an oxygen-containing functional group. These analogous sulfur- and oxygen-containing functional groups have somewhat similar properties. Thus, the —SH functional group, called the sulfhydryl group, is somewhat similar to the —OH group in an alcohol. Compounds containing the —SH group are therefore called thioalcohols or **thiols.** The prefix *thio-* indicates that a sulfur atom has replaced an oxygen atom in the characteristic functional group of these compounds. Thiols are also referred to as mercaptans (Table 18-1). We shall see that many reactions of thiols are similar to those of alcohols. For example, thiols can react with carboxylic acids to form **thioesters** (Table 18-1).

Compounds containing the functional group —S—S— are called **disulfides** (Table 18-1). Disulfides are important structural components of protein molecules.

Exercise 18-1

Draw structural formulas for all of the amines with the molecular formula C_3H_9N. Label each isomer as a primary, secondary, or tertiary amine.

Exercise 18-2

When ethanethiol reacts with acetic acid, a thioester is formed. Without looking ahead in the text, see if you can draw the structural formula for this thioester based on your knowledge of the reaction of alcohols and acids to form esters.

18-4 AMINES

Structure and Nomenclature

Table 18-2 shows the structural formulas and common names of several amines. The IUPAC names of amines are rather complicated, especially those for secondary and tertiary amines, and we shall not use them. In the common name for an amine, the alkyl groups attached to the nitrogen atoms are generally indicated by prefixes before the ending *-amine*. For example, ethyldimethylamine has two methyl groups and one ethyl group attached to the nitrogen atom. It is therefore a tertiary amine. The common names of some amines, such as aniline and alanine, are very old and do not indicate structural features. Notice in Table 18-2 that the name for the secondary amine, N-ethylaniline, contains the prefix N- to indicate that the ethyl group is bonded to the nitrogen and not to one of the carbon atoms in this molecule. This prefix is likewise used in the names of other amines where it is necessary to specify that an R group is bonded to the nitrogen and not another atom. The amine alanine also contains a carboxyl group. Compounds with both an amine and a carboxyl group are called **amino acids.** Amino acids that contain α-carboxyl groups, such as alanine (see Figure 18-1), are especially interesting because they are used to make proteins in the human body (Chapter 21).

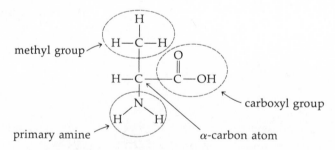

Figure 18-1   The α-amino acid alanine has a methyl group bonded to the α-carbon atom. Other α-amino acids have   different groups of atoms, called side chains, bonded to the α-carbon atom.

Quaternary ammonium ions are also very important in the human body. Compounds containing quaternary ammonium cations and anions such as Cl^-, OH^-, SO_4^{2-}, and so on, are ionic compounds.

Table 18-2 Some Common Amines

Structural Formula	Common Name
Primary amines:	

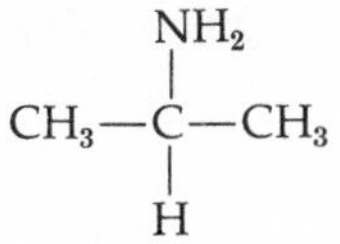

$$CH_3—\underset{\underset{H}{|}}{\overset{\overset{NH_2}{|}}{C}}—CH_3$$

Isopropylamine

$\langle \bigcirc \rangle —NH_2$

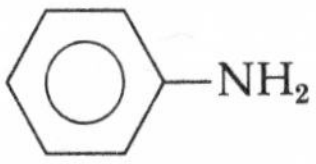

Aniline
(an aromatic amine)

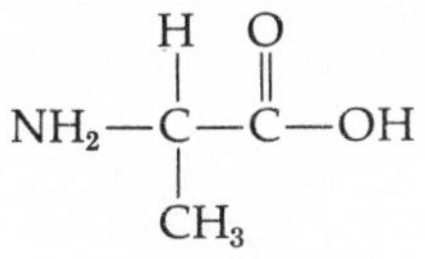

$$NH_2—\underset{\underset{CH_3}{|}}{\overset{\overset{H}{|}}{C}}—\overset{\overset{O}{\|}}{C}—OH$$

Alanine
(an amino acid)

Secondary amines:

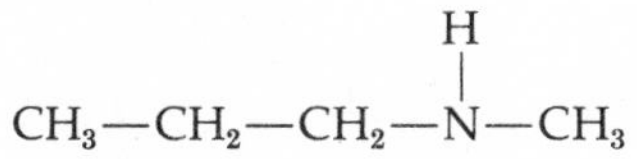

$$CH_3—CH_2—CH_2—\overset{\overset{H}{|}}{N}—CH_3$$

Methyl-*n*-propylamine

$\langle \bigcirc \rangle —\underset{\underset{H}{|}}{N}—CH_2—CH_3$

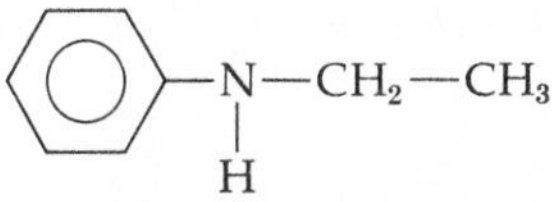

N-ethylaniline
(phenylethylamine)

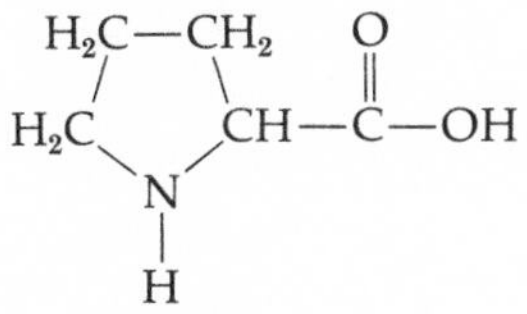

$$\underset{H_2C}{\overset{H_2C—CH_2}{\diagdown}} \quad \overset{O}{\|} \\ \underset{\underset{\underset{H}{|}}{N}}{\diagup} CH—C—OH$$

Proline
(an amino acid)

Tertiary amine:

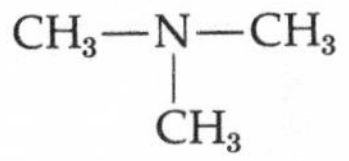

$$CH_3—\underset{\underset{CH_3}{|}}{N}—CH_3$$

Trimethylamine

Quaternary ammonium compounds:

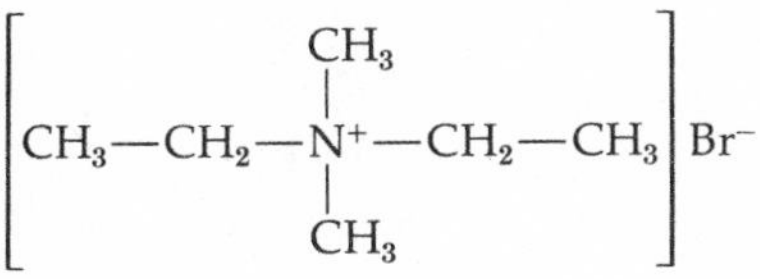

$$\left[CH_3—CH_2—\underset{\underset{CH_3}{|}}{\overset{\overset{CH_3}{|}}{N^+}}—CH_2—CH_3 \right] Br^-$$

Diethyldimethylammonium bromide

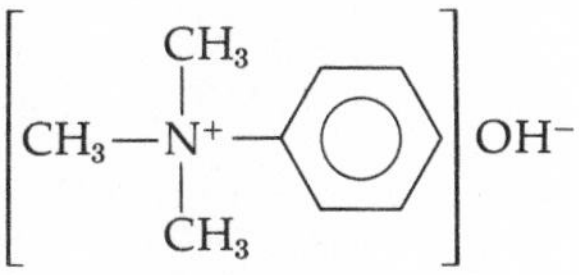

$$\left[CH_3—\underset{\underset{CH_3}{|}}{\overset{\overset{CH_3}{|}}{N^+}}—\langle \bigcirc \rangle \right] OH^-$$

Trimethylphenylammonium hydroxide

A number of important natural biomolecules as well as several drugs contain amine functional groups, often as part of a complicated structure that includes rings of atoms. The structural formulas of some of these complex amines are given in Table 18-3.

Table 18-3 Some Complex Amines

Drugs	Naturally Occurring in Humans

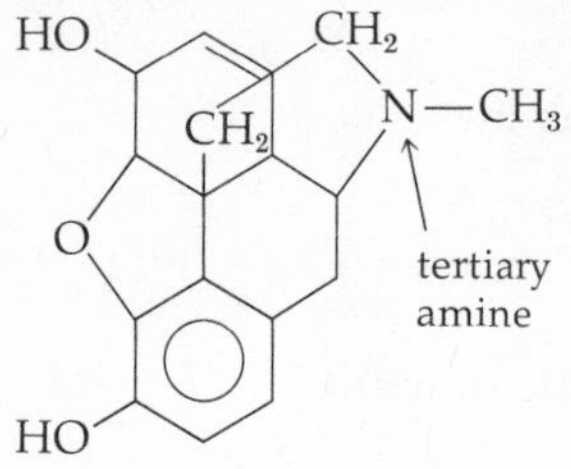

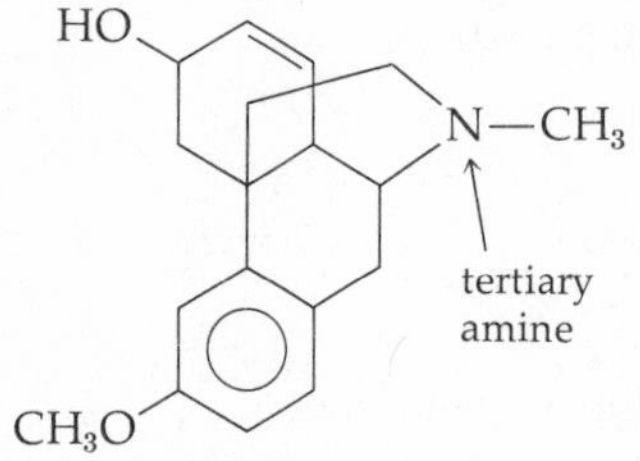

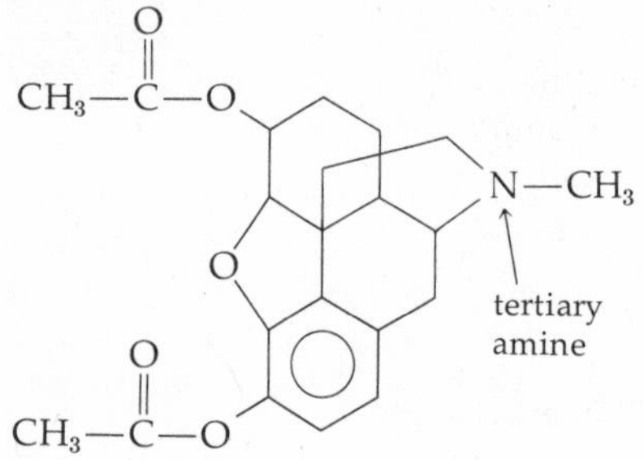

Morphine

Adenine (a component of nucleic acids)

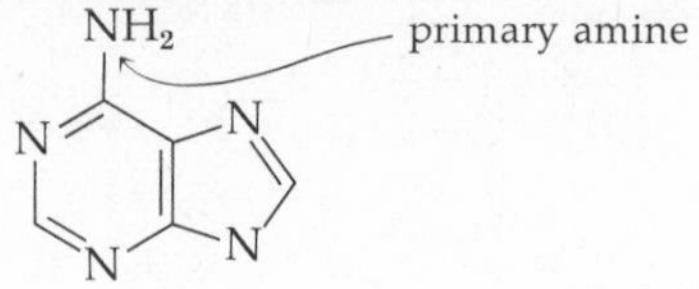

Codeine

Tryptophan (an amino acid)

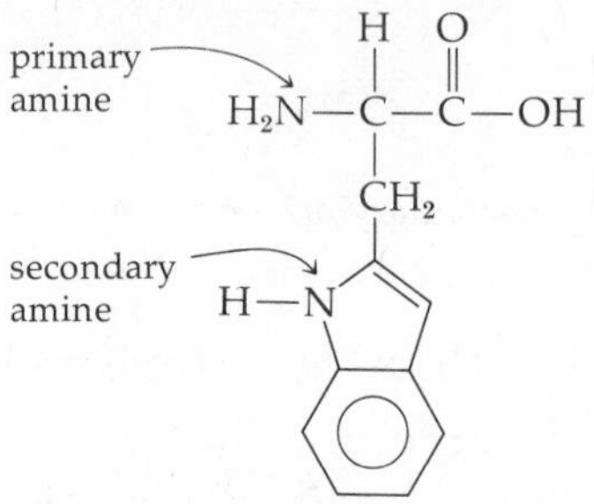

Heroin

Epinephrine (adrenaline)

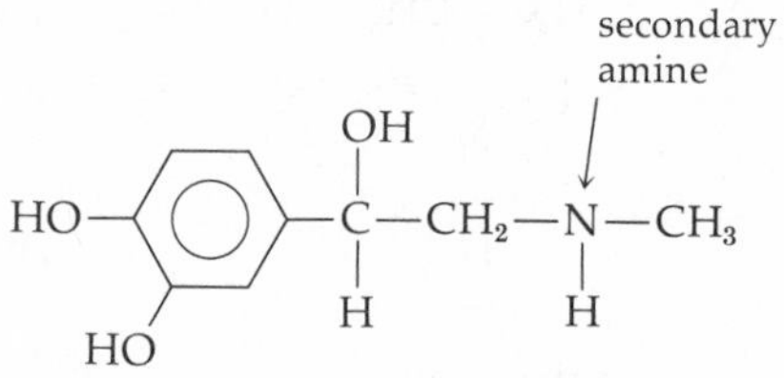

Amphetamine

Histamine

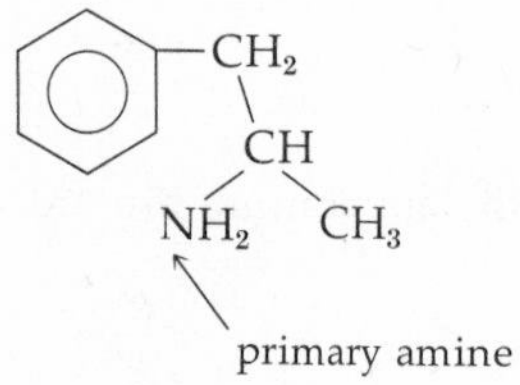

Nicotine (a poison)

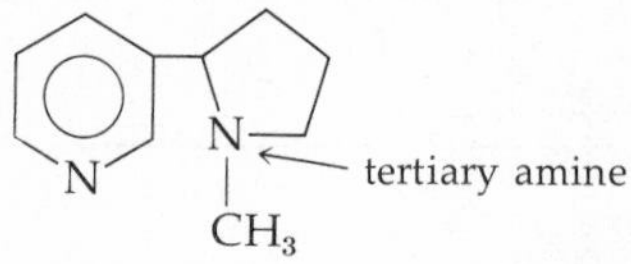

Procaine (Novocaine)

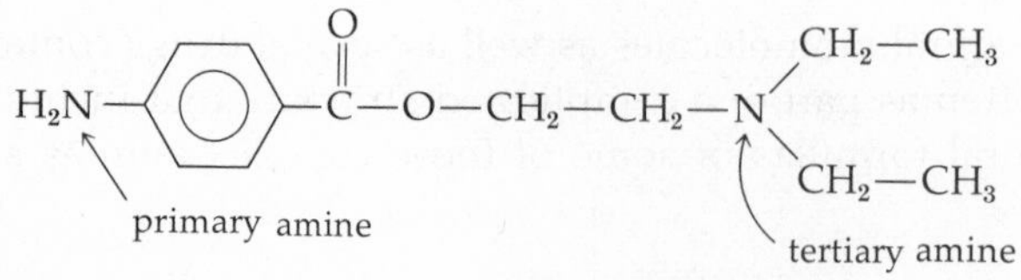

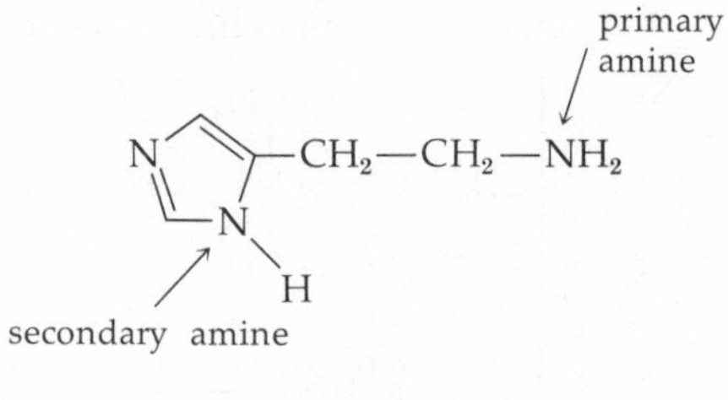

Physical Properties

An **amine** functional group is a very polar group that can hydrogen bond with water molecules. The nitrogen atom in a primary, a secondary, or a tertiary amine can act as a hydrogen bond acceptor; hence these amines can all form hydrogen bonds with water and other molecules (see Figure 18-2).

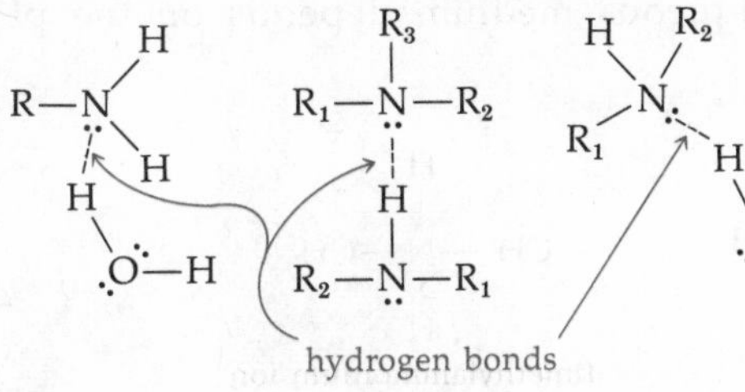

Figure 18-2   The amine nitrogen is a hydrogen atom acceptor in hydrogen bonds between an amine and water (left), between different amine molecules (center), and between an amine and an alcohol (right).

In addition, the hydrogen atom (or atoms) bonded to the nitrogen atom in primary or secondary amines can act as the hydrogen donor in the formation of a hydrogen bond with the oxygen atom in a water molecule or a very electronegative atom in another molecule (see Figure 18-3). Low-molecular-weight amines are therefore quite soluble in water. However, high-molecular-weight amines with large nonpolar parts are quite insoluble in water.

Figure 18-3   Primary (left) and secondary (right) amines are hydrogen atom donors in hydrogen bonds with molecules containing very electronegative atoms, such as the oxygen atom in water.

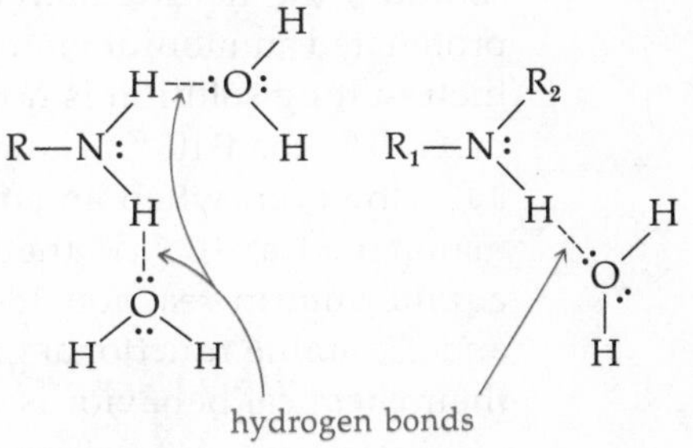

Chemical Reactivity

Ammonia can act as a proton acceptor, and therefore is a weak Brønsted-Lowry base. Likewise, primary, secondary, and tertiary amines can accept protons and are also weak bases. Thus, a solution of dimethylamine in water, for example, will be alkaline as a result of the acid-base reaction between water and the amine, producing the dimethylammonium ion and a hydroxide ion:

(18-1)

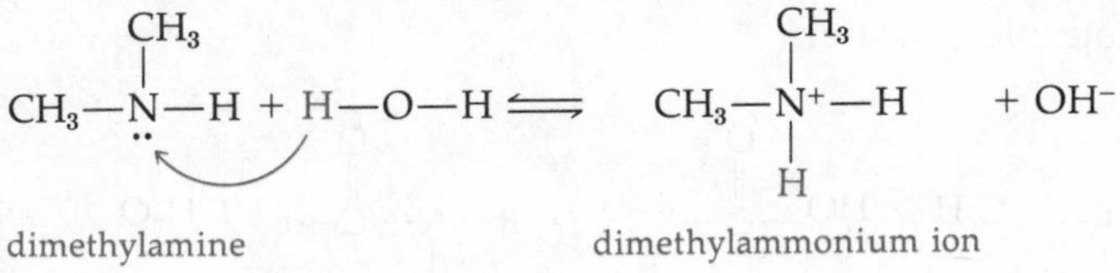

Since amines are weak bases, they can react with strong acids and form an ammonium salt. For example,

(18-2)

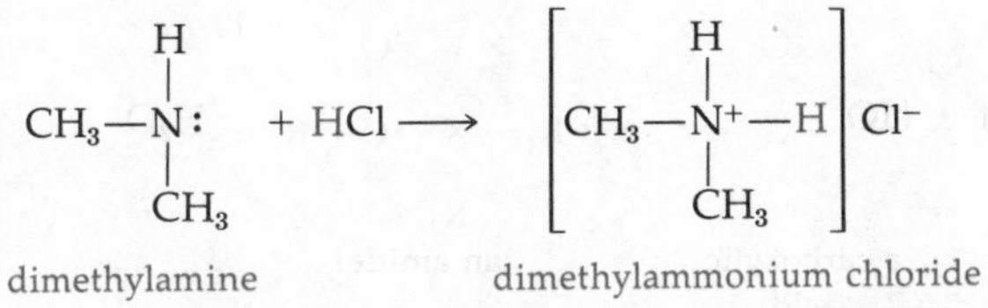

The salt formed in reaction 18-2, dimethylammonium chloride, is an ionic compound and is very soluble in water. It is often advantageous to convert an insoluble amine to the more soluble ammonium salt. For example, the solubility of morphine in water is 1 g/5000 ml, whereas 1 g of the ammonium sulfate salt of morphine will dissolve in only 15.5 ml of water.

Whether an amine will exist predominantly as a neutral molecule or as a protonated ion in an aqueous medium depends on the pH. Consider the equilibrium:

$$(18\text{-}3) \qquad CH_3-\overset{\cdot\cdot}{N}-CH_3 + H^+ \rightleftharpoons CH_3-\overset{\overset{\displaystyle H}{|}}{\underset{\underset{\displaystyle H}{|}}{N^{\pm}}}-CH_3$$

$$\qquad\qquad\qquad \text{dimethylamine} \qquad\qquad \text{dimethylammonium ion}$$

Dimethylamine, like all amines, is a weak base, and in basic solutions, where $[H^+]$ is very small, the equilibrium is shifted to the left and the predominant form is the neutral molecule $(CH_3)_2NH$. On the other hand, in neutral or acidic solutions, $[H^+]$ is large enough so that the equilibrium is shifted to the right and the predominant form is the protonated ammonium ion $(CH_3)_2\overset{+}{N}H_2$. Since the predominant form of an amine depends on the pH, the solubility of an amine can be drastically altered by changing the pH. Amines with large nonpolar groups are virtually insoluble in solutions of high pH because they exist predominantly in the form of neutral insoluble molecules. But in a neutral or acidic solution, the neutral amine molecule is converted to the much more soluble protonated ammonium ion form. Note that when an amine is dissolved in water, the resulting solution is not neutral—it is basic. The $[H^+]$ in such a solution is much less than 10^{-7} M, and the equilibrium in reaction 18-3 is shifted to the left. However, when an amine is dissolved in a buffer solution where $[H^+]$ is maintained at 10^{-7} M (neutral pH), the $[H^+]$ is sufficiently large to shift the equilibrium in reaction 18-3 to the right. As we shall see in Chapters 21, 22, and 23, amine functional groups are important parts of protein molecules, and their chemical behavior is important to the biological activity of proteins.

Quaternary ammonium salts are ionic compounds (see Table 18-2). Several substances containing quaternary (four-bonded) nitrogen ions have special functions in the human body. For example, we have frequently referred to a coenzyme called NAD^+. The positive charge of this coenzyme results from the presence of a quaternary nitrogen ion. Quaternary nitrogens have a strong tendency to attract electrons, which allows NAD^+ to function as a coenzyme in many oxidation-reduction reactions in the human body (see Figure 18-4).

The most important chemical property of primary and secondary amines is their ability to react with carboxylic acids to form amides, with the elimination of a molecule of water:

$$(18\text{-}4) \qquad R-\underset{\underset{\displaystyle H}{|}}{N}-H + HO-\overset{\overset{\displaystyle O}{\|}}{C}-R' \longrightarrow R-\underset{\underset{\displaystyle H}{|}}{N}-\overset{\overset{\displaystyle O}{\|}}{C}-R' + H_2O$$

$$\qquad\quad \underset{\text{amine}}{\text{a primary}} \qquad \underset{\text{acid}}{\text{a carboxylic}} \qquad\qquad \text{an amide}$$

$$(18\text{-}5) \qquad R_1-\underset{\underset{\displaystyle R_2}{|}}{N}-H + HO-\overset{\overset{\displaystyle O}{\|}}{C}-R' \longrightarrow R_1-\underset{\underset{\displaystyle R_2}{|}}{N}-\overset{\overset{\displaystyle O}{\|}}{C}-R' + H_2O$$

$$\qquad\quad \underset{\text{amine}}{\text{a secondary}} \qquad \underset{\text{acid}}{\text{a carboxylic}} \qquad\qquad \text{an amide}$$

Figure 18-4   The coenzyme nicotinamide
adenine dinucleotide (NAD^+).

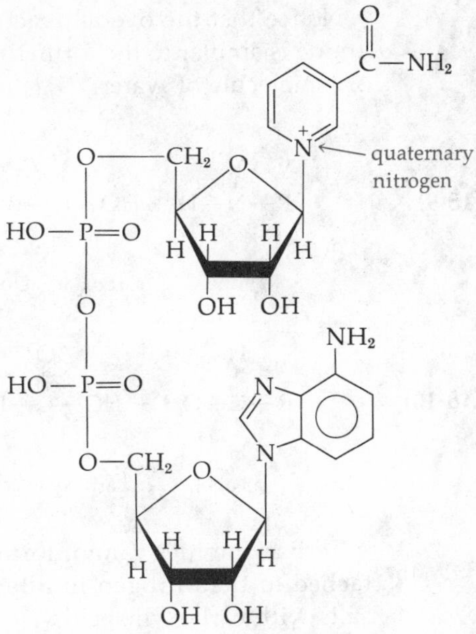

The reaction of an amine and a carboxylic acid can be viewed as an acid-base reaction to form a salt, followed by the loss of a molecule of water to form an amide. In the laboratory this is accomplished by heating the mixture to boil off the water:

(18-6)

$$R-\overset{\overset{H}{|}}{\underset{\underset{H}{|}}{N}}: + HO-\overset{O}{\overset{\|}{C}}-R' \longrightarrow \left[R-\overset{+}{N}H_3\right]\left[^-O-\overset{O}{\overset{\|}{C}}-R'\right]$$
$$\text{salt}$$

(18-7)

$$\left[R-\overset{+}{N}H_3\right]\left[^-O-\overset{O}{\overset{\|}{C}}-R'\right] \xrightarrow{\text{heat}} R-\underset{\underset{H}{|}}{N}-\overset{O}{\overset{\|}{C}}-R' + H_2O$$
$$\text{salt} \qquad\qquad\qquad\qquad \text{amide}$$

In the industrial manufacture of amides, high temperature is used to drive off the water and produce the amide product. The synthetic polymer nylon (Section 14-9) is made by a polymerization reaction that involves the formation of amide bonds between alternating dicarboxylic acid and diamine monomer units:

(18-8)

$$n\left(HO-\overset{O}{\overset{\|}{C}}-(CH_2)_x-\overset{O}{\overset{\|}{C}}-OH\right) + n(H_2N-(CH_2)_y-NH_2) \longrightarrow$$
$$\text{a dicarboxylic acid} \qquad\qquad\qquad \text{a diamine}$$

$$\left(-\underset{\underset{H}{|}}{N}-(CH_2)_y-\underset{\underset{H}{|}}{N}-\overset{O}{\overset{\|}{C}}-(CH_2)_x-\overset{O}{\overset{\|}{C}}-\right)_n + 2nH_2O$$
$$\text{nylon}$$

where x and y refer to the number of carbon atoms between the functional groups in these monomers. Nylon 66, produced from the six-carbon-atom dicarboxylic acid ($x = 4$) and the six-carbon-atom diamine ($y = 6$), is a typical commercial nylon polymer.

Notice that the overall reaction of an amine and a carboxylic acid to form an amide is similar to the formation of an ester in that both involve the elimination of a molecule of water:

(18-9)

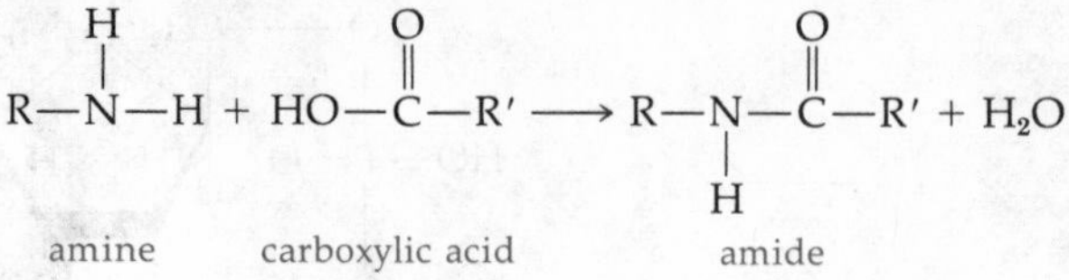

$$\underset{\text{amine}}{R-\overset{H}{\underset{H}{N}}-H} + \underset{\text{carboxylic acid}}{HO-\overset{O}{\overset{\|}{C}}-R'} \longrightarrow \underset{\text{amide}}{R-\overset{H}{\underset{H}{N}}-\overset{O}{\overset{\|}{C}}-R'} + H_2O$$

(18-10)

$$\underset{\text{alcohol}}{R-\overset{H}{\underset{H}{C}}-OH} + \underset{\text{carboxylic acid}}{HO-\overset{O}{\overset{\|}{C}}-R'} \longrightarrow \underset{\text{ester}}{R-\overset{H}{\underset{H}{C}}-O-\overset{O}{\overset{\|}{C}}-R'} + H_2O$$

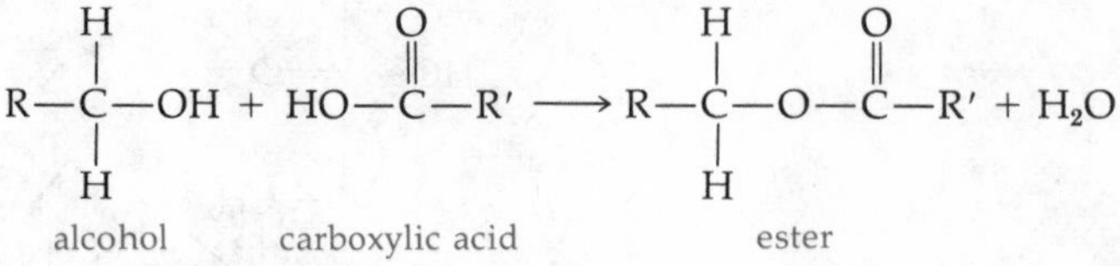

Tertiary amines cannot form amides, because there is no hydrogen atom attached to the nitrogen in a tertiary amine. However, tertiary amines do form salts with carboxylic acids.

Amides can also be formed by the reaction of an ester and an amine:

(18-11)

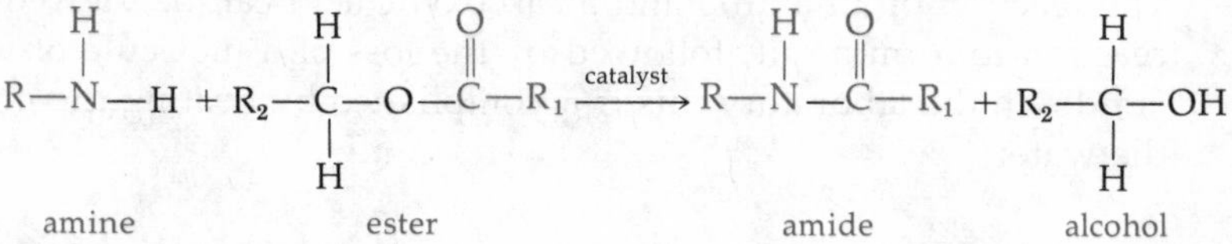

$$\underset{\text{amine}}{R-\overset{H}{\underset{H}{N}}-H} + \underset{\text{ester}}{R_2-\overset{H}{\underset{H}{C}}-O-\overset{O}{\overset{\|}{C}}-R_1} \xrightarrow{\text{catalyst}} \underset{\text{amide}}{R-\overset{H}{\underset{}{N}}-\overset{O}{\overset{\|}{C}}-R_1} + \underset{\text{alcohol}}{R_2-\overset{H}{\underset{H}{C}}-OH}$$

All of the enzymes and the other proteins in the human body are composed of amino acids held together by amide bonds (Chapter 21). In Chapter 24 we shall see that these amide bonds in proteins are formed from amines and esters in enzyme-catalyzed reactions similar to reaction 18-11.

Exercise 18-3

Indicate which of the following (1) can form a hydrogen bond by donating a hydrogen, (2) can form a hydrogen bond by accepting a hydrogen, (3) is a base, (4) is an acid.

(a) CH_3-NH_2 (b) $CH_3-\overset{+}{N}H_3$ (c) $CH_3-CH_2-\overset{}{\underset{CH_3}{N}}-H$

(d) $(CH_3)_3N$ (e) $(CH_3)_4N^+$

Exercise 18-4

Draw structural formulas for the products of the following reactions:

(a) 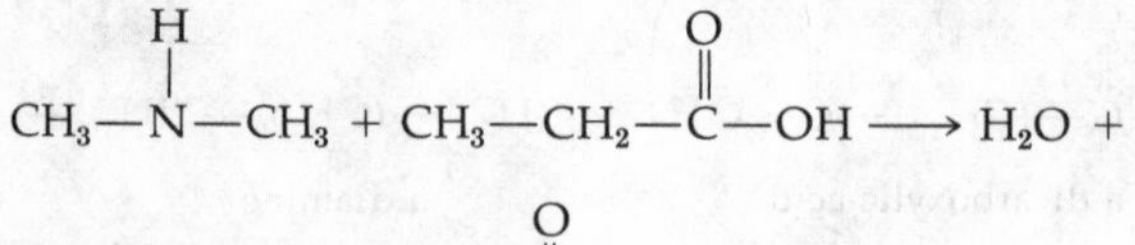 $CH_3-\overset{H}{\underset{}{N}}-CH_3 + CH_3-CH_2-\overset{O}{\overset{\|}{C}}-OH \longrightarrow H_2O +$

(b) $(CH_3)_3N + CH_3-CH_2-\overset{O}{\overset{\|}{C}}-OH \longrightarrow$

(c) $CH_3-\overset{NH_2}{\underset{CH_3}{C}}-H + H-\overset{O}{\overset{\|}{C}}-OH \longrightarrow H_2O +$

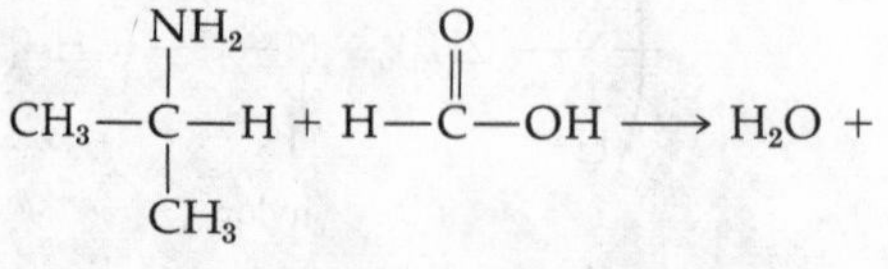

(d) 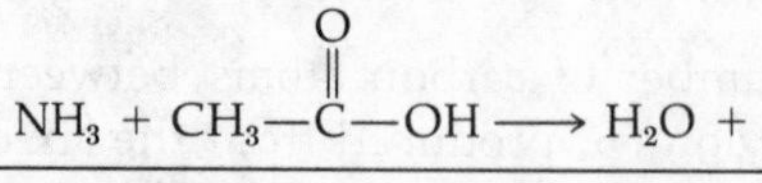 $NH_3 + CH_3-\overset{O}{\overset{\|}{C}}-OH \longrightarrow H_2O +$

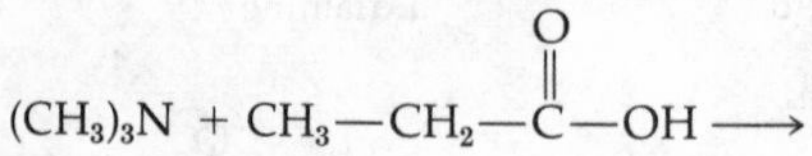

18-5 AMIDES

Structure and Nomenclature

The $\overset{\overset{\displaystyle O}{\|}}{-C-N-}$ group is called the **amide** functional group. When the nitrogen

atom of the amide group is bonded to two hydrogen atoms, the molecule is

called a **simple amide,** with the general formula $R'\overset{\overset{\displaystyle O}{\|}}{-C}-NH_2$.

Table 18-4 Some Simple Amides

Structural Formula	Common Name
$H\overset{\overset{\displaystyle O}{\|}}{-C}-NH_2$	Formamide
$CH_3\overset{\overset{\displaystyle O}{\|}}{-C}-NH_2$	Acetamide
$CH_3-CH_2\overset{\overset{\displaystyle O}{\|}}{-C}-NH_2$	Propionamide
$\begin{array}{c} OH \\ \| \\ C=O \quad\quad O \\ \| \quad\quad\quad \| \\ H-C-CH_2-C-NH_2 \\ \| \\ NH_2 \end{array}$	Asparagine (an amino acid)
$\begin{array}{c} OH \\ \| \\ C=O \quad\quad\quad O \\ \| \quad\quad\quad\quad \| \\ H-C-CH_2-CH_2-C-NH_2 \\ \| \\ NH_2 \end{array}$	Glutamine (an amino acid)
Benzamide (benzene ring with $\overset{\overset{\displaystyle O}{\|}}{-C}-NH_2$)	Benzamide
Nicotinamide (pyridine ring with $\overset{\overset{\displaystyle O}{\|}}{-C}-NH_2$)	Nicotinamide (a B vitamin)

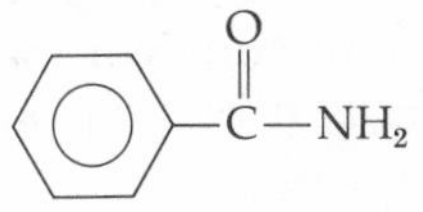

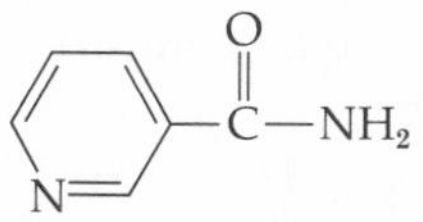

The structural formulas and common names of several simple amides are given in Table 18-4. The common name for a simple amide is derived from the common name of the parent carboxylic acid by dropping the suffix -ic (which denotes an acid) and adding the suffix -amide. Glutamine and asparagine are two amino acids that contain simple amide groups. Another important simple amide is nicotinamide, which is the active form of the B vitamin niacin, and is used in the formation of the coenzyme NAD^+ (see Figure 18-4).

N-substituted amides contain one or two R groups attached to the N atom of the amide group. The structural formulas of a few N-substituted amides are given in Table 18-5. Notice in Table 18-5 that the prefix N- is used to indicate that an R group is attached to the nitrogen atom in an N-substituted amide. We shall study N-substituted amides in more detail in Chapter 21, when we consider the amide bonds that join amino acids together to form proteins.

Table 18-5 Some N-Substituted Amides

Structural Formula	Name
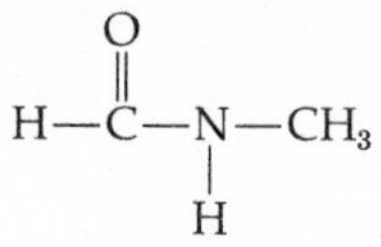 $H\!-\!\overset{\displaystyle O}{\overset{\|}{C}}\!-\!\underset{\displaystyle H}{\overset{\displaystyle}{N}}\!-\!CH_3$	N-methylformamide
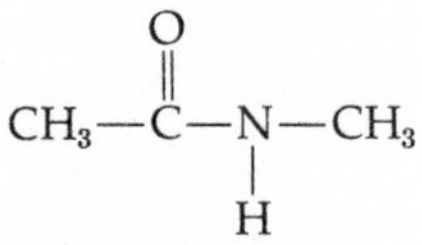 $CH_3\!-\!\overset{\displaystyle O}{\overset{\|}{C}}\!-\!\underset{\displaystyle H}{\overset{\displaystyle}{N}}\!-\!CH_3$	N-methylacetamide
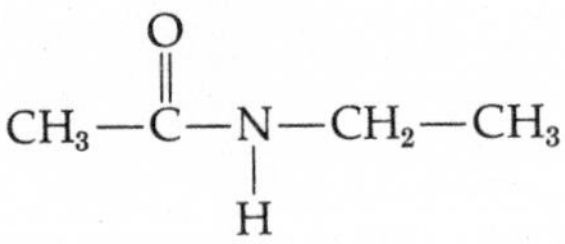 $CH_3\!-\!\overset{\displaystyle O}{\overset{\|}{C}}\!-\!\underset{\displaystyle H}{\overset{\displaystyle}{N}}\!-\!CH_2\!-\!CH_3$	N-ethylacetamide
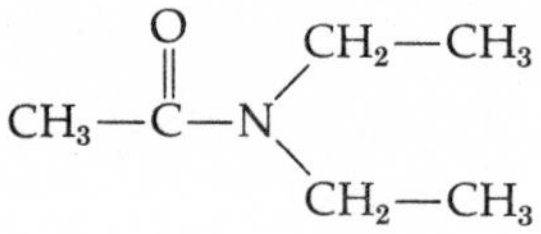 $CH_3\!-\!\overset{\displaystyle O}{\overset{\|}{C}}\!-\!N\!\!\begin{smallmatrix}\diagup\,CH_2-CH_3\\[4pt]\diagdown\,CH_2-CH_3\end{smallmatrix}$	N,N-diethylacetamide
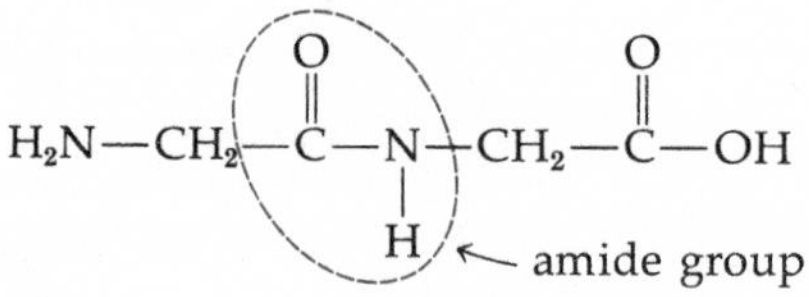 $H_2N\!-\!CH_2\!-\!\overset{\displaystyle O}{\overset{\|}{C}}\!-\!\underset{\displaystyle H}{\overset{\displaystyle}{N}}\!-\!CH_2\!-\!\overset{\displaystyle O}{\overset{\|}{C}}\!-\!OH$ ← amide group	Glycylglycine (two amino acids bonded by an amide linkage)

Properties of Simple Amides

Simple amides are very polar, and low-molecular-weight amides are quite soluble in water because they readily form hydrogen bonds with water (see Figure 18-5). Simple amides also have higher melting and boiling points than many

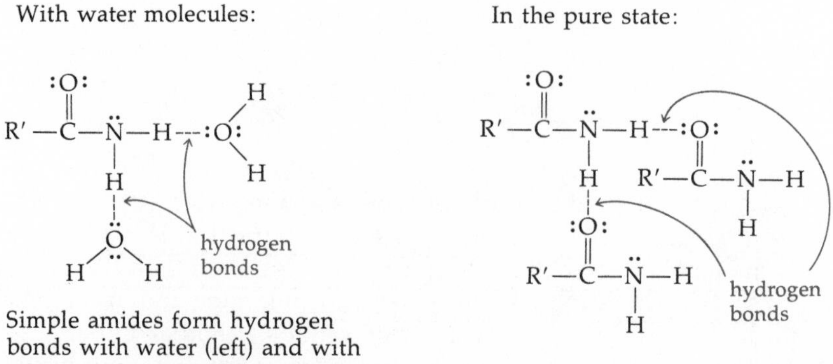

With water molecules: In the pure state:

Figure 18-5   Simple amides form hydrogen bonds with water (left) and with other amide molecules in the pure state (right).

other compounds of similar molecular weight, because of their ability to form hydrogen bonds in the pure state (see Figure 18-5). However, whereas amines are weak bases, amides are not, even though the amide group, like an amine, contains a nitrogen atom. <u>It has been determined experimentally that the carbon-nitrogen bond in amides is shorter than a typical carbon-nitrogen single bond, and that there is no free rotation about the amide carbon-nitrogen bond</u>. One representation of the amide bond that accounts for these observations is a composite picture of two resonance structures, one of which involves a C=N double bond (Figure 18-6). Note that in the structure on the right side of this figure, there are no unshared electrons on the nitrogen atom. The electronegative oxygen atom tends to pull these electrons away from the nitrogen atom, making it unable to serve as a proton acceptor (base).

Figure 18-6   Resonance structures for an amide. This composite representation accounts for the experimental observations that (1) an amide carbon-nitrogen bond is shorter than a typical carbon-nitrogen single bond; (2) the amide nitrogen atom is not a proton acceptor (a base); and (3) there is no free rotation around the carbon-nitrogen bond in amides.

$$R'-\overset{:\overset{\textstyle\ddot{O}}{\|}}{C}-\overset{\textstyle\ddot{N}}{\underset{H}{|}}-R \longleftrightarrow R'-\overset{:\overset{\textstyle\ddot{O}:^-}{|}}{C}=\overset{\textstyle+}{\underset{H}{N}}-R$$

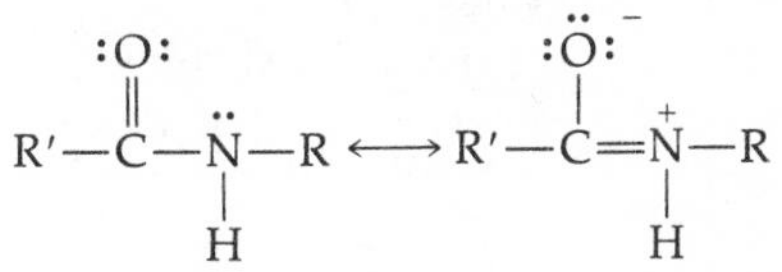

We have already seen how N-substituted amides are formed by the reaction of a carboxylic acid with a primary or a secondary amine (reactions 18-4 and 18-5). Simple amides can likewise be formed by the reaction of a carboxylic acid and ammonia:

(18-12) $$R'-\overset{\textstyle O}{\overset{\|}{C}}-OH + NH_3 \xrightarrow{\text{heat}} R'-\overset{\textstyle O}{\overset{\|}{C}}-NH_2 + H_2O$$

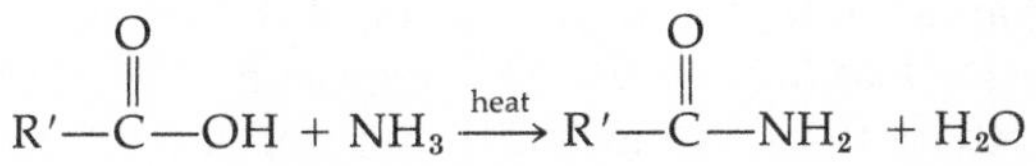

Note that the formation of both a simple and an N-substituted amide involves dehydration—removal of a molecule of water.

For our purposes, the most important reaction of amides is their hydrolysis to form amines (or ammonia) and carboxylic acids. Hydrolysis of an amide is exactly opposite to amide formation. For example,

(18-13)   Hydrolysis of a simple amide:

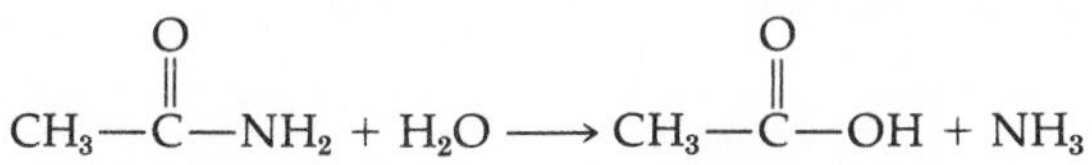

$$CH_3-\overset{\textstyle O}{\overset{\|}{C}}-NH_2 + H_2O \longrightarrow CH_3-\overset{\textstyle O}{\overset{\|}{C}}-OH + NH_3$$

(18-14)   Hydrolysis of an N-substituted amide:

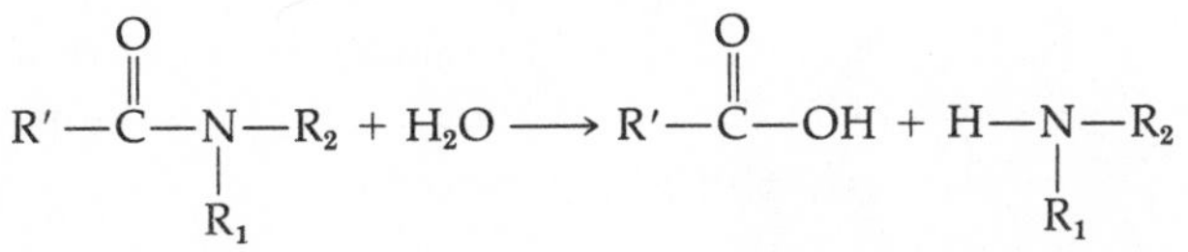

$$R'-\overset{\textstyle O}{\overset{\|}{C}}-\overset{N}{\underset{R_1}{|}}-R_2 + H_2O \longrightarrow R'-\overset{\textstyle O}{\overset{\|}{C}}-OH + H-\overset{N}{\underset{R_1}{|}}-R_2$$

Hydrolysis of the amide bonds in proteins is the major reaction that occurs during the digestion of proteins in the human body (Chapter 28).

Exercise 18-5
Draw structural formulas for the following compounds: (a) α-Methylpropionamide; (b) N,N-dimethylpropionamide; (c) N-cyclopentylformamide.

Exercise 18-6
Draw structural formulas for the products of each of the following reactions:

(a) 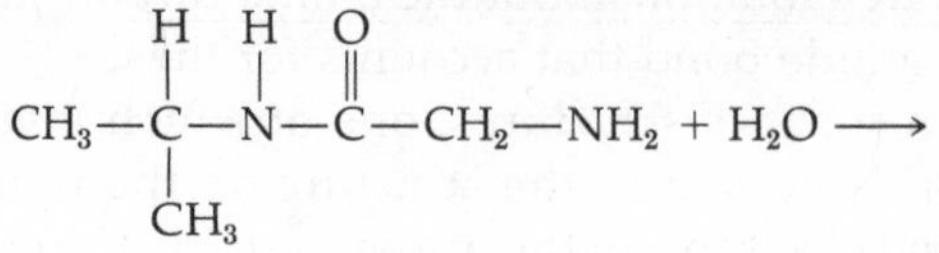
$$CH_3-\overset{\overset{\displaystyle H}{|}}{\underset{\underset{\displaystyle CH_3}{|}}{C}}-\overset{\overset{\displaystyle H}{|}}{N}-\overset{\overset{\displaystyle O}{\|}}{C}-CH_2-NH_2 + H_2O \longrightarrow$$

(b) $NH_3 + H_2O \longrightarrow$

(c) 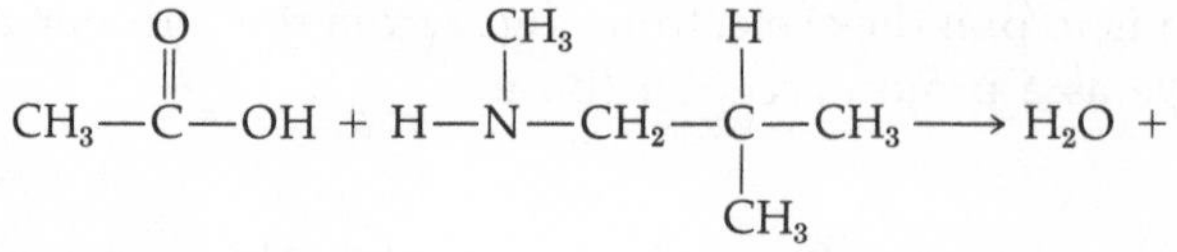
$$CH_3-\overset{\overset{\displaystyle O}{\|}}{C}-OH + H-\overset{\overset{\displaystyle CH_3}{|}}{N}-CH_2-\overset{\overset{\displaystyle H}{|}}{\underset{\underset{\displaystyle CH_3}{|}}{C}}-CH_3 \longrightarrow H_2O \ +$$

Exercise 18-7
For the reactions in Exercise 18-6, indicate which of the organic reactants and products can act as an acid, as a base, and as both an acid and a base.

18-6   THIOLS, DISULFIDES, AND THIOESTERS

The —SH functional group is called the **sulfhydryl** group, and is similar in many ways to the —OH hydroxyl group of alcohols. The class of organic molecules that contain only the sulfhydryl group are called **thiols.** The IUPAC name for a thiol is formed in the same manner as that of the corresponding alcohol, except the suffix -*ethiol* is used instead of the alcohol suffix -*ol*. The common name **mercaptan** is sometimes used. The names and structural formulas of a few thiols are given in Table 18-6. Cysteine is one of the amino acids used to make proteins. Ethyl mercaptan (ethanethiol) has the dubious distinction of being listed in the *Guinness Book of Records* as the foulest-smelling compound known! Minute amounts of ethanethiol are added to natural gas in order for gas leaks to be detectable (pure natural gas is odorless). 1-Butanethiol was long thought to be the malodorous molecule produced by skunks, but this was disproven in 1974. The odor of skunks is actually due to a mixture of sulfur-containing compounds, including 1-propanethiol.

Although the sulfhydryl group, —SH, and the alcohol group, —OH, are similar, remember that sulfur atoms are much less electronegative than oxygen atoms. Consequently, thiols do not form hydrogen bonds as readily as alcohols, have lower boiling points than alcohols of comparable weight (see Table 18-7), and are less soluble in water than alcohols.

Table 18-6 Some Thiols (Mercaptans)

Structural Formula	IUPAC Name	Common Name
CH_3-SH	Methanethiol	Methyl mercaptan
CH_3-CH_2-SH	Ethanethiol	Ethyl mercaptan
$CH_3-CH_2-CH_2-SH$	1-Propanethiol	n-Propyl mercaptan
$CH_3-CH_2-CH_2-CH_2-SH$	1-Butanethiol	n-Butyl mercaptan
$HO-\overset{\overset{\displaystyle O}{\|}}{C}-\overset{\overset{\displaystyle H}{|}}{\underset{\underset{\displaystyle NH_2}{|}}{C}}-CH_2-SH$		Cysteine (an amino acid)

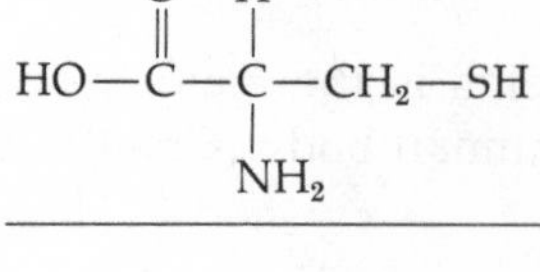

Table 18-7 Boiling Points of Some Alcohols and Thiols

Alcohol	Boiling Point (°C)	Thiol	Boiling Point (°C)
Methanol	65	Methanethiol	6
Ethanol	78	Ethanethiol	36
1-Butanol	117	1-Butanethiol	98

Two reactions involving sulfhydryl groups are of particular interest because of their importance in the human body. First, thiols can be easily oxidized to form disulfides in the reaction

(18-15) $$R_1—SH + HS—R_2 + X \longrightarrow R_1—S—S—R_2 + XH_2$$
a disulfide

In reaction 18-15, the two thiols each lose a hydrogen atom and couple together to form a disulfide. Since the thiols lose hydrogen, they are oxidized and the compound (X) that accepts the hydrogen atoms is reduced. Disulfides can be reduced back to the thiols in the reverse of reaction 18-15. Disulfide bonds between the sulfhydryl groups on cysteine amino acids are very important in determining the structure of proteins (Chapter 21).

The other important reaction of thiols is their combination with carboxylic acids to form **thioesters**:

(18-16)

$$\underset{\substack{\text{carboxylic} \\ \text{acid}}}{R'-\overset{\displaystyle O}{\overset{\|}{C}}-OH} + \underset{\text{thiol}}{HS-R} \longrightarrow \underset{\text{thioester}}{R'-\overset{\displaystyle O}{\overset{\|}{C}}-S-R} + H_2O$$

The $-\overset{\displaystyle O}{\overset{\|}{C}}-S-$ group is called a **thioester** group, and reaction 18-16 is similar to the reaction of alcohols and carboxylic acids to form esters.

Thioesters formed between a substance called coenzyme A (see Figure 18-7) and carboxylic acids are vital to energy production in the human body. We shall

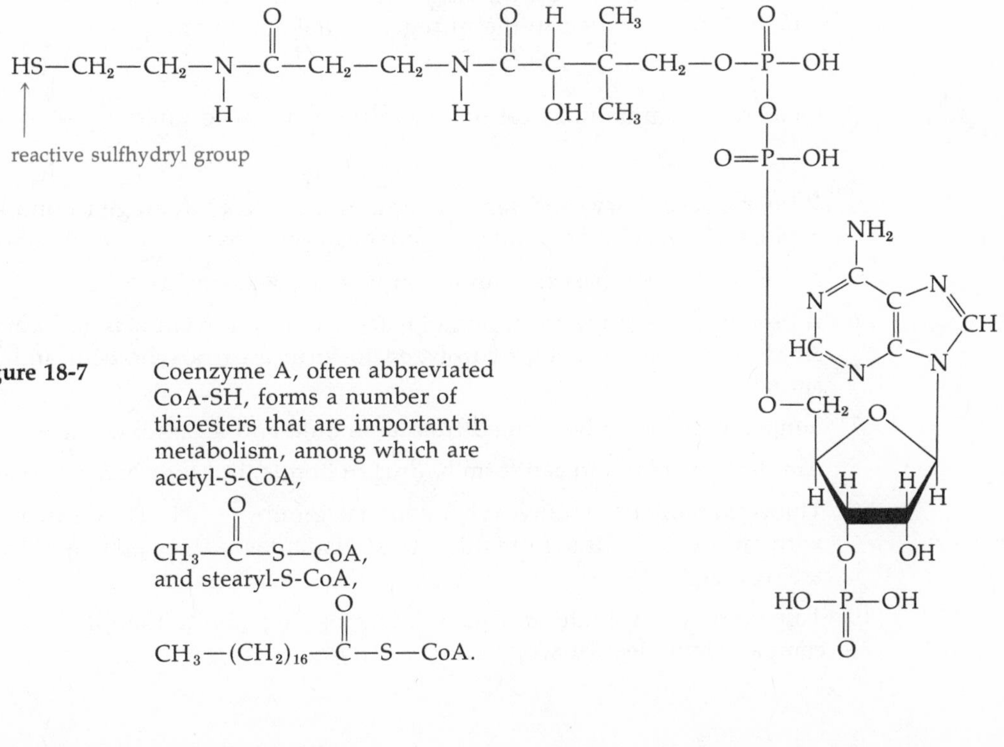

Figure 18-7 Coenzyme A, often abbreviated CoA-SH, forms a number of thioesters that are important in metabolism, among which are acetyl-S-CoA,

$$CH_3-\overset{\displaystyle O}{\overset{\|}{C}}-S-CoA,$$

and stearyl-S-CoA,

$$CH_3-(CH_2)_{16}-\overset{\displaystyle O}{\overset{\|}{C}}-S-CoA.$$

see that the thioester acetyl-S-coenzyme A is of central importance in metabo-

lism (Section 27-3). The CH_3—$\overset{\overset{O}{\|}}{C}$— portion of acetyl-S-coenzyme A is called an **acetyl** group. The general term **acyl** group refers to the atoms that remain when a hydroxyl group is removed from a carboxylic acid.

Hydrolysis of thioesters is similar to hydrolysis of esters:

(18-17)

$$R'—\overset{\overset{O}{\|}}{C}—S—R + H_2O \longrightarrow R'—\overset{\overset{O}{\|}}{C}—OH + HS—R$$

Thioester hydrolysis is highly exergonic. In fact, hydrolysis of acetyl-S-CoA and other thioesters provides the energy needed for some endergonic reactions in the human body.

Exercise 18-8

Draw structural formulas for the products of each of the following reactions:

(a) Methanethiol + ethanethiol + X (where X is an oxidizing agent) →
(b) Acetic acid + 1-butanethiol → H_2O +
(c) Methylthiopropionate + H_2O →

18-7 SUMMARY

1. Amines can be viewed as organic derivatives of ammonia. Primary amines, secondary amines, and tertiary amines have these respective general formulas: $R—NH_2$, $R_1—\overset{\overset{}{\underset{\overset{|}{H}}{N}}}{}—R_2$, and $R_1—\overset{\overset{}{\underset{\overset{|}{R_2}}{N}}}{}—R_3$.

2. The common names for amines use the names of the alkyl group attached to the nitrogen atom as prefixes followed by the ending -*amine*. The prefix N- is also used when it is necessary to specify that a given substituent is bonded to the nitrogen of the amine group.

3. Quaternary ammonium cations have the general formula $R_1—\overset{\overset{R_2}{\overset{|}{\underset{\overset{|}{R_4}}{\overset{+}{N}}}}}{}—R_3$.

4. Primary, secondary, and tertiary amines can act as hydrogen bond acceptors; primary and secondary amines can act as hydrogen bond donors.

5. Amines are weak bases; ammonium ions are weak acids.

6. A primary or secondary amine can react with a carboxylic acid to form an amide; an amide can be hydrolyzed to form a carboxylic acid and an amine.

7. Simple amides can be formed from ammonia and a carboxylic acid.

8. Amides are polar and can form hydrogen bonds, but they are not bases.

9. Thiols contain the sulfhydryl functional group, —SH. They can react with carboxylic acids to form thioesters. Disulfides are formed when thiols are oxidized.

10. Thiols are less soluble and have lower boiling points than alcohols of comparable molecular weight.

PROBLEMS

1. Identify and name each functional group in the following structural formulas:

(a) 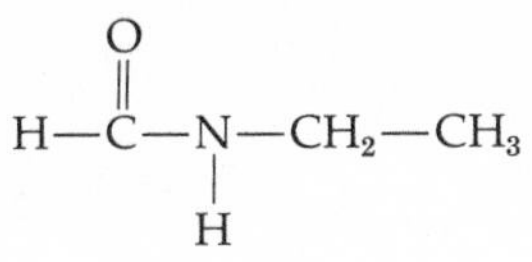 $H-\overset{\overset{\displaystyle O}{\|}}{C}-\underset{\underset{\displaystyle H}{|}}{N}-CH_2-CH_3$

(b) $H_2N-CH_2-\overset{\overset{\displaystyle O}{\|}}{C}-CH_3$

(c) $CH_3-\underset{\underset{\displaystyle H}{|}}{N}-\overset{\overset{\displaystyle O}{\|}}{C}-CH_3$

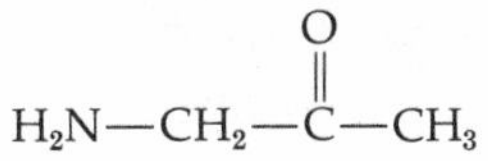

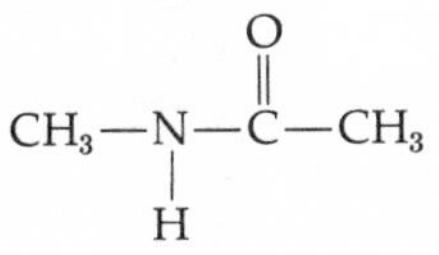

(d) $CH_3-S-S-CH_3$

(e) 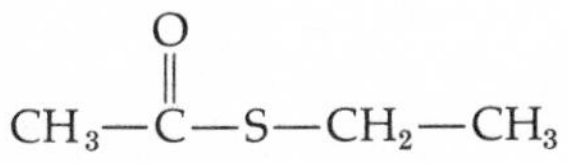 $CH_3-\overset{\overset{\displaystyle O}{\|}}{C}-S-CH_2-CH_3$

(f) 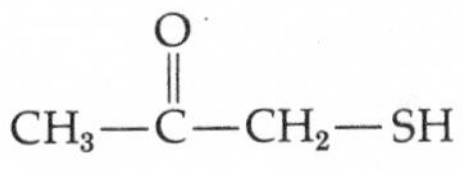 $CH_3-\overset{\overset{\displaystyle O}{\|}}{C}-CH_2-SH$

(g) 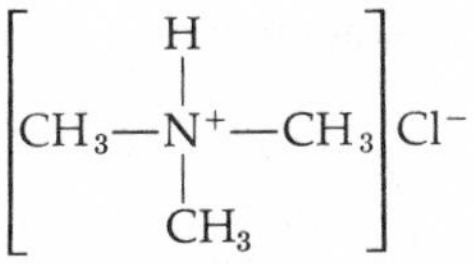 $\left[CH_3-\underset{\underset{\displaystyle CH_3}{|}}{\overset{\overset{\displaystyle H}{|}}{N^+}}-CH_3 \right] Cl^-$

2. Write the names for the compounds with the structural formulas (a), (c), (d), (e), and (g) in Problem 1.

3. Draw structural formulas for each of the following:
   (a) Tetraethylammonium bromide
   (b) *n*-Propylthioacetate
   (c) N-methylbutyramide
   (d) 2-Propanethiol
   (e) N,N-diethylbutyramide

4. Would you expect compound (a) or (e) in Problem 3 to be more soluble in aqueous solution? Explain.

5. Write structural formulas for the products of each of the following reactions:

(a) 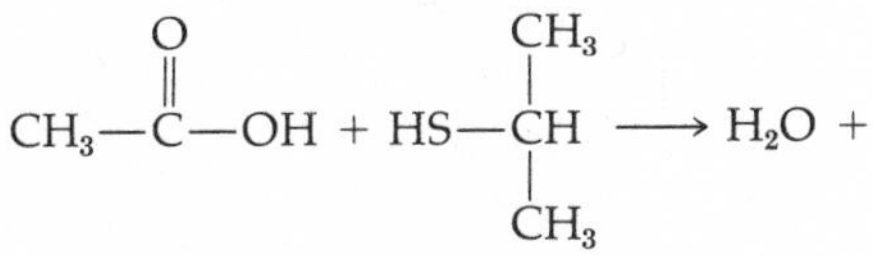 $CH_3-\overset{\overset{\displaystyle O}{\|}}{C}-OH + HS-\underset{\underset{\displaystyle CH_3}{|}}{\overset{\overset{\displaystyle CH_3}{|}}{CH}} \longrightarrow H_2O +$

(b) 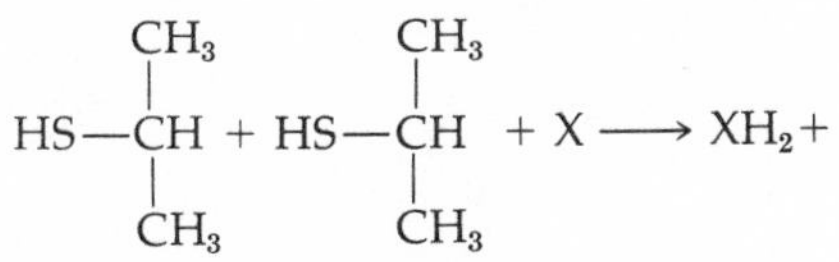 $HS-\underset{\underset{\displaystyle CH_3}{|}}{\overset{\overset{\displaystyle CH_3}{|}}{CH}} + HS-\underset{\underset{\displaystyle CH_3}{|}}{\overset{\overset{\displaystyle CH_3}{|}}{CH}} + X \longrightarrow XH_2 +$

(c) $H_2N-CH_2-CH_3 + H_2O \rightarrow$

(d) 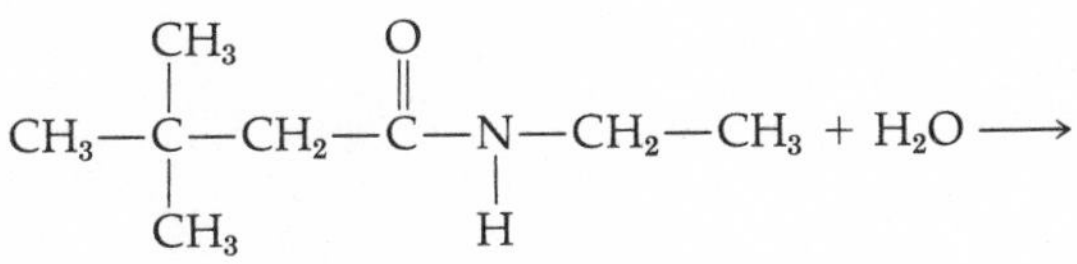 $CH_3-\underset{\underset{\displaystyle CH_3}{|}}{\overset{\overset{\displaystyle CH_3}{|}}{C}}-CH_2-\overset{\overset{\displaystyle O}{\|}}{C}-\underset{\underset{\displaystyle H}{|}}{N}-CH_2-CH_3 + H_2O \longrightarrow$

(e) 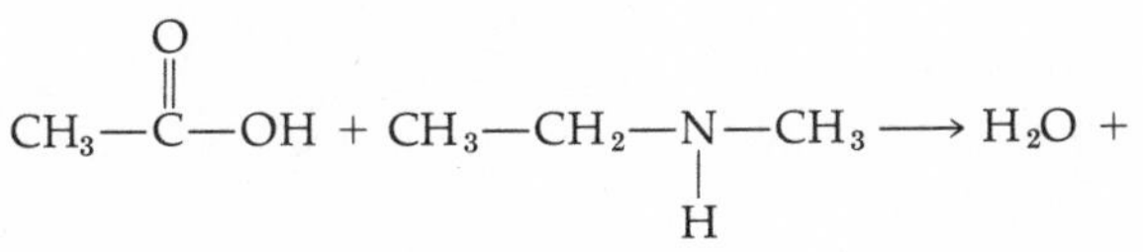 $CH_3-\overset{\overset{\displaystyle O}{\|}}{C}-OH + CH_3-CH_2-\underset{\underset{\displaystyle H}{|}}{N}-CH_3 \longrightarrow H_2O +$

(f) $CH_3-S-S-CH_3 + XH_2 \longrightarrow X +$

6. Which has the lower boiling point, ethanol or ethanethiol? Explain your answer.

SOLUTIONS TO EXERCISES

18-1 $CH_3—CH_2—CH_2—NH_2$ (*n*-propylamine), primary amine.

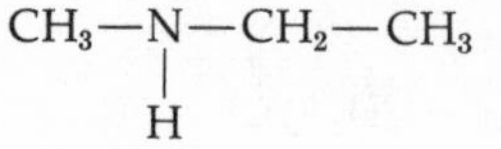

$CH_3—\overset{\displaystyle |}{\underset{\displaystyle H}{N}}—CH_2—CH_3$ (methylethylamine), secondary amine.

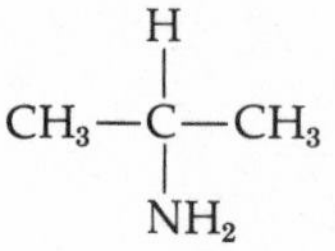

$CH_3—\overset{\displaystyle H}{\underset{\displaystyle NH_2}{C}}—CH_3$ (isopropylamine), primary amine.

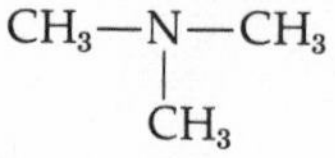

$CH_3—\underset{\displaystyle CH_3}{N}—CH_3$ (trimethylamine), tertiary amine.

18-2 $CH_3—\overset{\displaystyle O}{\overset{\displaystyle \|}{C}}—S—CH_2—CH_3$ is the thioester formed. Compare this reaction to the formation of ethyl acetate:

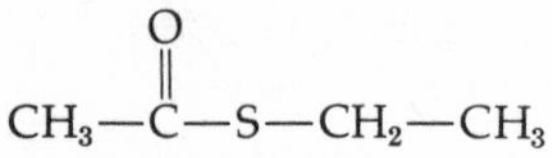

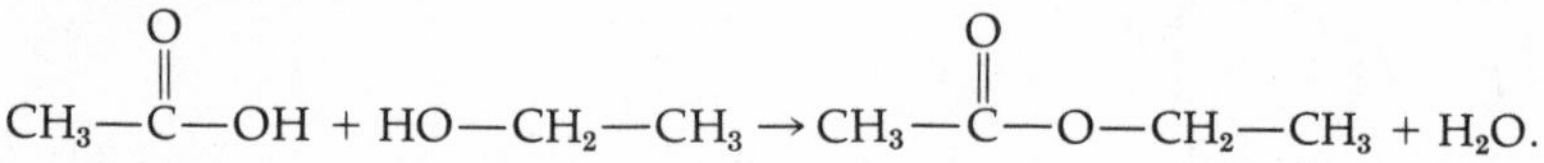

$$CH_3—\overset{O}{\overset{\|}{C}}—OH + HO—CH_2—CH_3 \rightarrow CH_3—\overset{O}{\overset{\|}{C}}—O—CH_2—CH_3 + H_2O.$$

18-3 (a) 1, 2, 3
    (b) 1, 4
    (c) 1, 2, 3
    (d) 2, 3
    (e) None of these

18-4 (a) 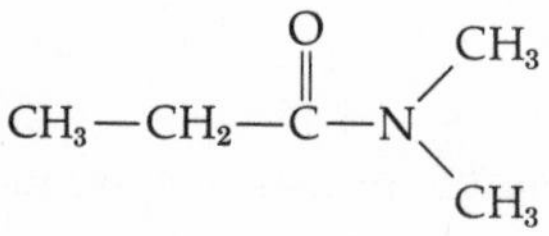 $CH_3—CH_2—\overset{O}{\overset{\|}{C}}—N\overset{CH_3}{\underset{CH_3}{\big<}}$

    (b) $\left[CH_3—CH_2—\overset{O}{\overset{\|}{C}}—O^-\right]\left[H—\overset{CH_3}{\overset{|}{\overset{+}{N}}}—CH_3 \atop CH_3\right]$

    (c) $H—\overset{O}{\overset{\|}{C}}—\overset{}{\underset{H}{N}}—\overset{CH_3}{\underset{H}{C}}—CH_3$

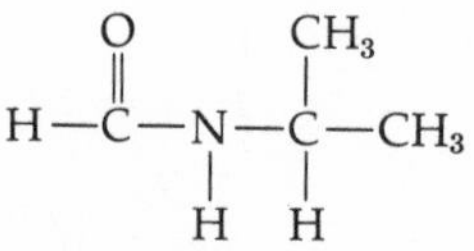

    (d) 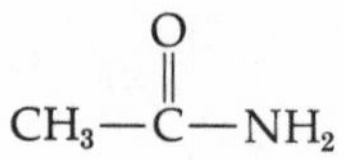 $CH_3—\overset{O}{\overset{\|}{C}}—NH_2$

18-5 (a) 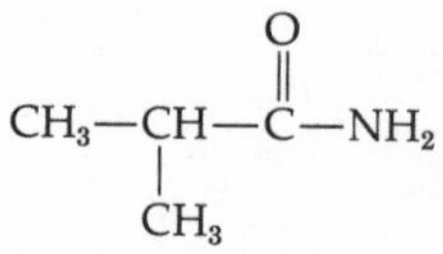 $CH_3—\underset{CH_3}{CH}—\overset{O}{\overset{\|}{C}}—NH_2$

    (b) $CH_3—CH_2—\overset{O}{\overset{\|}{C}}—\underset{CH_3}{N}—CH_3$

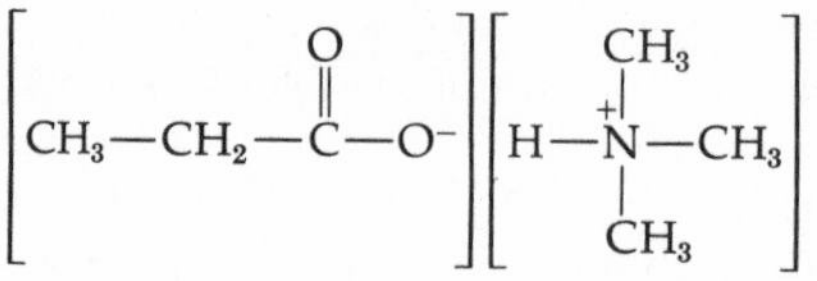

    (c) $H—\overset{O}{\overset{\|}{C}}—\underset{H}{N}—\bigcirc$

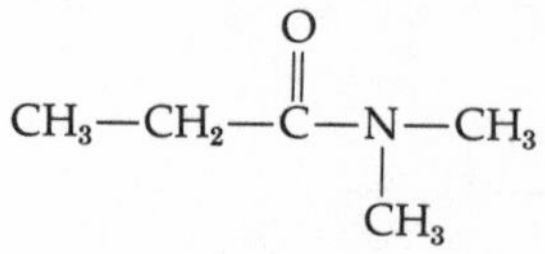

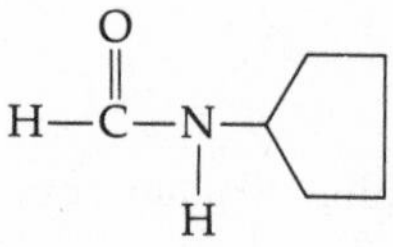

18-6 and 18-7

(a) $CH_3-\overset{\overset{\displaystyle H}{|}}{\underset{\underset{\displaystyle CH_3}{|}}{C}}-\overset{\overset{\displaystyle H}{|}}{N}-\overset{\overset{\displaystyle O}{\|}}{C}-CH_2-NH_2 + H_2O \longrightarrow CH_3-\overset{\overset{\displaystyle H}{|}}{\underset{\underset{\displaystyle CH_3}{|}}{C}}-NH_2 + HO-\overset{\overset{\displaystyle O}{\|}}{C}-CH_2-NH_2$

              base   base   acid and base

(b) $NH_3 + H_2O \longrightarrow NH_4^+ + OH^-$

(c) $CH_3-\overset{\overset{\displaystyle O}{\|}}{C}-OH + H-\overset{\overset{\displaystyle CH_3}{|}}{N}-CH_2-\overset{\overset{\displaystyle H}{|}}{\underset{\underset{\displaystyle CH_3}{|}}{C}}-CH_3 \longrightarrow H_2O + CH_3-\overset{\overset{\displaystyle O}{\|}}{C}-\overset{\overset{\displaystyle CH_3}{|}}{N}-CH_2-\overset{\overset{\displaystyle H}{|}}{\underset{\underset{\displaystyle CH_3}{|}}{C}}-CH_3$

      acid   base

18-8 (a) $XH_2 + CH_3-S-S-CH_2-CH_3$

(b) $CH_3-\overset{\overset{\displaystyle O}{\|}}{C}-S-CH_2-CH_2-CH_2-CH_3$

(c) $CH_3-CH_2-\overset{\overset{\displaystyle O}{\|}}{C}-OH + CH_3-SH$

CHAPTER 19

Stereochemistry

19-1 INTRODUCTION

Did you ever try to put your right foot into your left shoe? If you did, you probably found that your right foot fits a lot better in your right shoe, and that walking is a great deal easier when your shoes are on the correct feet.

Stereochemistry deals with the shapes of molecules. A molecule's shape is a significant factor in determining how that molecule can react with other molecules. We shall see in this chapter that stereoisomers can be subdivided into two classes—enantiomers and diastereomers. Enantiomers come in pairs, like shoes and feet. The difference in shape between a pair of enantiomers is analogous to the difference in shape between your right and left shoes or your right and left feet. Enantiomers are mirror images of one another. We are all familiar with objects that differ in shape but are mirror images, such as hands, feet, gloves, shoes, right-handed and left-handed threads on nuts and bolts, and so on.

For many compounds that can exist as a pair of enantiomers, only one of the enantiomers is found naturally. When enantiomers of two different compounds interact, there can be a large difference in the rate of reaction, depending on which specific enantiomer of each compound is involved in the reaction. The difference in the chemical properties of enantiomers is an essential feature of the enzyme-catalyzed reactions that take place in living cells. The reactants for most of these reactions are compounds that can exist as a pair of enantiomers. We shall see that, in most cases, the enzymes produced by a cell are capable of catalyzing reactions involving only one enantiomer of a pair.

In this chapter we also develop some useful ways to represent the shapes of molecules. We shall see how to describe the difference in shape between two enantiomers. It is interesting to note that while this difference in shape may result in quite different chemical properties, the only difference in the physical properties of enanatiomers is the manner in which they interact with polarized light.

A ballet dancer uses her reflection to fine-tune her movements. Bear in mind the difference between the original and its mirror image as you study the structures and properties of enantiomer pairs.

19-2 STUDY OBJECTIVES

After studying the material in this chapter, you should be able to:

1. Define the terms isomer, structural isomer, stereoisomer, enantiomer, diastereomer, superimposable, nonsuperimposable, chiral molecule, and asymmetric carbon atom.

2. Determine whether two given perspective drawings or Fischer projections represent (a) the same molecule, (b) enantiomers, (c) diastereomers, or (d) different molecules that are not isomers.

3. Draw perspective drawings or Fischer projections for all the stereoisomers of a compound, given a structural formula for the compound, and label which stereoisomers are enantiomers and which are diastereomers.

4. Explain why the two enantiomers of a chiral molecule can react differently with a single enantiomer of another chiral molecule, but will react identically with a nonchiral molecule.

5. Define the terms linearly polarized light, unpolarized light, transmission axis, and rotation.

6. Describe how the rotation of linearly polarized light by a solution can be measured with a polarimeter.

7. Explain what the designations (+) and (−) for enantiomers of a chiral molecule mean, and the distinction between the labels (+) and (−) and the labels D- and L- for a pair of enantiomers.

19-3 CLASSIFICATION OF STEREOISOMERS

As you read this chapter it is important to keep in mind the following points, which we have discussed previously:

1. The **shape,** or molecular geometry, of a molecule refers to the relative positions of all the atomic nuclei in the molecule.

2. A **structural formula** is a representation of the bonding arrangement in the molecule that indicates which atoms in the molecule are bonded together.

3. **Isomers** are different compounds that have the same molecular formula.

4. For two compounds to be different, there must be some differences in their chemical and physical properties that will allow their separation from a mixture.

5. Isomers are classified as either structural isomers or stereoisomers.

Structural isomers are compounds that have the same molecular formula but a different bonding arrangement (they have different structural formulas), such as the functional group isomers, ethyl alcohol and dimethyl ether, or the positional isomers, n-butane and isobutane. **Stereoisomers,** on the other hand, are isomers that have the same structural formula but differ in the way the bonds in the molecule are oriented in space. We previously described one class of stereoisomers, geometric or cis-trans stereoisomers. 1,2-Dimethylcyclohexane (Section 14-5) and 2-butene (Section 14-6) are examples of compounds that have cis-trans stereoisomeric forms.

Cis-trans stereoisomerism, however, is not the only type of stereoisomerism. For example, glyceraldehyde (see Figure 19-1) is a compound that exists in two stereoisomeric forms that are not of the cis-trans type. The two stereoisomers of glyceraldehyde are labeled D-glyceraldehyde and L-glyceraldehyde. In Section 19-7 we shall see why we label them D and L. The differences between the stereoisomers D- and L-glyceraldehyde are more subtle than the differences between, say, the cis-trans stereoisomers of 2-butene. In contrast to *cis*- and *trans*-2-butene, D- and L-glyceraldehyde have the same melting point, boiling point, solubility in water and other solvents, and so on. In fact, the physical properties of D- and L-glyceraldehyde are identical except for the way in which they interact with linearly polarized light (Section 19-6). There are, however, significant and important differences in the chemical properties of D- and L-glyceraldehyde, which we shall discuss shortly (Section 19-5).

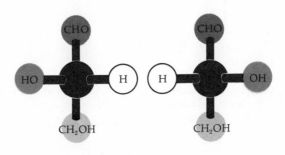

L-glyceraldehyde   D-glyceraldehyde

Figure 19-1   Ball-and-stick schematic models of the two stereoisomers of glyceraldehyde. Note that the H atom and the —OH group point toward you, and that the —CHO and —CH$_2$OH groups point away from you.

Look at the representations of D- and L-glyceraldehyde in Figure 19-1 closely. Do you see that they are **mirror images** of one another? Your left hand and your right hand function differently in certain circumstances because they have different shapes. Your right hand, for example, can only be used to shake hands properly with another right hand. In an analogous manner, D- and L-glyceraldehyde are examples of a class of stereoisomers called enantiomers. **Enantiomers** are stereoisomers that are mirror images of one another.

An actual three-dimensional model is the best way to visualize the shape of a molecule, to show correctly the spatial relationship between the atoms in the molecule, and to test whether or not two molecules are enantiomers. A variety of commercially available kits can be used to construct molecular models.

Several types of pictorial representations, such as those in Figure 19-1, are used to indicate molecular shapes, but none of these is as good as an actual model. Perspective drawings are a simpler way of indicating molecular shape than the ball-and-stick pictures in Figure 19-1. Since we are concerned primarily with the shapes of molecules that contain carbon atoms, we shall describe the way perspective drawings of carbon-containing compounds are interpreted, using D- and L-glyceraldehyde as examples.

Recall that for any carbon atom with four single bonds, the four bonds are oriented approximately toward the corners of a tetrahedron with the carbon atom in the center of the tetrahedron. The central carbon atom in glyceraldehyde has four single bonds and, as shown in Figure 19-2:

1. The central carbon atom is represented by a circle lying in the plane of the paper.

2. The H atom and the —OH group bonded to this carbon atom are represented as being closer to you, *in front* of the plane of the paper that contains the central carbon atom.

3. The —CH$_2$OH group and the aldehyde group, —$\overset{\overset{\displaystyle O}{\|}}{C}$—H (shown as —CHO), that are bonded to the central carbon atom are represented as being *behind* the plane of the paper.

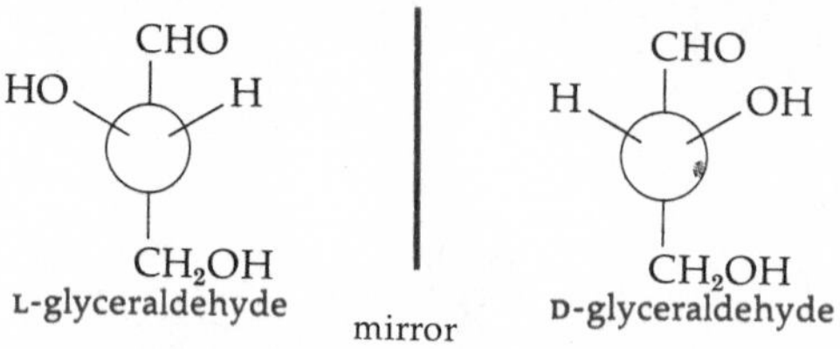

Figure 19-2   Perspective drawings of D- and L-glyceraldehyde, which are mirror images of one another. Visualize the H atom and the —OH group as pointing toward you, and the —CHO and —CH$_2$OH groups as pointing away from you.

Look at the perspective drawings of D- and L-glyceraldehyde in Figure 19-2 closely. It should be clear to you that they represent mirror images.

Figure 19-3 shows perspective drawings of two additional examples of enantiomers: 2-butanol and *cis*-[Co(H$_2$N—CH$_2$—CH$_2$—NH$_2$)$_2$Cl$_2$]$^+$, which is a transition metal complex ion containing cobalt.

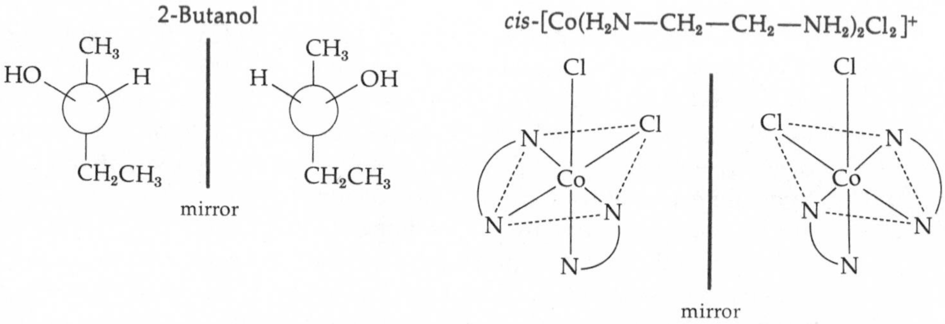

Figure 19-3   Perspective drawings of the enantiomers of 2-butanol and *cis*[Co(H$_2$N—CH$_2$—CH$_2$—NH$_2$)$_2$Cl$_2$]$^+$

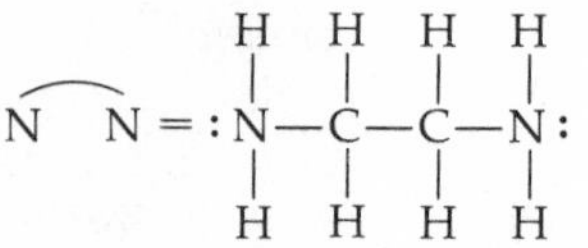

There is a fairly simply way of telling if a given compound can exist in two enantiomeric forms using both a model for a molecule of the compound and a model for its mirror image: When the molecule and its mirror image have the same shape, *all* the chemical and physical properties of the molecule and its mirror image are identical and there are no enantiomeric forms for the compound. If, and only if, the molecule and its mirror image have different shapes can a pair of enantiomeric stereoisomers exist.

We can test whether an object and its mirror image have the same or different shape by a *superposition procedure* in which we perform the following steps:

1. Think of the object and its mirror image as capable of penetrating one another without distorting their shapes.

2. Carry out this mental interpenetration procedure.

3. If all of the atoms in the resulting combination of the object and its penetrated mirror image are in positions identical to the original object, we say that the object and the mirror image are **superimposable.** If the resulting combination is not identical to the original object, we say that the object and its mirror image are **nonsuperimposable.**

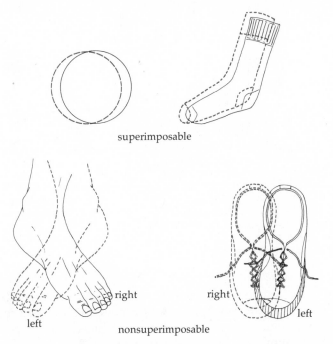

Figure 19-4   A ball and its mirror image, and a sock and its mirror image, are superimposable, whereas right and left feet, or right and left shoes, are nonsuperimposable.

Figure 19-4 illustrates several common examples of superimposable and nonsuperimposable objects. D-Glyceraldehyde and L-glyceraldehyde are nonsuperimposable mirror images of one another (see Figure 19-5). When a molecule and its mirror image are nonsuperimposable, the molecule is called a **chiral** molecule. Enantiomers can exist only for those compounds whose molecules are chiral.

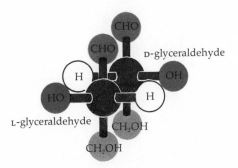

Figure 19-5   Ball-and-stick schematic models of D- and L-glyceraldehyde illustrate that these enantiomers are nonsuperimposable.

The term chiral comes from the Greek word *cheir,* meaning "hand." Glyceraldehyde, 2-butanol, and the *cis*-[Co(H$_2$N—CH$_2$—CH$_2$—NH$_2$)$_2$Cl$_2$]$^+$ ion are examples of chemical species that are chiral, and these species have enantiomeric
forms. Methane, dichloromethane, chlorodifluoromethane, and 2-chloropropane
(see Figure 19-6) are examples of compounds whose molecules are not chiral,
and these compounds do not have enantiomers.

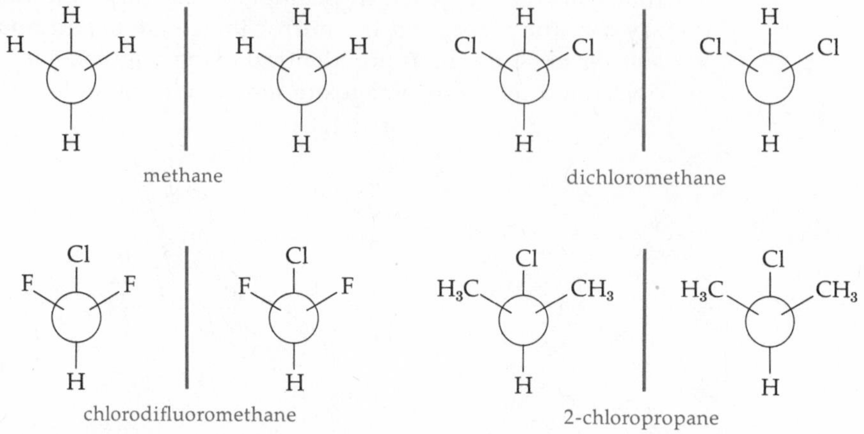

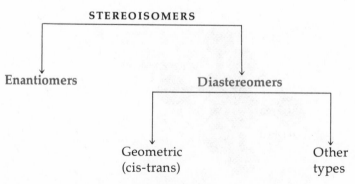

Figure 19-6   Perspective drawings of nonchiral   atoms attached to the carbon atom
molecules. Note that there are four   in chlorodifluoromethane, and
similar atoms attached to the   one pair of similar methyl groups
carbon atom in methane, two   attached to the central carbon atom
pairs of similar atoms attached to   in 2-chloropropane.
the carbon atom in dichloro-
methane, one pair of similar

An equivalent way of describing stereoisomers that are enantiomers is to call
them **mirror-image stereoisomers.** As we have mentioned previously, not all
stereoisomers are mirror images of each other. Stereoisomers that are *not* mirror
images of one another are called **diastereomers.** Thus, there are two classes of
stereoisomers (see Figure 19-7):

1. Mirror-image stereoisomers or enantiomers
2. Nonmirror-image stereoisomers or diastereomers

Geometric isomers, such as *cis-* and *trans*-2-butene, are only one of several types
of diastereomers. We shall discuss other types presently.

STEREOISOMERS

Enantiomers   Diastereomers

Geometric   Other
(cis-trans)   types

Figure 19-7   Enantiomers and diastereomers are the
two major classes of stereoisomers.

Many different types of compounds can exist as enantiomers, but since we are interested primarily in organic compounds, in our further discussion we shall consider only enantiomers that are carbon-containing compounds.

Exercise 19-1

Which of the following molecules are chiral? For each chiral molecule, draw a perspective drawing of the other enantiomer. The symbol —COOH is an alternative way of repre-

senting the carboxyl group, $-\overset{\overset{\text{O}}{\|}}{\text{C}}-\text{OH}$.

    (a) The amino acid alanine,

$$
\begin{array}{c}
\text{COOH} \\
\text{H}_2\text{N} \diagdown \underset{\bigcirc}{\big|} \diagup \text{H} \\
\text{CH}_3
\end{array}
$$

    (b) Chloroform,

$$
\begin{array}{c}
\text{Cl} \\
\text{Cl} \diagdown \underset{\bigcirc}{\big|} \diagup \text{H} \\
\text{Cl}
\end{array}
$$

    (c) Lactic acid,

$$
\begin{array}{c}
\text{COOH} \\
\text{H} \diagdown \underset{\bigcirc}{\big|} \diagup \text{OH} \\
\text{CH}_3
\end{array}
$$

    (d) 1,2-Dichloroethene

19-4 ENANTIOMERS AND DIASTEREOMERS OF CARBON-CONTAINING COMPOUNDS

Notice that in the chiral molecules glyceraldehyde and 2-butanol (Figures 19-2 and 19-3) there is a carbon atom that is bonded to four different atoms or groups of atoms. This is not true of the compounds in Figure 19-6, which are not chiral.

The central carbon atom in glyceraldehyde is bonded to a H atom, an —OH group, a —CH₂OH group, and a —CHO group, and the number 2 carbon in 2-butanol is bonded to a H atom, an —OH group, a —CH₃ group, and a —CH₂CH₃ group. Even though the number 2 carbon atom in 2-butanol is bonded to a carbon atom in a methyl group and another carbon atom in an ethyl group, the methyl group and the ethyl group are different.

Asymmetric Carbon

A carbon atom that is bonded to four different groups of atoms is called a **chiral carbon** or an **asymmetric carbon.** Usually (but not always) a molecule that has one or more asymmetric carbons is a chiral molecule. As our examples of glyceraldehyde and 2-butanol illustrate, in most cases carbon-containing molecules are chiral if they contain an asymmetric carbon; and molecules such as methane, dichloromethane, chlorodifluoromethane, and 2-chloropropane, which do not contain an asymmetric carbon, are not chiral. There are, however, exceptions to these generalizations: (1) Some carbon-containing compounds that are chiral do not have an asymmetric carbon; and (2) some carbon-containing compounds that have asymmetric carbons are not chiral. Thus, although the presence or absence of asymmetric carbons is a good indication of whether or not a molecule is chiral, the proof for chirality is nonsuperimposable mirror images.

We shall discuss only chiral molecules that contain one or more asymmetric carbons. Sugars and most amino acids possess asymmetric carbons. We shall not need to discuss chiral molecules that do not have an asymmetric carbon.

For larger molecules, with more than one asymmetric carbon, the perspective drawings we have used so far are tedious to draw and to use. It is more convenient to use planar representations called **Fischer projections.** Let us first illustrate how Fischer projections are drawn and used with a familiar example—glyceraldehyde.

Fischer Projections

To draw the Fischer projections for D- and L-glyceraldehyde we proceed as follows (see Figure 19-8a):

1. Consider the perspective drawing for D-glyceraldehyde and draw a cross to represent the asymmetric carbon with its four bonds.

2. Attach the H atom and the —OH group, which are coming toward you in the perspective drawing, to the left and right horizontal lines of the cross, respectively.

3. Attach the —CHO group and the —CH₂OH group, which are going away from you in the perspective drawing, to the top and bottom vertical lines of the cross, respectively.

Note that when you see a planar Fischer projection, it is very important that you visualize the horizontal lines as bonds coming toward you, and the vertical lines as bonds going away from you. As illustrated in Figure 19-8b, the Fischer projection for L-glyceraldehyde can be obtained by reflecting the Fischer projection for D-glyceraldehyde in a mirror.

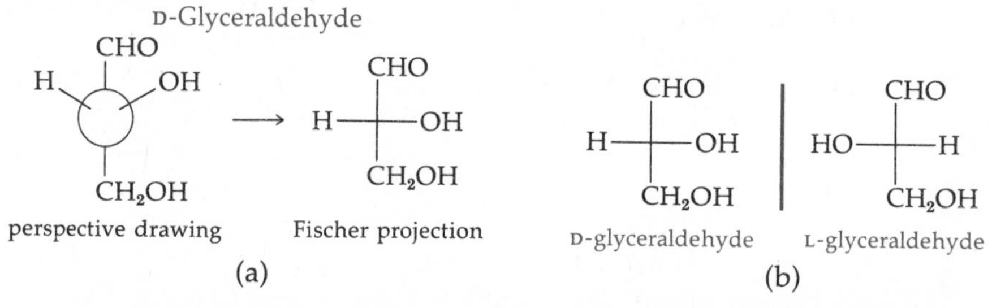

(a) perspective drawing — Fischer projection

D-glyceraldehyde L-glyceraldehyde

(b)

Figure 19-8 (a) Perspective drawing and Fischer projection for D-glyceraldehyde. Note that you should view the Fischer projection as if the H atom and the —OH group point toward you, and the —CHO and —CH₂OH groups point away from you. (b) The Fischer projections of D- and L-glyceraldehyde are mirror images of one another.

We know that D- and L-glyceraldehyde are nonsuperimposable mirror images, and this fact can be inferred from the Fischer projections, provided that the Fischer projections are used appropriately. Notice that the Fischer projection for L-glyceraldehyde *cannot* be superimposed upon the Fischer projection for D-glyceraldehyde by moving the L-projection to the left and then rotating the L-projection while keeping it in the plane of the paper (see Figure 19-9). We could superimpose the Fischer projections for D- and L-glyceraldehyde if we lifted the L-projection out of the plane of the paper, flipped it over, and put it on top of the D-projection, but *this operation is not permissible.* Fischer projections are two-dimensional representations; they can be used to test for superimposability only if you keep them in the projection plane and do not raise them from the surface of this plane and try to flip them over.

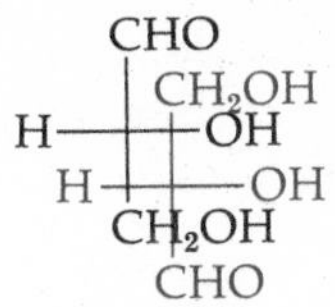

Figure 19-9   The Fischer projections for
D-glyceraldehyde (black) and
L-glyceraldehyde (blue) are
nonsuperimposable.

$$CHO$$
$$CH_2OH$$
$$H—|—OH$$
$$H—|—OH$$
$$CH_2OH$$
$$CHO$$

Let us now construct Fischer projections for the stereoisomers of a compound with two asymmetric carbons, for example, 2-bromo-3-chlorobutane. In Figure 19-10, the two asymmetric carbons are labeled with an asterisk (*). The perspective drawing and corresponding Fischer projection for one of the possible shapes for 2-bromo-3-chlorobutane is given in Figure 19-10a. The Fischer projection is labeled I and its mirror image is labeled II. To determine if projections I and II represent a pair of enantiomers, we have to test if I and II are superimposable. We cannot superimpose I and II (remember to keep them in the same plane), and therefore I and II do correspond to a pair of stereoisomers that are enantiomers.

These enantiomers are not the only stereoisomers of 2-bromo-3-chlorobutane. Another possible shape for this compound is represented in the perspective drawing and corresponding Fischer projection III given in Figure 19-10b. Fischer projection III is not superimposable on either projection I or II, and therefore Fischer projection III represents the shape of a third stereoisomer of our compound. The stereoisomer represented by Fischer projection III is not the mirror image of the shapes represented by Fischer projection I or II. Therefore, I and III represent diastereomers, as do II and III. When we consider Fischer projection IV, the mirror image of III, we find that III and IV are not superimposable. Therefore, III and IV correspond to a second pair of enantiomers. Thus there are two pairs of enantiomers, for a total of four stereoisomeric forms of 2-bromo-3-chlorobutane.

2-Bromo-3-chlorobutane:
$$CH_3—\overset{H}{\underset{Br}{\overset{*}{C}}}—\overset{H}{\underset{Cl}{\overset{*}{C}}}—CH_3$$

Perspective drawing   Fischer projection I   Perspective drawing   Fischer projection III

I   II   III   IV

I and II are an enantiomer pair   III and IV are an enantiomer pair

(a)   (b)

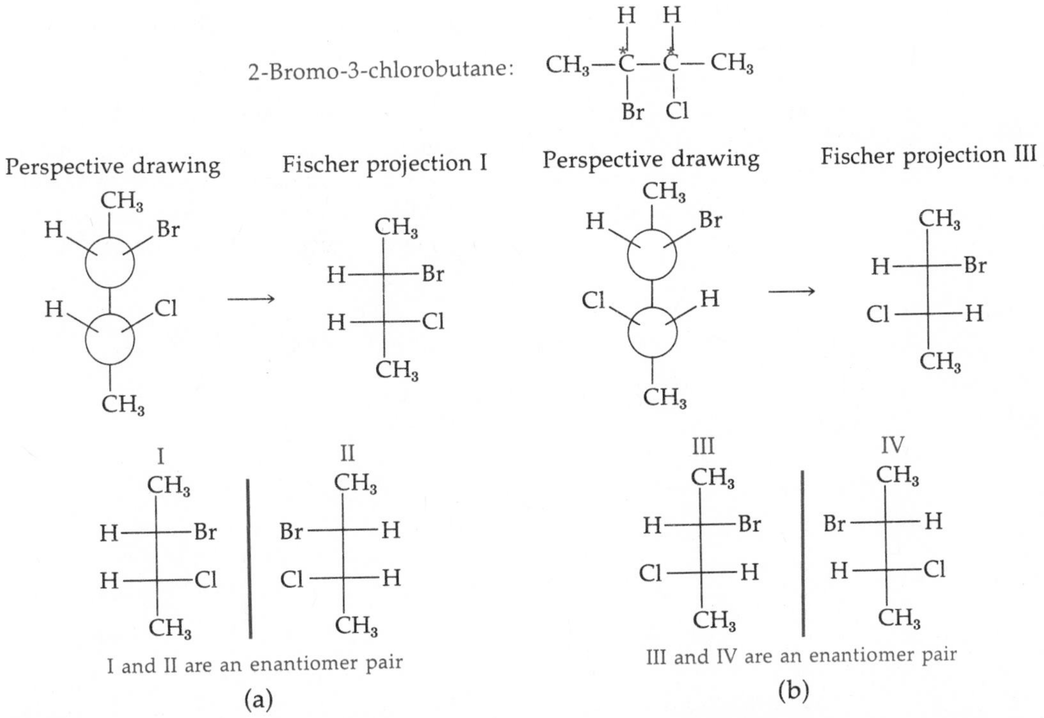

Figure 19-10   2-Bromo-3-chlorobutane has two asymmetric carbon atoms and four stereoisomers. The Fischer projections I and II correspond to one enantiomer pair (a), and the Fischer projections III and IV correspond to the other enantiomer pair (b).

Notice that there are two stereoisomers for glyceraldehyde, which has one asymmetric carbon, and four stereoisomers for 2-bromo-3-chlorobutane, which has two asymmetric carbons. The compound

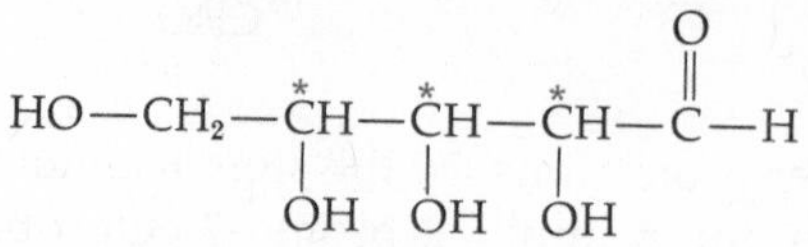

$$HO—CH_2—\overset{*}{C}H—\overset{*}{C}H—\overset{*}{C}H—\overset{\overset{\displaystyle O}{\|}}{C}—H$$
$$\underset{OH}{|}\quad\underset{OH}{|}\quad\underset{OH}{|}$$

has three asymmetric carbons, and there are a total of eight stereoisomers of this compound. As a general rule, for a compound with n asymmetric carbon atoms, the number of stereoisomers is no larger than 2^n ($2 \times 2 \times 2 \times \cdots n$ times). For most compounds, such as the ones we have just described, the actual number of stereoisomers is equal to 2^n. For some compounds, however, the number of stereoisomers is less than 2^n. Tartaric acid, for example,

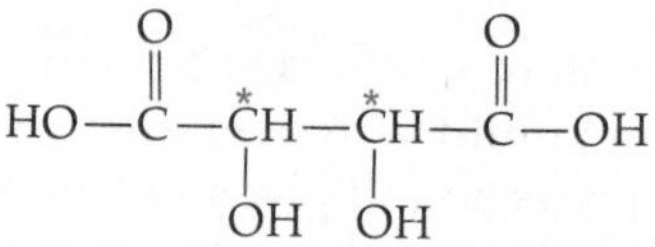

$$HO—\overset{\overset{\displaystyle O}{\|}}{C}—\overset{*}{C}H—\overset{*}{C}H—\overset{\overset{\displaystyle O}{\|}}{C}—OH$$
$$\underset{OH}{|}\quad\underset{OH}{|}$$

has two asymmetric carbons but only three stereoisomers. We can use Fischer projections to see why this is so (see Figure 19-11). Fischer projections A and B are nonsuperimposable mirror images, so they correspond to an enantiomeric pair, called L-tartaric acid (A) and D-tartaric acid (B). The first pair of enantiomers to be discovered involved a salt of tartaric acid. This discovery of enantiomers, which marked the beginning of the study of molecular shapes, was made by Louis Pasteur in 1848. Fischer projection C is not superimposable on either Fischer projection A or B, so Fischer projection C is a representation for another stereoisomer of tartaric acid, called *meso*-tartaric acid. *meso*-Tartaric acid and either D- or L-tartaric acid are diastereomers. Fischer projections C and D are superimposable, and therefore they do *not* represent an enantiomeric pair. Fischer projections C and D are thus *equivalent* representations for *meso*-tartaric acid, which has the shape indicated by the perspective drawing in Figure 19-12. Notice that the shape of *meso*-tartaric acid consists of two halves that are mirror images of one another—one half above and the other half below the mirror plane indicated in the figure. *meso*-Tartaric acid does not exist as an enantiomeric pair because of this *internal* mirror plane of symmetry.

$$\text{Tartaric acid: } HO—\overset{\overset{\displaystyle O}{\|}}{C}—\overset{*}{C}H—\overset{*}{C}H—\overset{\overset{\displaystyle O}{\|}}{C}—OH$$
$$\underset{OH}{|}\quad\underset{OH}{|}$$

A	B	C	D
COOH	COOH	COOH	COOH
H—⊢—OH	HO—⊢—H	H—⊢—OH	HO—⊢—H
HO—⊢—H	H—⊢—OH	H—⊢—OH	HO—⊢—H
COOH	COOH	COOH	COOH
L-tartaric acid	D-tartaric acid	*meso*-tartaric acid	*meso*-tartaric acid

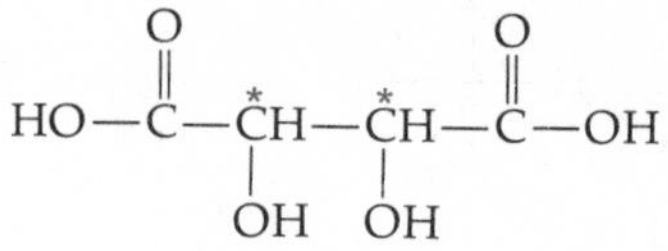

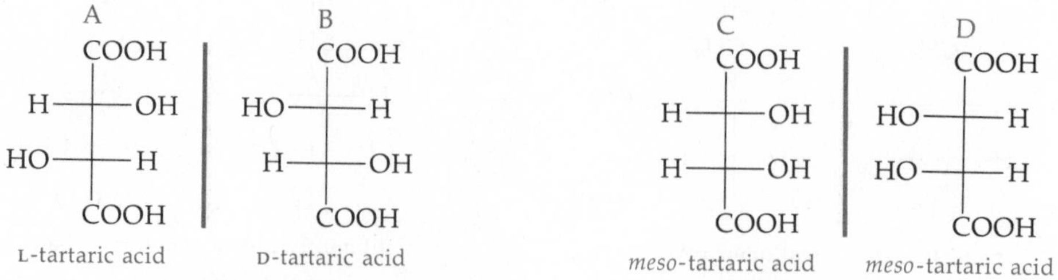

Figure 19-11   Fischer projections A and B are nonsuperimposable, and correspond to the enantiomer pair L- and D-tartaric acid, respectively. Fischer projections C and D are superimposable, and do *not* correspond to an enantiomer pair. C and D are equivalent representations for *meso*-tartaric acid.

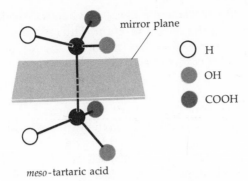

mirror plane

○ H

● OH

● COOH

meso-tartaric acid

Figure 19-12   Perspective drawing of *meso*-tartaric acid.

Exercise 19-2
Draw structural formulas for each of the following, and indicate with an asterisk (*) any asymmetric carbons.

    (a) 2-Methyl-1-butanol   (b) 2,3-Dichlorobutane
    (c) 2-Methyl-2-butanol   (d) 1,2-Dibromopropane

Exercise 19-3
Draw Fischer projections for all the stereoisomers in Exercise 19-2 and indicate which are enantiomers and which are diastereomers.

19-5   REACTIONS OF STEREOISOMERS

We have seen that chiral molecules have different shapes from their mirror images. Because of these differences in shape, a chiral molecule and its mirror image can have significantly different chemical reactivity.

    A chemical reaction involving a *single* enantiomer of a chiral molecule X (which we shall label D-X) and a *pair* of enantiomers, such as D- and L-glyceraldehyde, is analogous to trying to put your left foot into your right and left shoes. The fit between the enantiomer D-X and D-glyceraldehyde can be very different from the fit between D-X and L-glyceraldehyde (see Figure 19-13). Recall that

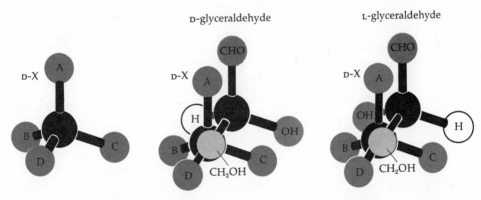

Figure 19-13   Ball-and-stick model of the chiral molecule D-X (left). The fit between D-X and D-glyceraldehyde (center) is different than the fit between D-X and L-glyceraldehyde (right).

the relative orientation of two reacting molecules when they collide (the fit) is a crucial factor in determining the rate of reaction. Thus if X and glyceraldehyde are molecules that react, it is possible that the rate of the reaction between D-X and D-glyceraldehyde is so slow that effectively no reaction takes place, whereas the rate of reaction between D-X and L-glyceraldehyde is very rapid. As we shall see, for most biologically important enzyme-catalyzed reactions, the enzyme fits with only one enantiomer of the reactant substance.

As we have just seen, there is a difference in the way D-X, a single enantiomer of the chiral molecule X, fits with D-glyceraldehyde or L-glyceraldehyde. On the other hand, there is an *identical* fit between any nonchiral molecule and D-glyceraldehyde or L-glyceraldehyde. This fact is illustrated in Figure 19-14, where Y is used as a label for the nonchiral molecule. A sock is analogous to a molecule that is not chiral, whereas your feet are analogous to a pair of enantiomers. The identical fit between the molecule Y, which is not chiral, and D- and L-glyceraldehyde is analogous to the identical fit between a sock and your left foot and the same sock and your right foot.

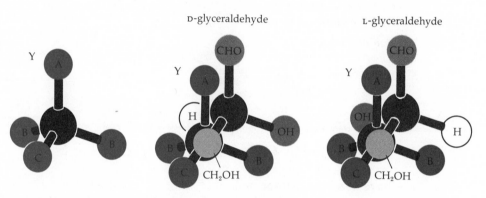

D-glyceraldehyde   L-glyceraldehyde

Figure 19-14   Model of the nonchiral molecule   same as that between Y and
Y (left). The fit between Y and   L-glyceraldehyde (right).
D-glyceraldehyde (center) is the

It is generally true that (1) a single enantiomer of a chiral molecule will fit differently, and react differently, with the two enantiomers of another chiral molecule; and that (2) a molecule that is not chiral will have an identical fit, and react identically, with the two enantiomers of a chiral molecule.

Exercise 19-4

Would you expect the enantiomers of 2-butanol (Figure 19-3) to react identically with the following molecules?

(a)

H_3C — OH, H — COOH

(b)

H_3C — H, H — COOH

19-6 OPTICAL ACTIVITY

The physical properties of two enantiomers of a chiral molecule are identical except for the way in which they interact with polarized light. Let us see what is meant by polarized light.

Light is an electromagnetic wave with an electric field and a magnetic field perpendicular to the direction of wave propagation. The magnitudes of both the electric and magnetic fields of any electromagnetic wave increase and decrease in a regular manner. For the electromagnetic wave in Figure 19-15, the direction of propagation is the y direction, the electric field is in the z direction, and the magnetic field (not shown) is in the x direction. The electric field direction is called the **direction of polarization.** We say that the electromagnetic wave in Figure 19-15 is **linearly polarized** in the z direction.

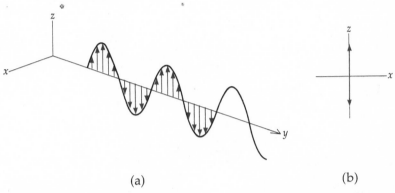

(a)   (b)

Figure 19-15   (a) Schematic representation of a linearly polarized electromagnetic wave propagating in the y direction, with the electric field (represented by arrows) oscillating in the z direction, perpendicular to the direction of propagation. There is also a magnetic field (not shown) oscillating in the x direction. The electric field, viewed with the y axis coming directly toward you, is shown in (b).

Other electromagnetic waves, with different directions of polarization, can propagate in the y direction. We can think of these waves as being obtained from the wave pictured in Figure 19-15 by rotating that wave by varying amounts about the y axis. Thus, the electric field of an electromagnetic wave propagating in the y direction can be in the z direction, or in the x direction, or in any direction in the xz plane (see Figure 19-16). Light coming from the sun, an electric light bulb, or any usual light source, is a combination of a very large number of electromagnetic waves with the electric field of the waves in all directions in the xz plane; light of this nature is called **unpolarized light.**

Figure 19-16   Representation of unpolarized light propagating in the y direction. The blue arrows indicate the electric fields of the component electromagnetic waves.

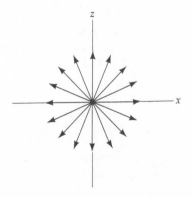

Certain substances transmit only those electromagnetic waves with the electric field in a single direction called the **transmission axis.** One of these substances is the commercial material Polaroid, which is used in sunglasses. When unpolarized light is passed through a piece of Polaroid, only those electromagnetic waves with the electric field in a single direction emerge, and this transmitted light is **linearly polarized light** (see Figure 19-17). If this linearly polarized light emerging from one piece of Polaroid is passed through a second piece of Polaroid, the light emerging from the second piece of Polaroid has a maximum intensity when the two pieces of Polaroid are aligned so that their respective transmission axes are in the same direction (Figure 19-18a); and the light has a minimum intensity when the two pieces of Polaroid are aligned so that their respective transmission axes are perpendicular to one another (Figure 19-18b). In an arrangement such as this, the first piece of Polaroid is called a polarizer, and the second piece of Polaroid is called an analyzer. An instrument with both a polarizer and an analyzer is called a **polarimeter.**

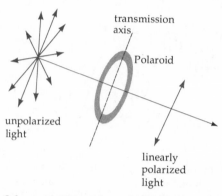

Figure 19-17   Schematic illustration of one way by which linearly polarized light can be generated. Unpolarized light is passed through a piece of Polaroid, which has a unique transmission axis. The emergent light is linearly polarized with its electric field in the same direction as the transmission axis of the Polaroid.

A solution consisting of one enantiomer of a chiral molecule in a nonchiral solvent rotates the direction of linearly polarized light. For example, if an aqueous solution of D-tartaric acid (Figure 19-11) is placed between the polarizer and the analyzer of a polarimeter, the direction of polarization of the linearly polarized light emerging from the polarizer is changed by an angle θ (see Figure 19-19). The angle θ is called the **rotation** of the solution. Compounds that rotate linearly polarized light are said to be **optically active.**

The rotation of the solution is easy to measure, since the light emerging from the analyzer will be at a maximum when the analyzer is rotated by the angle θ. When this is done, the direction of polarization of the light entering the analyzer is parallel to the transmission axis of the analyzer. When the analyzer is rotated to the right by an angle θ, we designate the rotation of the solution by $(+)\theta$, and when the analyzer is rotated to the left by an angle θ, we designate the rotation of the solution by $(-)\theta$.

The magnitude of the rotation of a D-tartaric acid solution (the size of the angle θ) is directly proportional to the number of D-tartaric acid molecules that interact with the light as it passes through the solution. Thus, the higher the concentration of the D-tartaric acid solution and the longer the length of the tube in the polarimeter containing the solution, the larger the rotation.

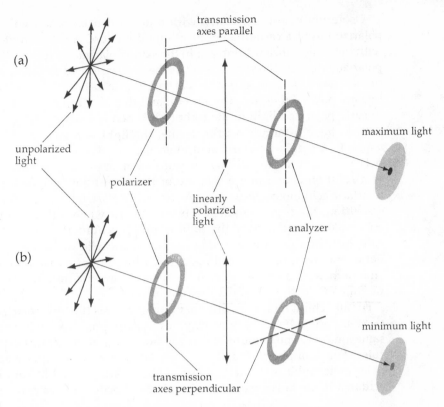

Figure 19-18   (a) When unpolarized light passes through two pieces of Polaroid (a polarizer and an analyzer) that have parallel transmission axes, the intensity of the transmitted light is at a maximum. (b) The intensity of the transmitted light is at a minimum when the transmission axes of the polarizer and the analyzer are perpendicular.

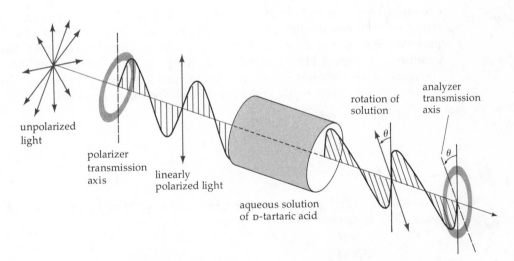

Figure 19-19   When linearly polarized light is passed through a solution of D-tartaric acid, the direction of polarization is rotated by an angle θ to the left. Thus, the light emerging from the analyzer has a maximum intensity when the analyzer is also rotated θ to the left, so that the transmission axis of the analyzer and the direction of polarization of the light emerging from the D-tartaric acid solution are parallel.

A solution of D-tartaric acid with a given concentration will rotate linearly polarized light a certain amount in one direction. A solution of L-tartaric acid with the same concentration put in the same polarimeter tube will rotate linearly polarized light by the *same amount* as the D-tartaric acid solution, *but in the opposite direction*. It is an experimental fact that a D-tartaric acid solution rotates linearly polarized light to the left (−), and that an L-tartaric acid solution rotates linearly polarized light to the right (+). When linearly polarized light interacts with a single D-tartaric acid molecule, the light is rotated to the left by a tiny amount. A single L-tartaric acid molecule, on the other hand, rotates linearly polarized light to the right by the same small amount. The measurable rotation of a solution containing a mixture of the *same* number of D-tartaric acid and L-tartaric acid molecules is zero. As linearly polarized light passes through such a solution, the light is rotated a bit to the left when it encounters a D-tartaric acid molecule and a bit to the right when it encounters an L-tartaric acid molecule, but since the number of D- and L-tartaric acid molecules is the same, the net overall rotation is zero. In general, a mixture of equal amounts of the two enantiomers of a chiral molecule is called a **racemic mixture,** and solutions of racemic mixtures have a net rotation of zero.

For any chiral molecule, one enantiomer rotates linearly polarized light to the right and the other enantiomer rotates linearly polarized light to the left by the same amount. The amount of this rotation is different for different chiral molecules. For example, the rotation of a 1 M solution of D-tartaric acid in a 10-cm polarimeter tube is (−)1.80°, whereas the rotation of a 1 M solution of D-lactic acid in a 10-cm polarimeter tube is (−)0.34°. Lactic acid is a chiral molecule with

$$\overset{\displaystyle H \quad O}{\underset{\displaystyle OH}{the\ formula\ CH_3-\overset{|}{\underset{|}{C}}-\overset{\|}{C}-OH.}}$$

19-7   WHICH ENANTIOMER IS WHICH?

Chemists have adopted certain labeling conventions, so that when they communicate with one another, they can refer to a specific enantiomer of a chiral molecule. For example, chemists have agreed to use the labels D- and L- for the enantiomers of glyceraldehyde indicated in Figure 19-1. Thus, if you know the conventions that chemists use, you know that the name "D-glyceraldehyde" refers to the glyceraldehyde enantiomer with the shape represented by

$$\begin{array}{c} CHO \\ | \\ H-\!\!\!-\!\!\!|-\!\!\!-\!\!\!-OH \\ | \\ CH_2OH \end{array}$$

in which the —OH group is on the right. Likewise, if you want to refer to the enantiomer of tartaric acid that has the following shape:

$$\begin{array}{c} COOH \\ | \\ H-\!\!\!-\!\!\!|-\!\!\!-\!\!\!-OH \\ | \\ HO-\!\!\!-\!\!\!|-\!\!\!-\!\!\!-H \\ | \\ COOH \end{array}$$

with the bottom —OH group on the left, you would use the label "L-tartaric acid."

The D- and L-convention is one way chemists designate enantiomers. In Chapter 20, we shall discuss the D- and L-labeling conventions for the enantiomers of glucose and other sugars.

In the early twentieth century, chemists were aware that there were two enantiomeric forms of glyceraldehyde, that one enantiomer rotated linearly polarized light to the right (+), and that the other enantiomer rotated linearly polarized light to the left (−), but they did not know which enantiomer was (+) and which was (−). The (+) enantiomer was arbitrarily assigned the shape D-glyceraldehyde, even though based on the information known at the time, the (+) enantiomer was equally likely to have the shape of L-glyceraldehyde. In 1949, however, experiments were performed which showed that this arbitrary choice was in fact correct. The (+) enantiomer of glyceraldehyde *does* have the shape shown above for D-glyceraldehyde.

It is important that you realize that it is a matter of *experimental fact* that the enantiomer with this shape rotates linearly polarized light to the right (+), whereas it is a *convention* to designate this enantiomer D-glyceraldehyde. At the present time, chemists have determined experimentally which enantiomer is (+) and which is (−) for many other chiral molecules.

Whether a given enantiomer of a chiral molecule is (+) or (−) depends on the specific atoms in the molecule and how they are bonded together. There is no general way of predicting which enantiomer of the chiral molecule is (+) and which is (−) by looking, for example, at the Fischer projection for that enantiomer. Thus, unless you have memorized the appropriate experimental fact, you cannot look at the Fischer projection

$$
\begin{array}{c}
\text{COOH} \\
\text{H} \!-\!\!|\!-\! \text{OH} \\
\text{HO} \!-\!\!|\!-\! \text{H} \\
\text{COOH}
\end{array}
$$

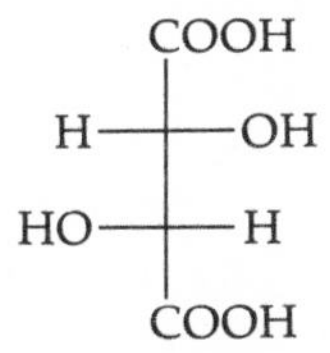

for an enantiomer of tartaric acid and call it (+)-tartaric acid. More important, even if you know this experimental fact, were you to refer to (+)-tartaric acid in your conversation with someone else, they would not know which enantiomer of tartaric acid you were talking about unless they also knew the same experimental fact. On the other hand, if you and another person both use the same labeling conventions, you could (1) look at the Fischer projection above and (2) notice that the bottom —OH group is on the left, and therefore call it L-tartaric acid; (3) refer to L-tartaric acid in conversation; and (4) expect the other person to know which enantiomer of tartaric acid you were talking about.

There is no relationship between the labels D- and L- assigned on the basis of certain conventions and the direction in which an enantiomer rotates linearly polarized light, (+) or (−). For example, we previously stated the experimental fact that L-tartaric acid is (+). On the other hand, D-glyceraldehyde is also (+).

Exercise 19-5

The Fischer projection for the amino acid L-alanine is

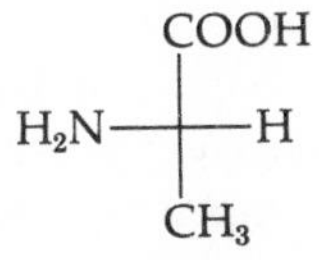

$$
\begin{array}{c}
\text{COOH} \\
\text{H}_2\text{N} \!-\!\!|\!-\! \text{H} \\
\text{CH}_3
\end{array}
$$

(a) Can you tell from the information given whether L-alanine is (+) or (−)?

(b) If you cannot tell whether L-alanine is (+) or (−) on the basis of the given information, what information is needed and how can it be obtained?

19-8   SUMMARY

1. Isomers are different compounds that have the same molecular formula. Structural isomers have different structural formulas. Stereoisomers have the same structural formula but differ in the way the bonds in the molecule are oriented in space.

2. Enantiomers are stereoisomers that are mirror images of one another, and diastereomers are stereoisomers that are not mirror images of one another. Geometric, or cis-trans stereoisomers, are one type of diastereomers.

3. Whether an object and its mirror image have the same shape or different shapes can be tested by a mental procedure called superposition. If an object and its mirror image have the same shape, then they are superimposable; if the object and its mirror image have different shapes, they are nonsuperimposable.

4. Molecules that are nonsuperimposable on their mirror images are called chiral molecules.

5. Enantiomers can exist only for those compounds whose molecules are chiral.

6. A carbon atom that is bonded to four different groups of atoms is called an asymmetric carbon.

7. Generally, but not always, carbon-containing compounds are chiral if they contain an asymmetric carbon, and molecules that do not contain an asymmetric carbon are not chiral.

8. Fischer projections are two-dimensional representations for the shapes of carbon-containing molecules that possess one or more asymmetric carbons.

9. For most, but not all, compounds with n asymmetric carbons, the total number of stereoisomers is 2^n.

10. In general, a single enantiomer of a chiral molecule will react differently with the two enantiomers of another chiral molecule, and a molecule that is not chiral will react identically with the two enantiomers of a chiral molecule.

11. An electromagnetic wave that is propagating in the y direction and that has the electric field in the z direction is said to be linearly polarized in the z direction.

12. Light coming from usual light sources is unpolarized.

13. Polaroid and certain other materials can convert unpolarized light to linearly polarized light.

14. A solution containing a single enantiomer of a chiral molecule can rotate the direction of linearly polarized light.

15. The physical properties of the two enantiomers of a chiral molecule are identical except for the way that they rotate linearly polarized light. For any chiral molecule, one enantiomer rotates linearly polarized light to the right and is known as the $(+)$ enantiomer, whereas the other enantiomer rotates linearly polarized light to the left and is known as the $(-)$ enantiomer.

16. Which enantiomer of a chiral molecule is $(+)$ and which enantiomer is $(-)$ is determined experimentally.

PROBLEMS

1. Describe the differences between structural isomers and stereoisomers. Draw structural formulas for pairs of molecules that are examples of each type of isomer.

2. Two substances, A and B, have the same molecular formula and the same structural formula but have slightly different melting points, boiling points, and solubilities in water. A and B are which of the following:
   (a) Structural isomers (b) Diastereomers
   (c) Enantiomers (d) Not isomers

3. Which of the following Fischer projections represent diastereomers? Which represent enantiomers?

(I) (II) (III) (IV)

Problems 4 and 5 refer to the following Fischer projections:

(I) (II) (III) (IV)

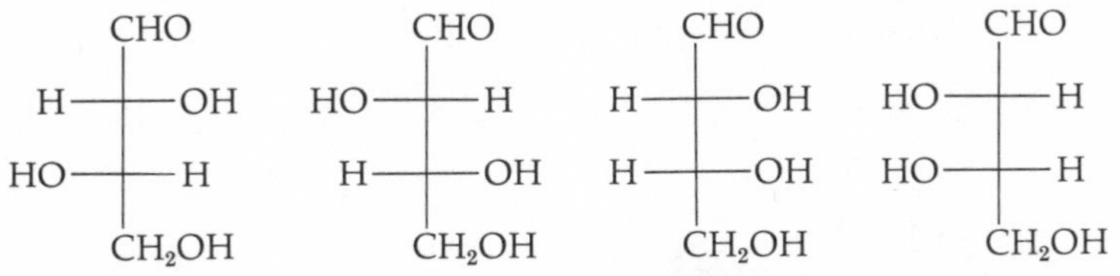

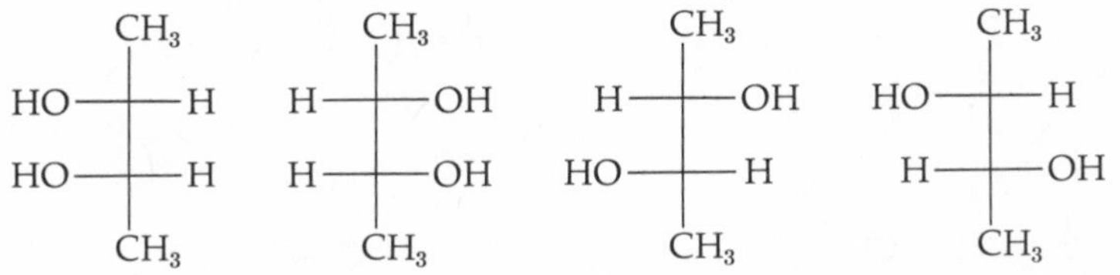

4. Which of these Fischer projections represent diastereomers? Which represent enantiomers?

5. Which of the following solutions would rotate linearly polarized light?
   (a) An equal mixture of III and IV (b) An equal mixture of I and II
   (c) Only I (d) Only III
   (e) None of these solutions

6. If substances C and D are enantiomers, which of the following statements are true?
   (a) The mirror image of C is identical with D.
   (b) C and D rotate linearly polarized light the same amount but in opposite directions.
   (c) If X is *any* substance that reacts with C, then X will react with D to a different degree.
   (d) C and D have slightly different melting points, boiling points, and solubility in water.

7. Draw structural formulas for each of the following and indicate with an asterisk (*) any asymmetric carbon atoms. Draw Fischer projections for all the stereoisomers (if any) of each, and indicate which are enantiomers and which are diastereomers.
   (a) α-Hydroxypropionamide (b) 2,2-Dichloro-3-methylbutane
   (c) 2,3-Hexanediol (d) α-Aminobutyric acid

8. There are four dimethylcyclopropane isomers (structural and stereoisomers). (a) Draw structural formulas for these four isomers. (b) Indicate which are structural isomers and which are stereoisomers. (c) For the stereoisomers, indicate which are enantiomers and which are diastereomers.

9. Describe the difference between linearly polarized light and unpolarized light.

10. Suppose that you have two bottles, one of which contains D-glyceraldehyde and the other contains L-glyceraldehyde. What experimental test could you perform and what information about the isomers could you utilize that would enable you to determine which isomer is in which bottle?

SOLUTIONS TO EXERCISES

19-1   (a)   Chiral,

$$\underset{CH_3}{\overset{COOH}{H\diagup\!\!\!\!\diagup NH_2}}$$

(b) Nonchiral. (Note that although the molecule and its mirror image may at first seem different, when you carry out the mental superposition procedure, the mirror image can be turned so that it is superimposable. Thus, the mirror image is a representation of the same molecule.)

(c)   Chiral,

$$\underset{CH_3}{\overset{COOH}{HO\diagup\!\!\!\!\diagup H}}$$

(d)   Both *cis*-1,2-dichloroethene and *trans*-1,2-dichloroethene are nonchiral.

19-2   (a) $HO-CH_2-\overset{\overset{\displaystyle CH_3}{|}}{\underset{\underset{\displaystyle H}{|}}{\overset{*}{C}}}-CH_2-CH_3$ (b) $CH_3-\overset{\overset{\displaystyle H}{|}}{\underset{\underset{\displaystyle Cl}{|}}{\overset{*}{C}}}-\overset{\overset{\displaystyle H}{|}}{\underset{\underset{\displaystyle Cl}{|}}{\overset{*}{C}}}-CH_3$

(c) $CH_3-\overset{\overset{\displaystyle OH}{|}}{\underset{\underset{\displaystyle CH_3}{|}}{C}}-CH_2-CH_3$ (d) $H-\overset{\overset{\displaystyle Br}{|}}{\underset{\underset{\displaystyle H}{|}}{C}}-\overset{\overset{\displaystyle Br}{|}}{\underset{\underset{\displaystyle H}{|}}{\overset{*}{C}}}-CH_3$

(no asymmetric carbon)

19-3   (a)

$$CH_3-\!\!\!\underset{\underset{\displaystyle CH_2CH_3}{|}}{\overset{\overset{\displaystyle CH_2OH}{|}}{\rule{0pt}{1pt}}}\!\!\!-H \qquad H-\!\!\!\underset{\underset{\displaystyle CH_2CH_3}{|}}{\overset{\overset{\displaystyle CH_2OH}{|}}{\rule{0pt}{1pt}}}\!\!\!-CH_3$$

enantiomer pair

(b)

$$\begin{array}{ccc}
\overset{\displaystyle CH_3}{|} & \overset{\displaystyle CH_3}{|} & \overset{\displaystyle CH_3}{|} \\
Cl-\!\!\!-H & H-\!\!\!-Cl & Cl-\!\!\!-H \\
H-\!\!\!-Cl & Cl-\!\!\!-H & Cl-\!\!\!-H \\
\underset{\displaystyle CH_3}{|} & \underset{\displaystyle CH_3}{|} & \underset{\displaystyle CH_3}{|} \\
(1) & (2) & (3)
\end{array}$$

(1) and (2) are an enantiomer pair
(1) and (3) are diastereomers, as are (2) and (3)
(3) is a *meso*-type compound

(d)

$$Br-\!\!\!\underset{\underset{\displaystyle CH_3}{|}}{\overset{\overset{\displaystyle CH_2Br}{|}}{\rule{0pt}{1pt}}}\!\!\!-H \qquad H-\!\!\!\underset{\underset{\displaystyle CH_3}{|}}{\overset{\overset{\displaystyle CH_2Br}{|}}{\rule{0pt}{1pt}}}\!\!\!-Br$$

enantiomer pair

19-4   (a) This is a chiral molecule, so it should react differently with the two enantiomers of 2-butanol.
(b) This is not a chiral molecule, so it should react identically with both enantiomers of 2-butanol.

19-5   (a) No. There is no relationship between the labels D- and L- and the direction in which an enantiomer rotates linearly polarized light.
(b) To determine whether L-alanine is (+) or (−), you can measure the rotation of an L-alanine solution with a polarimeter.

BIOLOGICAL MOLECULES

Overview

You have seen how the basic principles of chemistry can be applied to the study of simple molecules. Now you are ready to consider the much more complicated **biomolecules** found in living cells. In order to understand the function of biomolecules in healthy human cells, you must recognize that they are composed of simple parts whose properties you have already studied.

We can divide biomolecules into four categories: (1) carbohydrates, (2) amino acids and proteins, (3) nucleic acids, and (4) lipids. The structures and functions of the first three types of biomolecules will be presented in this part of the book. We shall begin with a study of carbohydrates, proceeding from small carbohydrates such as glucose to some very large carbohydrates such as starch and cellulose. We shall see that both starch and cellulose are composed of long chains of glucose building blocks bonded together in a particular arrangement.

Although proteins and nucleic acids are large and complex, they are also composed of smaller building blocks. We shall study these building blocks and how they are assembled in much the same way as you might attempt to understand how a radio operates by first learning about its component parts. Then we can consider the function of a protein molecule, keeping in mind that the function of any molecule, no matter how big, depends on its structure.

The relationship between the structure and the function of biomolecules will be heavily emphasized in our discussion. There is no magic involved—their function simply depends on the nature and orientation of their functional groups. We have already studied virtually all of the functional groups we shall encounter, as well as the types of chemical reactions that these functional groups can undergo.

One major property of all biomolecules is that they are found in living cells. Cells are very complicated, but we believe that an entire cell is no more than the sum of its parts. **Biochemistry** is the study of the component parts of cells and the interactions that take place between them. It is not enough to understand the structure of a protein if you do not understand its role in the cell. Yet there are thousands of proteins in each cell, and many other kinds of molecules as well. How can we possibly keep track of all of these molecules and what they are doing? To help out, we shall make extensive use of an analogy between a living cell and a factory. Factories have walls and cells have membranes. Factories have skilled workers and cells have enzymes. Factories have assembly lines and cells have metabolic pathways (a series of enzyme-catalyzed reactions leading to the synthesis or breakdown of a particular molecule).

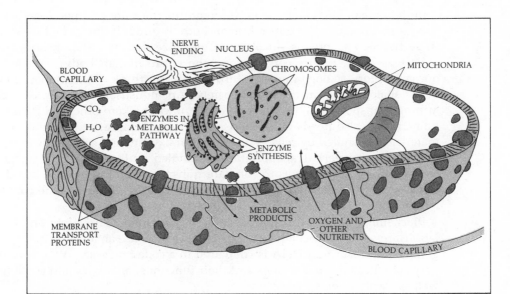

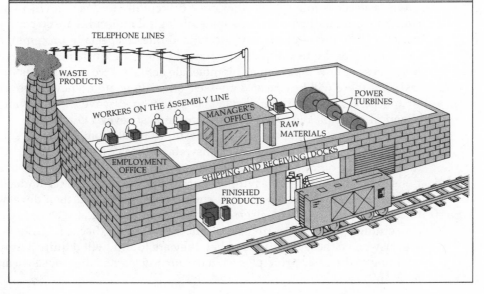

In the human body, cells with different functions are grouped together into tissues and organs. The entire human body is a highly integrated network of these tissues and organs. In order to visualize the interrelationships among molecules, cells, tissues, organs, and the entire human body, we can carry our cell-factory analogy one step further. We can think of the human body as being analogous to an industrial society. An industrial society must include mining, transportation, production, communication, and other industries. The human body also needs these kinds of services. Mining, or the intake of raw materials, is handled by cells in the lungs (for oxygen) and cells in the intestines (for food and water). Transportation is handled by the circulatory system. In fact, we might go so far as to compare transport proteins and red blood cells to trucks. Communication is handled by nerve cells (the telephone company) and hormones (Western Union).

Biochemists use several other analogies to describe the functions and interrelationships of biomolecules. In Chapter 24 we shall use an analogy in which the genetic information in DNA is compared to a coded message. When we study very complicated biomolecules and their functions, analogies add some fun as well as clarity to our discussions.

One aspect of biomolecules that must be kept in mind is that we do not know everything about them. Because some biological molecules are so complex; and because there are a lot of very specialized ones (especially proteins); and because many of the techniques for studying them are quite new—often we can only guess at how a particular biomolecule actually accomplishes its functional task. But these will be *educated* guesses, since there are no new rules, no magic in biochemistry. No biomolecule has yet been discovered that violates the chemical principles we have studied in the past several chapters.

Not having all the answers to our questions about biomolecules is sometimes hard to live with, especially when answers to these questions could be helpful to the practice of modern medicine. A great many of the answers to past questions have resulted in the saving of lives. For example, a better understanding of how enzymes work has allowed chemists to design and then develop drugs to combat certain specific diseases. Eventually, we shall also answer questions concerning the causes of diabetes, cancer, and other diseases. Understanding the answers to these questions, when they are found, will require a firm understanding of the basic principles we have already learned and those we are about to study.

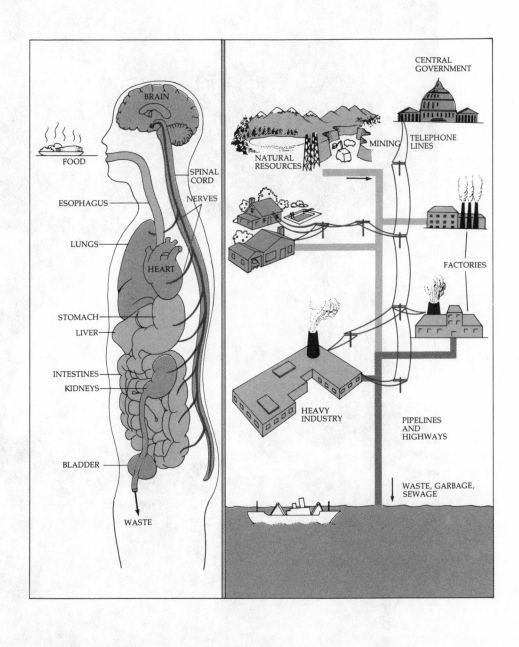

CHAPTER 20

Carbohydrates

20-1 INTRODUCTION

Compounds that contain an aldehyde or ketone functional group and two or more alcohol functional groups (i.e., polyhydroxy aldehydes or ketones), as well as compounds that can be hydrolyzed to give polyhydroxy aldehydes or ketones, are called carbohydrates. Carbohydrates are the most abundant organic compounds in nature, and they have been adapted for a wide variety of uses. Table sugar, the starch in potatoes and other vegetables, and the cellulose in wood are all carbohydrates.

Starch and cellulose are both large polymers that are assembled by plants from the same building block, D-glucose, a simple carbohydrate with the molecular formula $C_6H_{12}O_6$. The primary difference between starch and cellulose, as we shall see in this chapter, is simply a difference in the orientation of the bonds that join together the component D-glucose units.

In cellulose the D-glucose units form linear chains that are held together by hydrogen bonds to form long bundles. These bundles of cellulose molecules provide the structural form for all green plants, from blades of grass to huge trees. The availability and structural strength of trees make wood an important building material. In addition, wood pulp is the primary raw material in the manufacture of paper products. Cellulose-containing parts of other plants also have widespread applications; for example, cotton, which is about 90% cellulose, is a cornerstone of the textile industry. Synthetically prepared cellulose derivatives are also components of a large variety of commercial products. For example, nitrocellulose is used to make some lacquers and explosives, and cellulose acetate is used to make photographic films and phonograph records.

In starch the component D-glucose units are joined together differently than in cellulose. As a result, starch molecules have a spiral structure, rather than the linear structure of cellulose. Starch made by plants for energy storage is, in turn, used by humans and many other organisms for food. Most organisms possess enzymes that catalyze the breakdown of starch into its component glucose units, whereas relatively few organisms possess enzymes that catalyze the breakdown of cellulose.

For centuries, people throughout the world have used cotton to make clothing. Even today, despite the availability of a wide variety of synthetic fabrics, cotton remains a major raw material for the textile industry.

Like cellulose, starch is insoluble in water, but in boiling water it does form a colloidal dispersion that gels upon cooling. This property of starch is used in the preparation of jellies. Likewise, cornstarch is frequently used to thicken gravy, and clothing is occasionally starched to make it more resistant to wrinkling. Interestingly, bread generally becomes hard, not because it has lost water, but because a large number of starch chains have associated by forming additional hydrogen bonds. Hard, stale bread can be softened by warming it because the increased temperature breaks some of the hydrogen bonds. Starch is also used in the commercial manufacture of pastes and adhesives, as well as in the formation of drug tablets.

The human body produces several important carbohydrates. Glycogen, a polymer very similar to starch, is one of the compounds used for energy storage. An individual's blood type is determined by a specific carbohydrate on the surface of his or her red blood cells. An important component of the genetic material DNA is also a carbohydrate. In later chapters, especially Chapter 25, we shall study the specific roles of carbohydrates in the human body. In this chapter we shall explore the structure, nomenclature, and chemical reactions of carbohydrates in general, and study in detail those carbohydrates that are most important for humans.

20-2 STUDY OBJECTIVES

After studying the material in this chapter, you should be able to:

1. Define the terms carbohydrate, monosaccharide, disaccharide, polysaccharide, aldose and ketose, anomeric carbon, and glycosidic linkage.

2. Identify a monosaccharide as an aldose or ketose, and classify it to indicate the number of carbon atoms, given its structural formula.

3. Draw a Haworth projection for the hemiacetal or hemiketal form of a monosaccharide, given the Fischer projection for the free aldehyde or ketone form.

4. Identify the anomeric carbon atom in the hemiacetal or hemiketal form of a monosaccharide, tell if it is D or L, and specify whether the sugar is α or β, given its Haworth projection.

5. Draw Haworth projections for the monosaccharides glucose, galactose, ribose, and fructose.

6. Draw structural formulas for the products of reactions in which monosaccharides form internal hemiacetals or hemiketals, acetals, disaccharides, and esters, and for reactions in which acetals, disaccharides, and esters are hydrolyzed.

7. Classify the glycosidic bond in a disaccharide as α or β, given its Haworth projection.

8. Draw Haworth projections for the disaccharides maltose, lactose, and sucrose.

9. Determine if a disaccharide is a reducing sugar, given its Haworth projection.

10. Describe the structural differences between the polysaccharides starch (amylose and amylopectin), glycogen, and cellulose, and know which the human body is capable of digesting.

20-3 STRUCTURE AND CLASSIFICATION OF CARBOHYDRATES

We have defined **carbohydrates** as polyhydroxy aldehydes or ketones, or compounds that can be hydrolyzed to produce polyhydroxy aldehydes or ketones. Many carbohydrates have the general formula $C_n(H_2O)_n$, hence the name "carbo(n)-hydrate." Carbohydrates are also called sugars or saccharides. Many of the carbohydrates found in nature are very large molecules called polysaccharides. Starch and cellulose are examples of polysaccharides. These large polysaccharides are polymers. In Chapter 14 we saw that polymers are large molecules composed of simpler units called monomers. The simple units that form polysaccharides are called monosaccharides (see Figure 20-1). **Monosaccharides are carbohydrates that cannot be hydrolyzed into simpler compounds.** Polysaccharides can be hydrolyzed by acids or specific enzymes into monosaccharide units. For example, acid-catalyzed hydrolysis of starch or cellulose yields the monosaccharide glucose.

$$\text{polysaccharides} + H_2O \xrightarrow[\text{hydrolysis}]{\text{catalyst}} \text{monosaccharides}$$

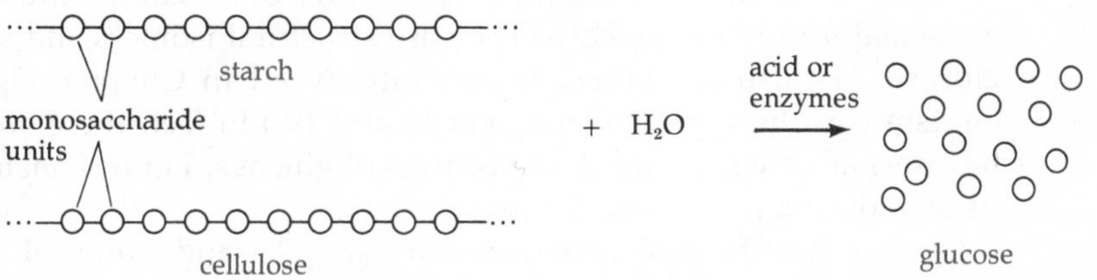

Figure 20-1   Polysaccharides can be hydrolyzed into their component monosaccharide units in reactions catalyzed by acids or specific enzymes.

The hydrolysis of either starch or cellulose yields the monosaccharide glucose.

Starch and cellulose are both polymers of glucose; however, the human body can digest starch but not cellulose. What is the basic difference between these two glucose polymers? In order to answer this question, and to explain other properties of polysaccharides, we must first study the structures and properties of monosaccharides. Glucose, the monosaccharide unit in cellulose and starch, is the most abundant carbohydrate in nature. In addition, glucose, which is also known as dextrose or "blood sugar," is vital to the human body. We shall therefore frequently use glucose as our example when discussing the chemistry of monosaccharides.

Classification of Monosaccharides

The general formula for many simple monosaccharides is $C_n(H_2O)_n$, where n is 3 or larger. For example, $n = 6$ for glucose. The molecular formula for glucose is thus $C_6H_{12}O_6$. The ending -*ose* is often used in the names of carbohydrates. Monosaccharides are classified in two ways: first, as aldehydes or ketones; and second, by the number of carbon atoms (n). A monosaccharide that contains the aldehyde functional group is called an **aldose,** whereas a monosaccharide that contains the ketone functional group is called a **ketose.**

Table 20-1 Classification of Monosaccharides
by Number of Carbon Atoms (n)

General Formula	n	Name
$C_3H_6O_3$	3	Triose
$C_4H_8O_4$	4	Tetrose
$C_5H_{10}O_5$	5	Pentose
$C_6H_{12}O_6$	6	Hexose
$C_7H_{14}O_7$	7	Heptose

Glucose is an aldose. Table 20-1 shows the names used to indicate the number of carbon atoms in a monosaccharide. For example, glucose has six carbon atoms and is therefore a hexose. In order to specify both the number of carbon atoms and the nature of the carbonyl-containing functional group, the names in Table 20-1 are combined with the name aldose or ketose. For example, glucose is both an aldose and a hexose; hence, glucose is an **aldohexose.** Other sugars important in the human body are aldopentoses, ketohexoses, aldotrioses, and so on.

Stereoisomers of Monosaccharides

Glucose is not the only aldohexose. Several other compounds have the same molecular formula as glucose. They are structural isomers and stereoisomers of glucose. (Structural isomerism was introduced in Chapter 13, and stereoisomerism was discussed in Chapters 14 and 19.) In fact, there are 31 other aldohexoses, all of which are stereoisomers of glucose. Let us consider glucose and its stereoisomers in more detail.

Glucose has the molecular formula $C_6H_{12}O_6$, and chemical evidence shows that it has one aldehyde group and five hydroxyl groups. The following structural formula is consistent with this experimental data:

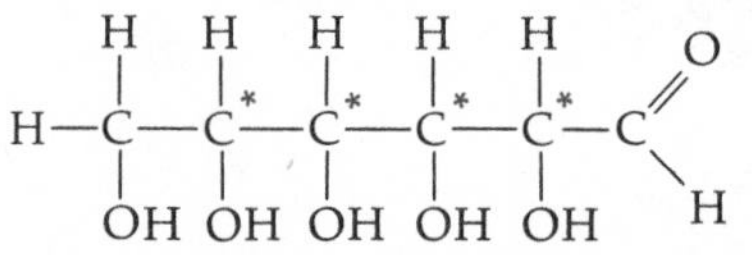

This structural formula has four asymmetric carbon atoms, which are indicated by asterisks. We therefore expect 16 stereoisomers or 8 pairs of enantiomers. Each pair of enantiomers has a different name, and the prefixes D- and L- are used to distinguish between an enantiomer pair (see Chapter 19). The Fischer projections for one of these enantiomer pairs, D-glucose and L-glucose, are shown in Figure 20-2. The D- and L-enantiomers of a monosaccharide are designated as follows: Draw the Fischer projection as a vertical carbon atom chain with the carbonyl group at the top of the projection, and then look at the *lowest* asymmetric carbon atom. If the hydroxyl group is to the right in the projection, the sugar is the D-enantiomer. If it is to the left, it is the L-enantiomer. Cells in the human body can use only sugars in the D-configuration, such as D-glucose. D-Galactose is another D-aldohexose that is important in the human body. Recall that the symbol D- indicates the configuration around only one particular carbon atom of a monosaccharide, not the rotation of polarized light by that compound. If the rotation of light needs to be specified, the symbols (+) and (−) are used to designate rotation to the right and left, respectively. What about D-glucose? D-Glucose happens to be dextrorotatory; it rotates polarized light to the right. Thus, D-(+)-glucose is often called **dextrose.**

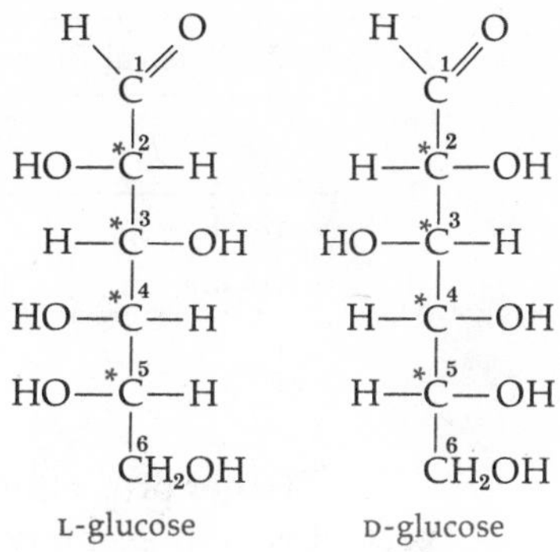

Figure 20-2   Fischer projections for the enantiomers L-glucose and D-glucose. The four asymmetric carbon atoms in each enantiomer are indicated by asterisks.

L-glucose   D-glucose

Formation of Internal Hemiacetals: Haworth Projections

The structure of D-(+)-glucose is not as simple as is indicated by the Fischer projection in Figure 20-2. This representation for aldohexoses shows only four asymmetric carbon atoms, and accounts for 16 stereoisomers. But we have stated that there are 32 stereoisomeric aldohexoses, which would indicate the presence of five asymmetric carbon atoms.

Recall from Chapter 16 that aldehyde and alcohol groups on the same molecule can react to form internal hemiacetals, and that internal hemiacetals with rings of six atoms are favored. D-Glucose can form *two* different hemiacetals, each containing a ring of six atoms, by a reaction of the aldehyde group and the alcohol group involving carbon atom number 5. The same is true for all the other 16 aldohexoses. Thus there are a total of 32 aldohexose stereoisomers. Fischer projections for the two different internal hemiacetals of D-glucose are given in Figure 20-3. Notice that each of these internal hemiacetals of D-glucose has a total of five asymmetric carbon atoms, because the carbon atom from the aldehyde group, carbon 1, is also asymmetric in the hemiacetal form. The additional carbon atom in an internal hemiacetal form is called an **anomeric** carbon atom. These two forms of D-(+)-glucose can be obtained as pure solids with different properties (see Table 20-2) and are called the α- and β-anomers of glucose. In general, **anomers** are diastereomers that differ only in the bonding arrangement of the anomeric carbon atom. The α-D-(+)-glucose anomer is the naturally occurring solid form of glucose. In addition to glucose and its stereoisomers, most other monosaccharides have α- and β-anomeric forms.

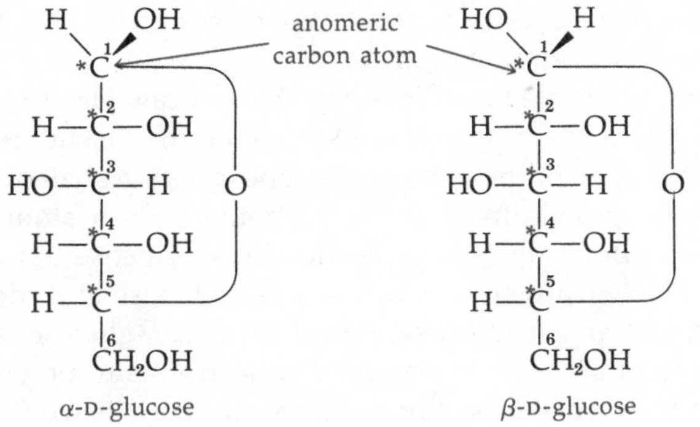

α-D-glucose β-D-glucose

Figure 20-3   Fischer projections for the two internal hemiacetal forms of D-glucose, the α- and β-anomers of D-glucose.

Table 20-2 Physical Properties of α- and β-D-Glucose Anomers

	Melting Point (°C)	Specific Rotation
α-D-Glucose	146	+112°
β-D-Glucose	150	+19°

The Fischer projections for α- and β-D-glucose, shown in Figure 20-3, are not very good representations of these molecules. Structural differences between the α- and β-forms, for example, are not easy to see from Fischer projections. Models would be much better for this purpose.

The structure of α-D-glucose, β-D-glucose, and the other stereoisomers of glucose resemble the structure of cyclohexane. Recall from Chapter 14 that cyclohexane, and other compounds containing rings of six carbon atoms, exist preferentially in a shape called the chair form, which allows all of the bond angles in the ring to be 109.5°. Thus the chair form is the best representation for the structures of the α- and β-forms of glucose (see Figure 20-4).

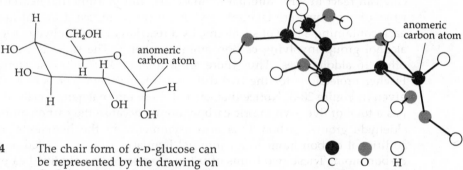

Figure 20-4   The chair form of α-D-glucose can be represented by the drawing on the left or the model on the right.

However, drawing the chair form of a monosaccharide is tedious, so a compromise method, called a **Haworth projection,** was developed in which all of the atoms in the ring are represented as being in one plane. Although Haworth projections are not the best representations of the shapes of monosaccharides, they are superior to Fischer projections for monosaccharides that exist in the internal hemiacetal or hemiketal form.

By applying a few simple rules, it is possible to draw Haworth projections for the α- and β-anomers of a monosaccharide, given the Fischer projection for the open-chain form. For all of the pentoses and hexoses we shall discuss, the hemiacetal or hemiketal group involves the carbonyl carbon atom and the lowest asymmetric carbon in the Fischer projection. Take glucose, for example. Figure 20-5 shows the Fischer projections for D- and L-glucose and the corresponding Haworth projections for α-D-glucose, β-D-glucose, α-L-glucose, and β-L-glucose. Note that the ring of atoms in a Haworth projection is to be visualized as projecting out from the plane of the paper. To emphasize this we usually use darker lines to represent the three bonds in the ring that are visualized as closer to the reader. Hydrogen atoms and hydroxyl groups are visualized as projecting up or down from the carbon atoms in the horizontal ring.

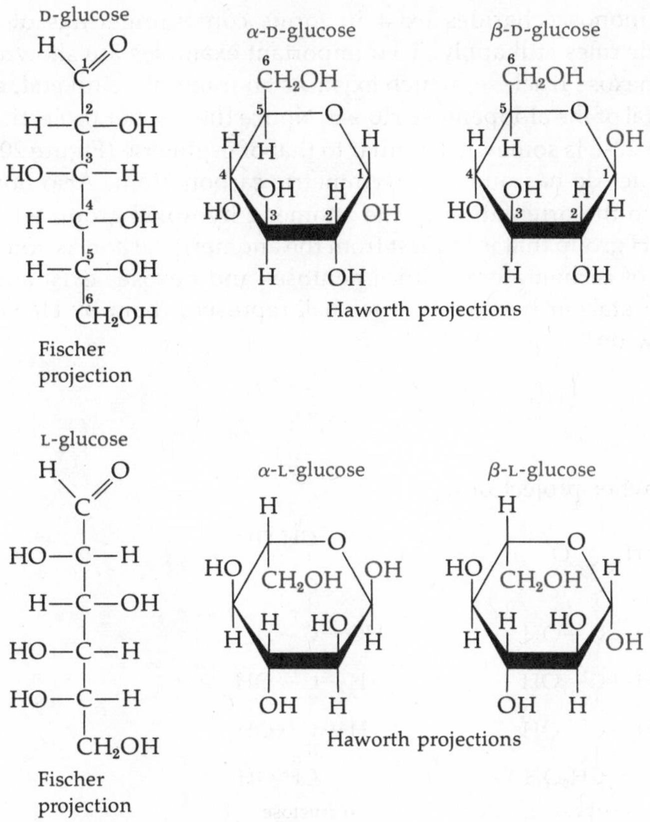

Figure 20-5   Fischer projections for D-glucose and L-glucose and the Haworth projections for their corresponding α- and β-anomeric forms.

The rules for drawing a Haworth projection are as follows:

1. Draw the anomeric carbon to the right and continue to draw the remaining atoms in the ring in a clockwise direction, with the ring oxygen at the top of the drawing.

2. Hydroxyl groups located on the *right* of the chain in the Fischer projection are drawn *down* in the Haworth projection. Hydroxyl groups drawn to the *left* in the Fischer projection are drawn *up* in the Haworth projection.

3. The last —CH₂OH group is drawn *above* the ring (up) in the Haworth projection of a D-sugar and *below* the ring (down) for an L-sugar.

4. The hydroxyl group attached to the anomeric carbon atom (the anomeric hydroxyl group) is drawn on the *same* side of the ring as the last —CH₂OH group for the β-anomer, and the *opposite* side of the ring for the α-anomer. When the α- or β-anomer is not specified, as in the designation D-glucose, the Haworth projection for the anomeric hydroxyl group is written as

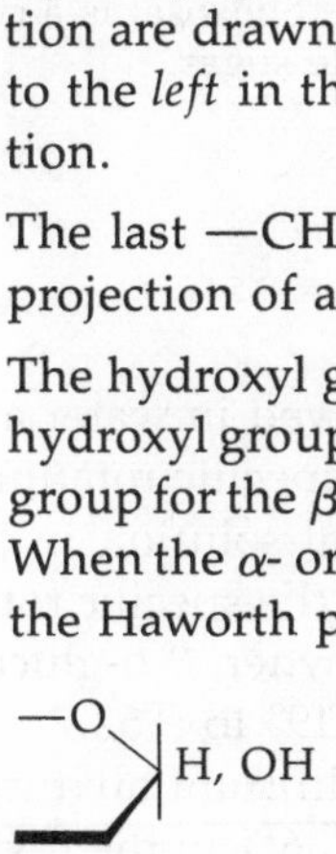

to indicate that the sugar is either α or β.

Some monosaccharides exist in forms containing rings of five atoms, for which our rules still apply. Two important examples are shown in Figure 20-6: the ketohexose fructose, which exists as an internal hemiketal, and the internal hemiacetal of the aldopentose ribose. Notice the Fischer projection for D-fructose in Figure 20-6 is somewhat similar to that of D-glucose (Figure 20-5), but that the ketose fructose has one less asymmetric carbon atom. Also notice that α- and β- refer to the orientation of the anomeric hydroxyl group with respect to the —CH₂OH group that is *farthest* from the anomeric carbon as you proceed around the ring of carbon atoms. Since pentoses and hexoses exist almost exclusively as hemiacetals or hemiketals, we shall represent them by Haworth projections from now on.

Fischer projections:

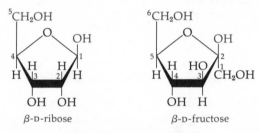

Haworth projections (for the predominant anomers):

Figure 20-6   Fischer and Haworth projections for two important monosaccharides that exist in forms containing rings of five atoms. β-D-Ribose, an aldopentose (left), is a component of DNA and other complex biomolecules, whereas β-D-fructose (right), a ketohexose, is a component of table sugar.

When naturally occurring α-D-glucose is dissolved in water and the rotation of polarized light is measured immediately, the specific rotation is +112°. The specific rotation is the rotation for a 1.0 g/1.0 ml solution in a 1.0-decimeter polarimeter tube. However, over a period of time the specific rotation decreases until it finally reaches +52.7°. On the other hand, when β-D-glucose is dissolved in water, the specific rotation increases from +19° to +52.7°. These data indicate that, in solution, glucose exists as an equilibrium mixture of the α-form and the β-form (see Figure 20-7). At equilibrium, 36% of the D-glucose exists in the α-form, 64% in the β-form, and only 0.02% exists in the free aldehyde or open-chain form.

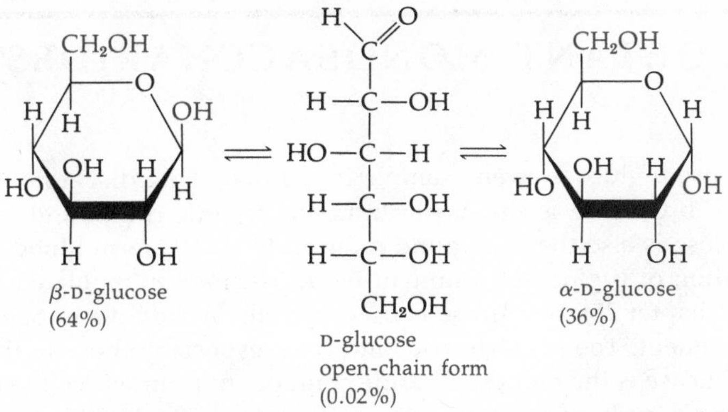

Figure 20-7   In aqueous solution almost all D-glucose molecules exist in the β- or α-anomer forms.

However, these anomers are in equilibrium with a small amount of the open-chain aldehyde form.

Exercise 20-1
Draw Fischer projections for all 8 of the D-aldohexoses in the open-chain form.

Exercise 20-2
The Fischer projection for D-galactose is given below. Draw the Haworth projection for β-D-galactose.

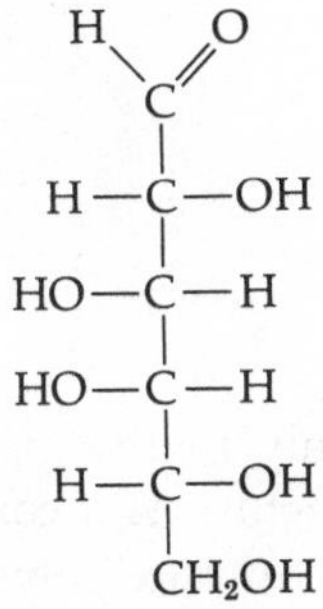

Exercise 20-3
The Haworth projections for three monosaccharides are given below. (a) Name these monosaccharides to indicate whether they are aldehydes or ketones and the number of carbon atoms (e.g., D-glucose is an aldohexose). (b) Indicate which carbon atom is the anomeric carbon atom, and determine if these are D- or L- and α- or β-anomers.

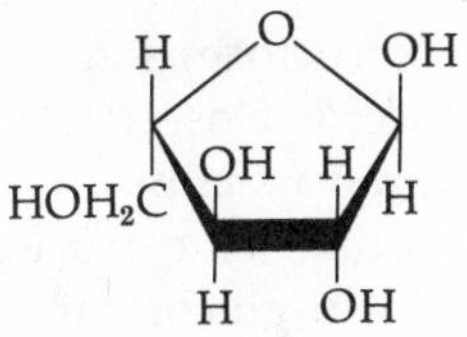

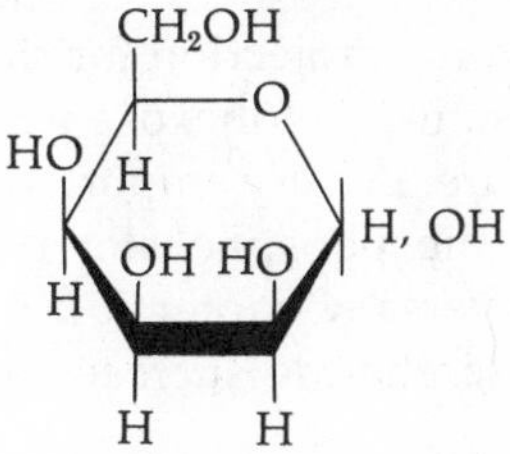

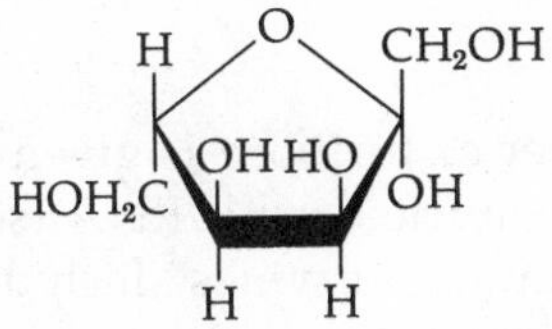

20-4   IMPORTANT MONOSACCHARIDES

D-Glucose

We chose D-glucose as an example in the preceding discussion for a very good reason. D-Glucose is the most abundant organic compound found in nature. D-Glucose is also the compound referred to by the term **blood sugar.** The concentration of D-glucose in human blood is carefully regulated, as we shall see later (Chapter 25). D-Glucose is used by cells in the human body as a source of nourishment. Too little glucose, and cells, especially those in the brain, starve.

D-Glucose is the monosaccharide component from which the polysaccharides cellulose, starch, and glycogen are built (see Section 20-7), and it is a component of several more complex polysaccharides. It is also a component of the important disaccharides sucrose, maltose, and lactose (see Section 20-6).

D-Galactose

The Haworth projection for the β-anomer of D-galactose, a stereoisomer of D-glucose, is

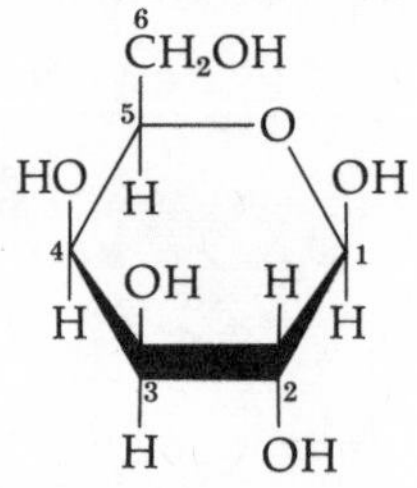

Notice that D-galactose differs from D-glucose only in the orientation of the groups bonded to carbon atom number 4. D-Galactose is a component of the disaccharide lactose found in milk, and of some complex polysaccharides. Ingested D-galactose is normally converted to D-glucose in the human body. The inability to perform this isomerization (conversion of one isomer to another) results in the disease galactosemia (Chapter 25).

D-Fructose

The Haworth projection for the β-anomer of D-fructose is given in Figure 20-6. D-Fructose is a ketohexose which, like galactose, is closely related to D-glucose in structure. A very important enzyme-catalyzed reaction in the human body converts the phosphoester D-glucose 6-phosphate to D-fructose 6-phosphate and vice versa (Chapter 25). β-D-Fructose and α-D-glucose are the components of the disaccharide sucrose (see Section 20-6), which is common table sugar.

D-Ribose

The Haworth projection for the β-anomer of D-ribose is given in Figure 20-6. This aldopentose is a component of ribonucleic acid, RNA (see Chapter 24). β-D-Ribose is also a component of several coenzymes, including NAD^+ (see Figure 18-4) and the compound ATP (Figure 25-4), which supplies energy for reactions in the human body.

The sugar cane plant is one of our major sources of sucrose (common table sugar), a disaccharide composed of the monosaccharide units β-D-fructose and α-D-glucose.

Many other monosaccharides, ranging in size from trioses to heptoses, are also found in nature. We shall encounter some of these in later chapters when we discuss human metabolism.

Exercise 20-4

D-Galactose and D-fructose can be converted to D-glucose by isomerization reactions. Can D-ribose be converted to D-glucose by such a reaction? Explain.

20-5 PROPERTIES OF MONOSACCHARIDES

Physical Properties

Pure monosaccharides are solids at room temperature. Monosaccharides contain alcohol groups and either a carbonyl group or a hemiacetal or hemiketal group. Like alcohols, they readily form hydrogen bonds with water or other molecules containing hydrophilic functional groups. Because a monosaccharide molecule such as glucose contains several alcohol groups, it can hydrogen bond with many water molecules and thus is extremely soluble in pure water and in aqueous solutions such as blood serum.

Reactions of Monosaccharides

In addition to the formation of internal hemiacetals or hemiketals, monosaccharides can participate in other reactions typical of compounds containing both alcohol and carbonyl groups. The most important specific reactions involving monosaccharides are those that occur in the human body. We shall discuss these reactions in detail in Chapter 25.

Let us consider some of the general reactions of monosaccharides, one of which is an oxidation reaction that can be used to determine the amount of a sugar present in a solution. Again, let us use glucose as our example. The α- and β-anomers of glucose in aqueous solution are in equilibrium with the free aldehyde form of glucose (see Figure 20-7). This equilibrium in aqueous solution between a free aldehyde and internal hemiacetal(s) is typical of all aldoses and other compounds that form internal hemiacetals. In solution, an internal hemi-ketal, such as fructose, is also in equilibrium with its free ketone form. Recall from Chapters 15 and 16 that aldehydes can be oxidized to form carboxylic acids, and that alcohols can also be oxidized to form aldehydes or ketones.

Benedict's, Fehling's, and Tollen's reagents are specific oxidizing agents that can oxidize aldoses with a free aldehyde group. Consider glucose. When one of these reagents reacts with the free aldehyde form of glucose, the oxidizing agent is reduced and glucose is converted to a complex mixture of products. Both Benedict's and Fehling's reagents contain Cu^{2+}, which is reduced, forming red Cu_2O. In solution, very little glucose (less than 1%) exists as a free aldehyde. The free aldehyde, however, is in equilibrium with the internal hemiacetal forms. As the free aldehyde form of glucose reacts with one of these reagents, more and more of the cyclic hemiacetal form is converted to the free aldehyde form and oxidized, until finally all of the glucose has been oxidized (see Figure 20-8). By measuring the amount of color change upon reaction with Benedict's or Fehling's reagents, one can determine the amount of aldoses such as glucose in a solution.

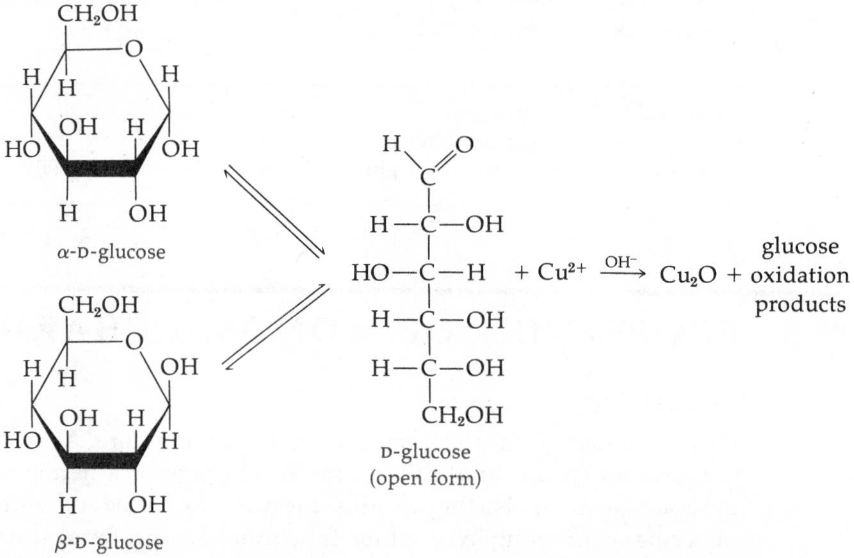

Figure 20-8   As the open-chain aldehyde form of glucose is oxidized, the two equilibria between this form and the α- and β-anomers shift, generating more of the open-chain aldehyde. In this manner all of the glucose in a solution can be oxidized.

Sugars that can reduce Fehling's, Benedict's, or Tollen's reagents are called **reducing sugars.** All aldoses are reducing sugars. Ketoses are also reducing sugars because they have a ketone functional group immediately adjacent to an alcohol functional group. The neighboring ketone group gives these adjacent alcohol groups slightly different properties than most alcohols. In particular, they can be more easily oxidized, and consequently D-fructose and other ketoses will also reduce Benedict's, Fehling's, or Tollen's reagents. We shall see that polysaccharides and some disaccharides are not reducing sugars.

Glucose is the only reducing sugar that is normally present in any appreciable concentration in human blood. Thus, Benedict's or Fehling's reagent can be used to measure "blood sugar." In patients with diabetes, these reagents can also be used to test for glucose in the urine. However, Benedict's and Fehling's reagents are not specific for glucose; diseases in which other reducing sugars accumulate in blood or urine can be misdiagnosed as diabetes when these reagents are used. Because of this possible difficulty, most clinical laboratories now use more specific reagents, usually enzymes, to determine the amount of glucose in a solution.

Another important reaction of carbohydrates is ester formation. Recall from Chapter 17 that alcohols can react with carboxylic acids or phosphoric acid to produce esters. There are many enzymes in the human body that catalyze the formation of phosphoesters from monosaccharides. For example, although glucose is present in fairly high concentration in the bloodstream, almost no free glucose exists within cells. In the cells of the human body, glucose is rapidly converted to the phosphoester called glucose 6-phosphate, where the phosphate group is bonded to carbon atom number 6 of glucose:

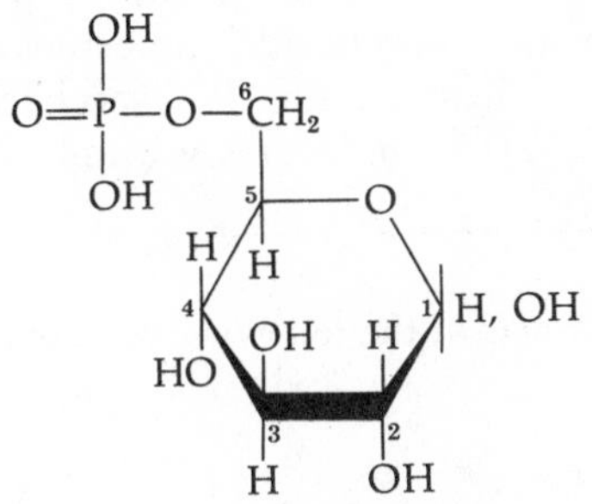

Glucose 6-phosphate is then oxidized in a series of reactions to provide energy for cellular functions. We shall encounter several other phosphoesters of monosaccharides when we study the metabolism of carbohydrates.

In Chapter 16 we saw that hemiacetals and hemiketals can react with alcohols to yield acetals and ketals, respectively. Monosaccharides that exist primarily as internal hemiacetals, such as glucose and ribose, or as internal hemiketals, such as fructose, also form acetals or ketals. When sugars react with alcohols to form acetals or ketals, the bond formed between the anomeric carbon atom of the sugar and the oxygen atom from the alcohol is called a **glycosidic bond,** and the term **glycoside** is used to refer to an acetal derivative of a carbohydrate. Glycosides of glucose are given the more specific name of **glucosides.** For example, under suitable conditions, α-D-glucose can react with methyl alcohol to produce the acetal called methyl-α-D-glucoside:

(20-1)

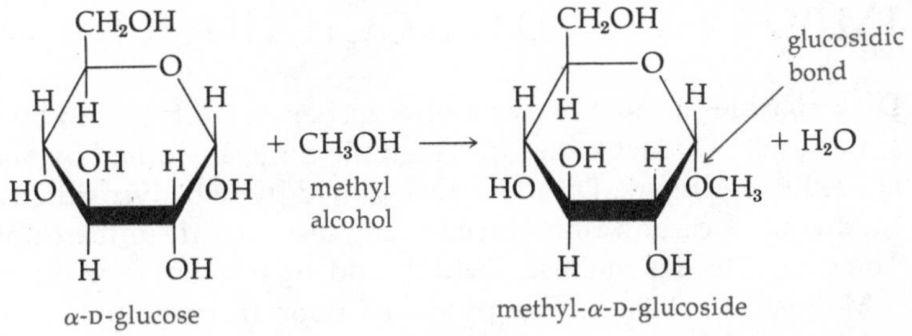

Notice that a glycosidic bond is a bond between the anomeric carbon and an oxygen atom that is not in the ring. Glycosidic linkages may be α or β. When β-D-glucose and methyl alcohol react, methyl-β-D-glucoside is formed.

Recall that acetals are much more stable in solution than are hemiacetals. The glycoside derivatives of monosaccharides, such as methyl-α-D-glucoside, do not exist as equilibrium mixtures of aldehydes and alcohols, and the carbon atoms in glycosidic bonds cannot be oxidized by Benedict's, Fehling's, or Tollen's reagents. However, glycosides can be hydrolyzed and the glycosidic bond broken in reactions catalyzed by an acid or by specific enzymes. An example is the acid-catalyzed hydrolysis of ethyl-β-D-riboside:

(20-2)

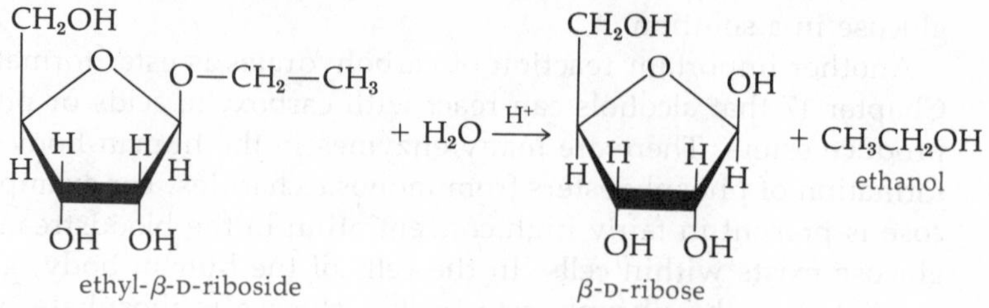

ethyl-β-D-riboside β-D-ribose

Glycosidic bonds can likewise be formed when the anomeric hydroxyl group of one monosaccharide and an alcohol functional group of another monosaccharide react. The joining of two monosaccharide molecules by a glycosidic bond(s) results in the formation of a disaccharide. Polysaccharides are formed by joining three or more monosaccharide molecules with glycosidic bonds.

Exercise 20-5

Draw Haworth projections for the products of the following reactions.

(a)

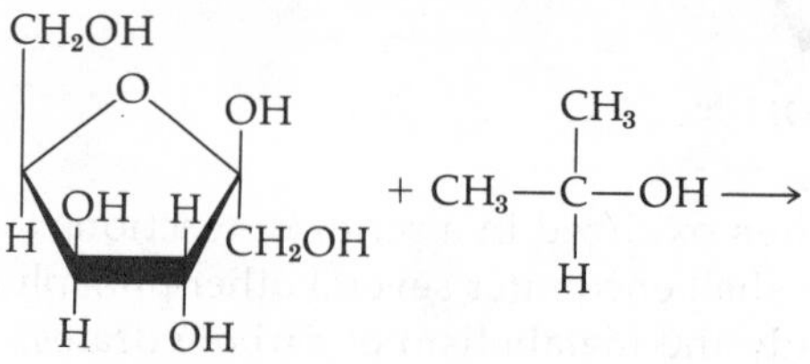

(b)

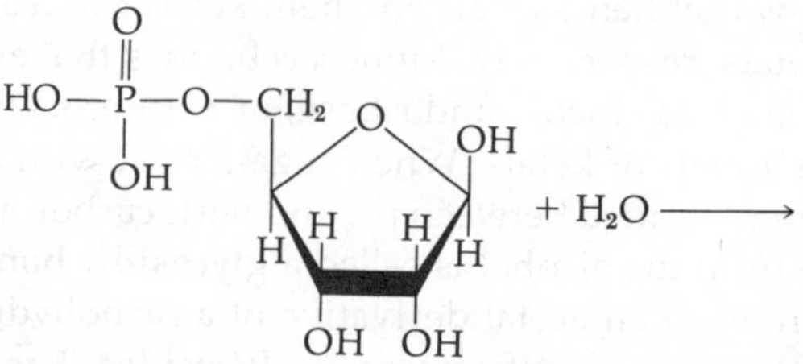

20-6 IMPORTANT DISACCHARIDES

Disaccharides consist of two monosaccharide units joined by one or two glycosidic bonds. Disaccharides in which the units are joined by one glycosidic bond are reducing sugars. Disaccharides joined by two glycosidic bonds that involve the anomeric carbon atoms on each monosaccharide unit are not reducing sugars. The disaccharides maltose, lactose, and sucrose are very important to humans.

Maltose is a disaccharide produced upon incomplete hydrolysis of the polysaccharide starch. It is found in germinating seeds. It is also produced commercially for use in the production of beer. Maltose is composed of two D-glucose units joined by an α-glucosidic bond between the anomeric carbon of one glucose unit and the number 4 carbon of the other glucose unit. This specific bond, termed an **α-1,4-glucosidic bond,** is also found in starch and glycogen:

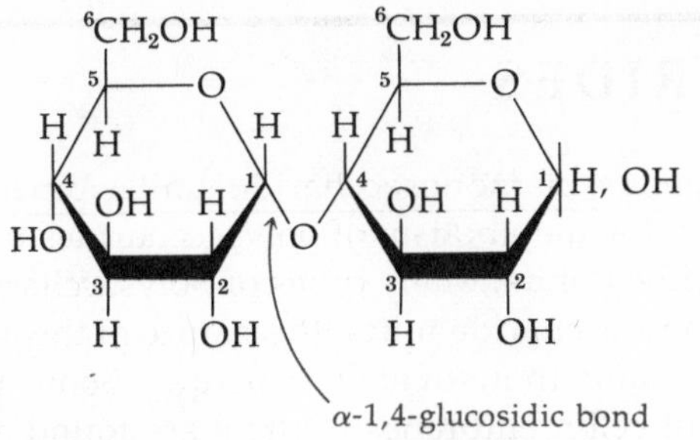

α-1,4-glucosidic bond

Note that the anomeric hydroxyl group of one of the glucose units participates in the glucosidic bond and therefore cannot be easily oxidized. However, the anomeric hydroxyl of the other glucose unit is not so occupied, and this glucose unit exists in equilibrium with the free aldehyde in solution. Thus maltose is oxidized by Benedict's, Fehling's, or Tollen's reagent and is a reducing sugar.

Lactose constitutes some 3% to 5% of the milk of mammals, including cows and humans. This disaccharide is composed of one galactose unit and one glucose unit joined by a glycosidic bond between the β-anomer of galactose and the number 4 carbon of glucose, a β-1,4-glycosidic bond:

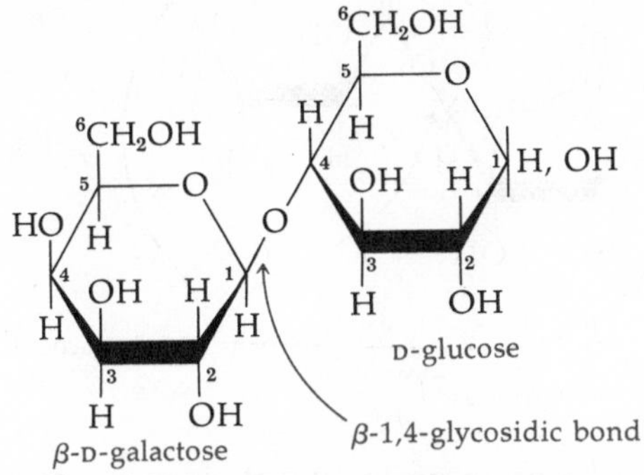

The glucose unit of lactose still exists as an equilibrium mixture of the α- and β-anomers and the free aldehyde in solution. Lactose is thus a reducing sugar.

Sucrose is found in honey, fruits, and vegetables. Sugar cane and sugar beets are the commercial sources for sucrose used as table sugar. Sucrose is composed of one glucose and one fructose unit, joined by two glycosidic bonds involving the anomeric carbons of both α-D-glucose and β-D-fructose:

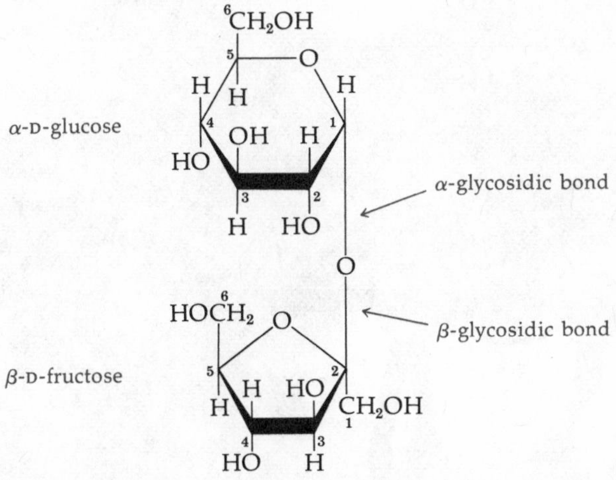

Sucrose is not a reducing sugar, since both anomeric carbons participate in the glycosidic bonds and thus no free aldehyde or ketone exists in solution. The metabolism of lactose and sucrose will be discussed in Chapter 25.

20-7   POLYSACCHARIDES

Polysaccharides are polymers of monosaccharide units joined by glycosidic bonds, and the first step in the metabolism of polysaccharides by humans is the hydrolysis of their glycosidic bonds. Many types of polysaccharides exist in nature, differing in their monosaccharide units, the nature of the glycosidic bonds that join the sugar units, and their overall structure. Some polysaccharides function as components of cell membranes. Others are found attached to proteins. Three of the most important polysaccharides found in nature are cellulose, starch, and glycogen. All three of these polysaccharides are built up from a single monosaccharide component, D-glucose. The difference between them is in the type of glucosidic bonds and in their overall structures.

Cellulose is a very large polymer of glucose units joined by β-1,4-glucosidic bonds in long chains:

$$\beta\text{-1,4-glucosidic bonds}$$

cellulose

One cellulose molecule may contain up to 10,000 glucose units. Because of the very large molecular weight of cellulose molecules, cellulose is insoluble in water. Cellulose is used by plants to form rigid cell walls. As we have already seen, there are many commercial uses for plant cellulose.

The cellulose found in grasses can be digested by herbivorous animals such as this horse, but the men and their dog show no appetite for this vegetation, which they cannot digest.

Humans cannot digest cellulose. We do not have any enzymes capable of hydrolyzing a β-1,4-glucosidic bond between two glucose units. Thus, the cellulose in our diet passes through the digestive tract intact. Note, however, that humans do possess an enzyme capable of hydrolyzing the β-1,4-glycosidic bond between the galactose and glucose units of lactose.

Starch is also a very large polymer of glucose units. However, the bonds that hold the glucose units together are α-1,4- and not β-1,4- as in cellulose. Starch is used by plants for food storage. There are two basic types of starch, amylose and amylopectin (see Figure 20-9). **Amylose** is a linear polymer of α-1,4-linked glucose units, and is water soluble. **Amylopectin** is a branched polymer. The main chains of amylopectin are joined by α-1,4-glucosidic bonds, as in amylose. However, about every 20 to 30 glucose units there are branches joined by α-1,6-glucosidic bonds. Amylopectin is not soluble in water. About 80% to 90% of the starch obtained from plants is amylopectin. These values, 20 to 30 glucose units and 80% to 90% amylopectin, are averages from a wide variety of plants. Different specific starches are made by different plants.

Humans possess the enzyme **amylase,** which catalyzes the hydrolysis of the α-1,4-glucosidic bonds in starch, and another enzyme that can break the α-1,6-bonds. As a consequence of the action of these enzymes, starch is completely hydrolyzed to individual glucose units during digestion. The salivary glands and the pancreas produce amylase, releasing it into the mouth and intestines, respectively. The free glucose is then absorbed by cells in the small intestine and transferred to the bloodstream.

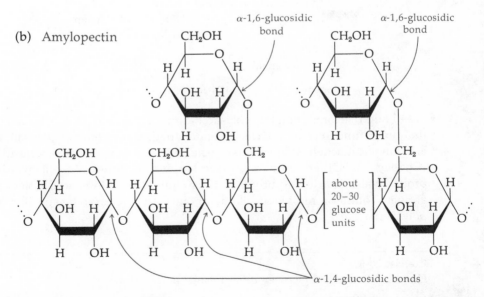

Figure 20-9   Amylose (a) and amylopectin (b) are two different forms of starch. Both have linear chains of glucose units joined by α-1,4-glucosidic bonds. However, amylopectin also has branch chains that are joined to the main chain by α-1,6-glucosidic bonds.

Glycogen is very similar to amylopectin in structure. It is a polymer of glucose units joined by α-1,4-bonds, with α-1,6-bonded branches:

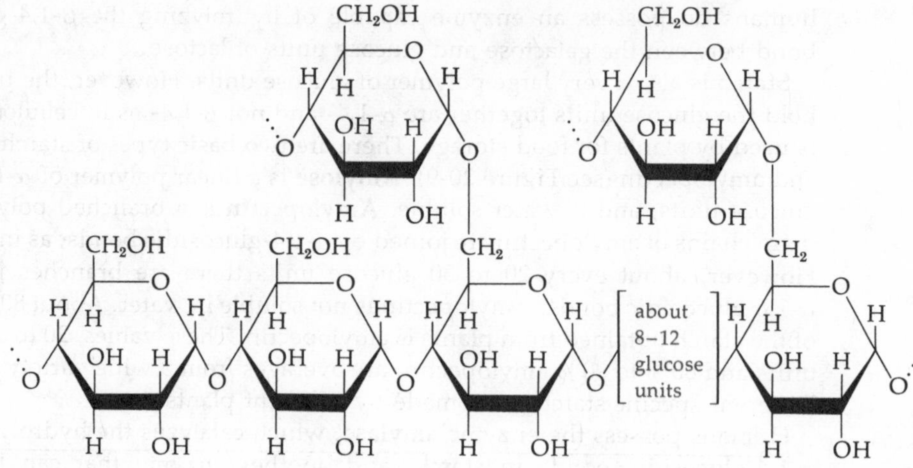

glycogen

However, it is more highly branched than amylopectin, with one branch point about every 8 to 12 glucose units. Glycogen is used for food storage in animal cells. The normal human body has about a one-day supply of stored glycogen. We shall study the metabolism of glycogen in Chapter 25.

Complex polysaccharides are carbohydrate polymers composed of more than one type of monosaccharide unit joined by glycosidic bonds. Some of the individual monosaccharide units that make up complex polysaccharides can also be fairly complex. For example, glucosamine and N-acetylgalactosamine are monosaccharides found in several complex polysaccharides:

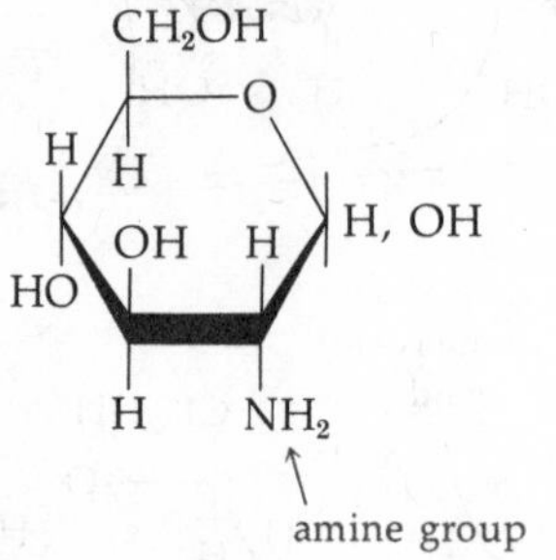

glucosamine 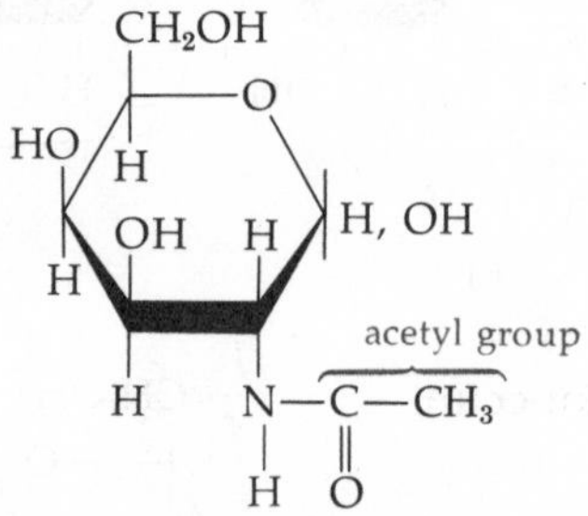   N-acetylgalactosamine

Although several complex polysaccharides are found in the human body, their exact functions are not clearly understood. Some complex polysaccharides are found attached to proteins, including some enzymes and antibodies (Chapter 22). Other complex polysaccharides are found on cell membranes, probably also attached to proteins. For example, the ABO and Rh blood-group antigens found on red blood cell membranes are known to be specific complex polysaccharides.

Exercise 20-6

How do cellulose and glycogen differ in structure?

Exercise 20-7

Would you expect amylopectin to reduce Benedict's reagent?

20-8 SUMMARY

1. Carbohydrates (also called saccharides or sugars) are polyhydroxy aldehydes or ketones or compounds that can be hydrolyzed to give polyhydroxy aldehydes or ketones.

2. Carbohydrates that cannot be hydrolyzed into simpler compounds are called monosaccharides.

3. Aldehyde monosaccharides are called aldoses, whereas ketone monosaccharides are called ketoses.

4. Many monosaccharides have the general formula $C_n(H_2O)_n$, where n is 3 or larger; in general, there are many stereoisomers of a particular monosaccharide.

5. The prefixes D- and L- are used to distinguish between the enantiomers of an enantiomeric pair of monosaccharides. Humans can metabolize only D-sugars.

6. Pentoses and hexoses form cyclic internal hemiacetals or hemiketals whose shapes can be represented by Haworth projections.

7. When a monosaccharide forms an internal hemiacetal or hemiketal, two different diastereomers, α- and β-anomers, can form. The carbonyl carbon used to form the hemiacetal or hemiketal becomes asymmetric and is called the anomeric carbon.

8. A monosaccharide exists in solution as an equilibrium mixture of the α-anomer and the β-anomer, with a small percentage in the open-chain form.

9. The monosaccharide D-glucose is the most abundant organic compound in nature. It is also called dextrose or blood sugar. D-Glucose is the monomer unit (building block) for the polysaccharides cellulose, starch, and glycogen, and is a component of the disaccharides maltose, sucrose, and lactose.

10. D-Galactose, D-fructose, and D-ribose are other monosaccharides vital to living organisms.

11. Monosaccharides are hydrophilic and are very soluble in water.

12. Monosaccharides and disaccharides that can be oxidized by Benedict's, Fehling's, or Tollen's reagent are called reducing sugars.

13. Monosaccharides can form esters with carboxylic or phosphoric acids. Phosphate esters of monosaccharides are very important in the human body.

14. Monosaccharides can react with alcohols to produce acetals or ketals that are called glycosides. Glycosides of glucose are called glucosides. Glycosidic bonds join monosaccharide units together to form disaccharides and polysaccharides.

15. Three important disaccharides are maltose (two α-1,4-bonded glucose units), lactose (a galactose joined to the number 4 carbon of glucose by a β-1,4-glycosidic bond), and sucrose (α-glucose joined to β-fructose).

16. Cellulose, starch, and glycogen are three important polysaccharides. Cellulose is a polymer of glucose units joined by β-1,4-glucosidic bonds. The glucosidic bonds in the polymers starch and glycogen are primarily α-1,4-bonds. In addition, amylopectin and glycogen contain α-1,6-branch points.

PROBLEMS

1. Draw Haworth projections for the indicated anomers of the following sugars:

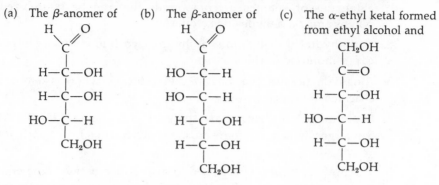

(a) The β-anomer of

(b) The β-anomer of

(c) The α-ethyl ketal formed from ethyl alcohol and

2. Draw Haworth projections for the following:
   (a) β-D-Fructose 1,6-diphosphate   (b) Methyl-α-D-riboside

3. Draw Fischer projections for all of the tetroses.

4. Draw Haworth projections for a short (three or four glucose units) segment of
   (a) Amylose   (b) Cellulose

5. What is an anomeric carbon atom?

6. Draw Haworth projections for maltose and sucrose. Identify the glycosidic bonds and indicate if these substances are reducing sugars.

7. Draw structural formulas or Haworth projections for the products of the following reactions. If no reaction occurs, write N.R.

(a)

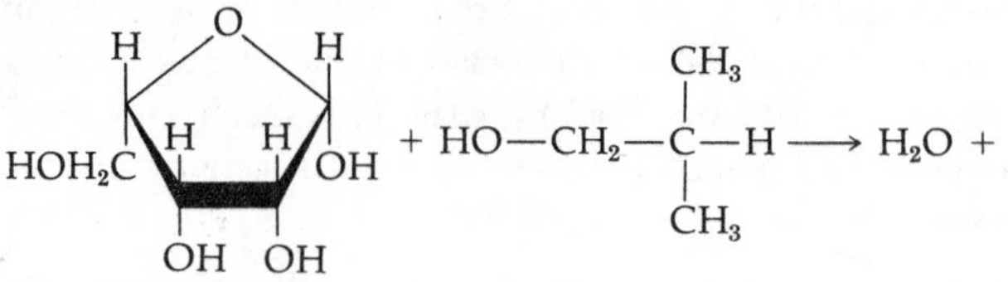

HOH_2C ... $+ HO—CH_2—\overset{\underset{|}{CH_3}}{\underset{|}{CH_3}}{\overset{|}{C}}—H \longrightarrow H_2O +$

(b)

$H—\overset{|}{\underset{|}{C}}—OH$ $+ H_2O \longrightarrow$
$\underset{\overset{|}{CH_2OH}}{}$

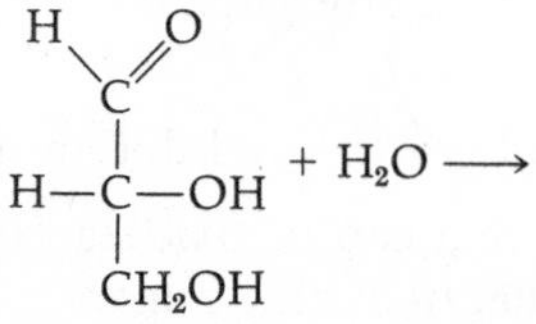

8. Draw Haworth projections for all of the D-aldopentoses.

9. Draw Fischer projections for all of the monosaccharide isomers of D-glyceraldehyde.

SOLUTIONS TO EXERCISES

20-1

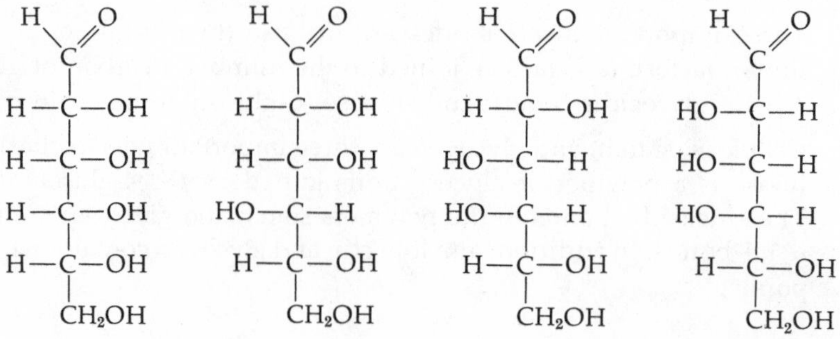

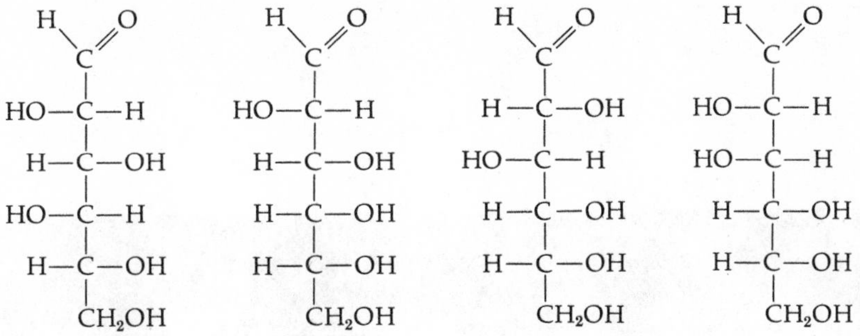

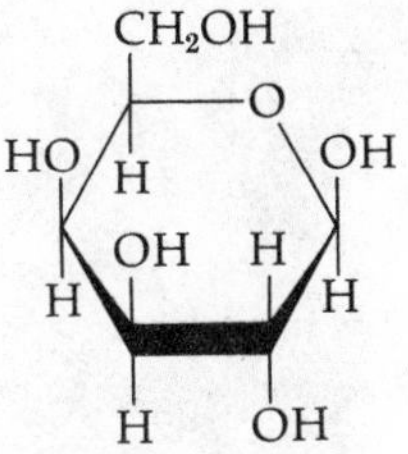

20-3 (a) (1) Aldopentose (2) Aldohexose (3) Ketohexose

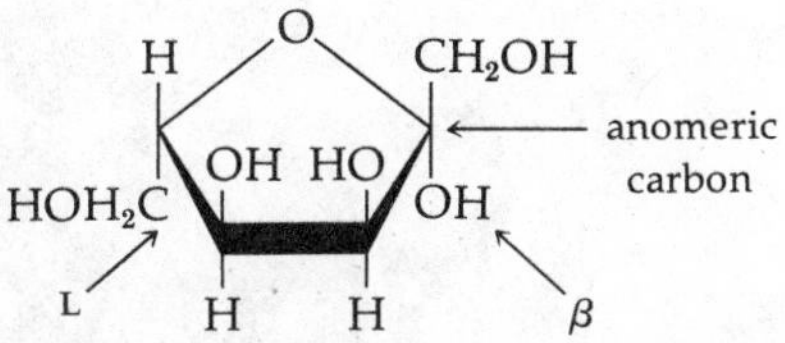

20-4 No, because D-ribose is an aldopentose, not an aldohexose; it is therefore not an isomer of glucose.

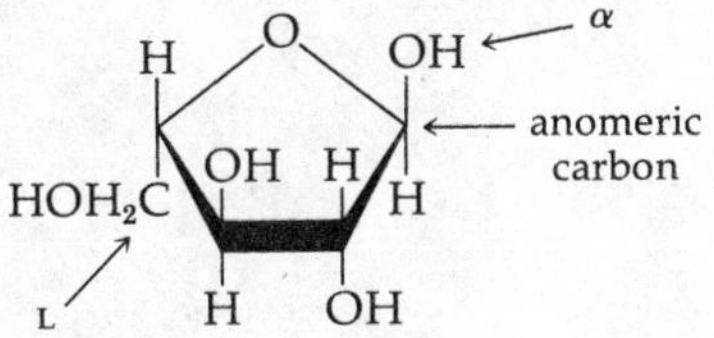

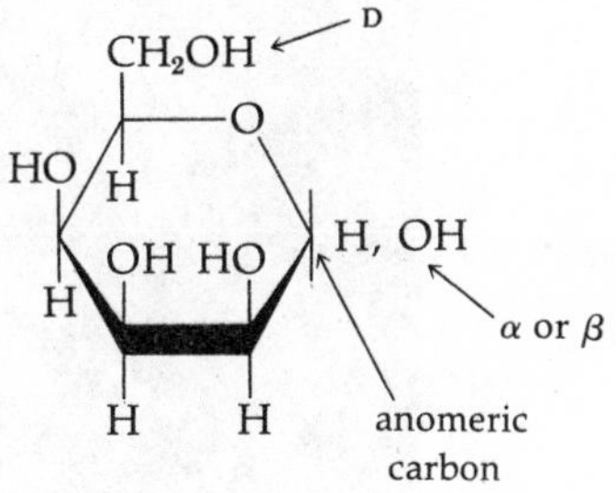

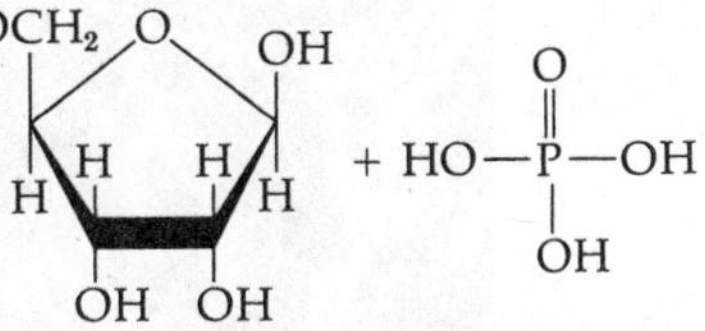

20-6 Cellulose contains only β-1,4-glucosidic bonds. Glycogen contains α-1,4-glucosidic bonds with α-1,6-branch points.

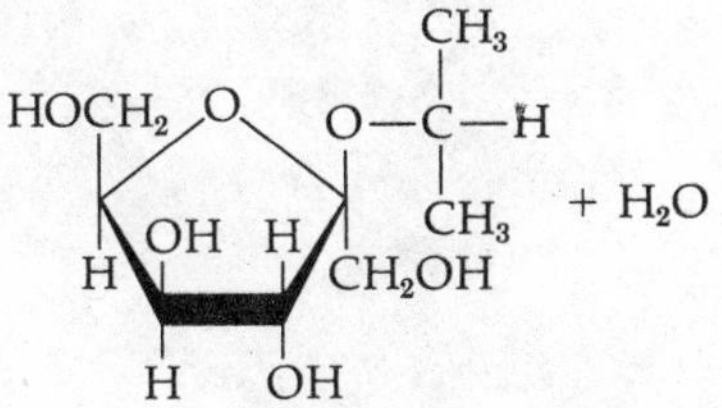

20-7 One molecule of amylopectin is composed of a very large number of glucose units. However, only one of them has a free anomeric hydroxyl group. Therefore amylopectin is not a reducing sugar, since the sole glucose unit that can be oxidized is negligible compared to the entire molecule.

CHAPTER 21

Amino Acids and Protein Structure

21-1 INTRODUCTION

The human body is largely made up of proteins. In fact, second only to water, proteins are the most plentiful chemical compounds in the human body. Protein molecules are complex—much more complex than water—and the body contains thousands of different ones. Hair, connective tissue, fingernails, and tendons are composed of proteins. Enzymes, the body's catalysts, are proteins. Antibodies, the molecules of immunity, are also proteins, as are a very large number of other important molecules. Proteins are vital to the functioning of cellular factories and to the integrated operation of all body organs.

It is not feasible to examine the detailed structure and function of each of the thousands of proteins in the human body. In fact, little is known about the structure of many proteins. In this chapter we shall examine the common structural features and properties of proteins. This will provide us with some insight into the relationship between protein structure and function.

Proteins are polymers. In the last chapter we saw that polysaccharides such as starch and cellulose are polymers composed of monosaccharide building blocks. The building blocks for proteins are called amino acids. Starch is composed of only a single building block, glucose, but there are 20 different amino acids used to assemble proteins. For example, hair consists predominantly of a type of protein called keratin. The properties of the keratin in hair are due to the unique arrangement, or sequence, of its component amino acids, which is different from the sequence of amino acids in every other protein in the body.

It is natural to begin our discussion of proteins by studying their amino acid building blocks. We shall then examine the way in which these amino acid building blocks are bonded together to form proteins. Only then can we study the forces that are responsible for the unique three-dimensional structure and properties of various proteins. We shall then be able to see why hair forms long strands, why it curls, and even the chemistry of permanent waving. More important, we shall have a basis for understanding the vital functions of enzymes, hemoglobin, protein hormones, and other proteins in the body.

The spatial relationship of the dewdrops on this spiderweb, intricate as it is, can only begin to convey the complexity of the arrangement of amino acid components in a protein. The structure and function of proteins are critically dependent on the precise arrangement of their component amino acids.

21-2   STUDY OBJECTIVES

After careful study of the material in this chapter, you should be able to:

1. Write the general formula for an α-amino acid.

2. Draw a peptide group and identify a peptide bond.

3. Define the terms peptide, protein, amino acid sequence, essential amino acid, zwitterion, N-terminal, C-terminal, hydrophobic core, oligomer, subunit, and isoenzyme.

4. Classify an amino acid as acidic, basic, neutral hydrophilic, or neutral hydrophobic, given the structural formula of its side-chain group.

5. Discuss the acid-base properties of a given amino acid and determine which form will predominate at low, neutral, or high pH.

6. Draw the structural formula for a peptide, given its abbreviated representation and the structural formulas for its component amino acids.

7. Define the terms primary, secondary, tertiary, and quaternary protein structure.

8. Describe the involvement of hydrogen bonds, disulfide bonds, hydrophilic and hydrophobic interactions, and salt bridges in maintaining protein shape.

9. Draw and describe an α-helix and a β-pleated sheet.

10. Discuss the process of protein denaturation and differentiate between reversible denaturation and coagulation of a protein.

21-3   PROTEIN POLYMERS

Proteins are polymers composed of amino acid building blocks that are joined together by amide bonds. The building blocks (monomers) for proteins are **α-amino acids,** so-called because each monomer possesses an amine functional group located on the carbon atom immediately adjacent (α) to a carboxyl functional group (see Figure 21-1). All 20 of the α-amino acids used to make proteins have the same general structure, but each has a different R' group. The R' group in an amino acid is often referred to as its **side-chain group.**

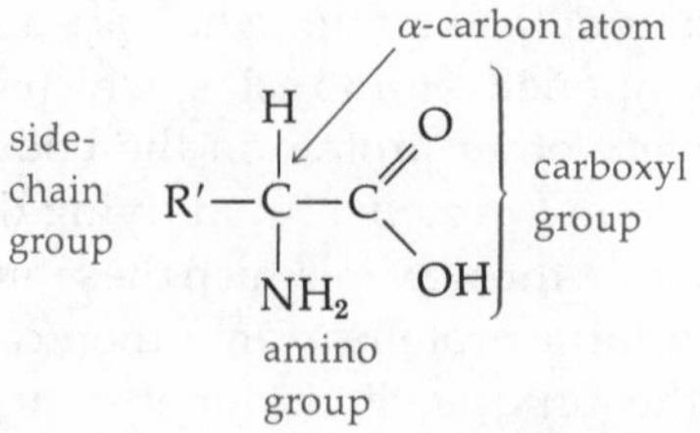

Figure 21-1   The general structural formula for an α-amino acid. The —NH_2 portion of a primary amine functional group is called an amino group.

In a protein, the carboxyl group of one amino acid is coupled to the amine group of the next amino acid by an **amide bond.** In Chapter 18 we saw that amides are formed by the reaction of a carboxylic acid and an amine:

(21-1)

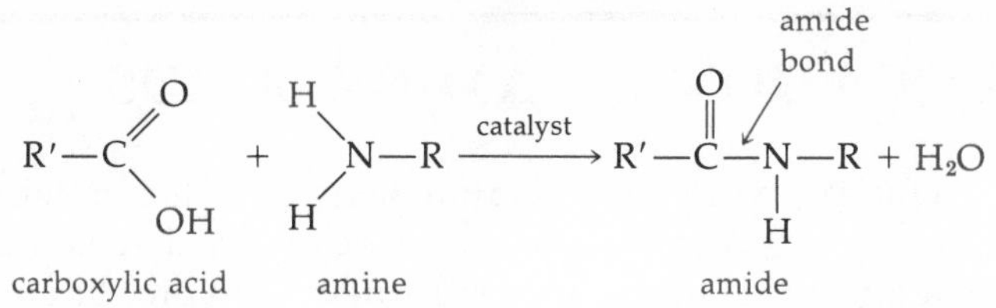

$$R'-C\overset{O}{\underset{OH}{\diagdown}} + \overset{H}{\underset{H}{\diagdown}}N-R \xrightarrow{\text{catalyst}} R'-\overset{O}{\overset{\|}{C}}-\overset{}{\underset{H}{N}}-R + H_2O$$

amide bond

carboxylic acid   amine   amide

We also saw in Chapter 18 that amide bonds join the diamine and dicarboxylic acid monomer units together in the synthetic polymer nylon. Nylon has an ordered sequence of monomer units and is made simply by catalyzing the formation of amide bonds in a vat filled with a diamine and a dicarboxylic acid (reaction 18-8). Recall that the monomer units in nylon can link up in only one way, that is, as alternate dicarboxylic acid and diamine units. Proteins, however, are quite different. Since each amino acid possesses *both* of the functional groups needed to form an amide bond, and 20 different amino acids can be used, an enormous number of different proteins—each with a different sequence of amino acids—is possible.

In a protein, the amide bond between the α-carboxyl group of one amino acid and the α-amino group of another amino acid is usually called a **peptide bond.** Figure 21-2 shows a short string of four amino acids joined by peptide bonds. Short strings of amino acids such as this are called *peptides* to distinguish them from the very long strings of amino acids called *proteins*. Peptides are sometimes classified according to the number of amino acids they contain. A peptide composed of two amino acids that are joined by a single peptide bond is called a *dipeptide*. A string of three amino acids joined by two peptide bonds is called a *tripeptide*. Figure 21-2 represents a *tetrapeptide*. Usually strings of fewer than 100 amino acids are called **peptides,** whereas longer strings are called **proteins.** The line separating peptides and proteins is somewhat arbitrary, but very long strings of amino acids are always called proteins.

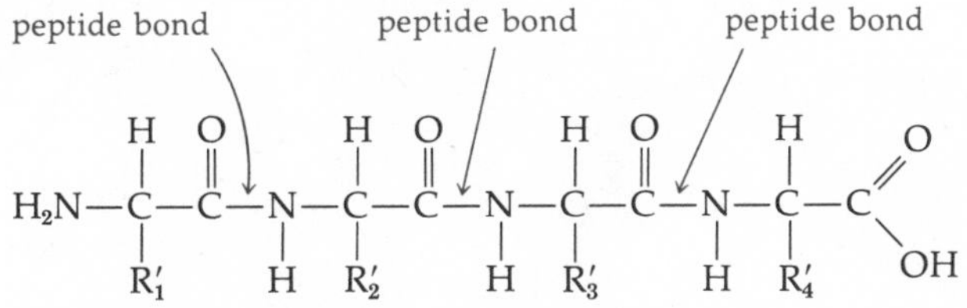

Figure 21-2   The general structural formula for a tetrapeptide.

The sequence of amino acids in a protein (or in a peptide) can be specified simply by indicating the order of the R' groups in the protein (or peptide). For example, $R_1'-R_2'-R_3'-R_4'$ refers to the sequence of amino acids in the tetrapeptide shown in Figure 21-2. These 4 different amino acids can form a total of 24 different tetrapeptides with different amino acid sequences (that is, $R_2'-R_1'-R_4'-R_3'$, $R_4'-R_3'-R_2'-R_1'$, etc.). We shall see that the exact sequence of amino acids in a peptide or a protein is crucial to its overall shape and therefore to its function. Because the sequence is so critical, the manufacture of each protein in the human body is carefully controlled so that the amino acids can be assembled in just the right sequence (see Chapter 24).

Exercise 21-1

A tripeptide composed of three different amino acids has the sequence $R_1'-R_2'-R_3'$. Write the sequence for all the other tripeptides containing R_1', R_2', and R_3'.

21-4 CLASSIFICATION OF AMINO ACIDS

It is very useful to divide the 20 amino acids used for protein synthesis into groups or classes based on the properties of their side-chain (R′) groups. We shall divide amino acids into four classes: neutral hydrophobic, neutral hydrophilic, acidic, and basic. Although other schemes are also used to classify these amino acids, this one is simple and will help us to understand how the amino acid side chains of a protein interact with each other and with other molecules. The name, abbreviation, structural formula, and classification of each of the 20 amino acids are given in Table 21-1.

Neutral Amino Acids

Fifteen of the 20 amino acids are classified as **neutral** because their side chains have no charge at the pH of body cells (which is about 7). These neutral amino acids are further subdivided according to how the side-chain group interacts with water molecules.

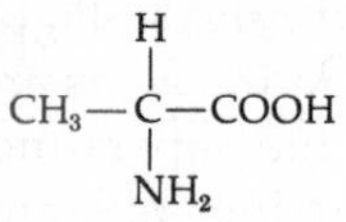

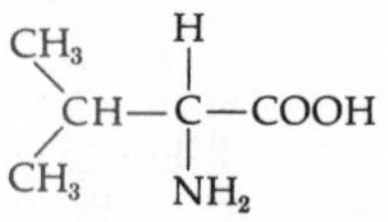

Table 21-1 The Amino Acids

A. Neutral Hydrophobic Amino Acids

	R′ groups
Alanine ala	
Valine val	
Leucine leu	
Isoleucine ile	
Proline pro	
Phenylalanine phe	
Tryptophan trp	

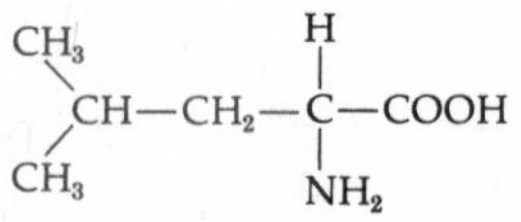

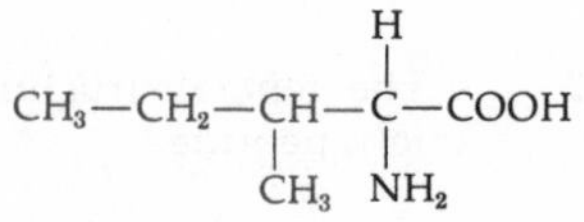

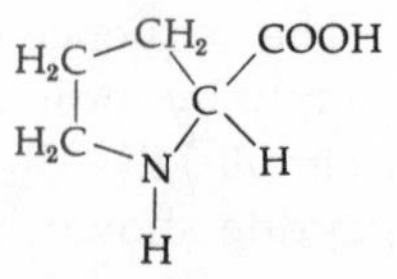

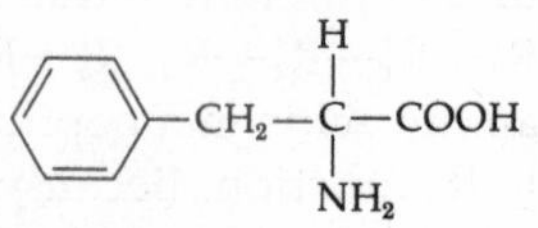

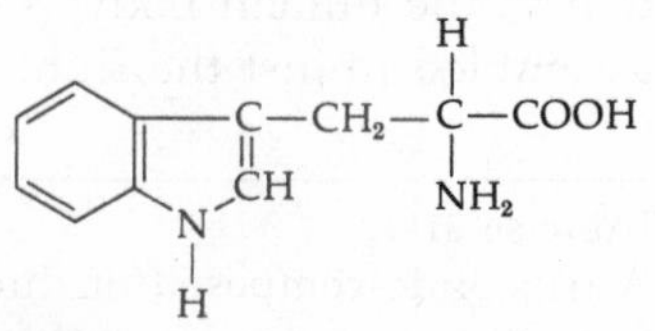

NEUTRAL HYDROPHOBIC AMINO ACIDS

Seven neutral amino acids have side chains (R') that are nonpolar or hydrophobic (Table 21-1A). These hydrophobic groups are either alkyl or aromatic in nature. The side chain of alanine (ala) is a methyl group, that of valine (val) is an isopropyl group, leucine (leu) has an isobutyl group, and isoleucine (ile) has a sec-butyl group. The amino acid proline (pro) is unusual in that the alkyl side chain is bonded to both the α-carbon and the α-amino group, forming a five-atom ring. The side chain of the amino acid phenylalanine (phe) is the aromatic benzyl group. The amino acid tryptophan (trp) has a rather complex aromatic side chain involving two rings.

NEUTRAL HYDROPHILIC AMINO ACIDS

Eight of the neutral amino acids are classified as hydrophilic (Table 21-1B). In general, these amino acids are more soluble in water than hydrophobic amino acids. The side chain of glycine (gly) is just a hydrogen atom. The other seven neutral hydrophilic amino acids have side chains that can form either strong or weak hydrogen bonds with water. Three have a hydroxyl group in their side chains: serine (ser), threonine (thr), and tyrosine (tyr). Two contain an amide functional group: asparagine (asn) and glutamine (gln). The remaining two contain a sulfur atom: cysteine (cys) and methionine (met).

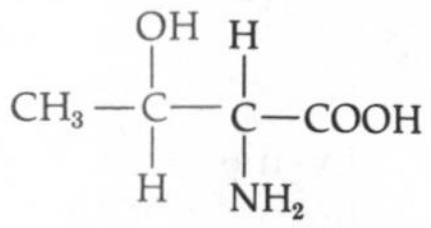

B. Neutral Hydrophilic Amino Acids

R' groups

Glycine
gly

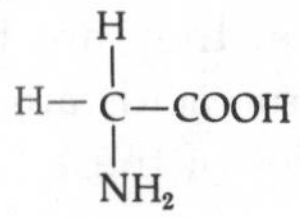

$$H-\underset{\underset{NH_2}{|}}{\overset{\overset{H}{|}}{C}}-COOH$$

Serine
ser

$$HO-CH_2-\underset{\underset{NH_2}{|}}{\overset{\overset{H}{|}}{C}}-COOH$$

Threonine
thr

$$CH_3-\underset{\underset{H}{|}}{\overset{\overset{OH}{|}}{C}}-\underset{\underset{NH_2}{|}}{\overset{\overset{H}{|}}{C}}-COOH$$

Tyrosine
tyr

$$HO-\!\!\left\langle\!\!\bigcirc\!\!\right\rangle\!\!-CH_2-\underset{\underset{NH_2}{|}}{\overset{\overset{H}{|}}{C}}-COOH$$

Asparagine
asn

$$\underset{O}{\overset{NH_2}{\diagdown}}C-CH_2-\underset{\underset{NH_2}{|}}{\overset{\overset{H}{|}}{C}}-COOH$$

Glutamine
gln

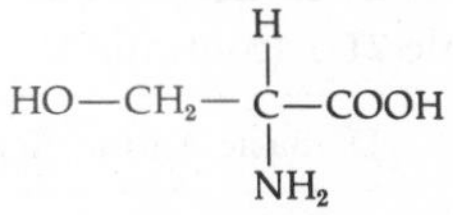

$$\underset{O}{\overset{NH_2}{\diagdown}}C-CH_2-CH_2-\underset{\underset{NH_2}{|}}{\overset{\overset{H}{|}}{C}}-COOH$$

Cysteine
cys

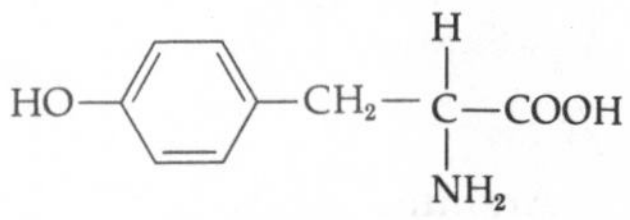

$$HS-CH_2-\underset{\underset{NH_2}{|}}{\overset{\overset{H}{|}}{C}}-COOH$$

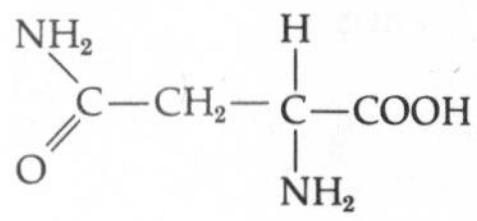

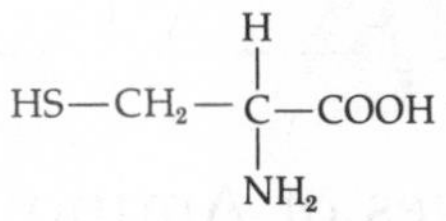

Methionine
met

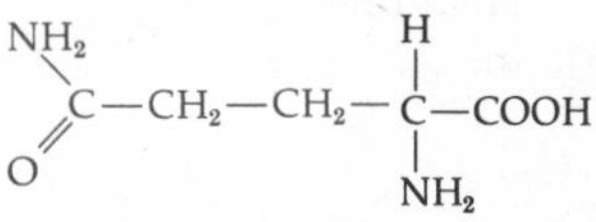

$$CH_3-S-CH_2-CH_2-\underset{\underset{NH_2}{|}}{\overset{\overset{H}{|}}{C}}-COOH$$

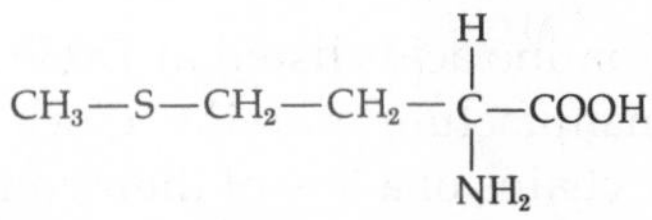

Acidic Amino Acids

Acidic amino acids have side chains that contain a second carboxyl group (Table 21-1C). The side chain of aspartic acid (asp) is $-CH_2-COOH$ and that of glutamic acid (glu) is $-CH_2-CH_2-COOH$. At the pH of cells in the body, these carboxyl groups exist primarily as negatively charged carboxylate ions and thus interact strongly with water molecules.

Table 21-1 (continued)

C. Acidic Amino Acids

R' groups

Aspartic acid asp	$\underset{O}{\overset{HO}{\diagdown}}C-CH_2-\underset{NH_2}{\overset{H}{\underset{|}{\overset{|}{C}}}}-COOH$
Glutamic acid glu	$\underset{O}{\overset{HO}{\diagdown}}C-CH_2-CH_2-\underset{NH_2}{\overset{H}{\underset{|}{\overset{|}{C}}}}-COOH$

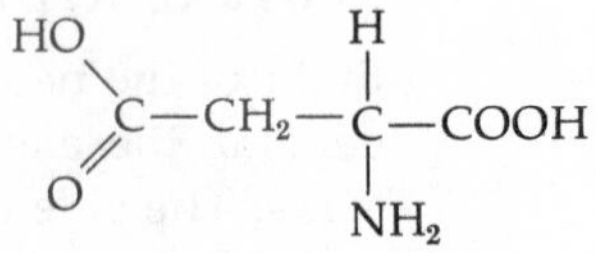

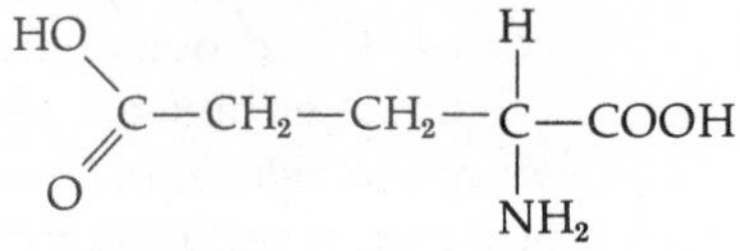

Basic Amino Acids

Three of the amino acids contain a side chain that can act as a proton acceptor or base; they are thus classified as basic amino acids (Table 21-1D). The basic amino acids are lysine (lys), arginine (arg), and histidine (his). The basic properties of these three amino acids will be discussed in Section 21-5.

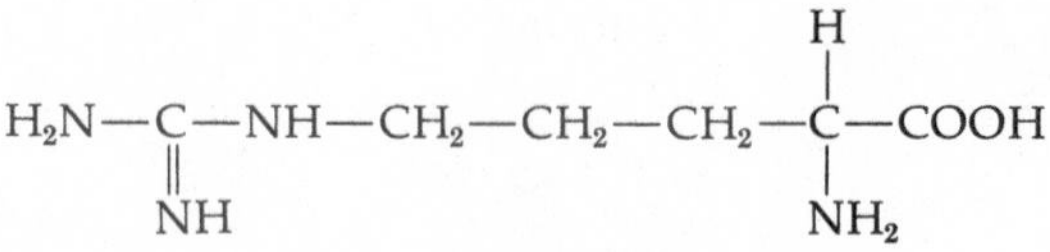

Table 21-1 (continued)

D. Basic Amino Acids

R' groups

Lysine lys	$H_2N-CH_2-CH_2-CH_2-CH_2-\underset{NH_2}{\overset{H}{\underset{|}{\overset{|}{C}}}}-COOH$
Arginine arg	$H_2N-\underset{NH}{\overset{||}{C}}-NH-CH_2-CH_2-CH_2-\underset{NH_2}{\overset{H}{\underset{|}{\overset{|}{C}}}}-COOH$
Histidine his	$HC=\!\!=\!\!C-CH_2-\underset{NH_2}{\overset{H}{\underset{|}{\overset{|}{C}}}}-COOH$

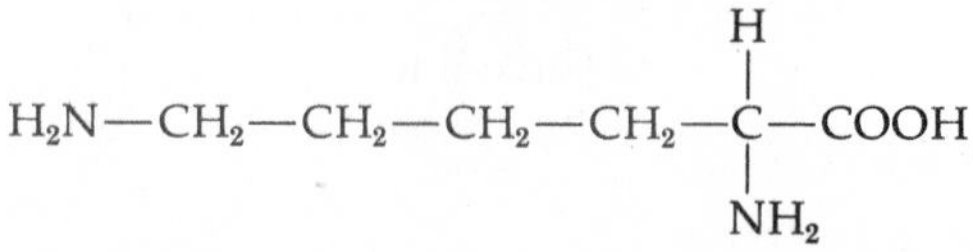

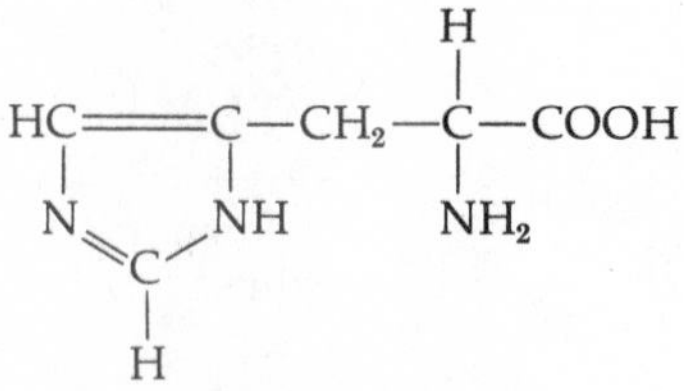

Derivatives of Amino Acids

The amino acids listed in Table 21-1 are the *only* amino acids used by living cells to manufacture proteins. Once certain proteins have been made, however, the side chains of a few of their component amino acids are sometimes modified by further chemical reactions.

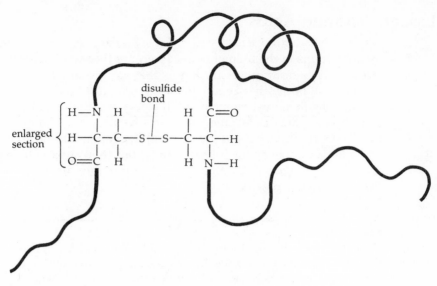

Figure 21-3   Schematic representation of a protein containing a disulfide bond between two component cysteines.

Recall that two thiols can be easily oxidized to form a disulfide bond (see Chapter 18). Frequently the thiol groups in the side chains of two cysteines $(R' = -CH_2-SH)$ in the same protein combine to form a covalent disulfide bond between the two sulfur atoms (see Figure 21-3). If two cysteine molecules that are not part of a protein link up by forming a disulfide bond, the product is called cystine:

$$\begin{array}{c} NH_2 \qquad\qquad\qquad NH_2 \\ | \qquad\qquad\qquad\qquad | \\ H-C-CH_2-S-S-CH_2-C-H \\ | \qquad\qquad\qquad\qquad | \\ C=O \qquad\qquad\qquad C=O \\ | \qquad\qquad\qquad\qquad | \\ OH \qquad\qquad\qquad\quad OH \end{array}$$

cystine

In some connective tissue proteins, the side chains of the amino acids lysine and proline are modified by specific enzymes to produce hydroxylysine and hydroxyproline:

hydroxylysine   hydroxyproline

The side-chain functional groups of some amino acids, especially the carboxyl groups of aspartic acid and glutamic acid and the hydroxyl groups of serine and threonine, can also react with a variety of nonprotein molecules, including carbohydrates and phosphoric acid.

A number of the α-amino acids used for protein synthesis are also used as starting materials for the manufacture of a variety of small molecules with very interesting functions, such as the hormone adrenaline (see Chapter 28).

Essential Amino Acids

All 20 of the α-amino acids listed in Table 21-1 are needed to make the many different proteins in the human body. Twelve of these amino acids can be synthesized by cells from other substances that are present in the body. The other eight cannot be synthesized by the body and must be included in the diet. The amino acids that must be included in a person's diet are called **essential amino acids.** They are isoleucine, leucine, lysine, methionine, phenylalanine, threonine, tryptophan, and valine. Arginine and histidine do not appear to be essential amino acids for human adults, but they apparently are essential for the normal growth of children. The dietary requirements for amino acids are discussed in Chapter 29.

Exercise 21-2
Draw the structural formulas for: (a) two acidic amino acids; (b) two basic amino acids; and (c) two neutral amino acids. In each case, indicate the functional group responsible for its classification as acidic, basic, or neutral.

Exercise 21-3
Classify each of the following amino acids as either neutral hydrophilic or neutral hydrophobic.

(a)

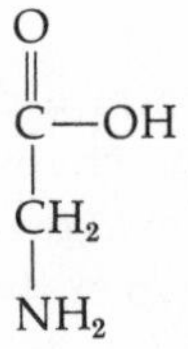

$$\begin{array}{c} O \\ \| \\ C-OH \\ | \\ CH_2 \\ | \\ NH_2 \end{array}$$

(b)

$$\begin{array}{c} O \\ \| \\ C-OH \quad O \\ | \quad\quad \| \\ H-C-CH_2-C-NH_2 \\ | \\ NH_2 \end{array}$$

(c)

$$\begin{array}{c} O \\ \| \\ C-OH \\ | \\ H-C-CH_2-\bigcirc \\ | \\ NH_2 \end{array}$$

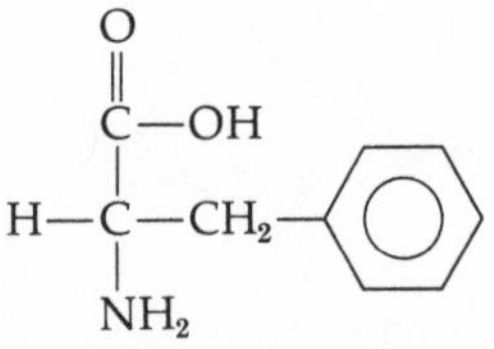

(d)

$$\begin{array}{c} O \\ \| \\ C-OH \\ | \\ H-C-CH_3 \\ | \\ NH_2 \end{array}$$

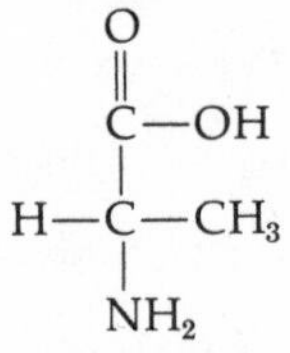

21-5 PROPERTIES OF AMINO ACIDS

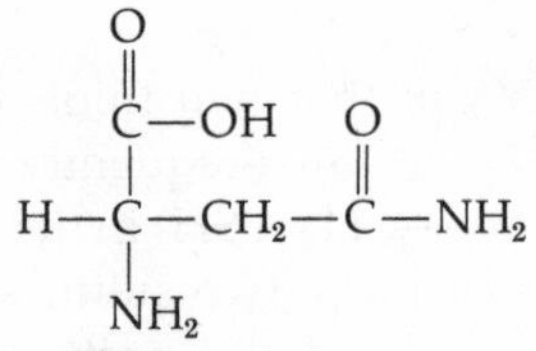

Optical Activity

All of the α-amino acids, with the exception of glycine, possess an asymmetric carbon (the α-carbon) and hence are optically active (see Figure 21-4). Only the L-isomers of the amino acids are synthesized and used by humans. In fact, all organisms in nature use only L-amino acids for protein synthesis. The few D-amino acids found in nature are used by some bacteria to form part of their cell wall.

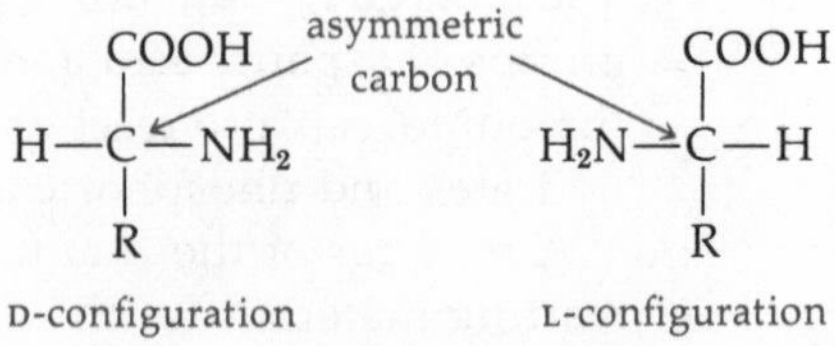

Figure 21-4   Fischer projections for the D- and L-enantiomers of amino acids. The L-enantiomers of the amino acids are the predominant ones found in nature. D-Amino acids are found only in some bacteria.

$$\begin{array}{c} COOH \\ | \\ H-C-NH_2 \\ | \\ R \end{array}$$

asymmetric carbon

$$\begin{array}{c} COOH \\ | \\ H_2N-C-H \\ | \\ R \end{array}$$

D-configuration   L-configuration

Acid-Base Properties

Every amino acid, regardless of its side chain, has an acidic carboxyl group and a basic amino group, so we shall consider first the acid-base properties of a typical neutral amino acid, such as alanine, which has no other acidic or basic functional groups. In the solid state, amino acids have saltlike properties because they have both a positively charged part and a negatively charged part. Such substances are called **zwitterions.** We can think of these zwitterions as being produced from the molecular form of the amino acids by an internal acid-base reaction:

(21-2)

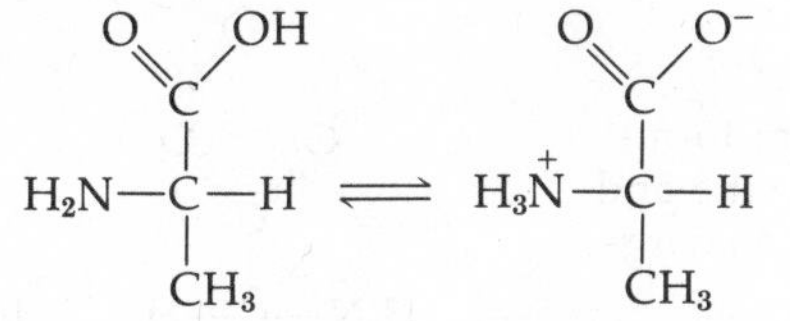

$$H_2N-\overset{\overset{\displaystyle O\diagdown\diagup OH}{\displaystyle C}}{\underset{\underset{\displaystyle CH_3}{|}}{C}}-H \rightleftharpoons H_3\overset{+}{N}-\overset{\overset{\displaystyle O\diagdown\diagup O^-}{\displaystyle C}}{\underset{\underset{\displaystyle CH_3}{|}}{C}}-H$$

L-alanine
(molecular form)

L-alanine
(zwitterion form)

Note in reaction 21-2 that neither the molecular form of alanine nor the zwitterion form has a net electrical charge. In aqueous solution these two forms are in equilibrium, but this equilibrium overwhelmingly favors the zwitterion at any pH.

Alanine in solution is also involved in two other equilibria,

(21-3)

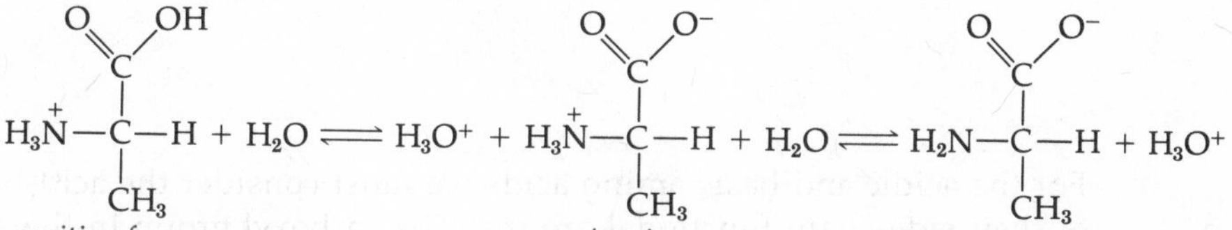

$$H_3\overset{+}{N}-\overset{\overset{\displaystyle O\diagdown\diagup OH}{\displaystyle C}}{\underset{\underset{\displaystyle CH_3}{|}}{C}}-H + H_2O \rightleftharpoons H_3O^+ + H_3\overset{+}{N}-\overset{\overset{\displaystyle O\diagdown\diagup O^-}{\displaystyle C}}{\underset{\underset{\displaystyle CH_3}{|}}{C}}-H + H_2O \rightleftharpoons H_2N-\overset{\overset{\displaystyle O\diagdown\diagup O^-}{\displaystyle C}}{\underset{\underset{\displaystyle CH_3}{|}}{C}}-H + H_3O^+$$

positive form

zwitterion
form

negative form

At any given pH some of the alanine in solution exists in the positive ion form, some of it in the negative ion form, some in the zwitterion form, and some in the molecular form. If the solution pH is very high—that is, $[H_3O^+]$ is very low—both of the equilibria in reaction 21-3 are shifted far to the right and the negative ion form of alanine predominates. On the other hand, if the solution pH is very low—that is, $[H_3O^+]$ is very high—both equilibria in reaction 21-3 are shifted far to the left and the positive ion form of alanine predominates. At the pH of human cells and fluids (pH ~7), alanine exists primarily as the zwitterion.

In solutions that are moderately basic (pH ~8.5 to 10.5), no single form of alanine predominates. In this pH range there are roughly comparable amounts of the zwitterion and the negatively charged forms. Similarly, in moderately acidic solutions there are roughly comparable amounts of the zwitterion and the positively charged forms of alanine. Note that the equilibria in reaction 21-3 can also be viewed as the successive dissociation of the diprotic acid

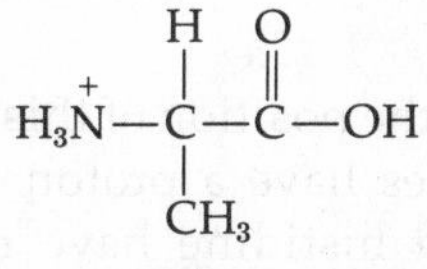

$$H_3\overset{+}{N}-\overset{\overset{\displaystyle H}{|}}{\underset{\underset{\displaystyle CH_3}{|}}{C}}-\overset{\overset{\displaystyle O}{\|}}{C}-OH$$

The other neutral hydrophilic or hydrophobic amino acids behave like alanine and are primarily zwitterions that have a net charge of zero at pH ~7.

Figure 21-5   At pH ~7, the predominant form of an acidic amino acid has a net negative charge.

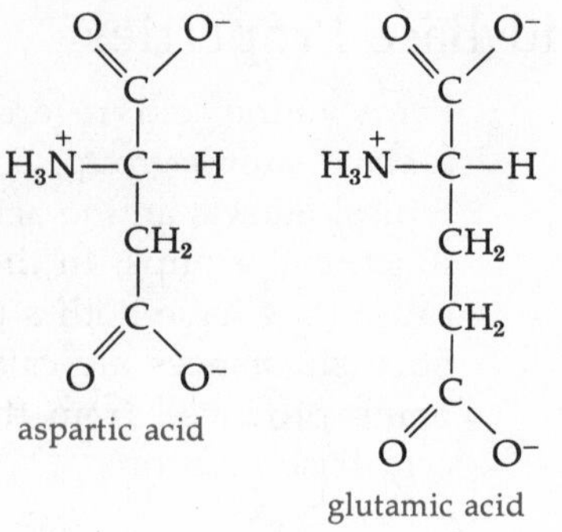

aspartic acid

glutamic acid

Figure 21-6   At pH ~7, the predominant forms of the basic amino acids lysine and arginine have a net positive charge.

lysine

arginine

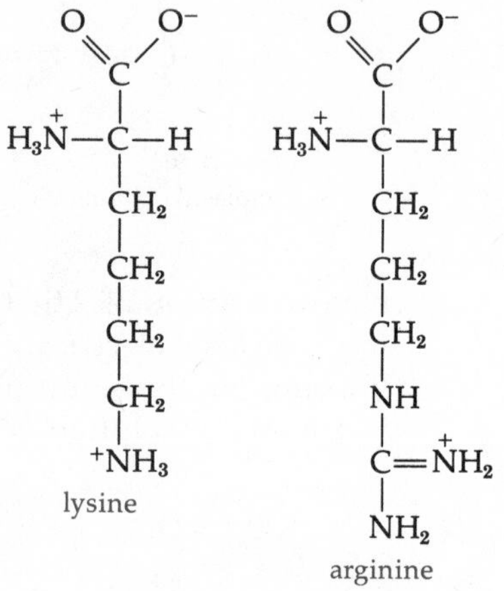

For the acidic and basic amino acids we must consider the acid-base behavior of their side-chain functional groups. The carboxyl group in the side chain of the acidic amino acids, aspartic acid and glutamic acid, is predominantly dissociated at neutral pH (see Figure 21-5). Two of the basic amino acids, lysine and arginine, exist mostly with a protonated nitrogen atom on their side chains at neutral pH and thus have a net positive charge (see Figure 21-6). The other basic amino acid, histidine, also contains a side-chain nitrogen that can serve as a proton acceptor.

(21-4)

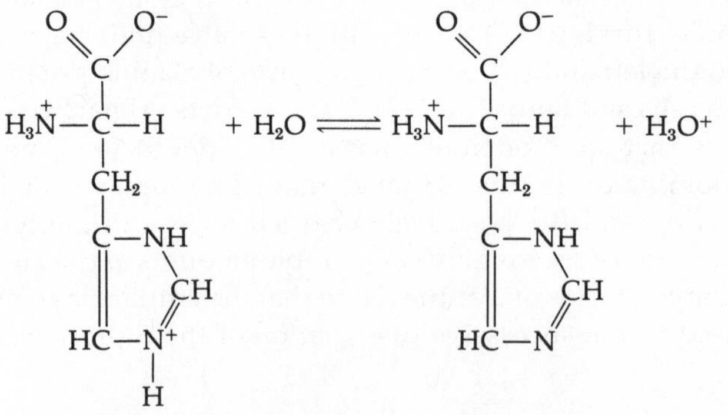

In an aqueous solution with a pH of about 6.5, the position of this equilibrium is such that about half of the histidine molecules have a protonated side-chain nitrogen and half do not. Both arginine and histidine have more than one nitrogen atom in their side chains, but they can accept only one proton per side chain. It is not necessary for us to consider why the particular nitrogen atoms shown in Figure 21-6 and reaction 21-4 are the ones that accept the protons.

Exercise 21-4

Draw structural formulas for the predominant forms of the following amino acids at pH values of 1, 7, and 14:

    (a) Valine   (b) Aspartic acid   (c) Serine   (d) Lysine

21-6 AMINO ACID SEQUENCE

As we mentioned at the beginning of the chapter, the sequence of amino acids in a given peptide or protein is crucial to its function, in much the same way as the arrangement of words in a sentence determines its meaning. For example, the two sentences, "Ali beat Frazier" and "Frazier beat Ali" contain the same three words but have quite different meanings. Similarly, there are two different dipeptides with different properties composed of the amino acids glycine and alanine (see Figure 21-7).

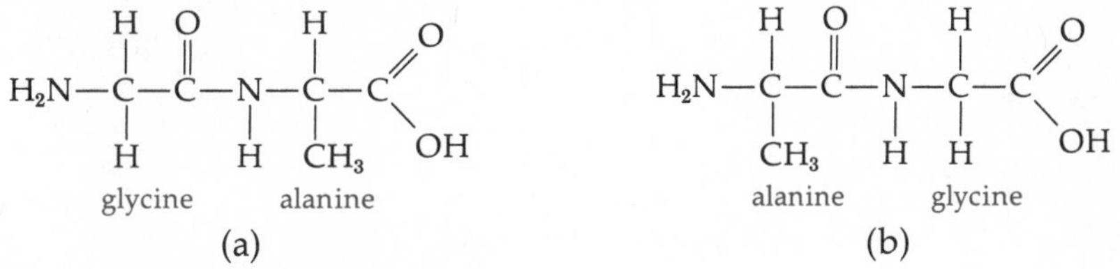

Figure 21-7   The two possible dipeptides composed of the amino acids glycine and alanine are gly-ala (a) and ala-gly (b).

Peptides and proteins usually have an amino acid with an uncombined α-amino group—called the **N-terminal amino acid**—at one end, and an amino acid with an uncombined α-carboxyl group—the **C-terminal amino acid**—on the other end. To avoid confusion, the structural formulas for peptides and proteins are always drawn with the N-terminal amino acid on the left and the C-terminal amino acid on the right, as we have done in Figure 21-7. The same convention is used when the structural formula for a protein or peptide is written in a condensed manner using the abbreviation for the names of the amino acids. Thus, for example, tyr-ala-ser is a condensed way of referring to the tripeptide with the structural formula

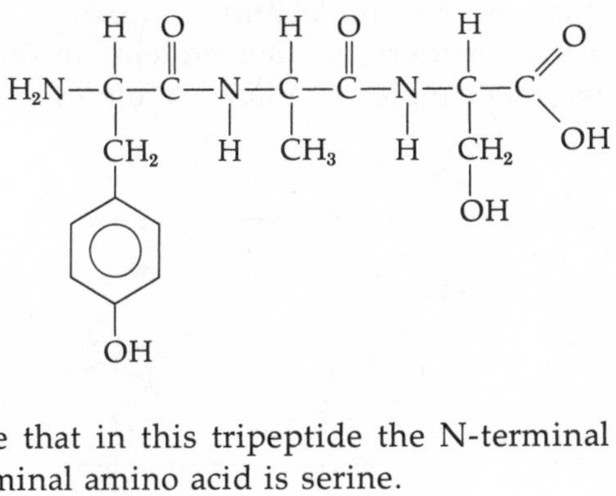

Notice that in this tripeptide the N-terminal amino acid is tyrosine and the C-terminal amino acid is serine.

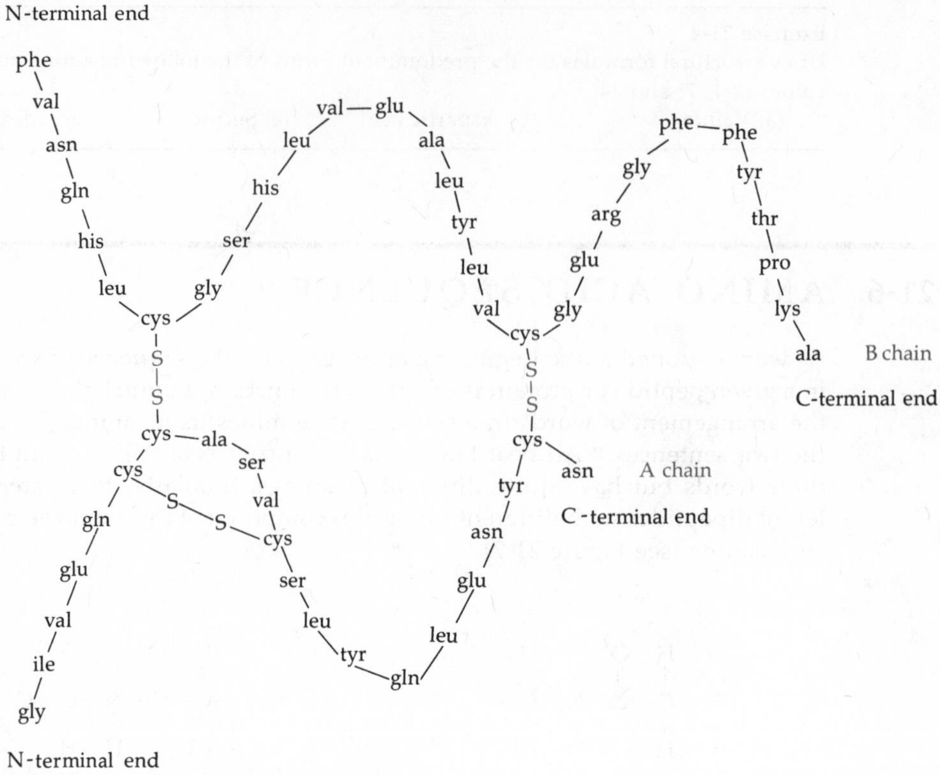

N-terminal end

Figure 21-8   A representation of the complete amino acid sequence of the polypeptide hormone insulin (bovine). Note that the insulin molecule is composed of two peptide chains (A and B) held together by two disulfide bonds between cysteine side chains.

For very long complex sequences of amino acids, it is sometimes necessary to emphasize further which amino acids are at the ends. This is done by writing the labels "N-terminal end" and "C-terminal end" next to the corresponding ends of the chains. The amino acid sequence in the peptide hormone insulin is shown in Figure 21-8. Notice that insulin is a single molecule consisting of two polypeptides joined by two disulfide bonds between cysteine side chains. There is also one disulfide bond between two cysteine side chains of amino acids in the same chain. It is interesting to note that insulin is actually synthesized from a single large peptide called proinsulin. Proinsulin is made in the pancreas and then a piece of the large peptide is removed, leaving the two joined peptide chains of insulin. Certain other proteins in the body are also produced by reactions that remove pieces of larger proteins.

Exercise 21-5

Draw the structural formula for the tetrapeptide

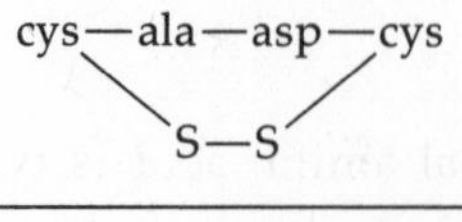

cys—ala—asp—cys
        \   /
         S—S

Optional

21-7 DETERMINATION OF AMINO ACID SEQUENCE

Until recently it was extremely difficult to determine the exact sequence of amino acids in peptides and proteins. Insulin (Figure 21-8), which has 51 amino acids, was the first large peptide whose amino acid sequence was determined. This was accomplished by Frederick Sanger and his colleagues in 1953, after several years of work. Basically, Sanger's technique involved, first, the hydrolysis of all the peptide bonds in insulin by strong acids, followed by separation and identification of the component amino acids, and finally, determination of the sequence of these amino acids.

When proteins are completely hydrolyzed, the constituent amino acids can be separated on the basis of their different chemical properties and identified. Today, a machine called an **amino acid analyzer** is routinely used for this purpose. This machine separates a mixture of amino acids according to their ability to interact with a highly charged insoluble matrix called an ion-exchange resin.

Determining the sequence of amino acids in a protein is more difficult. Sanger determined the sequence for insulin in a very laborious way. We can illustrate the principles he used by considering how the sequence of a tetrapeptide known to contain the amino acids tyrosine, alanine, valine, and serine could be determined. Twenty-four different tetrapeptides containing these four amino acids are possible. Suppose, however, that we react a sample of the tetrapeptide with an enzyme that can hydrolyze only those peptide bonds that contain the carbonyl group of tyrosine, and then determine that the products of this reaction are (1) the amino acid serine and (2) a tripeptide composed of alanine, valine, and tyrosine. Only two possible sequences are consistent with this information: val-ala-tyr-ser and ala-val-tyr-ser.

Now, suppose we take another sample of our tetrapeptide, react it with a reagent that can hydrolyze only the peptide bond involving the N-terminal amino acid, and determine that the products of this reaction are (1) a tripeptide composed of serine, tyrosine, and alanine, and (2) the amino acid valine.

The *only* sequence that is consistent with both experiments is val-ala-tyr-ser. Sanger performed a large number of these types of reactions on insulin, and after a great deal of effort was finally able to determine its amino acid sequence. Today, techniques are available whereby the amino acid sequence of small proteins can be determined relatively easily. Basically these methods involve the sequential hydrolysis and identification of amino acids, starting with the N-terminal amino acid. This new technique has even been automated, so that the sequence of small proteins such as insulin can now be determined in a matter of days, rather than years, as was the case for Sanger.

21-8 THE SHAPE OF PROTEINS

Proteins are very large molecules, and one could envision many possible shapes for a protein with a particular amino acid sequence. It is known, however, that every protein has its own definite, unique shape and that this shape must be maintained in order for the protein to function properly. Let us consider the factors that determine a protein's shape.

The Peptide Group

Thus far, we have considered peptides and proteins as strings of amino acids held together by peptide bonds (see Figures 21-2 and 21-3). Notice that these strings involve a repetitive pattern, called a **covalent backbone,** consisting of one nitrogen atom followed by two carbon atoms, and so on, which we can represent as

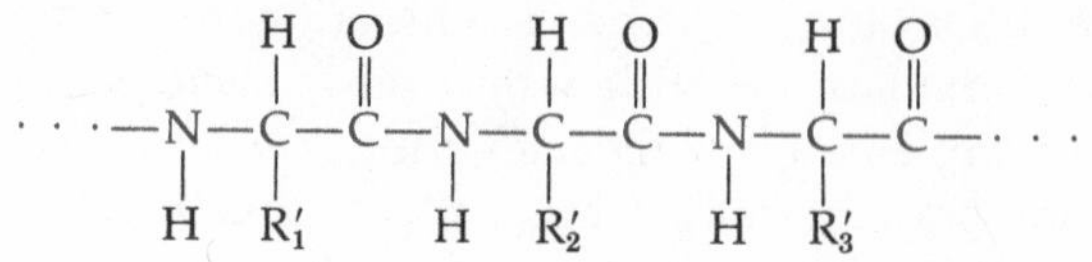

$$\cdots -\underset{\underset{H}{|}}{N}-\underset{\underset{R_1'}{|}}{C}-\overset{\overset{O}{\|}}{C}-\underset{\underset{H}{|}}{N}-\underset{\underset{R_2'}{|}}{C}-\overset{\overset{O}{\|}}{C}-\underset{\underset{H}{|}}{N}-\underset{\underset{R_3'}{|}}{C}-\overset{\overset{O}{\|}}{C}- \cdots$$

We would expect the covalent backbone to be extremely flexible, since there is usually free rotation about single covalent bonds. Starting at the N-terminal end, however, the third bond in the covalent backbone and every third bond thereafter is an amide bond. In Chapter 18 we saw that (1) the carbon-nitrogen bond in amides is shorter than a typical carbon-nitrogen single bond; (2) there is *not* free rotation about amide bonds; and (3) amides are best represented by a composite picture of two resonance structures, one of which involves a C=N double bond. Resonance structures for a peptide group are shown in Figure 21-9. This representation of peptide groups agrees with their experimentally determined properties, including the fact that all six atoms in a peptide group lie in the same plane.

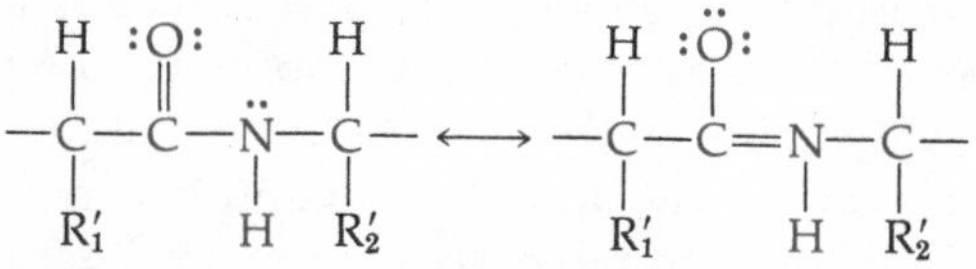

$$-\underset{\underset{R_1'}{|}}{C}-\overset{\overset{:O:}{\|}}{C}-\underset{\underset{H}{|}}{\overset{..}{N}}-\underset{\underset{R_2'}{|}}{\overset{H}{C}}- \longleftrightarrow -\underset{\underset{R_1'}{|}}{C}-\overset{\overset{:\overset{..}{O}:}{|}}{C}=\underset{\underset{H}{}}{N}-\underset{\underset{R_2'}{|}}{\overset{H}{C}}-$$

Figure 21-9   Resonance-structure representation for a peptide group. All six of the atoms in the peptide group (colored blue) lie in the same plane.

The rigid nature of peptide bonds restricts the possible shapes that a given protein can assume, but there is still free rotation of the group of atoms in one plane defined by one peptide bond relative to groups of atoms in other planes defined by other peptide bonds. The free rotation of these groups of atoms allows much flexibility in the covalent backbone of a protein (see Figure 21-10).

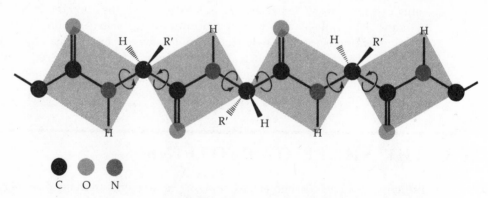

C   O   N

Figure 21-10   Flexible and nonflexible bonds in a polypeptide. The shaded areas represent portions of the backbone containing a peptide group in which atoms all lie in the same plane (i.e., there is nonflexible bonding around peptide bonds). Between the shaded portions the backbone is free to rotate. Arrows indicate this flexibility.

Interactions Between Peptide Groups

A very important factor in determining the shapes of proteins is the ability of two peptide groups to form a strong hydrogen bond. This hydrogen bond can form between the hydrogen atom in one peptide group and the carbonyl oxygen in another peptide group (Figure 21-11). This hydrogen bond is exceptionally strong because there is a particularly large fractional negative charge on the oxygen atom of the peptide group and a particularly large fractional positive charge on the hydrogen atom of this group. Compare the resonance structures in Figure 21-11. In the structure on the right there are three unshared pairs of electrons on the oxygen atom, whereas there are only two in the structure on the left, and there are no unshared electron pairs on the nitrogen atom on the right. The resonance structure on the right implies that some of the valence electrons in the amide group are pulled toward the oxygen atom and away from the nitrogen atom and the nitrogen-hydrogen bond.

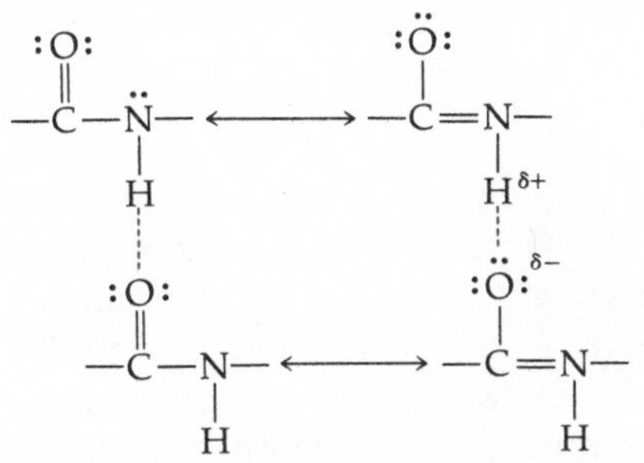

Figure 21-11   Hydrogen bonds (blue dashed lines) between peptide groups. These two resonance structures provide a composite representation that accounts for the particularly strong hydrogen bond between two peptide groups.

Interactions Between Side Chains

If the covalent backbone in a protein is folded or coiled, then the side-chain groups that extend out from the backbone can interact with one another. The shape of a particular protein is determined to a large degree by side-chain interactions. We can divide the attractive forces involved in side-chain interactions into four types: (1) hydrogen bonds, (2) disulfide bonds, (3) hydrophobic and hydrophilic interactions, and (4) salt bridges.

Hydrogen bonds can form between appropriate side-chain groups, and a disulfide bond can form between two cysteine side chains (Figure 21-3). Recall from Chapter 7 that hydrophilic parts of larger molecules interact strongly with water molecules and that hydrophobic parts of larger molecules tend to cluster together. Thus in an aqueous medium most proteins will tend to coil up like a ball of string so that the hydrophobic side chains extend inwards and the hydrophilic side chains are on the outside of the "ball," where they can interact with water molecules (see Figure 21-12).

Figure 21-12   The compact "glob" formed by a typical protein in an aqueous medium is schematically represented here. The interior, or hydrophobic core, of the protein is composed largely of nonpolar hydrophobic amino acids, whereas most hydrophilic side-chain groups are on the exterior, exposed to the aqueous environment.

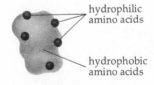

hydrophilic amino acids

hydrophobic amino acids

At the pH of body cells, the acidic side-chain groups of aspartic or glutamic acid are negatively charged, and there is a positive charge on the basic side-chain groups of lysine and arginine and on most of the side-chain groups of histidine. The attraction between a negatively charged acidic side chain and a positively charged basic side chain is called a **salt bridge** (see Figure 21-13).

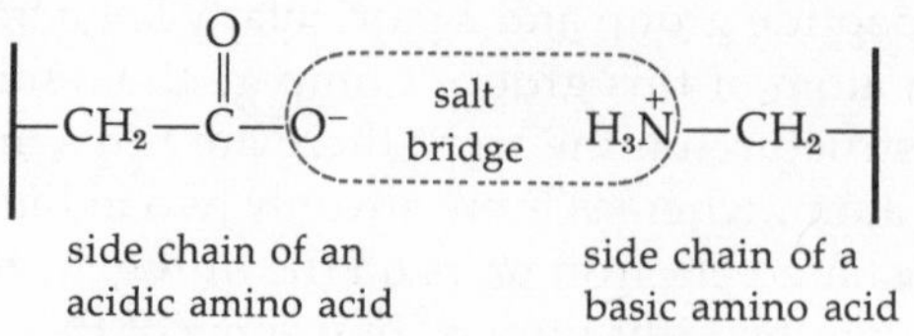

$$-CH_2-\overset{\overset{\displaystyle O}{\|}}{C}-O^-\ \underset{\text{bridge}}{\text{salt}}\ \overset{+}{H_3N}-CH_2-$$

side chain of an   side chain of a
acidic amino acid   basic amino acid

Figure 21-13   A salt bridge between charged side chains of nearby acidic and basic amino acids.

Levels of Protein Structure

It is extremely difficult to determine the exact shape of a protein, although this has been accomplished for several proteins by using a technique called X-ray diffraction analysis. When discussing the complicated shapes of large protein molecules, it is useful to specify four different levels of protein structure.

The amino acid sequence of a protein is called its **primary structure.** The term *secondary structure* refers to the spatial relationship between amino acids fairly close together in the protein's amino acid sequence. To clarify this, let us represent a protein as a string with knots corresponding to amino acids (see Figure 21-14). We could visualize several possible types of secondary structure, including a linear chain, a zigzag chain, a twisted spiraling chain, as well as a chain without a repetitive pattern. A large chain could consist of portions with different types of secondary structure (see Figure 21-15). We shall study the common types of protein secondary structure in Section 21-9. The term *tertiary structure* refers to the overall shape of a protein, including the spatial relationship

Figure 21-14   Representations for possible types of protein secondary structure. The peptide backbone can be viewed as a string and the amino acid components as knots in the string.

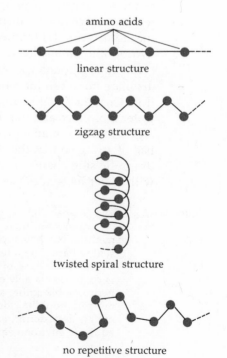

amino acids

linear structure

zigzag structure

twisted spiral structure

no repetitive structure

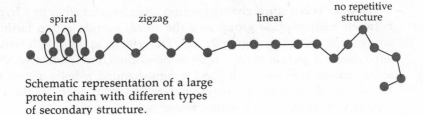

spiral   zigzag   linear   no repetitive
                                                structure

Figure 21-15   Schematic representation of a large
protein chain with different types
of secondary structure.

Figure 21-16   Schematic representation of the
tertiary structure of a protein. Note
that the amino acids labeled *a* and
b in the figure are not nearby in the
amino acid sequence but are close
together in space because of the
overall shape of this polypeptide.

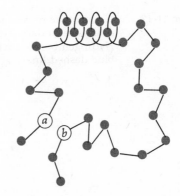

of amino acids that are far removed from each other in the amino acid sequence.
This is represented schematically in Figure 21-16 and will be discussed further
in Section 21-10.

Many proteins in the human body consist of more than one peptide chain.
The spatial relationship of component peptide chains in such a protein is called
the *quaternary structure* of the protein (see Section 21-11). For example, a protein
composed of two peptide chains is analogous to two balls of string that have
been glued together.

The precise shape of an individual protein is uniquely determined by its
amino acid sequence. Dramatic support for this assertion was recently obtained. In 1969, R. B. Merrifield and his colleagues completed the laboratory
synthesis of a protein with the same amino acid sequence as the naturally
occurring enzyme, ribonuclease. Their synthetic protein curled up into exactly
the same shape as that of natural ribonuclease.

Exercise 21-6
Using structural formulas, indicate side-chain interactions between the following pairs
of amino acids at pH 7.0. If there is no interaction, write none.

(a) Glycine and serine   (b) Lysine and aspartic acid
(c) Glutamine and serine   (d) Tryptophan and phenylalanine

21-9   SECONDARY STRUCTURE

Secondary structure of a protein refers to the spatial relationship of amino acids
that are fairly close together in the amino acid sequence. Several years ago,
Linus Pauling showed that two types of protein secondary structure, called the
α-helix and the β-pleated sheet, would allow a large number of hydrogen bonds
to form between nearby peptide groups. He postulated that many proteins have
an α-helix secondary structure or a β-pleated sheet secondary structure, and his
prediction was later verified by experimental observations.

The **α-helix** is a spiral chain of amino acids held together by a hydrogen bond between each peptide group and the third peptide group farther along the amino acid chain (see Figure 21-17). The α-helix spiral might twist in either a left-handed direction or in a right-handed direction. Naturally occurring proteins contain L-amino acids, and a right-handed α-helix allows for more effective hydrogen bonding between L-amino acids than would a left-handed α-helix. Therefore natural α-helices are right-handed.

Figure 21-17   A schematic representation of the α-helix secondary structure. Hydrogen bonds are indicated by blue dashed lines.

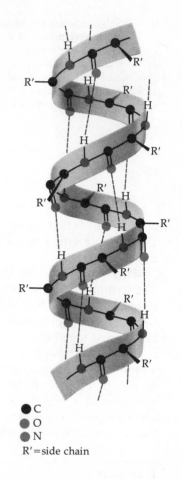

● C
● O
● N
R′ = side chain

The **β-pleated sheet** is a zigzag type of secondary structure in which two or more chains of amino acids are held together by hydrogen bonds between peptide groups on adjacent chains (see Figure 21-18). A single chain of amino acids can also curl around so that a β-pleated sheet forms between parts of the same chain (see Figure 21-19).

Notice in Figures 21-17 and 21-18 that in both the α-helix and the β-pleated sheet secondary structures the amino acid side chains (R′) stick out from the covalent backbone and that adjacent side chains are fairly close together. The stability of an α-helix or a β-pleated sheet is enhanced when salt bridges can form between negatively charged (acidic) and positively charged (basic) side-chain groups. On the other hand, if a portion of a peptide contains large, bulky side chains or adjacent side chains with the same electrical charge, then the repulsion between side-chain groups in this portion of the chain is likely to prevent an α-helix, a β-pleated sheet, or any other characteristic repetitive secondary structure. It has been determined experimentally that some proteins are entirely α-helical. One example is the keratin in human hair, which consists of three

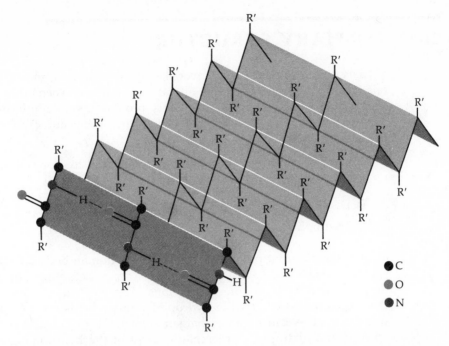

●C
●O
●N

Figure 21-18   A schematic representation of the β-pleated sheet type of secondary structure. The hydrogen bonds (blue dashed lines) between the three adjacent peptide chains are shown only in the first plane.

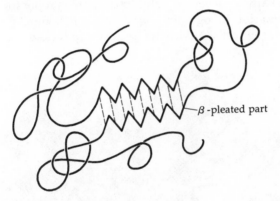

β-pleated part

Figure 21-19   Schematic representation of a β-pleated sheet formed between two parts of the same large protein. The blue dashed lines represent hydrogen bonds.

α-helices wound around each other as shown in Figure 21-20. Most proteins, however, contain only a part that is α-helical. For example, about 70% of the protein hemoglobin is helical, whereas only 7% of some other proteins is α-helical in shape.

Figure 21-20   The protein keratin, a component of hair, is a helix formed from three protein chains. Each component protein chain has an α-helical shape.

Exercise 21-7
Would you expect the peptide phe-tyr-ser-phe-tyr-glu-glu-glu-phe-tyr-lys-lys-lys to form an α-helix? Explain your answer.

21-10 TERTIARY STRUCTURE

Tertiary structure refers to the overall shape of a protein, including the spatial relationship of amino acids that are far removed from each other in the amino acid sequence (see Figure 21-16). Every protein has a definite unique overall (or tertiary) structure because of interactions involving side-chain groups. These interactions include:

1. Hydrogen bonding between pairs of neutral hydrophilic side chains
2. Salt bridges between oppositely charged side chains
3. Disulfide bonds between pairs of cysteines
4. Hydrophobic and hydrophilic interactions

Most of the different protein molecules in the human body are surrounded by water molecules and have a globular overall shape determined primarily by hydrophilic and hydrophobic interactions. The particular globular shape that these proteins adopt has the hydrophilic amino acids on the outside, where water molecules can hydrate charged side-chain groups and hydrogen bond with neutral hydrophilic side chains. Most of the hydrophobic amino acids, however, are packed very tightly together in the interior of the molecule, which is often called the **hydrophobic core.**

The tertiary structure of the protein myoglobin is represented in Figure 21-21. The myoglobin chain is packed so tightly that there is room for only four water molecules in its interior. Myoglobin is very similar to hemoglobin and is used to store oxygen in muscle cells. Notice the heme group tucked into a pocket on the

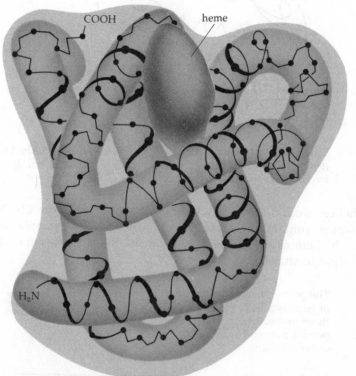

Figure 21-21   A schematic representation of the tertiary structure of whale myoglobin. The backbone is shown in black. The spaces between portions of the backbone are completely filled by side-chain groups, indicated here by pale blue shading. Tertiary structure is maintained by interactions between side-chain groups on different parts of the protein.

surface of myoglobin. This is where the oxygen molecule binds. Most other proteins, including enzymes, also have pockets where they can bind small molecules. We shall see that the existence and shape of such pockets is a crucial factor in determining how enzymes work (Chapter 23).

Exercise 21-8
Which of the following amino acids would you expect to find in the hydrophobic core of a protein?
(a) Lysine   (b) Isoleucine   (c) Glycine   (d) Arginine   (e) Histidine   (f) Tryptophan

---

21-11 QUATERNARY STRUCTURE

Proteins that contain more than one polypeptide chain are called **oligomeric** proteins. The component polypeptide chains are called **subunits,** and they may all be the same or they may differ. Oligomeric proteins with either two or four subunits are the most common types, but proteins with three or many more subunits are known to exist.

Hemoglobin, for example, is an oligomeric protein composed of four subunits, all of which are quite similar in structure to myoglobin. In normal human adults, hemoglobin has two different subunits, called α and β. A single hemoglobin molecule consists of two α-type and two β-type subunits. Each subunit can bind oxygen. When oxygen binds to one subunit, the shapes of all the subunits change slightly. The functional significance of this slight change in shape will be discussed in the next chapter.

The subunits of an oligomeric protein usually interact considerably with one another, and an oligomeric protein is thus one grand protein molecule. Subunits are held together by hydrogen bonds, salt bridges between positively and negatively charged side-chain groups, hydrophilic and hydrophobic interactions, and, in some cases, disulfide bonds between cysteines on different subunits.

In the next chapter we shall see how oligomeric proteins can have certain functional advantages over single-subunit (monomeric) proteins. One advantage of making large proteins out of smaller subunits is analogous to the advantage a component stereo sound system has over a single-unit system. If the cell makes a defective component (subunit), it does not have to scrap the entire system. It may simply replace the defective subunit with a good one.

Isoenzymes, a special class of oligomeric proteins, are enzymes that catalyze the same reaction but have slightly different subunits. The best-known examples of isoenzymes are those of the enzyme lactic acid dehydrogenase (LDH). LDH has two types of subunits, called M and H, which differ to some extent in their amino acid sequence. These subunits associate to form a tetrameric (four-subunit) quaternary structure. There are five LDH isoenzymes: M_4, M_3H, M_2H_2, MH_3, and H_4. It is relatively easy to separate a mixture of LDH isoenzymes in the laboratory, because M and H subunits differ in the number of charged amino acid side-chain groups.

It has been found that heart muscle cells make almost entirely H_4-type LDH (H stands for heart), whereas cells that make up skeletal muscle make predominantly M_4-type LDH (M stands for muscle). White blood cells and other tissues make both types of subunits, so they have a mixture of LDH isoenzymes. This information is used clinically to aid in the diagnosis of certain diseases. For example, if a person's heart cells are damaged, the H_4 type of LDH will leak out of the damaged cells and the resulting increase in H_4 concentration in that person's serum can be detected.

21-12 PROTEIN DENATURATION

Under certain conditions it is possible to disrupt the unique shape of a protein without breaking any of the bonds that hold the covalent backbone together. Any process that drastically alters the shape of a protein but leaves the primary structure intact is called **denaturation.** A protein denatures when an increase in temperature or a change in the composition of the solution disrupts the hydrogen bonds, salt bridges, and/or hydrophobic interactions responsible for its secondary, tertiary, and quaternary structure. A change in solution pH or a change in salt concentration can cause a protein to denature. Substances that reduce disulfide bonds can also denature proteins.

A denatured protein loses its ability to function properly. For example, in order to measure the activity of blood serum enzymes in the clinical laboratory, it is imperative that the serum samples be protected from conditions that would denature and thus inactivate them.

Some proteins that have been denatured by a change in the composition of the solution or the breaking of disulfide bonds can be *renatured* — that is, refolded into their original shape. In such a case, we say that the denaturation is *reversible.* For example, the enzyme ribonuclease can be denatured by the addition of substances that disrupt hydrogen bonds, hydrophobic interactions, salt bridges, and disulfide bonds. Denatured ribonuclease cannot function as a catalyst. If the substances responsible for the denaturation are slowly removed, however, ribonuclease will refold into its original shape and regain its enzyme activity (see Figure 21-22). The phenomenon of reversible denaturation is additional evidence for the fact that the primary structure of a protein determines its overall shape.

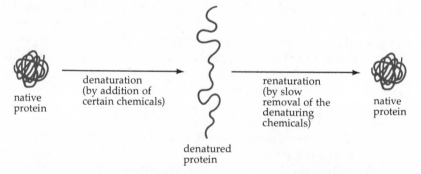

Figure 21-22   Schematic representation of the reversible denaturation of a protein. The shape and enzymatic activity of the renatured protein are the same as they are for the undenatured, native protein.

Some proteins cannot be reversibly denatured, and also, some methods of protein denaturation are not reversible (see Figure 21-23). A familiar example is cooking an egg. When an egg is cooked, the proteins in the egg are irreversibly denatured by the heat. When the egg cools, these proteins do not renature. Rather, large numbers of individual protein molecules clump together into an insoluble complex held together primarily by hydrophobic interactions. The irreversible denaturation of proteins that results in such insoluble complexes is called **coagulation.**

Partial denaturation and renaturation are involved in the process of hair curling. People commonly curl hair by using hot curlers or by a more involved process called permanent waving. We have seen that hair protein consists of three helices twisted around each other and held together by hydrogen bonds and disulfide bonds (see Figure 21-24). Hot curlers can disrupt some of these

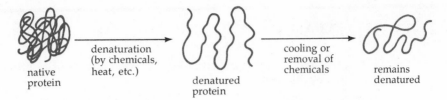

Figure 21-23 Schematic representation of irreversible protein denaturation.

hydrogen bonds (partial denaturation), and when the hair cools new hydrogen bonds form. Since the hair cools on the curlers, the new hydrogen bonds tend to stabilize the curled shape of the hair. However, hot, humid weather or a soapy shower can disrupt these hydrogen bonds and remove the curls.

In the permanent waving process, the disulfide bonds between the three interwoven helices of hair are broken by reducing agents in the wave treatment solution (Figure 21-24). The hair is then set on curlers and new disulfide bonds form, using oxygen in the air as the oxidizing agent; the new disulfide bonds hold the hair in a curled shape. These covalent disulfide bonds hold curls much more permanently than hydrogen bonds.

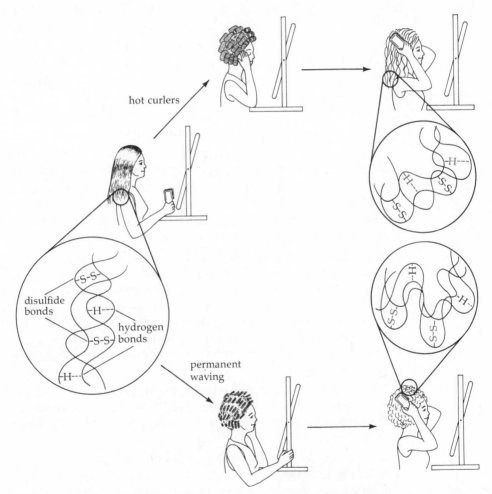

Figure 21-24 Curling hair. Hot curlers disrupt some hydrogen bonds. As the hair cools on the curlers, new hydrogen bonds form, maintaining the curled shape of the hair. However, by reducing and then re-forming disulfide bonds in the permanent waving process, more stable curls can be formed.

21-13 SUMMARY

1. Proteins are polymers composed of long chains of α-amino acids joined by peptide bonds.

2. A peptide bond is an amide bond formed between the carboxyl group of one amino acid and the amino group of another amino acid. Short chains of amino acids joined by peptide bonds are called peptides.

3. Each peptide or protein has a unique amino acid sequence, which is crucial to its overall shape and function. By convention, amino acid sequences are written with the N-terminal amino acid on the left and the C-terminal amino acid on the right.

4. The general formula for an α-amino acid is
$$H_2N-\underset{\underset{R'}{|}}{\overset{\overset{H}{|}}{C}}-\overset{\overset{O}{\|}}{C}-OH.$$
The 20 α-amino acids used to assemble proteins are classified according to the properties of their side-chain (R') groups.

5. The essential amino acids are those that are required for protein synthesis but that cannot be synthesized by the human body. They must be included in the diet.

6. Glycine is the only amino acid that is not optically active. Only the L-enantiomers of the other 19 amino acids are used in the synthesis of proteins.

7. In the solid state, amino acids exist as zwitterions. The predominant form of an amino acid in solution depends on the pH.

8. The unique overall shape of a protein is a result of the planar nature of its peptide groups, hydrogen bonds that occur between peptide groups, and various types of interactions among the side-chain groups of its component amino acids.

9. The attraction between a negatively charged and a positively charged side-chain group is called a salt bridge.

10. The shapes of proteins are discussed in terms of four levels of protein structure. The primary structure of a protein is its amino acid sequence. Secondary structure refers to the spatial relationship of amino acids that are fairly close together in the amino acid sequence of a protein. Tertiary structure refers to the overall shape of a protein, including the spatial relationship of amino acids that are far apart in the amino acid sequence. The spatial relationship of the subunits of an oligomeric protein is referred to as quaternary structure.

11. The α-helix and the β-pleated sheet are two types of secondary structure that are frequently found in proteins. These types of secondary structure are held together by hydrogen bonds between peptide groups of nearby amino acids.

12. Interactions among amino acid side-chain groups, including hydrogen bonding, salt bridges, disulfide bonds, and hydrophilic and hydrophobic interactions, stabilize the tertiary structure of proteins. These types of interactions also hold subunits together in oligomeric proteins.

13. Oligomeric enzymes that catalyze the same reaction but that have somewhat different subunits are called isoenzymes.

14. Any process that drastically alters the shape of a protein but leaves its primary structure intact is called denaturation. Under certain conditions the denaturation of some proteins is reversible.

15. Coagulation is the irreversible denaturation of a protein, which results in the formation of insoluble complexes.

PROBLEMS

1. Define the following terms: (a) zwitterion, (b) essential amino acid, (c) isoenzyme, (d) denaturation, (e) salt bridge.

2. Draw the general structural formula for α-amino acids.

3. What is the difference between a peptide and a protein?

4. The R group of the amino acid glutamic acid is $-CH_2-CH_2-\overset{\overset{\displaystyle O}{\|}}{C}-OH$. What is the net charge on the predominant form of glutamic acid at (a) pH = 1.0, (b) pH = 6.8, and (c) pH = 13.5?

5. What type of force is most important in stabilizing the α-helix form of a polypeptide chain?

6. The respective R groups of the amino acids glutamic acid, lysine, alanine, and serine are

$-CH_2-CH_2-\overset{\overset{\displaystyle O}{\|}}{C}-OH$, $-CH_2-CH_2-CH_2-CH_2-NH_2$, $-CH_3$, and $-CH_2OH$.

(a) Draw the structural formula for the tetrapeptide ala-glu-lys-ser in its predominant ionic form at pH 7.0.

(b) Identify the peptide bonds in ala-glu-lys-ser and indicate side-chain interactions if any occur.

7. Hydrophobic interactions are important in which levels of protein structure?

8. Which of the following changes is most likely to denature a globular protein?
   (a) A decrease in temperature from 37°C to −10°C
   (b) An increase in pH from 7.4 to 8.5
   (c) A decrease in pH from 7.4 to 6.5
   (d) An increase in temperature from 37°C to 85°C
   (e) None of the above

9. The predominant form of the amino acid histidine at pH 8.5 is

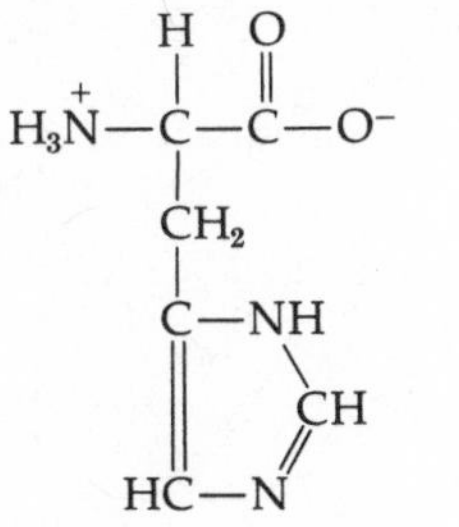

What is the predominant form of histidine at pH = 5.0?

10. Referring to Table 21-1, indicate which side chains of the following amino acid pairs will form (a) a hydrogen bond, (b) hydrophobic interactions, or (c) neither:
   (i) ser and ser   (ii) gln and thr   (iii) glu and phe
   (iv) phe and tyr   (v) tyr and thr

11. Define the terms (a) oligomer, (b) subunit, (c) coagulation, (d) hydrophobic core, and (e) β-pleated sheet, as they refer to proteins.

12. How are disulfide bonds involved in maintaining the shape of proteins?

13. Referring to Table 21-1, draw the structural formula for the predominant form of the tetrapeptide gly-ile-ser-val at pH 7.0.
    (a) Indicate which atoms in this tetrapeptide must lie in a common plane.
    (b) What is the net charge on this molecule at pH 1.0?

14. At what pH does the uncharged form of the amino acid arginine predominate?
    (a) pH = 1.0   (b) pH = 7.0
    (c) pH = 14.0   (d) None of these

SOLUTIONS TO EXERCISES

21-1 $R_1'—R_3'—R_2'$
      $R_2'—R_1'—R_3'$
      $R_2'—R_3'—R_1'$
      $R_3'—R_1'—R_2'$
      $R_3'—R_2'—R_1'$

21-2 (a) The structural formulas for the two acidic amino acids, aspartic acid and glutamic acid, are

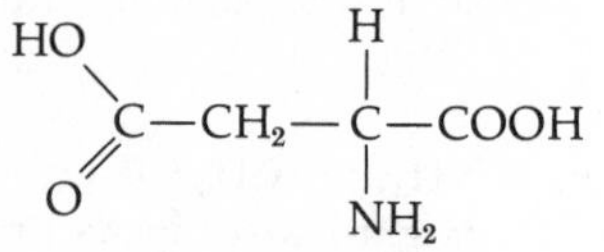

aspartic acid 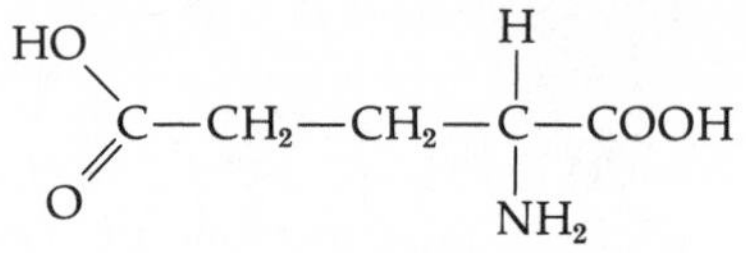   glutamic acid

(b) Structural formulas for the three basic amino acids, lysine, arginine, and histidine, are

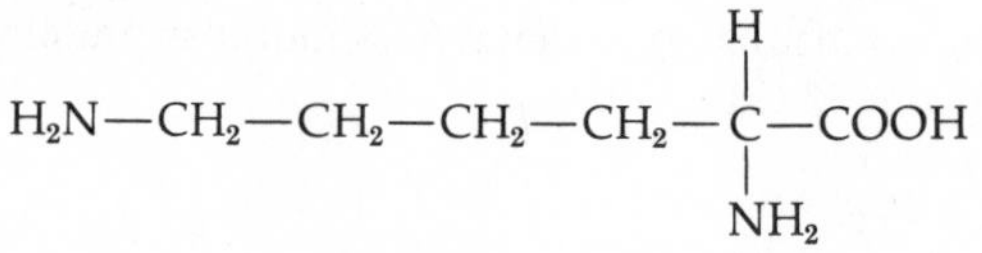

lysine

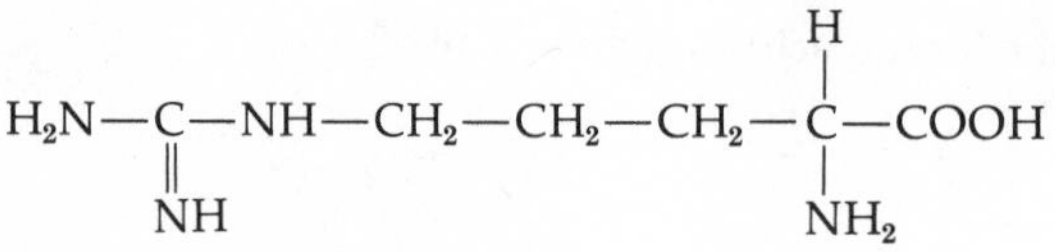

arginine

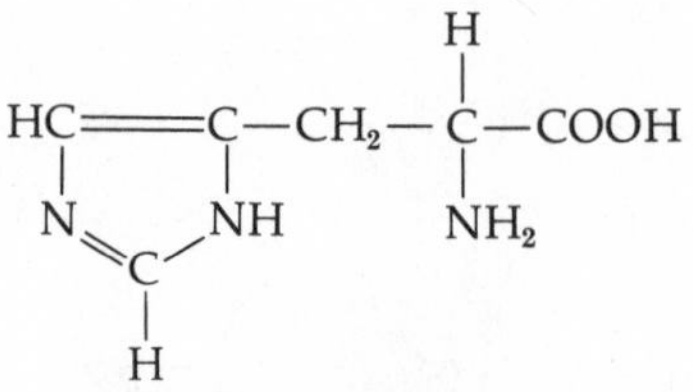

histidine

(c) The neutral amino acids (the other 15 amino acids used for protein synthesis) are shown in Table 21-1A and 21-1B. Note that this exercise does not specify hydrophilic or hydrophobic neutral amino acids, so either is correct.

The functional groups responsible for the classification of these compounds are as follows:

(a) Acidic: For both aspartic acid and glutamic acid, the side-chain carboxyl group.

(b) Basic: For lysine, the amino group (—NH$_2$) on the side chain.

For arginine, the —N—C—NH$_2$ group.
$\qquad$ | $\quad$ ||
$\qquad$ H $\;$ NH

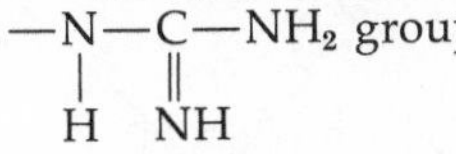

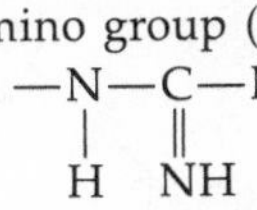

For histidine, the —CH$_2$—C===CH group.

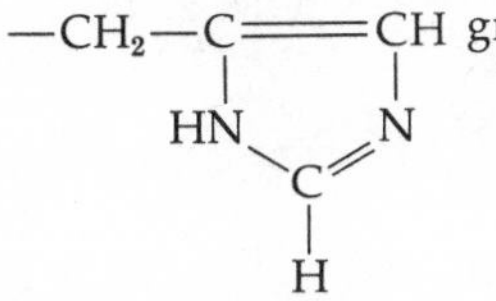

(c) Neutral: The side chains of the various neutral amino acids contain several types of functional groups, including —SH, —OH, and so on, but none that act as proton donors or acceptors, so they are classified as neutral.

21-3 (a) Neutral hydrophilic
$\qquad$ (b) Neutral hydrophilic
$\qquad$ (c) Neutral hydrophobic
$\qquad$ (d) Neutral hydrophobic

21-4 (a) Valine

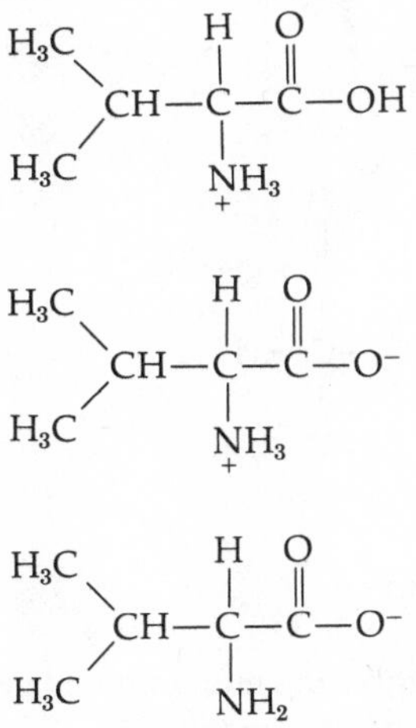

(b) Aspartic acid

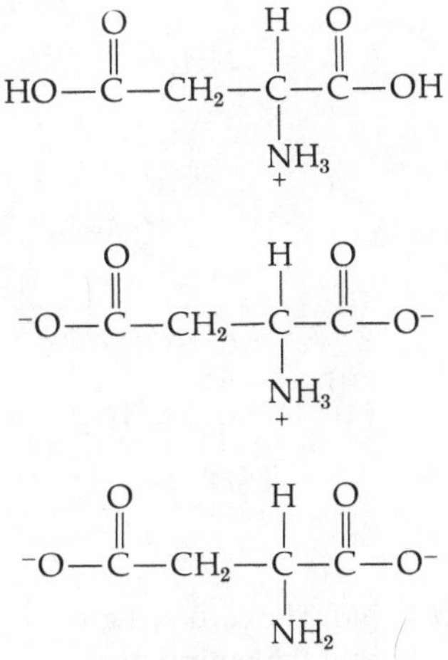

(c) Serine

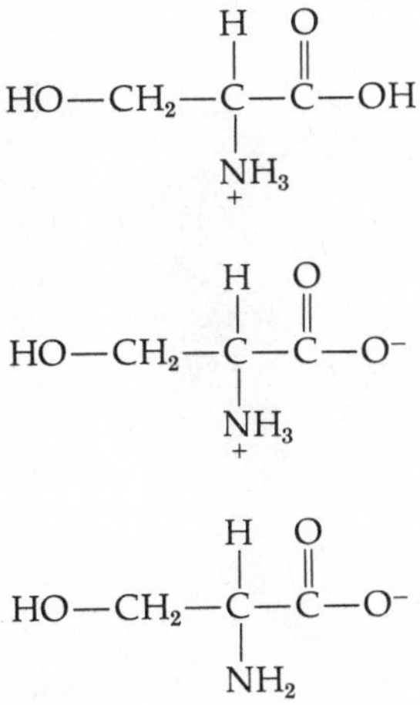

(d) Lysine

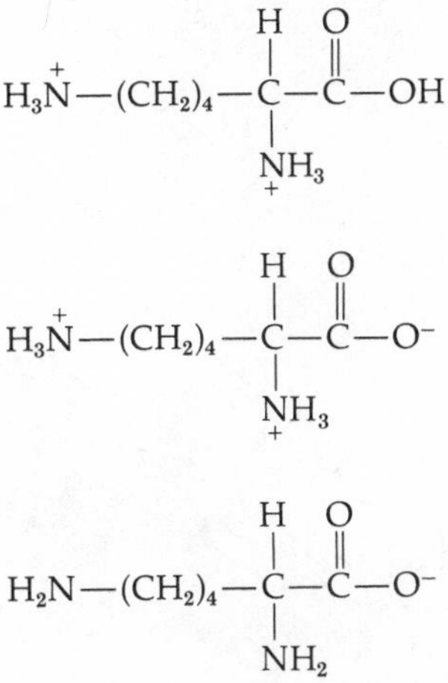

21-5

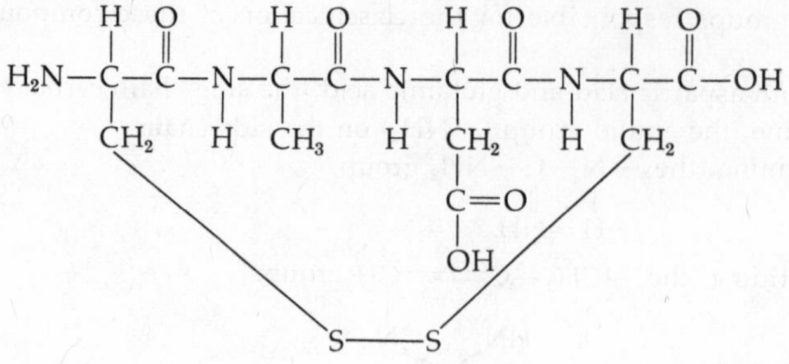

21-6 (a) None

(b)

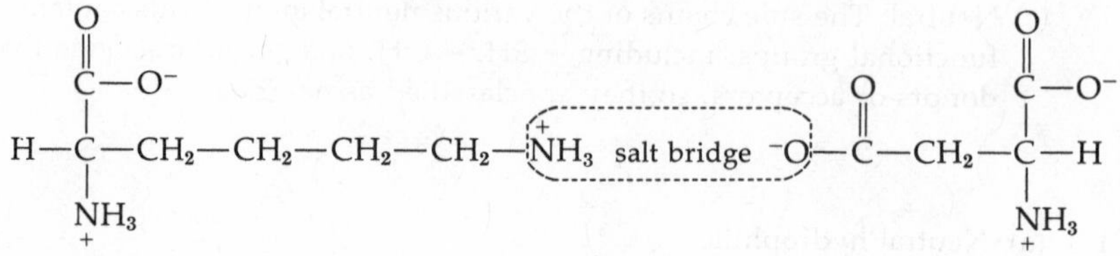

(c)

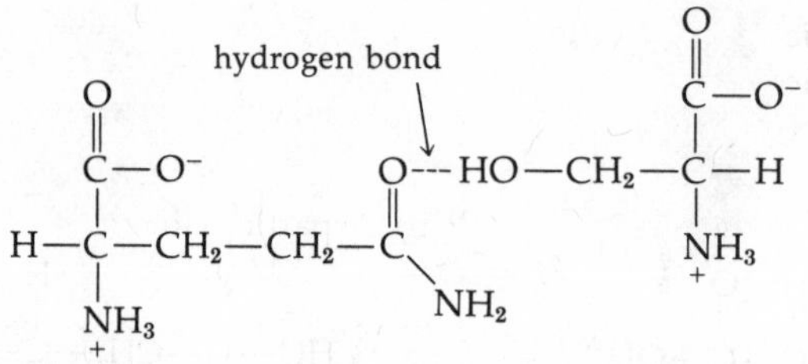

(d)

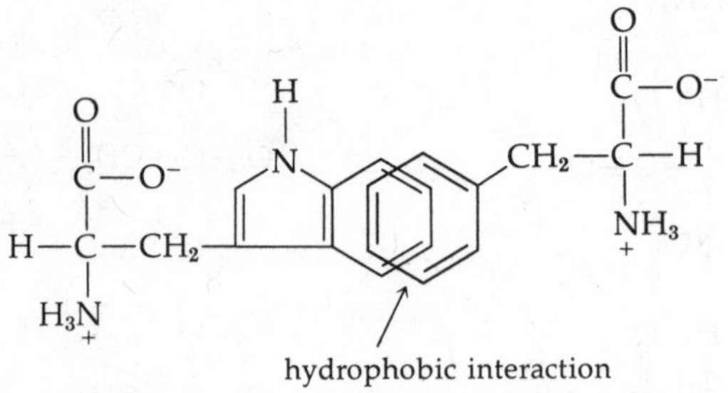

21-7 No. The carboxyl groups of the three neighboring glutamic acids will repel each other, as will the amino groups of the three neighboring lysines.

21-8 (b) Isoleucine, (f) tryptophan, and possibly (c) glycine, because the side chain of glycine (—H) does not interact strongly with water.

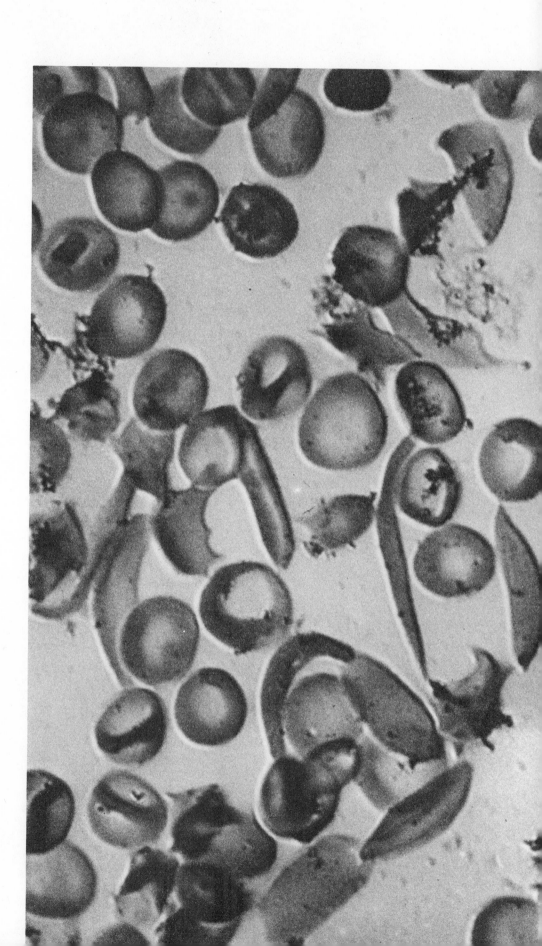

CHAPTER 22

Protein Function

22-1 INTRODUCTION

In the early years of this century, Henry Ford realized that a factory could operate much more efficiently if each worker assembled only a small portion of the total product. Of course, Ford was not the first person to realize the advantages of the assembly line, but he was extraordinarily successful at putting the idea into practice and so did as much as anyone to launch modern society into the age of specialization.

Long before Henry Ford promoted this industrial application of specialization, the proteins in living organisms had evolved their specialized functional roles. Each cell in the human body, for example, contains more than a thousand different proteins. Each protein has a specific, very limited role in the overall operation of the cell.

It would be impossible to study each protein individually; there are just too many. Instead, we shall concentrate on *functional categories* of proteins. The division of proteins into functional classes resembles the division of factory workers by classification. Many proteins function as *enzymes*, which work on cellular assembly lines called metabolic pathways. *Hormones* are chemical messengers and *carrier proteins* convey assembled products or nutrients between cells in different parts of the body. *Membrane transport proteins* work on the loading dock of the cellular factory, carrying molecules or ions into and out of cells. *Antibody molecules* guard the body against foreign invaders.

In the last chapter we saw that proteins are polymers of amino acids. Each different protein has a unique amino acid sequence that determines its overall structure. Its precise structure, in turn, determines its function in the cell. After we have classified proteins according to their roles in the cell, we shall look more closely at the structural and functional properties of protein hormones, transport proteins, and antibodies. Understanding how proteins function in the healthy individual will enable us to see how defects in protein function can lead to disease.

In the blood of a patient suffering from sickle cell disease, some of the red blood cells have a shape that is strikingly different from the normal doughnut-shaped red blood cell. The sickle shape is due entirely to a difference in a single amino acid in two of the four subunits of the hemoglobin molecule.

22-2 STUDY OBJECTIVES

After studying the material in this chapter, you should be able to:

1. Name the major functional classes of proteins and describe the role of each class in the cell and the whole body.

2. Describe how hormonal messages are amplified.

3. Differentiate among simple diffusion, passive transport, and active transport.

4. Define the terms prosthetic group and conjugated protein.

5. Describe specificity, saturation, and inhibition as they apply to membrane transport proteins.

6. Define the terms antigen, immunoglobulin, complement, agglutination, and vaccination.

7. Draw a diagram representing an IgG molecule and use the drawing to describe the structure and function of antibodies.

22-3 EVOLUTION OF PROTEIN FUNCTION

The complex and precise structure of each individual protein is essential to the survival of the living organism. During the course of evolution, the structures of individual proteins have been selected to optimize their specific functions. Even slight variations in the structure of a particular protein can lead to enormous consequences. For example, hemoglobin S, which occurs in people with sickle cell anemia, differs from normal hemoglobin A by only 1 amino acid out of 146 in half of the four hemoglobin subunits. The change of a single amino acid in the sequence of a protein is called an **amino acid substitution.** Such substitutions almost always lead to a functionally inferior or inactive protein, and thus to an organism with a poorer ability to survive and reproduce in a given environment.*

In our modern society, organisms with defects are not always selected against. Malfunctions such as diabetes or myopia (near-sightedness) can be compensated for. Given proper treatment, people with these and other disorders can survive and reproduce. In fact, the proportion of individuals with such disorders in the human population is actually increasing.

An amino acid substitution rarely leads to a functionally "better" protein. Yet these rare favorable changes are necessary, because eventually they lead to the evolution of organisms that are better adapted to their particular environment, organisms that have a better chance to survive and reproduce. These better adapted organisms are more likely to thrive and may even compete so successfully with those that do not possess the "better" protein that the latter organisms may become extinct.

---

* Interestingly, individuals possessing hemoglobin S have increased resistance to malaria. The **sickle cell trait,** in which a person has genes for, and synthesizes, both hemoglobin A and hemoglobin S, is quite common in individuals of African ancestry. Although most of the red blood cells of these individuals will sickle in laboratory tests, they have enough hemoglobin A to prevent this from occurring in their bodies. Individuals with **sickle cell disease** have genes for hemoglobin S only. Their red blood cells, which are devoid of any hemoglobin A, can sickle in their bodies—causing much pain and even death.

This process of evolution and selection has bestowed on modern organisms, including humans, a large number of functional proteins. Each of these proteins has evolved to the very specific structure it needs to function as it does. The general process of evolution of protein function has also made more highly evolved organisms different in appearance from their predecessors. The genetic information that each of us has inherited (such as height, a tendency for baldness, or the shape of our ears) is coded in our DNA. It is DNA that directs the synthesis of proteins in such a way as to give each of us our unique qualities.

Since virtually everything that happens in the body depends on the function of proteins, the study of human biochemistry can ultimately be viewed as the study of proteins and what they do. As we proceed to look at the functional roles and properties of proteins, we must always bear in mind that it is the highly evolved structure of a protein that determines its very specific function.

Exercise 22-1

In two subunits of hemoglobin S, the amino acid valine is substituted for a glutamic acid found in hemoglobin A. Hemoglobin S is much less soluble than hemoglobin A. How do you think a hypothetical hemoglobin, call it hemoglobin Z, with an aspartic acid substituted for this glutamic acid, would compare to hemoglobin A and hemoglobin S in solubility and function?

22-4 CLASSIFICATION OF PROTEINS

Proteins may be classified in two ways. They may be classified on the basis of their gross physical properties, such as solubility in water or coagulation by heat. Recall that coagulation is a special kind of irreversible denaturation in which protein precipitates out of solution, such as egg white solidifying upon cooking. Proteins may also be classified on the basis of their functional properties (that is, what they do).

Classification According to Physical Properties

The classification scheme based on physical properties was devised first, before the functions of most proteins were known. It is still a useful way to classify proteins because the physical properties of a protein are the result of its structure, as are its functional properties.

Table 22-1 shows the classification of proteins based on their physical properties. You will notice that **fibrous proteins** are important in the structure of hair, nails, and connective tissue. **Globular proteins,** on the other hand, include proteins with a wide variety of roles, such as enzymes and histones. For example, serum albumin is found in large quantities in human blood. One particular property, its high solubility, allows it to function as a carrier protein for molecules, such as fats, that would otherwise be insoluble in the aqueous medium of serum. Most other globular proteins are less soluble than albumin and are placed in the globulin subclass. Some globulins are antibodies, whose function is to bind and inactivate foreign substances and thus combat disease and infection. Most enzymes are also globulins. Enzymes work inside the cellular factories and are not usually found in the blood serum. Indeed, the presence of certain enzymes in serum is used to diagnose some diseases. For example, the injury of cardiac muscle cells caused by a myocardial infarction (heart attack) will cause these cells to release specific enzymes into the serum. Laboratory tests for these enzymes are used to detect the presence and severity of a heart attack.

Table 22-1 Classification of Proteins Based on Physical Properties

Class	Properties	Examples
Fibrous proteins	Insoluble in aqueous solutions; elongated molecules, often consisting of several coiled polypeptide chains	
Collagens	Can be converted into soluble gelatins by boiling; contain large amounts of hydroxy-proline and hydroxylysine but no cysteine or tryptophan	The major proteins of connective tissues
Elastins	Similar to collagens but cannot be converted to gelatins by boiling	Proteins of tendons and arteries
Keratins	Contain large amounts of cysteine	Hair, wool, nails (Hair is about 14% cysteine.)
Globular proteins	Soluble in aqueous solutions; spherical or ellipsoidal in shape	
Albumins	Readily soluble in pure water; coagulated by heat; function as carriers for hydro-phobic molecules	Serum albumin, egg albumin
Globulins	Insoluble or only slightly soluble in pure water; very soluble in aqueous salt solutions; can be coagulated by heat	Enzymes and antibodies
Histones	Basic proteins; contain large amounts of arginine and lysine; soluble in pure water	Histones in chromatin
Protamines	Very basic proteins; contain large amounts of arginine, but no tryptophan or tyrosine	Found in sperm cell chromosomes

Other subclasses of proteins in Table 22-1 also have physical properties that are necessary for their particular function. For example, proteins known as histones contain large amounts of the basic amino acids arginine and lysine. This basic property allows histones to interact with the acidic phosphate portions of DNA molecules. The combination of histones and DNA in the cells of higher organisms such as humans is called chromatin.

Another practical use of this classification scheme can be found in the clinical laboratory, where the difference in solubility between albumins and globulins is used to determine their concentrations. The relative amount of albumins and globulins in serum is usually expressed as the ratio of albumin to globulin concentration, or the A/G ratio. Variations in the A/G ratio indicate several disease states, such as hypogammaglobulinemia, which means that a patient has less than (*hypo-*) the normal amount of serum gammaglobulins. Since most gammaglobulins are antibodies, a patient with a high A/G ratio may have an impaired resistance to infection, that is, a malfunctioning immune system.

Exercise 22-2

An individual is suspected of having a type of cancer called multiple myeloma, in which some of the cells that produce antibody molecules have become malignant. Would this individual's A/G ratio be lower or higher than normal? Explain your answer.

Classification According to Functional Properties

Classification of proteins based on function is relatively recent. We have already mentioned several of these functions, including transport, catalysis, and communication. Major functional classes of proteins are listed in Table 22-2. This

Table 22-2 Protein Classification by Function

Class	Properties	Examples
Catalytic proteins		
Enzymes	Catalyze chemical reactions	Lactate dehydrogenase (LDH), amylase, pyruvate dehydrogenase
Noncatalytic proteins		
Carrier proteins	Carry molecules or ions through the bloodstream	Hemoglobin, albumin
Receptor proteins	Bind hormones and neurotransmitters to cell membranes	The insulin receptor
Membrane transport proteins	Carry molecules across cell membranes	Na^+K^+-ATPase, which transports K^+ ions into cells and pumps Na^+ ions out of cells
Structural proteins	Form extracellular structures such as hair and nails	Collagen, keratin
Contractile proteins	Extend or contract to produce movement of muscles, cells, or subcellular parts	Myosin, tubulin
Protein hormones	Messenger molecules that direct the activities of various cells and organs	Insulin, adrenal corticotropic hormone (ACTH), growth hormone
Antibodies	Bind to foreign substances and activate their elimination from the body	Anti-Rh, Anti-A (antibodies to Rh factor and to blood group A)

method for classification of proteins is extremely useful. We want to know what proteins do and how they work. When a protein does not work, we want to know how this defect will affect the body as a whole. We shall spend the rest of this chapter and the next studying protein function.

Conjugated Proteins

Many proteins are self-sufficient, that is, they function perfectly well without the help of other molecules. Many other proteins require the assistance of a nonprotein molecule in order to function. For example, as we have previously mentioned, many enzymes require small molecules called coenzymes (such as NAD^+) in order to function properly as catalysts. As we shall see in Chapter 23, coenzymes are altered in enzyme-catalyzed reactions. In most enzyme-catalyzed reactions the altered coenzyme (such as NADH) quickly dissociates from the enzyme. On the other hand, several proteins, including some enzymes, have non-amino acid molecules or ions bound very tightly to them. We call such proteins **conjugated proteins,** and the auxiliary molecules bound to them are called **prosthetic groups.**

There are several types of prosthetic groups, ranging from metal ions to the complex heme group in hemoglobin. Heme is a complex ion containing a central iron ion bonded to a complicated nitrogen-containing organic molecule. Some examples of conjugated proteins and their prosthetic groups are listed in Table 22-3. Notice the wide variety of molecules that serve as prosthetic groups and the rather arbitrary distinction between glycoproteins and mucoproteins. Prosthetic groups are bound to their proteins very tightly, frequently via covalent bonds to certain amino acid side chains. They can also be bound via multiple charge interactions or by hydrogen bonds as in nucleoproteins. The role of prosthetic

Table 22-3 Conjugated Proteins

Type	Prosthetic Group	Properties	Example
Nucleoproteins	Nucleic acid (DNA, RNA)	Large, compact complexes	Chromatin, ribosomes
Mucoproteins*	Carbohydrate	More than 4% carbohydrate by weight	Human chorionic gonadotropin, a hormone used to test for pregnancy
Glycoproteins*	Carbohydrate	Less than 4% carbohydrate	Antibodies
Lipoproteins	Lipid	Water soluble	Serum lipoproteins
Proteolipids	Lipid	Not very water soluble, soluble in nonpolar solvents	Cell membranes
Hemoproteins	Heme group	Characteristic color	Hemoglobin, cytochrome c
Metalloproteins	Metal ion (Fe^{3+}, Zn^{2+}, Mg^{2+}, Mn^{2+})	Require a metal ion to function	Carbonic anhydrase

* The distinction between mucoproteins and glycoproteins is somewhat arbitrary.

groups in the function of a particular protein is not always completely understood. In hemoglobin, for example, oxygen actually binds to the heme group (see Figure 22-1), but the function(s) of the carbohydrate prosthetic groups found on antibodies and several enzymes is unclear.

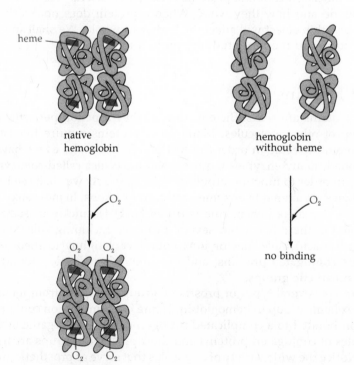

heme

native
hemoglobin

hemoglobin
without heme

O_2

O_2

O_2 O_2

no binding

O_2 O_2

Figure 22-1   Hemoglobin has four subunits, each of which normally contains one heme prosthetic group. An O_2 molecule binds to each of the heme groups in native hemoglobin (left), but cannot bind to hemoglobin if the heme groups have been removed (right).

22-5 NONCATALYTIC PROTEINS

Proteins can be divided broadly into two functional categories: those that serve as catalysts—the enzymes—and those that do not. The noncatalytic proteins can be further classified according to their various functional roles. Carrier proteins, receptor proteins, membrane transport proteins, structural proteins, contractile proteins, hormones, and antibodies are all functional classes of noncatalytic proteins.

Carrier Proteins

Carrier proteins carry nutrients and water-insoluble vitamins to cells in need of them. Carrier proteins are the transportation workers of the human body. Albumin, hemoglobin, and transferrin are examples of carrier proteins. Albumin is relatively nonselective and transports a variety of fats through the bloodstream. Hemoglobin is more selective, but can still bind several different kinds of molecules. It binds oxygen tightly and carries it to cells that need it. Unfortunately, hemoglobin can also bind carbon monoxide (CO), and it does so more readily than it binds O_2. Therefore, exposure to too much CO leads to oxygen starvation of cells. The cells suffocate and the person dies. Transferrin is a carrier protein that is very selective. It carries iron from the intestines to the liver.

Receptor Proteins

The chemical messages between cells in the body, in the form of hormones and neurotransmitters, are received by specific **receptor proteins** located in cell membranes. After "capturing" the specific chemical messenger that it binds, a given receptor protein may (1) help escort the messenger across the cell membrane and into the cell, as is the case with steroid hormone receptors; or (2) activate or deactivate an enzyme located on the interior side of the cell membrane in order to convey the message to the interior of the cell. In this case the original messenger (a hormone) does not itself enter the cell. Receptors for the hormones adrenaline and insulin work in this fashion. The receptor protein may also (3) generate an electrical response. This occurs when the receptor proteins on nerve cells bind small molecules called **neurotransmitters,** which carry messages between nerve cells.

Membrane Transport Proteins

The **membrane transport proteins,** located in cell membranes, carry molecules and ions from the aqueous environment that surrounds the cell across the cell membrane, which contains a large amount of water-insoluble fats, and into the aqueous cytoplasm. (Some chemists refer to carrier proteins as transport proteins, but we shall not in order to avoid confusion.) Membrane transport proteins are analogous to workers on the loading dock of a factory. They handle shipping as well as receiving operations for the cellular factory. Sometimes transport across the cell membrane must be accomplished against a concentration gradient. This is hard work and requires energy input. As you will recall, molecules have a natural tendency to move from an area of high concentration to an area of low concentration. Molecules concentrated inside cells have a natural tendency to leak out. A great deal of the energy used daily by the body goes to reverse this tendency, moving molecules to a place where they are already more concentrated. We shall discuss transport across cell membranes in greater detail shortly.

Structural Proteins

Structural proteins form large polymers. Their functional properties, again, are due to their specific amino acid sequences. Hair and nails (keratins), the collagen of connective tissue, and the elastins of tendons and arteries are present in large amounts in the body. In fact, more than 30% of the total weight of protein in the body is collagen. Structural proteins have a broad range of properties. Keratins in nails are hard and brittle, elastins make tendons tough and elastic. Fibrin, the protein in blood clots, is formed from fibrinogen, a soluble globular protein normally present in unclotted blood. The process of blood clotting involves the conversion of fibrinogen into insoluble fibrin polymers by the enzyme thrombin (see Figure 22-2). The major difference between serum and plasma is the presence of fibrinogen in plasma. **Plasma** is obtained from blood that has been prevented from clotting and thus still contains fibrinogen. **Serum** is the fluid obtained from clotted blood and thus has no fibrogen.

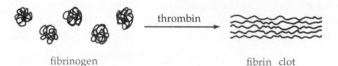

fibrinogen   fibrin clot

Figure 22-2   Certain stimuli, such as the injury   Thrombin in turn converts soluble
                   of a blood vessel, promote the   fibrinogen molecules into
                   conversion of prothrombin (an   insoluble fibrin clots.
                   inactive precursor) to thrombin.

Contractile Proteins

Contractile proteins function in the movement of cells and organs. A major example is muscle contraction. The contraction process in muscle cells is very complex and requires the association of several proteins into a network that contracts by the sliding of some proteins along a string of other proteins. Different contractile proteins are required for cells to divide, for sperm to swim, and so on. These contractile proteins, called *microtubules*, are long, hollow filaments that are built up from very large numbers of a subunit called tubulin. The spindle fibers of dividing cells, the tails of sperm cells, and the hairlike cilia on the cells lining the esophagus are examples of cellular components made from microtubules.

Hormones

Hormones are chemical messengers in the human body. Not all hormones are proteins. Some hormones are just modified amino acids, whereas others are lipid molecules. We shall discuss a number of these later in this text (see Chapters 26 and 28). However, several hormones are proteins or smaller polypeptides. Table 22-4 lists some of these polypeptide hormones and the messages they carry. By acting as chemical messengers, hormones allow cells in different parts of the body to communicate and thus to function in concert. They tell specific types of cells, called **target cells** (or target tissue), to speed up or slow down their assembly lines. One example is oxytocin:

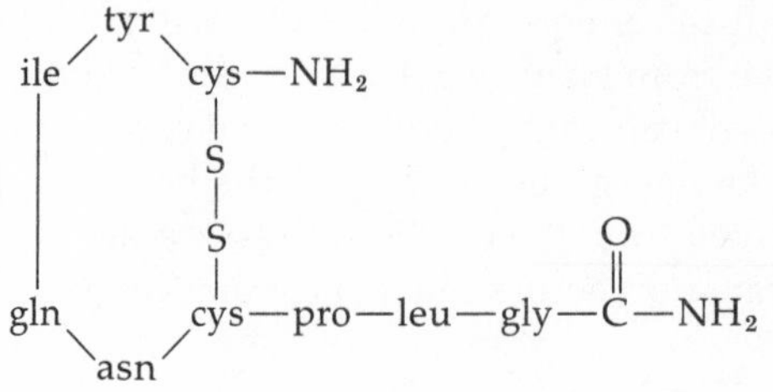

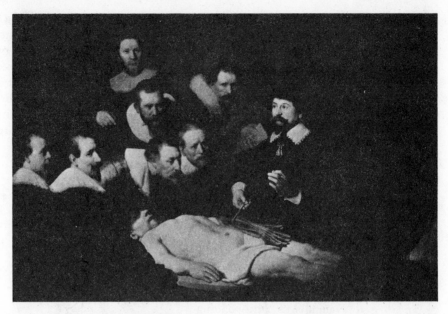

Tendons, which are composed to a large extent of elastic structural proteins, appear to be the subject of the anatomy lesson depicted in this Rembrandt painting.

The hormone oxytocin is released from the brain and travels through the bloodstream to its target tissue, the uterus. It tells the uterine muscle cells to contract, causing labor and childbirth. We can also use synthetic oxytocin or oxytocin obtained from another animal, such as a pig, to induce labor artificially.

Table 22-4 Some Polypeptide Hormones

Hormone	Producer Gland	Target Tissue	Effect
Luteotropin (prolactin)	Anterior pituitary	Mammary gland	Proliferation and initiation of milk secretion
		Corpus luteum of the ovary	Stimulation of progesterone secretion
Thyrotropin	Anterior pituitary	Thyroid	Formation and secretion of thyroid hormones
Somatotropin (growth hormone)	Anterior pituitary	Several	Several, including growth of bone and muscle
Follicle stimulating hormone	Anterior pituitary	Ovary	Secretion of estrogens and ovulation
		Testis	Development of seminiferous tubules, spermatogenesis
Oxytocin	Hypothalamus-posterior pituitary	Uterus	Contraction, parturition
		Mammary gland	Let-down of milk
Vasopressin	Hypothalamus-posterior pituitary	Arterioles	Regulates blood pressure
		Kidney tubules	Water resorption
Thymosin	Thymus	Lymphocytes	Maturation of T-lymphocytes
Insulin	Pancreas	General	Utilization of carbohydrate, increase in protein synthesis
Glucagon	Pancreas	Liver	Breakdown of glycogen to glucose
		Adipose tissue	Release of lipids

The pituitary gland, which sits at the base of the brain, makes a large number of polypeptide hormones. Pituitary hormones are released into the bloodstream upon orders from the brain. These orders are carried in the form of a small peptide, called a **releasing factor.** This multistep communication process between the brain and a target tissue is not wasteful. Each step in the process amplifies the signal. A minute quantity of a specific releasing factor from the brain causes the pituitary to release a large amount of a specific hormone. This specific pituitary hormone in turn stimulates an amplified response in its target tissue (see Figure 22-3).

Figure 22-3   The amplification that occurs in successive steps during the transmission of a hormonal message is represented by progressively larger arrows at each step. In this case, a stimulus causes the hypothalamus to release a tiny amount of a small peptide (a specific releasing factor), which in turn causes the pituitary gland to release a much larger amount of its corresponding specific hormone into the bloodstream. This hormone then binds to receptors on target cell membranes, causing the activation of a membrane-bound enzyme. Subsequent steps within the target cells result in an even greater amplification of the response.

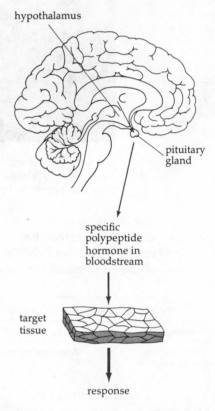

Hormones thus regulate the activities of cells throughout the body in response to several conscious and subconscious stimuli. Fright, for example, results in the brain's sending out the message: "Supply the muscle cells with more energy." The response to this hormonal message is the stimulation of enzymes in certain target cells. These enzymes break down stored glycogen polymers into their glucose components. In Chapter 25 we shall see that glucose is a very good source of energy.

Antibodies

Antibodies are the soldiers in the body's "defense department," which is called the immune system. Antibodies are fairly large proteins, containing more than 1400 amino acids each, and they are capable of specifically binding to certain molecules, called **antigens,** which are not normally found in an individual's body. The binding of antibody molecules to harmful antigens leads to their elimination from the body, as we shall see shortly.

Different individuals have different **blood types.** The most significant difference between blood types is the presence on an individual's red blood cell membranes of one of the three complex polysaccharides shown in Figure 22-4. These polysaccharides are the basis for the A, B, AB, and O blood groups. An individual with blood group A will have the group A polysaccharide on his or her red blood cell membranes but will not have the blood group B polysaccharide.

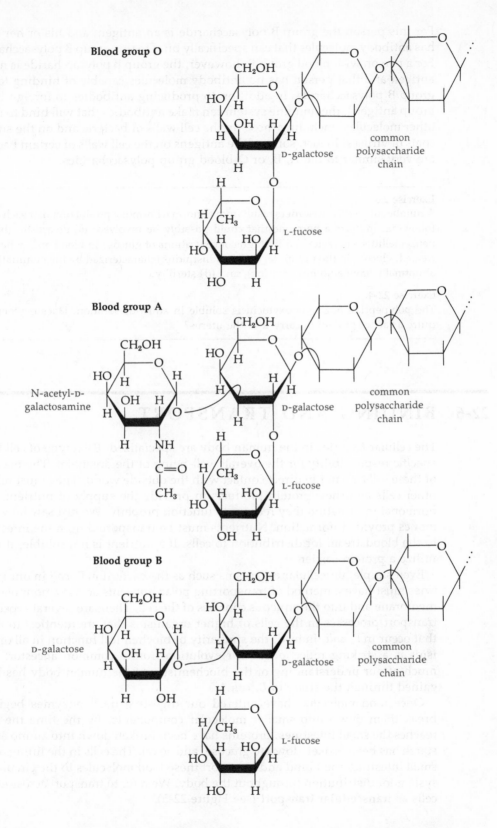

Blood group O

D-galactose

common polysaccharide chain

L-fucose

Blood group A

N-acetyl-D-galactosamine

D-galactose

common polysaccharide chain

L-fucose

Blood group B

D-galactose

D-galactose

common polysaccharide chain

L-fucose

Figure 22-4   The complex polysaccharides that determine A, B, or O blood groups all consist of a polysaccharide chain, which is attached to the red blood cell membrane. Notice that the only difference between these polysaccharides is the presence of an additional galactose (group B) or N-acetylgalactosamine (group A) at the end of a common polysaccharide chain. These additional components are absent in the group O polysaccharide. For people with blood type AB, both group A and group B polysaccharide chains are present.

For this person the group B polysaccharide is an antigen, and his or her body has antibody molecules that can specifically bind to this group B polysaccharide. For a person with blood group B, however, the group B polysaccharide is not an antigen and that person has no antibody molecules capable of binding to this group B polysaccharide. In addition to producing antibodies to foreign blood group antigens, the immune system can make antibodies that will bind to many other molecules, including those on the cell walls of bacteria and on the surface coats of viruses. In fact, some of the antigens on the cell walls of certain bacteria are very similar to the A, B, or O blood group polysaccharides.

Exercise 22-3
A number of medical disorders result from abnormal hormone production. For each of the following, indicate a hormone that could possibly be involved: (a) dwarfism, (b) diabetes mellitus (characterized by high concentrations of glucose in blood and, when untreated, glucose in the urine), (c) diabetes insipidus (characterized by the elimination of abnormally large amounts of urine), and (d) sterility.

Exercise 22-4
The polypeptide hormone oxytocin is soluble in aqueous solution. Does oxytocin require a carrier protein to carry it to the uterus?

22-6 BINDING AND TRANSPORT

The cellular factories in the human body are specialized. Each type of cell has a specific responsibility for the overall well-being of the organism. The majority of these cells are not in direct contact with the outside world. They must rely on other cells and their protein products to provide the supply of nutrients and hormonal information they require to function properly. We just saw how hormones provide information. Nutrients must be transported from the intestines to the bloodstream for distribution to cells. If a nutrient is not soluble, it must utilize a protein carrier.

Even simple, unicellular organisms such as the bacterium *E. coli* in our intestines must have a method of transporting polar nutrients across a nonpolar cell membrane and into the aqueous confines of the cell. There are several aspects of transport processes in the cells of higher organisms that are identical to those that occur in *E. coli*. Indeed, the similarity of biochemical function in all organisms is a striking reflection of our evolution from a common ancestor. And much of our understanding of the biochemistry of the human body has been gained through the study of *E. coli*.*

Once food molecules have entered our digestive tract, enzymes begin to break them down into smaller molecular components. By the time the food reaches the small intestines, proteins have been broken down into amino acids, starch has been broken down to glucose, and so on. The cells in the lining of the small intestines then bind and transport these food molecules to the circulatory system for distribution throughout the body. We refer to transport across entire cells as **transcellular transport** (see Figure 22-5).

---

* This has led some biochemists to suggest that *all* biochemical events in humans are identical to those in *E. coli,* or: "The elephant is a large *E. coli.*" This idea is true for many but not all of the biochemical events that occur in all organisms. The dissimilarities between humans and organisms as large and complex as mice have led to many difficulties in cancer research. Some treatments that kill tumors in mice have little or no effect on similar tumors in humans. Similarly, the drug thalidomide had no ill effects when tested on monkeys, but it produced disastrous birth defects in humans.

Figure 22-5 A schematic representation of transcellular transport. Molecules (indicated by blue circles and squares) are carried across the cells lining the small intestines by specific membrane transport proteins.

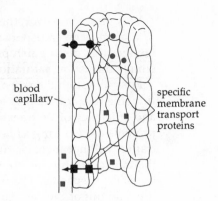

blood capillary

specific membrane transport proteins

Very little is known about transcellular transport. Some molecules may even pass in between the intestinal cells. However, specific membrane transport proteins appear to be more frequently employed. In the case of amino acids, specific membrane transport proteins are required for different groups of amino acids. For example, the amino acids serine and threonine are transported by the same transport protein.

Once food molecules have been transported to the bloodstream, those that are not soluble in the aqueous plasma must be transported by carrier proteins to the cells. One example is apo-β-lipoprotein, which is made by the intestinal cells and apparently picks up its lipid passengers at or near the site of its synthesis.

Characteristics of Membrane Transport Proteins

Membrane transport proteins are often required to allow food molecules to enter "hungry" cells. Some molecules may pass through the membrane by a **simple diffusion process** (due to a concentration gradient) without the help of another molecule in the membrane (see Figure 22-6).

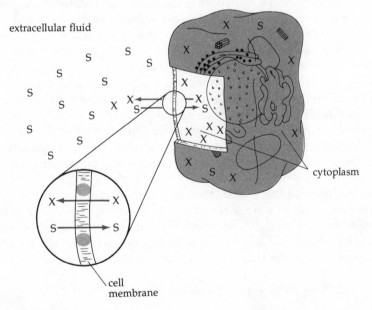

extracellular fluid

cytoplasm

cell membrane

Figure 22-6 Schematic representation of simple diffusion of molecules across a cell membrane. The molecules simply move through the cell membrane from a region of high concentration toward one of lower concentration, as shown in a magnified view of a portion of the cell membrane. Note that the [S] is higher outside the cell, whereas [X] is higher inside the cell.

Very often, however, the transport of a molecule or ion requires the assistance of a membrane transport protein. There are three readily identifiable characteristics that distinguish protein-assisted transport across cell membranes from simple diffusion: **saturation, specificity,** and **inhibition.**

First, although the rate of diffusion increases with increased solute concentration, the rate of transport of solutes into or out of a cell is limited by the number and efficiency of its transport proteins in the cell membrane. Once all of these proteins are working as fast as they possibly can, an increase in solute concentration will not increase the rate of transport. When this happens, we say that the transport proteins are **saturated.** Second, diffusion is nonspecific, whereas transport is **specific.** The transport proteins are selective in what they will carry into and out of the cell. For example, amino acid transport proteins in humans will only carry L-amino acids; D-amino acids are discriminated against. Third, membrane transport proteins can be **inhibited** by molecules that are structurally similar to the molecule they transport. The transport protein is fooled and binds the inhibitor. The inhibited transport protein is then unable to transport its usual solute. Saturation, specificity, and inhibition are also important properties of enzymes, as we shall see later.

Exercise 22-5

Carrier proteins in the blood also show the characteristics of saturation, specificity, and inhibition. Use hemoglobin and O_2, N_2, and CO to describe these characteristics.

Passive Transport Versus Active Transport

The proteins that transport molecules across cell membranes can be further divided into two types, depending on whether or not the transport process requires energy.

When transport does not require energy it is called **passive transport** (see Figure 22-7). Passive transport is necessary to transfer polar molecules across the nonpolar cell membrane. A good example is the transport of glucose into human erythrocytes (red blood cells). The rate of entry of glucose into erythrocytes increases as the concentration of glucose goes up, but the rate reaches a maximum when the transport proteins are saturated. The glucose transport proteins are also specific. They transport D-sugars, such as D-glucose, D-galactose, and D-ribose. They will not bind and transport L-isomers of these sugars. There are hundreds of thousands of glucose transport protein molecules on each erythrocyte membrane, and they can be inhibited by molecules that are similar to glucose in structure. Therefore, the glucose transport protein satisfies all of the characteristics of protein-assisted transport (saturation, specificity, and inhibition).

In addition to these characteristics, **active transport** proteins use energy to transport molecules from regions where they are relatively dilute to regions where they are already concentrated. The energy is supplied by an exergonic chemical reaction, such as the hydrolysis of ATP, which we shall discuss in Chapter 25. For the time being, let us symbolize this reaction as follows:

$$ATP \rightarrow ADP + \text{phosphate ion} + \text{energy}$$

One of the best-known examples of active transport is the Na^+K^+-ATPase pump. This protein pumps K^+ into cells while simultaneously pumping out Na^+. The sodium ion concentration in blood plasma is already much higher than in cells, so energy must be expended to pump more Na^+ into plasma. Potassium

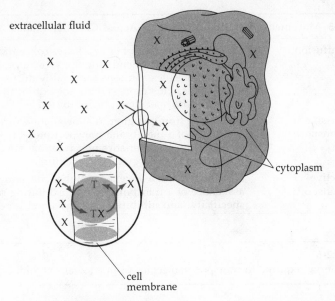

extracellular fluid

cytoplasm

cell
membrane

Figure 22-7   Schematic representation of passive transport of molecules via the membrane transport protein T. The molecules move from a region of high concentration to one of lower concentration, but they require the assistance of a membrane transport protein.

ions are more concentrated in cells than in plasma, so work is also needed to pump even more K^+ into cells. In both cases, the hydrolysis of ATP supplies the required energy (see Figure 22-8). Several other active transport proteins are known. They transport sugars, amino acids, and other molecules into or out of cells. Diffusion, active transport, and passive transport are compared in Table 22-5.

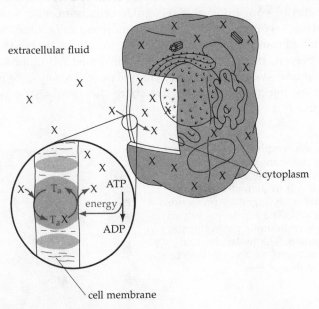

extracellular fluid

cytoplasm

cell membrane

Figure 22-8   Active transport of the substance X into a cell where it is already concentrated is accomplished by a specific active transport protein, represented by T_a. The energy required to carry even more X into the cell is supplied by the hydrolysis of ATP.

Table 22-5 Movement of Molecules and Ions Across Cell Membranes

Diffusion:	Movement from a higher to a lower concentration without the assistance of a membrane transport protein. Only certain molecules and ions can readily diffuse across cell membranes. This process does not exhibit the characteristics of saturation, specificity, or inhibition.
Passive transport:	Movement from a higher to a lower concentration with the assistance of a membrane transport protein. This process exhibits the characteristics of saturation, specificity, and inhibition. No energy input is required.
Active transport:	Movement from a lower to a higher concentration with the assistance of a transport protein, exhibiting saturation, specificity, and inhibition. Energy input is required.

Exercise 22-6

Why is energy required to transport molecules from a lower to a higher concentration?

22-7 ANTIBODIES AND IMMUNITY

In order for the human body to survive and thrive in a world filled with disease-causing viruses and bacteria, it must be capable of defending itself against invasions by these agents. As we mentioned previously, this defense is the responsibility of the **immune system,** and antibody molecules can be considered to be the soldiers of this defense system. The antibodies are produced by a certain type of white blood cell called a **lymphocyte.** Lymphocytes circulate through the body in search of foreigners such as bacteria. When a foreigner is encountered it is usually inactivated and removed from the body.

Lymphocytes in your body also recognize cells from other humans and differences between your normal cells and cancerous cells. One consequence of this is that blood must be typed and matched to a recipient before a transfusion in order to prevent the destruction of the transfused red blood cells by the recipient's immune system. Likewise, transplanted kidneys, hearts, and other organs may be rejected if they do not share the same tissue type as the recipient's cells.

A human lymphocyte as it appears under the scanning electron microscope. These cells measure about 5×10^{-6} m in diameter and are covered with fingerlike projections called villi. The functional role of these membranous projections is not known. The molecules that bind antigens are too small to be seen in this photo.

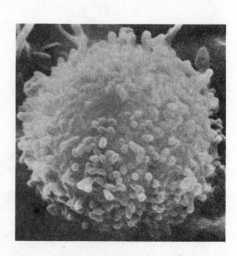

To understand the chemical basis for the recognition and elimination of such foreigners by the immune system, we must first examine which kinds of molecules are recognized as foreign, that is, which are antigens. Then we must examine the structure of the molecules that lymphocytes frequently employ as soldiers—antibodies. We shall then study how antibodies are formed, how they work, and how we can stimulate the immune system to produce antibodies against deadly organisms in the process of immunization.

Antigens

In a healthy individual viruses, bacteria, mismatched red blood cells, and other foreign "invaders" are distinguished from native body cells by the immune system. The term **antigen** is used to describe those molecules that can be recognized as foreign by the immune system of a given individual. Any molecule, such as a protein, that is synthesized by the cells of an individual does not normally act as an antigen in that person's body, but it may be recognized as foreign (act as an antigen) if introduced into the body of a second individual. A very large number of molecules, including most foreign proteins and certain polysaccharides, can be recognized as antigens by the immune system of a normal person.

A very slight difference between a foreign molecule and one native to an individual may be sufficient to be recognizable by that person's immune system. For example, in Figure 22-4 we saw that the only difference between the A and B blood groups was the identity of the last sugar of a common polysaccharide chain. Similarly, a protein that differs by only a few amino acids from a protein native to an individual may be an antigen for that individual. For example, the hormone insulin, which is administered to persons with diabetes, is usually obtained from pigs or cows. The insulins obtained from these animals have slightly different amino acid sequences than human insulin, and some patients make antibodies against them. Table 22-6 lists various types of antigens that are important to human health.

Table 22-6 Some Common Types of Antigens

Type	Specific Examples	Importance
Viral	Polio virus coat protein	Used for polio vaccination
Bacterial	Bacillus of Calumette and Guerin (BCG)	Being tested for cancer immunotherapy
Blood group	B substance, Rh substance	Blood typing for transfusions
Histocompatibility (tissue antigens)	Histocompatibility antigen HLA-A7	Involved in organ transplantation
Fetal	Alphafetoprotein, carcinoembryonic antigen	Produced by fetuses and by some tumors, but not found in normal adults

Antibodies

Let us consider what happens when a foreign antigen enters the human body, as, for example, in a bacterial infection. The invading bacteria grow and reproduce rapidly in the warm, nutrient-rich body. Some of them also come into contact with lymphocytes, which constantly circulate through the body (see Figure 22-9). Attached to the membranes of some of these lymphocytes are antibody molecules specific for these bacterial antigens, allowing the lymphocytes to bind to the bacteria. This binding causes the lymphocytes to divide, producing many more cells that are capable of making antibodies against these specific bacterial antigens. In fact, an increase in the number of white blood cells in the circulatory system is used in the diagnosis of certain illnesses, such as acute infection of the appendix (appendicitis). Eventually the rate of antibody production exceeds the rate of bacterial multiplication and these antibody molecules bind to all of the bacteria and cause their inactivation and elimination from the body. This entire process may require several days, during which time the individual may feel quite ill.

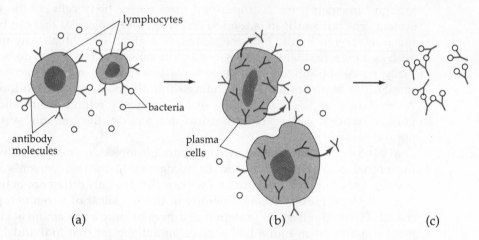

| (a) | (b) | (c) |

Figure 22-9   When the body is infected by bacteria, a few of the many circulating lymphocytes have antibody molecules on their membranes that specifically bind to antigens on the invading bacteria (a). These lymphocytes then divide several times—producing larger antibody-producing plasma cells and numerous identical lymphocytes. The plasma cells produce and secrete large amounts of antibodies directed against the bacterial antigens (b). These antibodies bind to the bacteria (c) and cause their elimination.

What special structural features enable antibody molecules to bind to antigens and cause their elimination from the body? Antibodies, which are also called **immunoglobulins,** are globular proteins. There are five types or classes of immunoglobulin in normal humans. The most prevalent one is immunoglobulin G, abbreviated IgG. IgG molecules are oligomers consisting of two large and two small polypeptide chains or subunits. The large chains are about twice as long as the small chains, and all four chains are held together by disulfide bonds and by hydrophilic and hydrophobic interactions (see Figure 22-10). Notice that an individual IgG molecule is symmetrical, so that it possesses two identical arms. At the tip of each arm is a binding site for the particular antigen with which it reacts. The binding sites of an antibody molecule are very specific for a particular antigen. The antigen is bound via hydrophilic and hydrophobic interactions with the side chains of individual amino acid components of the antibody.

Because one molecule of IgG can bind to two identical antigens, it is said to be **divalent.** This double-binding ability is crucial to the function of IgG. Consider the reaction between IgG antibody molecules that are specific for type B blood group antigen and type B red blood cells (see Figure 22-11). There are several type B antigen molecules on each type B red blood cell. Because the antibody molecules are divalent, some of them will bind to both an antigen on one red blood cell and to another antigen on a different red blood cell. When a large number of these divalent interactions occur, they cause the red blood cells to

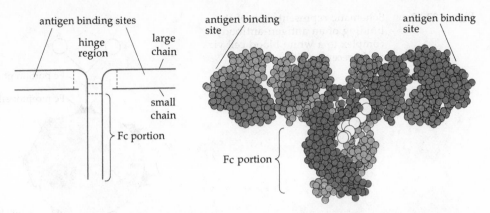

Figure 22-10   The structure of an IgG molecule. In the simplified schematic representation on the left, each component polypeptide chain is indicated by a solid line, and disulfide bonds by dotted blue lines. The figure on the right, a space-filling model of an actual human IgG molecule, shows the compact arrangement of the component polypeptide chains. The large chains are shaded in blue and the small chains in gray. The open circles represent a carbohydrate prosthetic group.

clump together or **agglutinate.** Agglutination of red blood cells can occur in the circulatory system of an individual given a transfusion of blood that is not matched to his or her blood type. This type of agglutination can produce severe clinical problems, and can lead to death. To prevent such problems, a few blood cells from any unit of blood considered for a transfusion are routinely mixed in a test tube with some serum from the patient who is to receive the blood. This procedure is called a **cross-match.** If agglutination occurs in the cross-match test, the unit of blood in question is not transfused into that patient.

Figure 22-11   Schematic representation of the agglutination of red blood cells. The binding of several antibody molecules to the antigens on individual red blood cells is accompanied by cross-linking of different cells by these divalent antibody molecules.

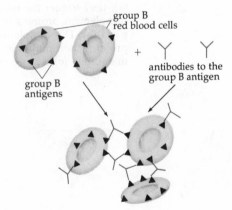

IgG and other immunoglobulins can also agglutinate invading bacteria. The human body can then easily eliminate the small clumps of bacteria.

An IgG molecule with a different amino acid sequence is required for each antigen that is to be recognized. The differences between these IgG molecules are restricted almost entirely to the antigen binding regions. The rest of the molecules are essentially identical, and they have common functions. The "hinge region" (see Figure 22-10) allows two antigens to be bound at varying angles. The Fc portion of an IgG molecule can function in a number of ways. It may allow an antibody and its antigen, an antigen-antibody complex, to be bound to different types of white blood cells via an **Fc receptor site** (see Figure 22-12). The binding of the complex to the Fc receptor site activates the destruction of certain types of antigens by mechanisms that are not well understood.

Figure 22-12   Schematic representation of the binding of an antigen-antibody complex to a white blood cell via an Fc receptor site.

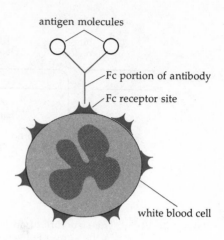

The Fc portions of antibodies are also involved in complement reactions. **Complement** is a term used to refer to a special group of proteins found in human blood. When several antibody molecules are bound to antigens on certain types of foreign cells, their Fc portions can bind to complement proteins (see Figure 22-13). The complement proteins then catalyze reactions that in turn cause the destruction of the foreign cell. Besides agglutination and complement reactions, there are a number of other ways in which foreign cells can be eliminated after binding by antibody molecules.

In addition to IgG, there are four other classes of immunoglobulins: IgA, IgM, IgD, and IgE. All are composed of small and large chains that combine to form molecules with more than one antigen binding site. IgA is found primarily in secretions (milk, saliva) and the intestinal tract, whereas IgG and IgM are found

Figure 22-13   When several antibody molecules attach to a foreign cell or organism (a), they trigger the binding of complement proteins to their Fc parts (b). The complement proteins then puncture the cell membrane (c), killing the cell.

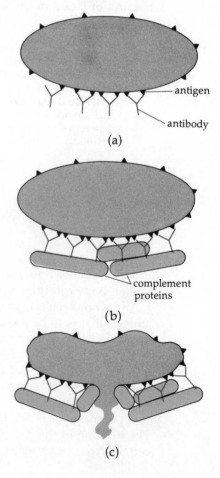

predominantly in blood. IgE antibodies are present in very low concentrations, but they play an important role in allergic reactions. Little is known of the IgD class of immunoglobulins. The properties of the five classes of immunoglobulins are summarized in Table 22-7. Note that an IgM molecule, for example, has 10 large and 10 small chains and can thus bind to 10 identical antigen molecules simultaneously.

Table 22-7 The Five Classes of Human Immunoglobulins

Immunoglobulin Class	IgG	IgA	IgM	IgD	IgE
Number of small chains per molecule	2	2 or 4	10	2	2
Number of large chains per molecule	2	2 or 4	10	2	2
Structural representation	(diagram)	(diagram)	(diagram)	(diagram)	(diagram)
Serum concentration (mg/ml)	12	1.8	1.0	0.03	0.0003
Functional role	Complement and agglutination reactions in serum	Some agglutination reactions	Complement and agglutination reactions in serum	Not known	Allergic reactions

Immunization

A cornerstone of modern medicine is the deliberate stimulation of lympho-cytes in normal individuals to produce antibodies against disease-causing bac-teria and viruses—a process called **immunization.** Basically, this process in-volves the injection of a specified dose of a bacteria or virus that has previously been killed or otherwise inactivated (to prevent illness). The antigens on the dead bacteria or virus stimulate lymphocytes to divide, just as the live infec-tious organisms would. The immunization results in the production of a large number of lymphocytes that produce antibody molecules. These antibodies do not remain in the bloodstream for very long. The larger number of lympho-cytes, however, do remain in circulation and are quickly able to produce very large amounts of antibody should the individual be infected by that organism at a later time.

In 1798, Edward Jenner achieved the first successful immunization, which was made against the smallpox bacillus. Today we immunize individuals against a wide variety of diseases. However, we still do not know exactly how a foreign antigen can trigger a lymphocyte to divide and synthesize antibodies. There are many other aspects of the human immune system that are not yet well understood but that are actively being investigated. Experiments currently being conducted on the immune system offer great promise for significantly im-proving medical care within our lifetimes.

Exercise 22-7
In addition to group A and group B antigens, an individual's blood type is determined by a number of other antigens. Hemolytic disease of the newborn can result in the death of an infant whose blood type is Rh$^+$ (i.e., whose red blood cells have the Rh antigen), if the child's mother is Rh$^-$. We shall not discuss the detailed pathology of this disease, but with proper prenatal care most cases of hemolytic disease of the newborn can now be avoided. Suggest two laboratory tests that should be a part of good prenatal care in order to prevent this disease.

22-8   SUMMARY

1. The unique amino acid sequence of a protein determines its structure and its function.

2. The change of a single amino acid in a protein is called an amino acid substitution.

3. Proteins can be classified according to their physical properties or according to their functional properties.

4. There are two main functional classes of proteins: catalytic proteins or enzymes and noncatalytic proteins.

5. The subclasses of noncatalytic proteins are carrier proteins, receptor proteins, membrane transport proteins, structural proteins, contractile proteins, hormones, and antibodies.

6. A conjugated protein has an auxiliary molecule, called a prosthetic group, tightly bonded to it.

7. Membrane transport proteins exhibit three basic features: saturation, specificity, and inhibition.

8. There are three different ways in which molecules cross cell membranes: simple diffusion, passive transport, and active transport. Active transport requires energy.

9. The human immune system contains lymphocyte cells, which can bind to antigens and produce antibodies.

10. An antigen is a molecule that can be recognized as foreign by lymphocytes and antibodies.

11. An antibody or immunoglobulin is a protein, produced by lymphocytes, that possesses two or more binding sites for a particular antigen.

12. Agglutination and complement reactions are two ways in which antigens can be eliminated after their recognition by antibody molecules.

PROBLEMS

1. A substance X moves from the outside of a cell to the inside under conditions where the concentration of X inside the cell is greater than the concentration of X outside. This process is an example of which of the following: (a) simple diffusion, (b) passive transport, (c) active transport, (d) allosteric activation, (e) none of these?

2. Which of the following are found as prosthetic groups on at least some proteins: (a) carbohydrates, (b) lipids, (c) hemes, (d) metal ions, (e) none of these?

3. Which of the following classes of proteins is most soluble in pure water: (a) collagens, (b) albumins, (c) globulins, (d) keratins, (e) elastins? Explain your answer.

4. Thrombin is which of the following: (a) a protein hormone, (b) a membrane transport protein, (c) a coenzyme, (d) an enzyme, (e) a structural protein?

5. Which of the following are not characteristic properties of a membrane transport protein: (a) specificity, (b) inhibition, (c) coenzyme requirement, (d) saturation, (e) solubility?

6. Consider an IgG antibody molecule that binds to the blood group B antigen. The molecular weights of the large and small peptide chains in this antibody are 50,000 and 25,000, respectively. What is the molecular weight of the entire antibody molecule? How many antigen binding and Fc parts does this molecule contain?

SOLUTIONS TO EXERCISES

22-1 Hemoglobin Z, with the acidic glutamic acid side chains replaced by other acidic side chains in aspartic acid, should be as soluble as hemoglobin A, and probably would be indistinguishable from hemoglobin A in function.

22-2 An individual with multiple myeloma has malignant antibody-producing cells, which tremendously increases the concentration of antibodies—globulins—in the serum. A patient suffering from this malignancy should have a lower-than-normal A/G ratio.

22-3 (a) Lack of sufficient somatotropin (growth hormone)
   (b) Insufficient, or inactive, insulin
   (c) Insufficient, or inactive, vasopressin
   (d) Insufficient, or inactive, follicle stimulating hormone, or inappropriate amounts of one of the other sex hormones

22-4 Carrier proteins are used to transport relatively insoluble substances, such as oxygen and fats. Since oxytocin is soluble, it does not require a carrier protein.

22-5 A hemoglobin molecule is saturated when it has bound four O_2 molecules. Even with a large excess of O_2, no more oxygen can bind. Hemoglobin normally binds and transports O_2. It does not bind to N_2. CO, however, binds more readily to hemoglobin than does O_2 and can thus inhibit O_2 binding.

22-6 Molecules have a natural tendency to move from an area of high concentration to one of lower concentration. This is an example of the natural tendency toward disorder (increased entropy). Energy is required to overcome this tendency.

22-7 Good prenatal care should include determination of the mother's Rh group. If she is Rh$^-$, then tests should be performed to determine if she has antibodies directed against Rh$^+$ cells.

CHAPTER 23

Enzymes

23-1 INTRODUCTION

Using our factory analogy, enzymes are skilled workers who are employed on the cellular assembly lines. Two important facts should be remembered: Virtually all of the reactions that occur in living cells are catalyzed by enzymes, and all enzymes are proteins. Enzymes are the most numerous and diverse of all the functional types of protein. Enzymes possess the same basic structural features as all other proteins; that is, they are polymers of amino acids whose side chains interact to form a unique three-dimensional structure. Enzymes also share certain functional features with other proteins, such as membrane transport proteins, in that they show *specificity* for the molecules on which they work and also exhibit the properties of *saturation* and *inhibition*, as we shall see shortly.

Enzymes are very good catalysts. Many reactions that occur rapidly within living cells proceed at an incredibly slow rate in the absence of the required enzymes. Not only do enzymes increase the rates of reactions within cells, but they catalyze only those reactions that produce desired products.

Some enzyme-catalyzed reactions occur in every living cell. Others occur only in certain types of organisms or only in certain types of cells. Our appearance, as well as the way we function, is due to our unique assortment of enzymes. Major breakthroughs in the understanding of the biochemical workings of the human body have often resulted from the discovery of new enzymes. Increasingly, enzymes are being used in modern medicine. For example, certain diseases are diagnosed by measuring the levels of specific enzymes in a patient's serum. Other diagnostic tests make use of the tremendous specificity of enzymes to measure the quantity of smaller molecules in serum.

Remember that all catalysts, including enzymes, increase reaction rates but do not alter the position of equilibrium in the reactions they catalyze. Why, then, you might ask, don't all of the reactions in cells quickly reach a state of equilibrium? This does not happen because the products of one enzyme-catalyzed reaction are the starting materials for other reactions. Almost all of the molecules within each living cell are constantly being synthesized and used up. The reactions in cells reach a state of equilibrium only when the cells die.

Droplets secreted by the very specialized leaves of the Venus flytrap contain enzymes that digest the insect prey of this carnivorous plant.

In this chapter we shall examine the unique functional features of enzymes, including their specificity and the rates at which enzyme-catalyzed reactions proceed. Finally, we shall look at the way enzyme function is controlled by cells. This is very important. Using our factory analogy, cellular control of enzyme activity is like supervisors telling assembly line workers to speed up or slow down the production of a product so that the supply is kept in line with the demand.

23-2   STUDY OBJECTIVES

After studying the material in this chapter, you should be able to:

1. Describe the active site of an enzyme and the process of induced fit.
2. Discuss the specificity, saturation, and inhibition of enzymes.
3. Describe the simplest two-step mechanism, $E + S \rightleftharpoons ES$, followed by, $ES \rightarrow E + P$, for enzyme-catalyzed reactions.
4. Define molar activity and maximum rate (V_{max}).
5. Use the relationship among molar activity, maximum rate (V_{max}), and enzyme concentration to calculate any one of these quantities, given numerical values for the other two.
6. Explain the difference between a competitive and a noncompetitive enzyme inhibitor.
7. Explain how pH and temperature affect the rates of enzyme-catalyzed reactions.
8. Describe the function of coenzymes in some enzyme-catalyzed reactions, as well as the connection between coenzymes and vitamins.
9. Describe how an enzyme-catalyzed reaction can be regulated by the substrate concentration, rate of enzyme synthesis, allosterism, and chemical modification.
10. Describe how a sequence of enzyme-catalyzed reactions can be regulated by a process of feedback inhibition.
11. Describe the relationship between competitive inhibitors and the action of many drugs.

23-3   ENZYME CATALYSTS

Thousands of enzymes have been identified, and several have been purified and studied in great detail. We shall not attempt to describe all of the known enzymes. Instead we shall examine the common properties and characteristics of some typical enzymes and the reactions they catalyze. Again, we must keep in mind that enzymes are proteins. An enzyme's unique function is a result of its unique protein structure.

Catalytic Power

One common property of enzymes is their tremendous **catalytic power**—their ability to increase greatly the rate of a chemical reaction. As we discussed in Chapter 9, several factors influence the rate of any chemical reaction. For example, the rate of many chemical reactions doubles with each 10°C rise in temperature. The observed rate of a chemical reaction depends on the overall activation energy for that reaction. Recall that for a single step in a reaction mechanism, the activation energy (E_a) is the amount of energy required to form an activated complex in which the reactants are in close contact and in the correct orientation to each other to favor formation of products. Catalysts, as you will recall, alter the reaction pathway to one that has a lower overall E_a and thereby increase the reaction rate.

An example of the decrease in activation energy for an enzyme-catalyzed reaction is the reaction $2H_2O_2 \rightarrow 2H_2O + O_2$. Without the presence of an enzyme catalyst, this reaction has an activation energy in aqueous solution of 18 kcal/mole. The enzyme *catalase* decomposes hydrogen peroxide in an enzyme-catalyzed pathway with an overall E_a of only 2 kcal/mole. The activation energy for the reaction pathway catalyzed by catalase is only one-ninth (1/9) that of the uncatalyzed pathway. Therefore, the rate of the enzyme-catalyzed reaction is much faster than the rate for the uncatalyzed pathway.

In enzyme-catalyzed reactions, the reactant on which an enzyme works is called a **substrate.** The simplest general model, which is applicable to many but not all enzyme-catalyzed reactions, is the following two-step mechanism:

Step 1 $E + S \rightleftharpoons ES$

Step 2 $ES \rightarrow E + P$

where E, S, and P represent enzyme, substrate, and product, respectively, and ES represents an intermediate in which the enzyme and substrate are bound together. Notice that, according to this model, product(s) can be formed only if an enzyme and a substrate first bind together to form ES. The intermediate ES can either revert back to unbound enzyme and substrate or can break down to form product(s) and regenerate free enzyme.

There are a number of ways whereby an enzyme can lower the activation energy for a chemical reaction. First, it provides a surface on which the reactants can come together in close proximity. Second, it binds its substrates in the correct orientation for the reaction. Third, it may actually stretch some of the bonds in the reactants (substrates), making them easier to break. [Many of these properties are also common to nonenzyme catalysts. For example, platinum (a nonenzyme catalyst) is often used by chemists to provide a surface on which reactants can come close together.] The tremendous catalytic power of enzymes, however, results from the exact combination of effects necessary for an individual reaction. Enzymes align their substrates in the right orientation and under the right conditions to favor a rapid reaction.

Nomenclature for Enzymes

Most enzymes are named and classified according to the reactions they catalyze. Their names are their job descriptions in the cellular factory. For example, *urease* catalyzes a reaction in which urea is hydrolyzed to ammonia and carbon dioxide:

(23-1) $H_2O + H_2N-\overset{\overset{\displaystyle \|}{}}{\underset{\underset{\displaystyle O}{}}{C}}-NH_2 \xrightarrow{\text{urease}} CO_2 + 2NH_3$

urea

In the early days of biochemistry, enzymes were often named by adding the suffix *-ase* to the name of the substrate. Thus the enzyme that hydrolyzes urea is called urease. An enzyme that hydrolyzes ATP would be an ATPase. However, we shall see that there are several enzymes that hydrolyze ATP, and, in order to be precise, names must be used that distinguish different enzymes that work on the same substrate.

Several enzymes were originally given nondescriptive names. Catalase is one example. The protein-splitting enzymes in our digestive tract were given the names trypsin, chymotrypsin, and pepsin. Even though the names of these enzymes do not indicate what reactions are catalyzed, nondescriptive names are still used for many common enzymes. As more enzymes were discovered, however, nondescriptive names became confusing. Therefore, an international commission on enzymes was convened in order to adopt a systematic enzyme nomenclature system. Each enzyme was given a precise name and a number. Although we shall not discuss this nomenclature system in great detail, a look at the major classes will be helpful.

Enzymes are divided into six major classes according to the type of reaction they catalyze. These classes are listed in Table 23-1, which illustrates the many types of reactions that can be catalyzed by enzymes. For example, the Enzyme Commission placed the enzymes that catalyze oxidation-reduction reactions in one class called **oxidoreductases.** One specific oxidoreductase is called glycerol dehydrogenase, which oxidizes glycerol to dihydroxyacetone (see Table 23-1) while reducing the coenzyme NAD^+. In Section 23-6 we shall discuss the role of hydrogen acceptors such as NAD^+, as well as acceptors for functional groups removed from substrates by enzymes. Another large class of enzymes, the **transferases,** transfer functional groups of atoms from one molecule to another. Glutamic-pyruvic transaminase is an example (Table 23-1). This enzyme catalyzes the exchange of an amine group on one molecule and a carbonyl oxygen of another molecule. Other classes of enzymes listed in Table 23-1 are hydrolases, lyases, isomerases, and ligases. **Hydrolases** break bonds with the addition of water. **Lyases** break bonds without the addition of water. Some lyases catalyze decarboxylation reactions. **Isomerases** interconvert isomers. **Ligases** join small molecules together into larger ones, with the energy for synthesis coming from hydrolysis of ATP (see Chapter 25). This latter reaction can be represented by $ATP + H_2O \rightarrow ADP + P_i + energy$, or $ATP + H_2O \rightarrow AMP + PP_i + energy$.

Table 23-1   Classification of Enzymes

Class 1. Oxidoreductases (catalyze oxidation-reduction reactions)
    Example: glycerol dehydrogenase

Class 2. Transferases (transfer groups of atoms)
    Example: glutamic-pyruvic transaminase

Class 3. Hydrolases (break bonds with addition of water)
Example: carboxypeptidase A

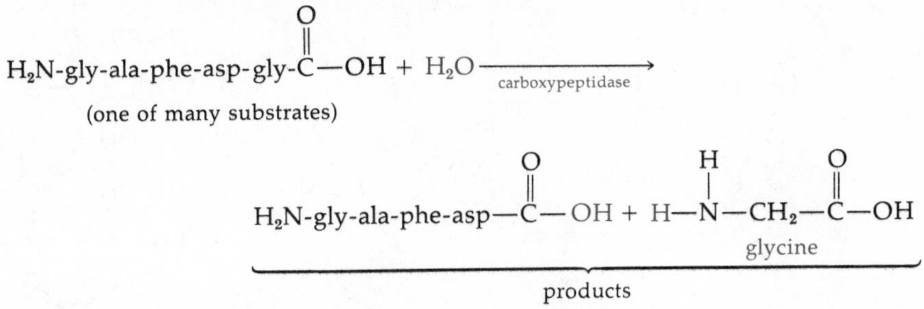

$$H_2N\text{-gly-ala-phe-asp-gly-}\overset{\overset{\displaystyle O}{\|}}{C}\text{—OH} + H_2O \xrightarrow[\text{carboxypeptidase}]{}$$

(one of many substrates)

$$H_2N\text{-gly-ala-phe-asp—}\overset{\overset{\displaystyle O}{\|}}{C}\text{—OH} + H\text{—}\overset{\overset{\displaystyle H}{|}}{N}\text{—}CH_2\text{—}\overset{\overset{\displaystyle O}{\|}}{C}\text{—OH}$$
glycine

products

Class 4. Lyases (cleave bonds without addition of water)
Example: pyruvate decarboxylase

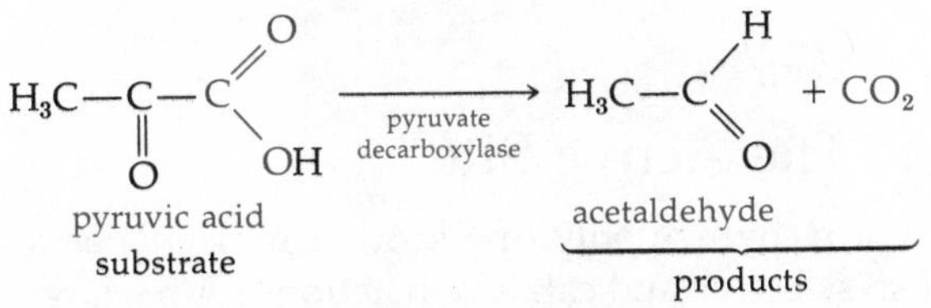

$$H_3C\text{—}\overset{\overset{\displaystyle O}{\|}}{C}\text{—}\overset{\overset{\displaystyle O}{\diagup}}{C}\diagdown_{OH} \xrightarrow[\text{decarboxylase}]{\text{pyruvate}} H_3C\text{—}C\overset{\diagup H}{\diagdown O} + CO_2$$

pyruvic acid   acetaldehyde
substrate

products

Class 5. Isomerases (catalyze isomerization reactions)
Example: triosephosphate isomerase

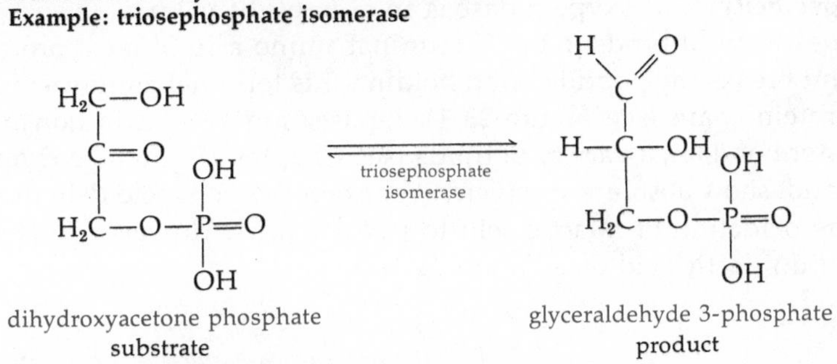

dihydroxyacetone phosphate   glyceraldehyde 3-phosphate
substrate   product

Class 6. Ligases (join molecules via ATP hydrolysis)
Example: pyruvate carboxylase

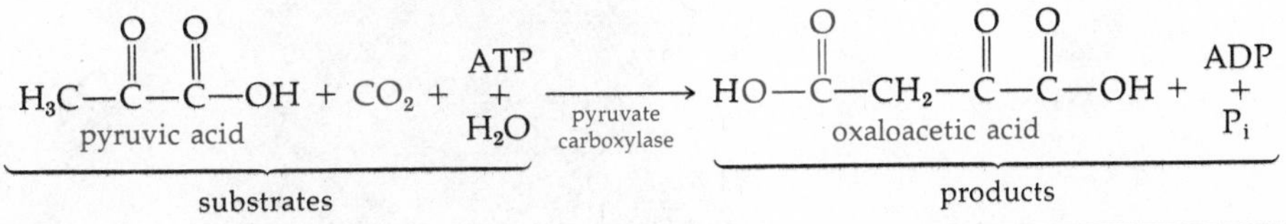

$$H_3C\text{—}\overset{\overset{\displaystyle O}{\|}}{C}\text{—}\overset{\overset{\displaystyle O}{\|}}{C}\text{—OH} + CO_2 + \overset{ATP}{\underset{H_2O}{+}} \xrightarrow[\text{carboxylase}]{\text{pyruvate}} HO\text{—}\overset{\overset{\displaystyle O}{\|}}{C}\text{—}CH_2\text{—}\overset{\overset{\displaystyle O}{\|}}{C}\text{—}\overset{\overset{\displaystyle O}{\|}}{C}\text{—OH} + \overset{ADP}{\underset{P_i}{+}}$$

pyruvic acid   oxaloacetic acid

substrates   products

Exercise 23-1
Several enzyme-catalyzed reactions follow. To which class does each of the enzymes E_a, E_b, E_c, and E_d belong?

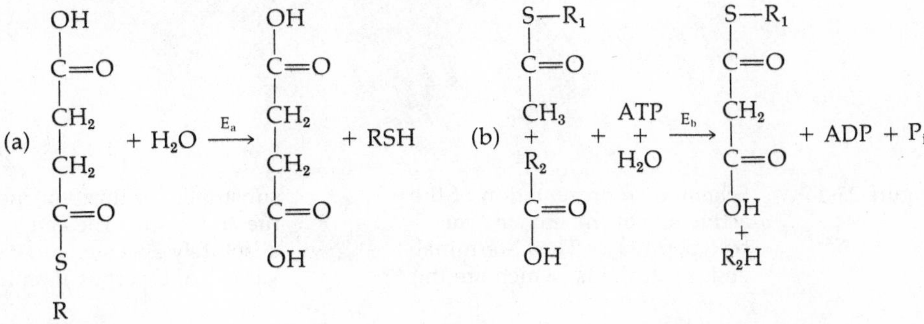

(c)
$$
\begin{array}{c}
\text{OH} \\
| \\
\text{C}=\text{O} \\
| \\
\text{CH}_2 \\
| \\
\text{CH}_2 \\
| \\
\text{C}=\text{O} \\
| \\
\text{OH}
\end{array}
\quad + \text{FAD} \xrightarrow{E_c}
\begin{array}{c}
\text{OH} \\
| \\
\text{C}=\text{O} \\
| \\
\text{CH} \\
|| \\
\text{HC} \\
| \\
\text{C}=\text{O} \\
| \\
\text{OH}
\end{array}
\quad + \text{FADH}_2
$$

(d)
$$
\text{HOCH} \xrightarrow{E_d}
$$

Enzyme Specificity: The Active Site

Some enzymes will recognize only one kind of molecule as a substrate. Other enzymes are not so selective and catalyze reactions with a larger number of substrates. The selectivity an enzyme shows in choosing a substrate is called its **specificity.** Carboxypeptidase is an example of an enzyme with a broad range of specificity. It binds to the C-terminal amino acid of most proteins or peptides, and breaks the peptide bond holding this terminal amino acid to the rest of the protein chain (see Figure 23-1). Lipases are also fairly nonspecific and break ester bonds in a variety of lipids (see Chapter 27). Some enzymes, on the other hand, show absolute specificity. For example, lactic acid dehydrogenase catalyzes the oxidation of L-lactic acid to pyruvic acid. However, this enzyme will not bind D-lactic acid (see Figure 23-2).

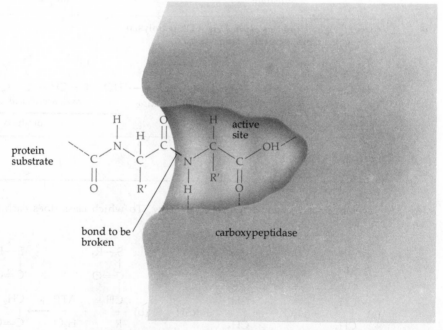

Figure 23-1   Schematic representation of the active site of the enzyme carboxypeptidase. The C-terminal ends of proteins, which are the substrates for this enzyme, bind at the active site. The active site is not absolutely specific, and can bind a variety of C-terminal amino acids.

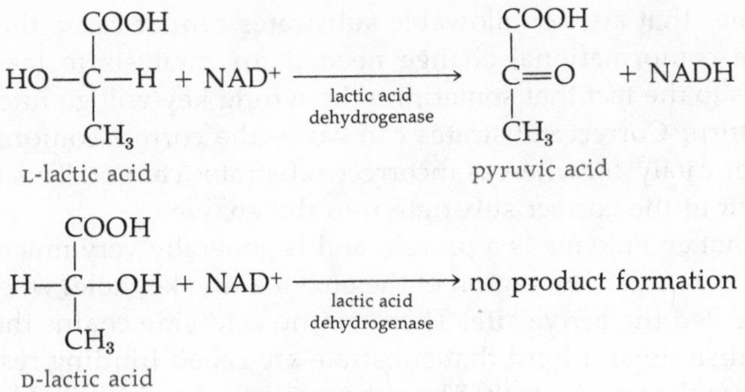

$$\underset{\text{L-lactic acid}}{\overset{\displaystyle \text{COOH}}{\underset{\displaystyle \text{CH}_3}{\text{HO}-\overset{|}{\underset{|}{\text{C}}}-\text{H}}}} + \text{NAD}^+ \xrightarrow[\substack{\text{lactic acid}\\\text{dehydrogenase}}]{} \underset{\text{pyruvic acid}}{\overset{\displaystyle \text{COOH}}{\underset{\displaystyle \text{CH}_3}{\overset{|}{\underset{|}{\text{C}}}=\text{O}}}} + \text{NADH} + \text{H}^+$$

$$\underset{\text{D-lactic acid}}{\overset{\displaystyle \text{COOH}}{\underset{\displaystyle \text{CH}_3}{\text{H}-\overset{|}{\underset{|}{\text{C}}}-\text{OH}}}} + \text{NAD}^+ \xrightarrow[\substack{\text{lactic acid}\\\text{dehydrogenase}}]{} \text{no product formation}$$

Figure 23-2   Lactic acid dehydrogenase is absolutely specific for L-lactic acid. D-Lactic acid is not a substrate for this enzyme.

This ability of an enzyme to discriminate between substrate and nonsubstrate is somewhat analogous to the fit of a key into a lock (see Figure 23-3). Not all of the parts of enzymes, or locks for that matter, are rigid. Recent experiments have shown that enzymes undergo a change in shape, called a **conformational change,** upon binding an allowable substrate. This is like the tumblers of a lock moving to accommodate the inserted key.

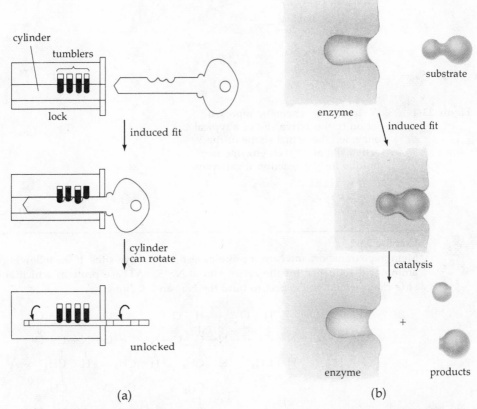

(a)   (b)

Figure 23-3   (a) The insertion of a key into its lock causes the tumblers of the lock to move into positions that allow the cylinder to rotate.

(b) The binding of a substrate to its enzyme induces an analogous change in the enzyme to a shape that allows for catalysis.

Molecules that are not allowable substrates cannot cause the enzyme to undergo the conformational change needed for catalysis to take place. This is analogous to the fact that sometimes the wrong key will go into a lock but then will not turn. Correct substrates can cause the correct conformational change needed for catalysis, whereas incorrect substrates cannot. This process is called **induced fit** of the correct substrate into the enzyme.

Recall that an enzyme is a protein and is generally very much larger than the substrate molecule. The region of the enzyme (the keyhole) where the substrate binds is called the **active site.** Those amino acid side chains that recognize the correct substrate and bind that substrate are called **binding residues.** They are like the tumblers of the lock. The substrate is recognized and bound by hydrophilic and hydrophobic interactions. Those amino acid side chains whose functional groups help to stretch bonds or otherwise participate in catalysis are called **catalytic residues.** Catalytic residues may also serve as proton donors or acceptors. They are like the cylinder of the lock, which turns to the unlocked position. The binding residues plus the catalytic residues constitute the active site (see Figure 23-4).

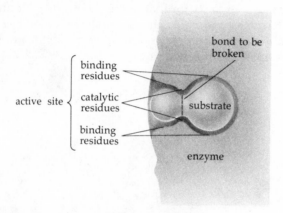

Figure 23-4   A simplified schematic representation of the active site of a typical enzyme. The actual shape of the active site of a given enzyme is unique for the reaction it catalyzes.

Exercise 23-2

Membrane transport and carrier proteins also have active sites. If the following were the amino acid sequence for the active site of Na^+K^+-ATPase protein, which amino acid side chain would you expect to bind the Na^+ and K^+ ions?

Exercise 23-3

The following amino acid sequence is part of the active center of an enzyme in its predominant form at pH 7. This enzyme interacts with its substrate by donating a proton. Which amino acid side chain can do this?

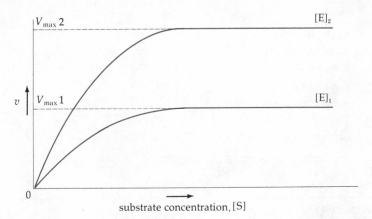

23-4   ENZYME ACTIVITY

Among the unique features of enzymes, there are three that are like those of transport proteins: *specificity, saturation,* and *inhibition.*

Enzyme Kinetics

Saturation of an enzyme with substrate means that all of the active sites of all of the available enzyme molecules are bound to substrate molecules. This is different from uncatalyzed reactions, where collisions between reactants and the formation of products continually increase as the concentration of the reactants increases. Figure 23-5 shows that the rate of an enzyme-catalyzed reaction, v, depends on the concentration of the substrate when relatively small amounts of substrate are present. In the presence of a relatively large amount of substrate, each of the enzyme molecules is saturated with substrate and the **maximum velocity** (V_{max}) is achieved. The value of V_{max} is directly proportional to the amount of enzyme present. If the amount of an enzyme is doubled, the V_{max} also doubles, all other things being equal.

Figure 23-5   The effect of substrate concentration on the rate of an enzyme-catalyzed reaction. The rate of product formation (v) increases with increasing substrate concentration until the enzyme is saturated, when it is working at its maximum velocity. Note that $[E]_2$ is twice $[E]_1$ and $V_{max}2$ is twice $V_{max}1$.

Figure 23-5 illustrates the most general change that occurs in the rate of an enzyme-catalyzed reaction as the substrate concentration is increased. This relationship is typical for enzyme-catalyzed reactions that follow the two-step process described on page 555, and in certain other cases as well. Note that this figure shows the kind of data that can be obtained for experiments conducted in the laboratory. In the human body, enzymes rarely work at maximum velocity, because substrate concentrations in cells are usually well below the saturation level.

The rate of an enzyme-catalyzed reaction also depends on such factors as the temperature and the pH, as well as on the inherent efficiency of the enzyme in question. Each enzyme has a set of preferred conditions (pH, temperature, and so on) for maximum activity. Enzymes also differ widely in their inherent efficiency. The term used to express this inherent efficiency is the molar activity of an enzyme. **Molar activity** is defined as the ratio of V_{max} to [E]:

(23-2) $$\text{Molar activity} = \frac{V_{max}}{[E]}$$

Thus, the molar activity of an enzyme can be determined by measuring V_{max} at a known concentration of enzyme. The molar activity is then obtained by dividing the V_{max} obtained by the enzyme concentration used. For example, if a $10^{-3}\ M$ solution of a particular enzyme exhibits a V_{max} of $1\ M/min$, then the molar activity of this enzyme is

(23-3) $$\text{Molar activity} = \frac{1\ M/min}{10^{-3}\ M} = 1000\ \text{min}^{-1}$$

Table 23-2 gives the molar activities of some specific enzymes. The large differences in molar activities between different enzymes is not surprising, since there is such a vast array of enzymes and reactions catalyzed. Remember that not all enzymes are specific for only one substrate. For an enzyme that is not specific for a single substrate, the values for V_{max} may be quite different for different substrates.

Table 23-2 Molar Activities of Some Enzymes

Enzyme	Molar Activity (min^{-1})
Carbonic anhydrase	36,000,000
Ketosteroid isomerase	17,100,000
Fumarase	1,200,000
β-Amylase	1,100,000
β-Galactosidase	12,500
Phosphoglucomutase	1,240
Succinate dehydrogenase	1,150
Aconitase	900

Exercise 23-4

A 1-liter solution that contains 10^{-5} mole of the enzyme catalase could theoretically produce 56 moles of H_2O from hydrogen peroxide, H_2O_2, per minute at substrate saturation. (The catalase reaction was discussed in Section 23-3.) What is the molar activity of the enzyme catalase?

Effect of pH

Many functional groups on the amino acid side chains of enzymes can lose protons or be protonated (see Chapter 21). The number of functional groups that are protonated depends on the pH. For most enzymes, salt bridges involving charged amino acid side chains are necessary to hold the enzyme together. In other words, the shape of an enzyme (its tertiary structure) depends on the pH. Now, since the activity of an enzyme depends critically on the shape of the enzyme, the activity of most enzymes varies with pH (see Figure 23-6). Trypsin is typical of most enzymes in the human body in that its activity varies substantially with pH, and it has a pH optimum very close to the actual pH of human cells. Some enzymes, however, are very active over a fairly broad range of pH values. Papain is an example of such an enzyme.

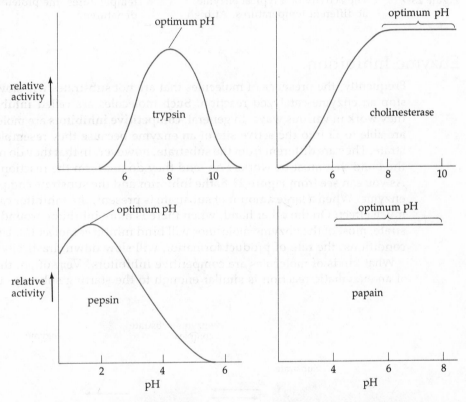

Figure 23-6   The effect of pH on the activity of some enzymes. Note the different pH optima of these enzymes and the broad pH optimum of papain. The behavior of most enzymes is similar to that of trypsin.

Effect of Temperature

The shape of an enzyme also depends on the temperature (Chapter 21). Very high temperatures are detrimental to enzymes. After all, they are proteins and can be denatured by heat. Unfolded, denatured enzymes are inactive (see Figure 23-7). Thus, as with noncatalyzed chemical reactions, the rate of an enzyme-catalyzed reaction first increases as the temperature increases, but then the rate decreases as the enzyme becomes denatured. Most enzymes in humans have an optimum temperature at which they are most active. This optimum temperature is close to 37°C (normal body temperature).

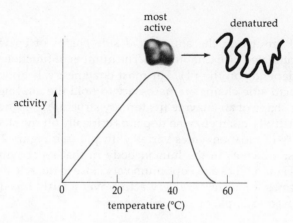

Figure 23-7   The activity of a typical enzyme at different temperatures. At high temperatures, the protein is denatured.

Enzyme Inhibition

Frequently, the presence of molecules that are not substrates will slow down or stop an enzyme-catalyzed reaction. Such molecules are called **inhibitors,** and they work in various ways. In general, **competitive inhibitors** are molecules that are able to fit into the active site of an enzyme because they resemble the substrate. They are different from the substrate, however, in that they do not contain the bond that must be worked on, and they do not form the reaction products. As you can see from Figure 23-8, the inhibitor and the substrate compete for the enzyme. When a large amount of substrate is present, the inhibitor cannot have much effect. On the other hand, when there is more inhibitor around than substrate, most of the enzyme molecules will bind inhibitor molecules. Under these conditions, the rate of product formation will slow down drastically.

What kinds of molecules are competitive inhibitors? Very often, the product of an enzymatic reaction is similar enough to the starting substrate that it can

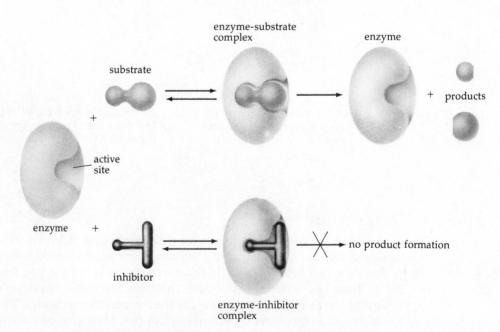

Figure 23-8   Schematic representation of competitive inhibition of an enzyme. The inhibitor competes with the substrate for the enzyme. When inhibitor is bound, substrate cannot bind.

act as a competitive inhibitor and prevent further production of itself. Thus, the product-inhibited enzyme reaction slows down when the amount of product is too large.

Noncompetitive Enzyme Inhibition

There are other inhibitors that do not usually compete with the substrate for the active site of the enzyme, and usually are not similar to the substrate in structure. **Noncompetitive inhibitors** bind to the enzyme at some point other than the active site (see Figure 23-9). The substrate can still bind to the enzyme-inhibitor complex, but it cannot be converted to product. Since the substrate does not bind to the same site as a noncompetitive inhibitor, the presence of a large amount of substrate cannot overcome noncompetitive inhibition.

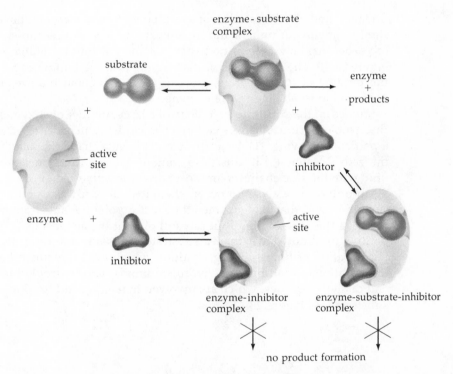

Figure 23-9   Noncompetitive inhibition of an enzyme. The inhibitor does not compete with the substrate for the active site. It can bind to either the free enzyme or the enzyme-substrate complex. When bound to the inhibitor, the enzyme is incapable of catalyzing the conversion of the substrate to products.

Exercise 23-5

The enzyme succinate dehydrogenase catalyzes the reaction

$$
\begin{array}{c}
\text{O} \\
\parallel \\
\text{C}-\text{OH} \\
\mid \\
\text{CH}_2 \\
\mid \\
\text{CH}_2 \\
\mid \\
\text{C}-\text{OH} \\
\parallel \\
\text{O}
\end{array}
\;+\; \text{FAD}
\;\underset{\substack{\text{succinate} \\ \text{dehydrogenase}}}{\rightleftharpoons}\;
\begin{array}{c}
\text{O} \\
\parallel \\
\text{C}-\text{OH} \\
\mid \\
\text{C}-\text{H} \\
\parallel \\
\text{H}-\text{C} \\
\mid \\
\text{C}-\text{OH} \\
\parallel \\
\text{O}
\end{array}
\;+\; \text{FADH}_2
$$

succinic acid   fumaric acid

Two of the following molecules are competitive inhibitors of succinate dehydrogenase. Which is not?

$$O=\overset{\underset{\displaystyle |}{OH}}{C}-CH_2-\overset{\underset{\displaystyle |}{OH}}{C}=O$$

malonic acid

$$O=\overset{\underset{\displaystyle |}{OH}}{C}-CH_2-\overset{\overset{\displaystyle O}{\|}}{C}-\overset{\underset{\displaystyle |}{OH}}{C}=O$$

oxaloacetic acid

$$I-CH_2-\overset{\overset{\displaystyle O}{\|}}{C}-NH_2$$

iodoacetamide

23-5 COENZYMES AND THE MECHANISMS OF ENZYME-CATALYZED REACTIONS

Chemical and physical studies of a number of typical enzymes have contributed significantly to our understanding of enzyme-catalyzed reactions. For example, the experimentally determined shape of the enzyme hexokinase is shown in Figure 23-10. The active site of this enzyme is at the bottom of a "pocket," and the binding of the substrate, glucose, to this enzyme induces a pronounced conformational change in the enzyme.

Studies on the substrate specificity of a large number of enzymes have shown that a molecule must possess certain characteristics in order to be a substrate for a particular enzyme: (1) A substrate must have a shape that allows it to fit into the active center; (2) it must also contain specific functional groups that can bind with the side chains of amino acids at the active center of the enzyme; and (3) in order for catalysis to occur, the substrate must also contain the type of chemical bond that the enzyme attacks. If a potential substrate molecule does *not* have this type of bond but can still bind at the active site, it will be a competitive inhibitor rather than a substrate. Thus enzyme catalysis requires the participation of both binding residues, which hold the substrate in place via combinations of salt bridges, hydrogen bonds, and hydrophobic interactions, and catalytic residues, which are involved in making and breaking bonds in the substrate (see Figure 23-4).

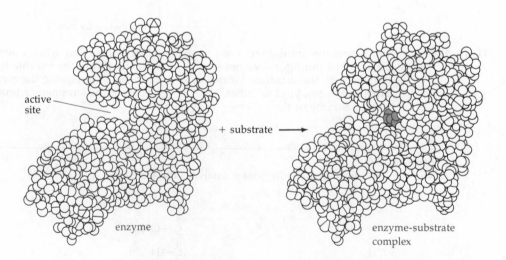

active site

enzyme

+ substrate ⟶

enzyme-substrate complex

Figure 23-10   The experimentally determined three-dimensional shape of the enzyme hexokinase isolated from yeast cells. The left figure shows the shape of the free enzyme. Note the deep pocket. The active site is at the bottom of this pocket.

Binding of substrate, glucose, induces hexokinase to undergo the conformational change necessary for it to catalyze its reaction. Note the molecule of glucose (in blue) bound at the active site in the right figure.

The nature of the catalytic residues (those amino acids required for bond making or breaking) depends on the reaction catalyzed. Some enzymes even form covalent bonds with their substrate as part of the reaction mechanism. For example, aldolase, one of the enzymes required for glucose oxidation, uses such a reaction pathway.

Many enzymes require small accessory molecules called **coenzymes** in order to function as catalysts. These accessory molecules supply groups of atoms to the substrate or accept groups of atoms from it. We have seen that the coenzyme NAD^+ accepts a hydride ion, $H:^-$. Some enzymes, called transaminases, remove amino groups from substrates. Since these enzymes must not be altered if they are to keep functioning as catalysts, transaminases require coenzymes that will accept these amino groups. Other coenzymes specialize in methyl groups, carboxyl groups, and so on. The number of different coenzymes in the human body is much less than the number of different enzymes, since many different enzymes can use the same coenzyme. Coenzymes are bound to enzymes at coenzyme binding sites in the same manner as substrates.

There is some similarity between coenzymes and prosthetic groups. However, prosthetic groups on proteins need not be altered in catalysis.

Let us consider one example of a reaction that requires a coenzyme:

(23-4)

In reaction 23-4, catalyzed by the enzyme alanine transaminase, two bonds in the substrate alanine are broken (the bond between the α-carbon atom and a hydrogen atom and the bond between the α-carbon atom and the amine group) and two bonds are formed between the α-carbon atom and an oxygen atom to give the product, pyruvic acid. The enzyme does not accept the amine group from alanine, nor does it donate the oxygen atom needed to form pyruvic acid. Rather, these atoms are exchanged between the substrate alanine and a coenzyme called pyridoxal phosphate. In a subsequent enzyme-catalyzed reaction (not shown), the pyridoxamine phosphate produced in reaction 23-4 is converted back to pyridoxal phosphate. Pyridoxal phosphate is also involved in a number of other enzyme-catalyzed reactions in which amine groups are transferred.

Humans cannot synthesize some parts of coenzymes. Over the many years of evolution, we have lost the ability to do so. Those parts of coenzymes that we cannot synthesize must be included in our diet and are called **vitamins.** Vitamins were recognized as a necessary component of our diet long before we really knew why. If people did not have an adequate supply of them, they became ill. Now, we know why we need them. We shall see some of the specific reactions for which vitamins are required when we study the pathways for the synthesis and degradation of molecules in the body. We shall talk about nutritional requirements of the human body in Chapter 29. Structural formulas for some coenzymes, with their vitamin portions indicated, are shown in Appendix 3.

Unlike enzymes, coenzymes such as pyridoxal phosphate and NAD^+ are altered in enzyme-catalyzed reactions. However, they can be changed back to their original form by other reactions. Since coenzymes can be recycled, we only need small amounts of them to satisfy our needs.

Exercise 23-6
Figure 23-10 shows the enzyme hexokinase bound to its substrate, glucose. (a) List some amino acids that might function as binding residues at the active site of this enzyme. (b) Notice how small glucose is compared to hexokinase. Is the size of this enzyme wasteful? In other words, couldn't a lot smaller protein do the same job? Explain.

23-6 REGULATION OF ENZYME ACTIVITY

It is one thing to possess a collection of enzymes that are capable of making or breaking apart a wide variety of molecules. It is quite another thing to control the activity of these enzymes. We do not want to make too much or too little of any substance. Enzyme activity can be influenced by a number of factors. We have already seen that it depends on pH and temperature. For warm-blooded humans, a change in temperature is not an important means of controlling enzyme activity. At 37°C, almost all enzymes are working at or near their optimum temperature, and our body temperature does not vary significantly. The pH may be important in controlling the activity of an enzyme, but we do not yet know to what extent human metabolism makes use of this means of regulation. Enzyme activities are controlled in several other ways that are well understood.

Regulation of Substrate Concentration

One sure way to stop an enzyme from making too much product is to starve it, that is, to limit the amount of substrate available to it. If we use our factory analogy, this would be equivalent to stopping the assembly line. The worker (the enzyme) is there and ready, but has nothing on which to work. Substrate limitation can be achieved in a number of ways. Intracellular enzymes require that their substrates get into the cell. The rate of entry of substrates is, in turn, controlled by their ability to penetrate the cellular membrane or by their rate of transport into the cell. By controlling these factors, the rate of product formation can be controlled. For those enzymes that work farther down the assembly line, the availability of substrate depends on how fast the enzymes on the beginning of the assembly line are working (see Figure 23-11).

Figure 23-11   In this series of reactions, the concentrations of substrates B and C available to enzymes E_B and E_C depend on the rate of reaction A → B catalyzed by E_A. If E_A is inhibited, the formation of D will decrease.

$$A \xrightarrow[E_A]{} B \xrightarrow[E_B]{} C \xrightarrow[E_C]{} D$$

Regulation of Enzyme Concentration

Clearly, the rate of product formation is limited by the amount of enzyme present. If no enzyme is present, no product will form. Cells control the amounts of the various enzymes they contain by controlling their rate of synthesis. Synthesis of enzymes that are not needed can be shut off completely. However, even if synthesis of a particular enzyme is turned off completely, there will be a lot of it around for a while. Using our factory analogy, after a factory stops hiring one type of employee, there is usually a long time interval before the number of employees of that type decreases because some have quit or retired. (Actually, enzymes are "retired" quite forcibly—they are cut apart by other enzymes!) We shall see how enzyme synthesis is controlled in the next chapter.

Allosteric Regulation of Enzyme Activity

A faster way to stop enzyme workers in a cellular factory is to lay them off. This is accomplished by allosteric effects. **Allosteric enzymes** have more than one site (*allo-* = other, *-steric* = site) to which substrates or other small molecules can bind. The other sites may also be catalytic sites. Hemoglobin is an example of a multisite protein. Each of its four subunits can bind oxygen. The binding of an O_2 molecule by one of the four subunits makes binding of O_2 easier for the other subunits. This is because binding of O_2 by induced fit causes a conformational change in hemoglobin, as shown in Figure 23-12. The new conformation makes it easier for the other O_2 molecules to bind.

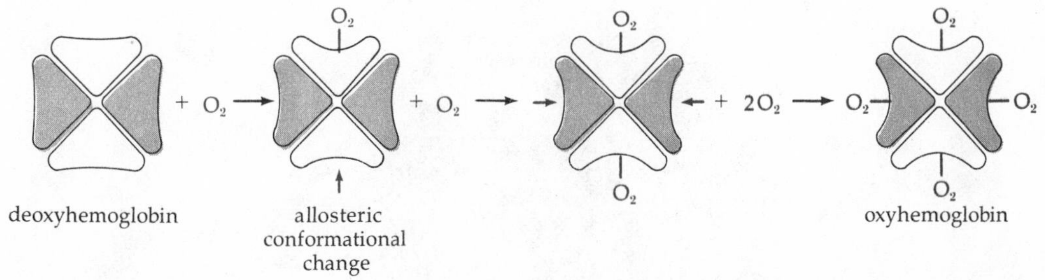

deoxyhemoglobin   allosteric   oxyhemoglobin
                    conformational
                        change

Figure 23-12   A model for the allosteric effect of oxygen binding to hemoglobin. The allosteric conformational change induced by the first molecule of O_2 to bind facilitates the binding of additional O_2 molecules. As indicated by the blue arrows, the subunits bound to O_2 exert pressure on the other subunits to change their shape as well.

Several allosteric proteins have binding sites for molecules that are not substrates. Many of these undergo an unfavorable conformational change when they bind this other molecule. In the unfavorable conformation it is more difficult for the substrate to bind to the active site. The other molecule is then called an **allosteric inhibitor** of the enzyme. End products of the assembly line on which the particular enzyme is working are frequently allosteric inhibitors. The enzyme that is inhibited is usually the first one on that assembly line. Thus when a product of the assembly line or pathway builds up, it stops production until it is used up. This type of control, called **feedback inhibition,** is illustrated in Figure 23-13. In this example, E_1D, the complex formed by the enzyme E_1 and the feedback inhibitor D, is in equilibrium with free enzyme and D, that is, $E_1D \rightleftharpoons E_1 + D$. When the concentration of D is large, most of the E_1 will be bound to D and inactivated; but when the concentration of D is small, most of the E_1 will be free to work. Thus, the formation of products, B, C, and D will slow down or speed up as the concentration of D increases or decreases.

Figure 23-13   Feedback inhibition of an enzyme. The final product of this series of reactions, D, binds to an allosteric site on the first enzyme in the series, E_1. The activity of enzyme E_1 is thus dependent on the concentration of D, and large amounts of D shift the equilibrium to favor the inactive E_1D complex.

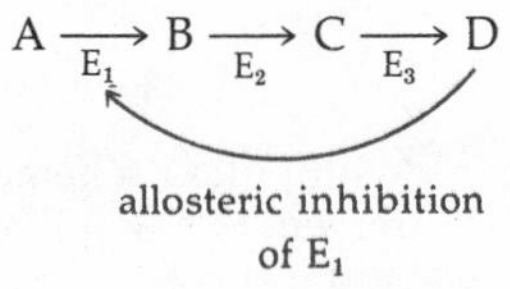

allosteric inhibition
of E_1

Regulation by Chemical Modification of Enzymes

Still another way of controlling enzyme activity is chemical modification. **Chemical modification** of an enzyme is achieved by the covalent attachment of a functional group to the enzyme. The attachment is catalyzed by yet another enzyme, and occurs at a specific site but usually not at the active site. For example, one of the serine residues on the enzyme glycogen phosphorylase can react to form a phosphate ester. This enzyme has a different shape when phosphate is attached to the serine residue, and is more active (has a higher molar activity).

Other enzymes can be modified by attachment of acetate or methyl or phosphate groups to them. The modified enzyme is either more or less active than the unmodified enzyme. Figure 23-14 shows how the control system works for glycogen phosphorylase. Here, a hormone reaches the cell and gives it the message to increase glucose production. The hormones glucagon and adrenaline (via their receptor proteins) are allosteric activators for a particular enzyme called adenyl cyclase, located in the cell membrane. These hormones change the con-

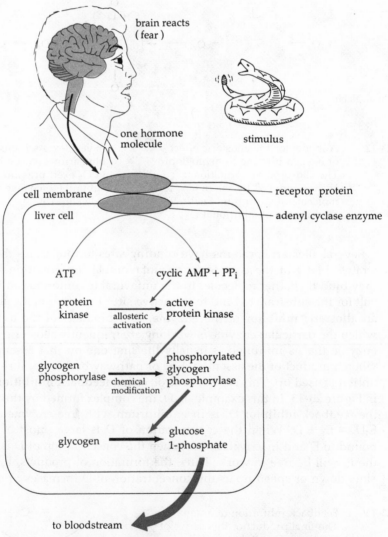

Figure 23-14   A simplified scheme showing the tremendous amplification achieved by the successive steps in the process whereby fright ultimately results, via the hormone adrena-line, in an increase in blood sugar. The amplification is represented by the increasing thickness of the blue arrows.

formation of adenyl cyclase to a highly active shape that can make a lot of cyclic AMP from ATP. The cyclic AMP then allosterically activates a second enzyme, a protein kinase, which results in the binding of phosphate groups to proteins, such as glycogen phosphorylase. Finally, the phosphorylated glycogen phosphorylase breaks glycogen down to glucose 1-phosphate. Thus, the activation of glycogen phosphorylase proceeds through several steps of allosteric activation and chemical modification. Each step also amplifies the message, so that about 3 million glucose molecules are produced for each messenger hormone molecule.

Exercise 23-7
Note that each successive step in the hormonal activation scheme in Figure 23-14 results in added amplification of the response. If each of the four steps within the cell resulted in only a 10-fold amplification, how many glucose 1-phosphate molecules would be produced as a result of the action of one hormone molecule at the cell membrane?

23-7 ON THE MECHANISM OF DRUG ACTION

Many drugs that are used to combat microbial infections or to fight cancer are competitive inhibitors of specific enzymes. For example, some enzymes that are vital to infectious organisms but are not found in healthy humans can be inhibited. Inhibitors of these enzymes can therefore kill the infectious organism but not affect human cells. Sulfanilamide was one of the first of these inhibitors to be identified. It is structurally similar to p-aminobenzoic acid, which bacteria use to make folic acid (see Figure 23-15).

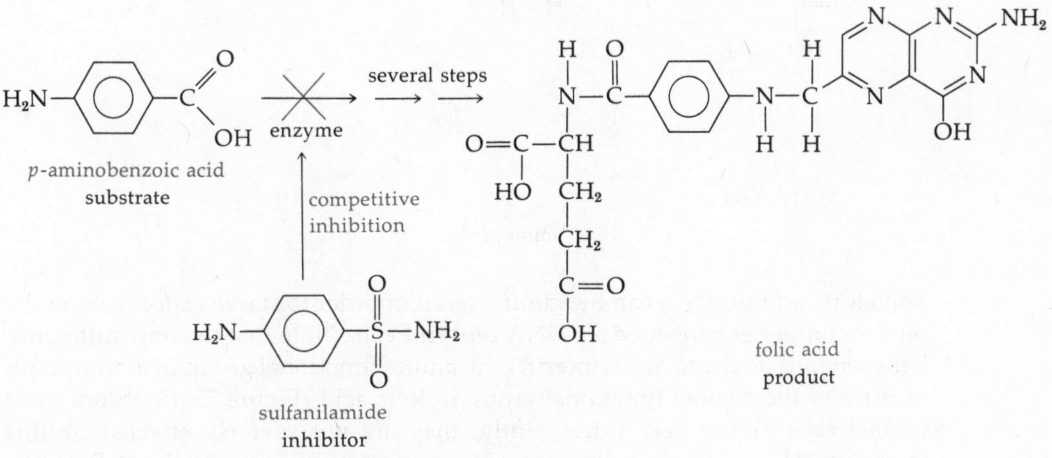

Figure 23-15   The drug sulfanilamide competitively inhibits the formation of folic acid from p-aminobenzoic acid in many microorganisms.

Humans cannot make folic acid. For us it is a vitamin. Thus, when we take sulfanilamide, bacteria die, but we are not affected. Several derivatives of sulfanilamide have now been synthesized, some of which are even better competitive inhibitors of folic acid synthesis. Together with sulfanilamide, these compounds are popularly referred to as the sulfa drugs, some of which have the following names and structures:

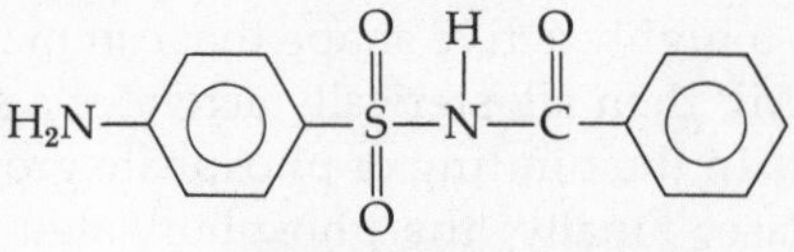

sulfabenzamide

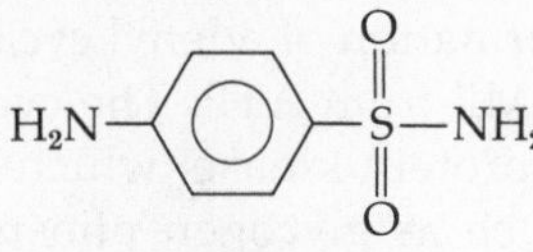

sulfanilamide

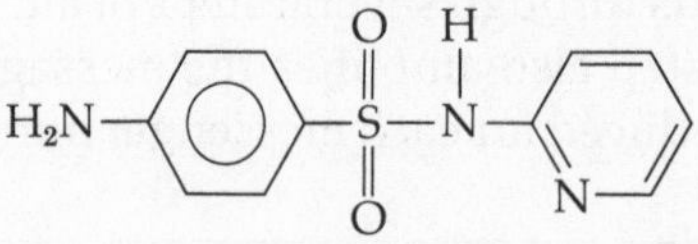

sulfapyridine

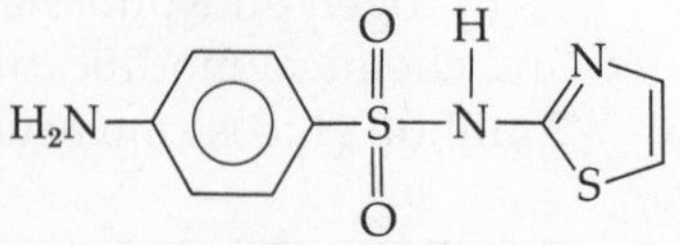

sulfathiazole

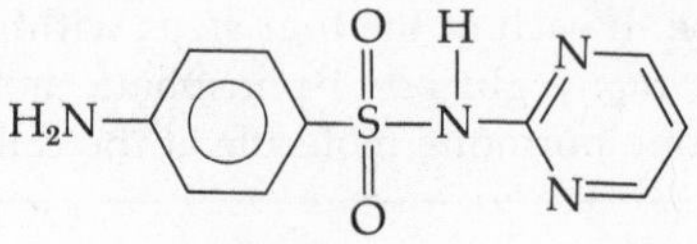

sulfadiazine

Potent competitive inhibitors of enzymes in important metabolic pathways, such as the sulfa drugs, are often referred to as **antimetabolites.** Other kinds of antimetabolites are used in the chemotherapy of cancer. They take advantage of the fact that cancer cells divide rapidly, and therefore need to make a lot of DNA, whereas normal cells do not. For example, aminopterin, an analog of folic acid, is a competitive inhibitor of purine biosynthesis.

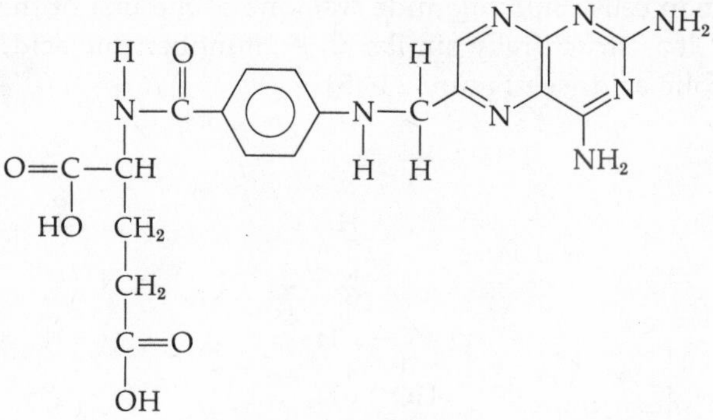

aminopterin

Therefore, aminopterin can be administered in order to starve cancer cells of the purine components needed for DNA biosynthesis. Note that the only difference between folic acid and aminopterin is an amine functional group in aminopterin in place of the alcohol functional group in folic acid (Figure 23-15). Since most normal cells divide very infrequently, they are not severely affected by this treatment. However, the cells of some tissues do divide frequently. Cells in the intestines and some other types of cells are also killed by aminopterin treatment. Since cancer cells are derived from normal cells, and thus contain the same enzymes, it is difficult to eliminate them selectively with antimetabolites. Since some human cancers now appear to be caused by viruses (Chapter 24), there may be a better chance of finding specific antimetabolites in these cases.

Other compounds used to treat microbial infections are the **antibiotics**—molecules made by one organism that are toxic to others. They are thought to act as antimetabolites, but the reactions they competitively inhibit are not always clearly understood. Penicillin was the first antibiotic to be discovered (accidentally, by Alexander Fleming in the 1920s); it interferes with cell wall biosynthesis in a variety of bacteria. The structural formulas of several antibiotics are shown at the top of the next page.

Penicillin G

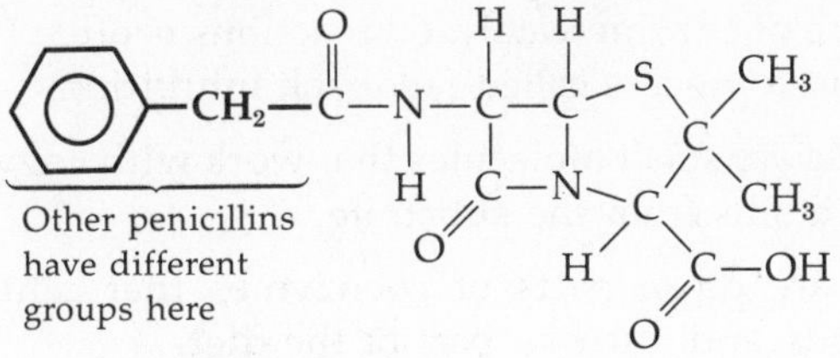

Other penicillins
have different
groups here

Penicillin F $CH_3-CH_2-CH=CH-CH_2-$

Penicillin K $CH_3-(CH_2)_6-$

Penicillin X $HO-$⟨○⟩$-CH_2-$

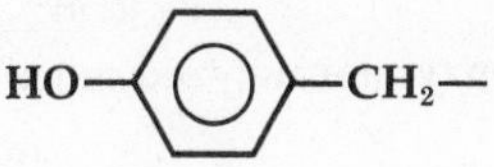

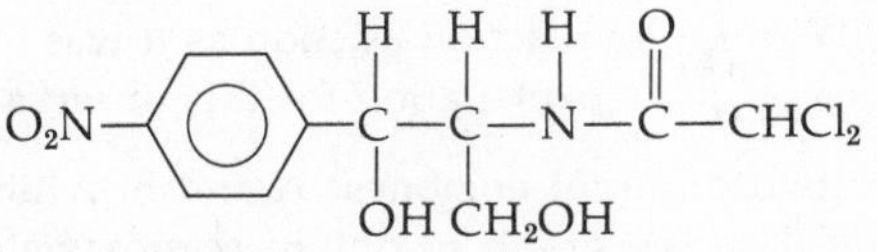

Chloramphenicol (chloromycetin)

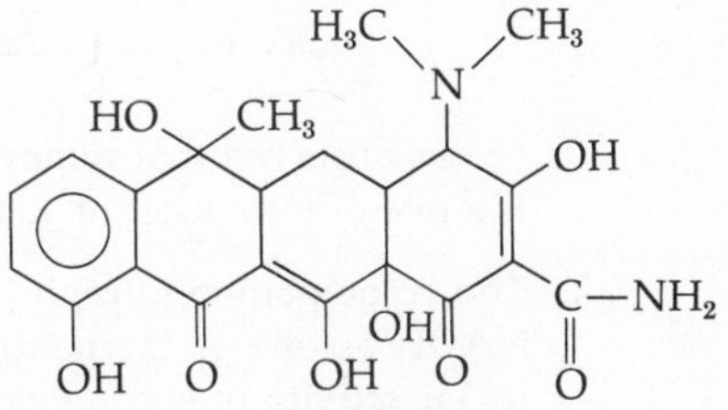

Tetracycline

Exercise 23-8
Would you expect aminopterin to be a useful drug for the treatment of bacterial infections?
Explain your answer.

23-8 SUMMARY

1. Enzymes are proteins that catalyze chemical reactions. They all exhibit the basic functional features of specificity, saturation, and inhibition.

2. Enzymes are named and classified according to the reactions that they catalyze. There are six major classes of enzymes: oxidoreductases, transferases, hydrolases, lyases, isomerases, and ligases.

3. The region of an enzyme where the substrate binds is called the active site, which consists of binding residues and catalytic residues.

4. When a substrate binds to an enzyme, a conformational change in the enzyme occurs. This process is called induced fit.

5. One general mechanism for enzyme-catalyzed reaction is the two-step process $E + S \rightleftharpoons ES$ followed by $ES \rightarrow E + P$, where E, S, and ES are enzyme, substrate, and enzyme-substrate complex, respectively.

6. The molar activity of an enzyme is a measure of the catalytic power of the enzyme.

7. The rate of an enzyme-catalyzed reaction is influenced by temperature, pH of the reaction medium, enzyme concentration, substrate concentration, chemical modification of the enzyme, allosteric effects, and the presence of inhibitors.

8. Two general types of inhibitors are competitive inhibitors and noncompetitive inhibitors.

9. A sequence of enzyme-catalyzed reactions is often regulated by allosteric enzymes in a process called feedback inhibition.

10. Coenzymes are small molecules that work with enzymes by supplying or accepting atoms from the substrate.

11. Vitamins are those parts of coenzymes that cannot be made by the human body and must be part of the diet.

12. Many drugs are competitive inhibitors of enzymes.

PROBLEMS

1. In a study of the following sequence of enzyme-catalyzed reactions,

$$S \xrightarrow{E_1} T \xrightarrow{E_2} U \xrightarrow{E_3} V \xrightarrow{E_4} W$$

it was observed that removing W from the reaction as soon as it was formed increased the rate of formation of T. Give a possible explanation for this observation.

2. The accompanying graph is for two different enzymes, A and B, which both produce NADH as one of their products. Use the graph to obtain approximate values for the molar activity of each enzyme. Which is more efficient in producing NADH?

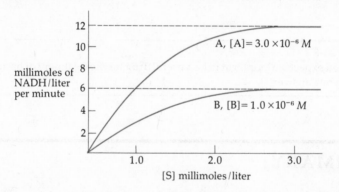

3. The molar activity of a certain enzyme, E, is 4800 min^{-1}. Calculate the rate of the reaction catalyzed by E when the concentration of E is 5.0×10^{-6} M and the substrate concentration is extremely large.

4. The enzyme amylase catalyzes the hydrolysis of starch to glucose. To which class of enzymes does amylase belong?

5. Explain what is meant by induced fit with regard to enzyme-catalyzed reactions.

6. The membrane transport protein Na$^+$K$^+$-ATPase, which is also an enzyme, has a binding site for ATP. The structural formula for the predominant form of ATP at a pH of about 7 is

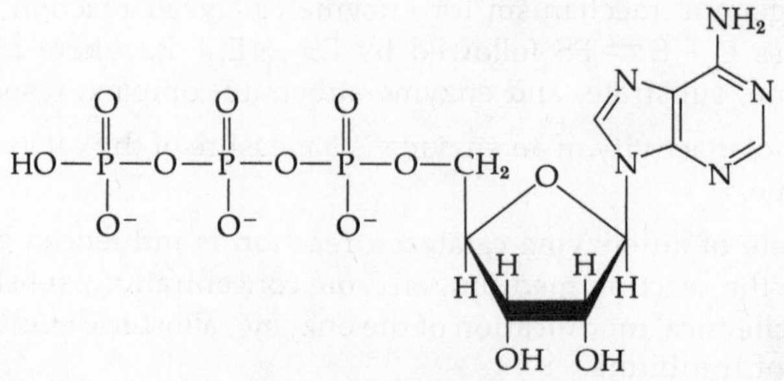

List some amino acids of Na$^+$K$^+$-ATPase whose side chains may be involved in binding ATP, and describe how each can interact with a portion of ATP.

SOLUTIONS TO EXERCISES

23-1 (a) A thioester bond is broken with the addition of water, so this enzyme is a hydrolase (Class 3).

(b) Two substrates are joined together via the exergonic hydrolysis of ATP, so this is a ligase reaction (Class 6).

(c) The $-CH_2-CH_2-$ portion of the substrate loses carbon-hydrogen bonds (an oxidation), while FAD is reduced. Therefore, this enzyme is classified as an oxidoreductase (Class 1).

(d) E_d catalyzes the interconversion of two isomers (Class 5).

23-2 The aspartic acid side chain contains a negatively charged carboxylate ion $\left(-\overset{\overset{O}{\|}}{C}-O^-\right)$, which will bind the positive Na^+ or K^+ ions.

23-3 At the pH of cells in the body (~7.4), the carboxyl group exists primarily as the negative ion and thus cannot donate a proton. However, the ammonium ion form of the amino acid lysine is a weak acid and can serve as a proton donor at this pH (see Chapter 18).

23-4 For catalase, $V_{max} = 56$ (moles/liter) min^{-1} at this enzyme concentration. Therefore,

$$\text{Molar activity of catalase} = \frac{V_{max}}{[E]} = \frac{56 \text{ (moles/liter) } min^{-1}}{10^{-5} \text{ (mole/liter)}} = 5.6 \times 10^6 \ min^{-1}$$

23-5 Malonic acid and oxaloacetic acid are similar to succinic acid in size and both possess two carboxyl groups. Iodoacetamide, however, is smaller and does not contain a carboxyl group and is not a competitive inhibitor of succinate dehydrogenase.

23-6 (a) Glucose, the substrate of this enzyme, contains several hydroxyl groups. We would therefore expect to find binding residues capable of forming hydrogen bonds with this substrate, possibly including serine, threonine, asparagine, glutamine, or tyrosine, as opposed to amino acids with hydrophobic side chains. (b) In addition to possessing the correct binding and catalytic residues, these residues must be correctly oriented in the active site of the enzyme. Hexokinase also undergoes a fairly substantial conformational change upon binding its substrate. The large sizes of enzymes are necessary in order for them to provide active sites that have very specific shapes as well as to undergo precise conformational changes.

23-7 There are four steps of amplification, so if we assume a 10-fold amplification per step, the overall amplification is $10 \times 10 \times 10 \times 10 = 10,000$ glucose 1-phosphate molecules.

23-8 Aminopterin is not useful for treating bacterial infections. Bacteria can make their own folic acid in the presence of aminopterin. A dose of aminopterin large enough to competitively inhibit the folic acid-dependent steps of DNA synthesis in bacteria would also severely hinder DNA synthesis in normal human cells.

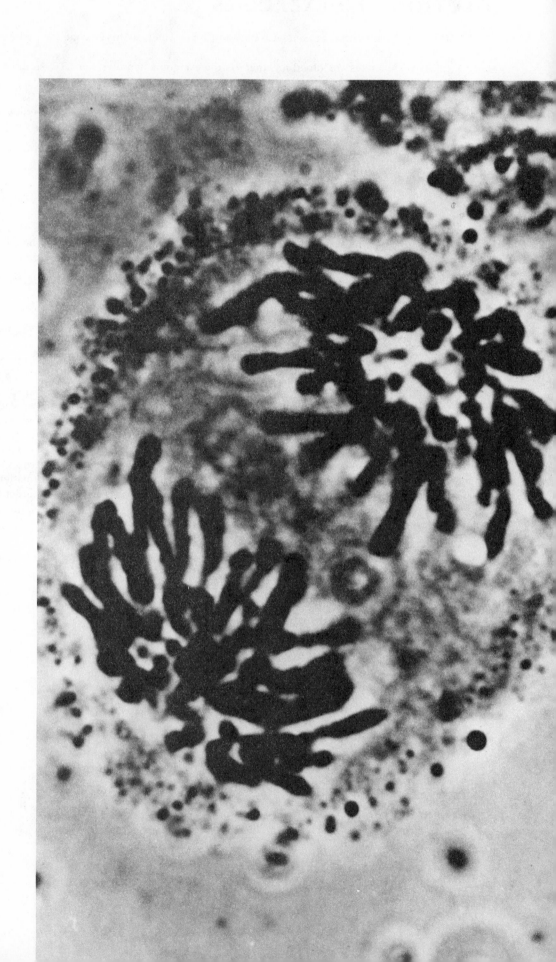

CHAPTER 24

Nucleic Acids and the Biosynthesis of Proteins

24-1 INTRODUCTION

In our analogy of cells in the human body as industrial factories, proteins play the roles of mail carriers, assembly-line workers, and so on. Obviously, someone needs to be in charge of an industrial factory, and there must be control of the cellular machinery as well. The task of directing the activities of the cellular factory begins with information stored in DNA molecules found in chromosomes located in the cell nucleus. Each chromosome consists of one DNA molecule and several proteins. There are 46 chromosomes (23 pairs) in all human cells except sperm and ova, which each have 23 unpaired chromosomes. The synthesis of a particular protein is directed by a definite portion of a DNA molecule called a gene. One DNA molecule contains hundreds of genes.

The total of all of the DNA present in all of the chromosomes in each cell is analogous to a set of master blueprints for the entire body. When a cell divides, each daughter cell must receive its own set of these master blueprints. Thus, prior to cell division, the DNA in a cell is reproduced. We shall see that the reproduction of DNA occurs in a process called *replication*, which produces an *exact* copy of a cell's DNA.

Using our analogy, managing a particular cellular factory is accomplished by determining which sections of the blueprint are to be read. We know quite a bit about how this blueprint is read, but we know relatively little about the process by which only certain portions are selected to be read by a given cell. As you study how cells read their DNA blueprints, you will see that the process of deciphering genetic information closely parallels the process of deciphering a code or a foreign language. Therefore, the steps in this process have been given names that suggest decoding, such as *transcription* and *translation*. The terms *code letters* and *codons* are also used in the description of protein synthesis. We shall also look at how the process of protein synthesis is controlled. Finally, we shall see how viruses invade our cells and use the cellular machinery to make their own proteins.

As a human cell divides, two sets of chromosomes (the oblong, dark objects in this micrograph) can be detected as they separate. Each chromosome consists of a single DNA molecule and several proteins.

24-2   STUDY OBJECTIVES

After careful study of this chapter, you should be able to:

1. Differentiate purine bases from pyrimidine bases, nucleosides from nucleotides, and DNA from RNA.

2. Discuss the three-dimensional structure of the nucleic acids and the relationship of their structure to their function.

3. Describe the role of hydrogen-bonded base pairs in the structure and function of nucleic acids.

4. Describe the process of DNA replication.

5. Compare and contrast transcription and replication.

6. Detail each of the steps in translation and its role in the overall process of protein synthesis.

7. Explain the necessity for regulating protein synthesis and the differences between induction and repression.

8. Describe what can happen when viruses infect cells.

24-3   GENES AND THEIR FUNCTION

A **gene** is a definite portion of a DNA molecule that codes for the synthesis of the specific sequence of amino acids in a particular protein. Since each cell in the human body (except sperm and ova) contains 23 pairs of chromosomes, each cell is, in theory, capable of producing all of the different proteins in the human body. Any *particular* cell, however, uses only some of its genes and synthesizes relatively few proteins. Which genes are used and which proteins are synthesized by a cell distinguishes one type of cell from another. As the human embryo matures, groups of cells become specialized as liver cells, brain cells, kidney cells, and so on. This process is called **differentiation.** We shall describe later what little is known about the complexities of differentiation.

Although we know relatively little about the mechanisms involved in differentiation, we do know a good deal about the everyday functioning of DNA molecules and their component genes. If we think of the DNA molecules in the nucleus as the master blueprints in the manager's office, then we realize that they must be kept safe for future reference. Therefore, in order to use the information contained in the DNA, the cell must make expendable copies of each portion of the blueprint (each gene) that it wants to use. Cells do not have Xerox machines, but they do have their own copying process, which we call **transcription.** The transcription process produces an expendable copy of a gene, called a **messenger RNA** molecule, that can be carried out into the factory proper. (As we shall see in Section 24-8, a messenger RNA molecule is not an exact copy of a DNA gene. A better, but still imperfect analogy of the relationship between messenger RNA and its DNA gene would be the relationship between a photographic negative and its positive print.)

Once outside of the nucleus, messenger RNA directs the assembly of the appropriate amino acids into the proper sequence to form the desired protein molecule. This process of protein assembly, called **translation,** takes place on ribosomes, specialized particles located in the cell's cytoplasm. The translation process involves several enzymes.

Basically, the management of cellular factories is accomplished according to the following model, reverently referred to as the **central dogma of molecular biology.**

$$\text{DNA} \xrightarrow[\text{transcription}]{} \text{messenger RNA} \xrightarrow[\text{translation}]{} \text{protein}$$

master   copy   product
blueprint

The arrows here represent information transfer, not simple chemical reactions. In order to understand the details of these processes, you must first become familiar with the structures and properties of the nucleic acid polymers DNA and RNA, and the components, building blocks, from which these nucleic acid polymers are assembled.

24-4 COMPONENTS OF NUCLEIC ACIDS

Nucleic acids, like proteins, are large polymers made up of a small number of different building blocks. The building blocks of nucleic acids are called **nucleotides.** Each nucleotide is in turn composed of three smaller parts: a phosphate group, a monosaccharide, and a nitrogen-containing base. The term **nucleoside** is used to refer to a nitrogenous base bound to a monosaccharide, and a nucleotide is a nucleoside phosphate (see Figure 24-1).

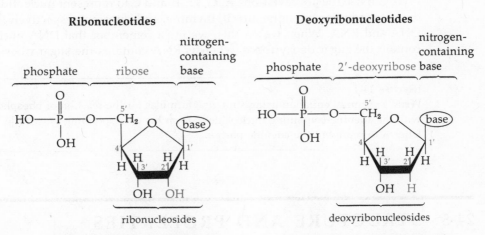

Figure 24-1   The components of nucleotides. Note the different sugar components in ribonucleotides and deoxyribonucleotides. Carbon atoms of the sugar component of nucleotides are numbered with primes (1′, 2′, etc.) to distinguish them from the atoms on the base component, which are numbered 1, 2, 3, and so forth.

There are two major types of nucleic acids: **deoxyribonucleic acid (DNA)** and **ribonucleic acid (RNA).** There are only two differences between the deoxyribonucleotide components of DNA and the ribonucleotide components of RNA. The first difference is in the sugar component. Both DNA and RNA have the sugar D-ribose, but in DNA there is no hydroxyl group on the number 2′ carbon of ribose, so the prefix *deoxy-* is used to denote the absence of oxygen at this position. Second, each DNA and RNA nucleotide contains one of four possible bases. Three of them—adenine, guanine, and cytosine—are the same in both DNA and RNA. The fourth base is thymine in DNA and uracil in RNA. The structural formulas for these bases, as well as their names, are given in Figure 24-2. Adenine and guanine have two nitrogen-containing rings. They are called **purines** because of their similarity to the molecule purine. Likewise, cytosine, thymine, and uracil are called **pyrimidines,** because of their similarity to pyrimidine. Notice that thymine is structurally identical to uracil except that there is a methyl group on thymine and none on uracil.

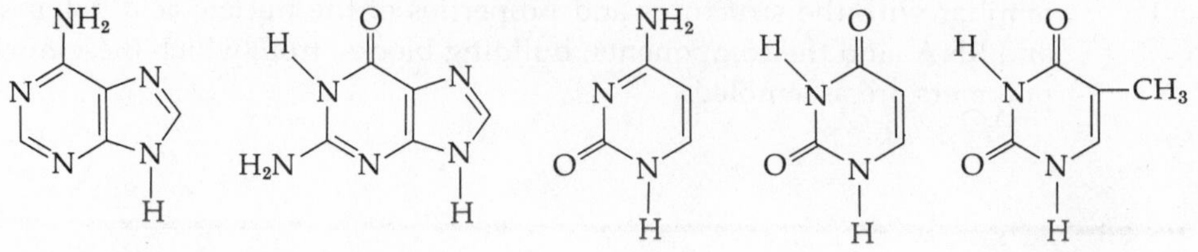

adenine (A)   guanine (G)   cytosine (C)   uracil (U)   thymine (T)

Figure 24-2   The nitrogenous bases of DNA and RNA. In a nucleoside, either ribose or deoxyribose replaces the hydrogen atom shown in blue. Note the similarity of adenine and guanine to the molecule purine, and of uracil, thymine, and cytosine to the molecule pyrimidine.

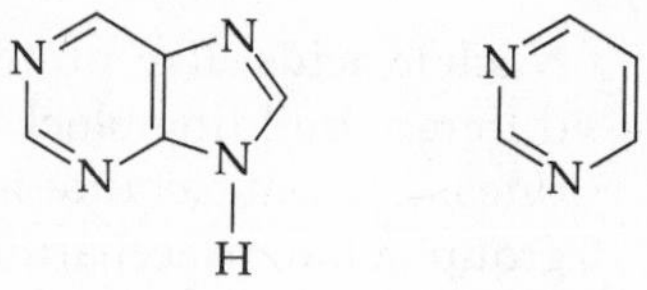

purine   pyrimidine

We can use the abbreviations A, G, U, T, and C to represent nucleotides with the bases adenine, guanine, uracil, thymine, and cytosine, respectively, in both DNA and RNA. When we do this, we must remember that DNA nucleotides contain the sugar deoxyribose, whereas RNA contains the sugar ribose.

Exercise 24-1

Write a chemical equation using structural formulas for the reaction of phosphoric acid with the adenine-containing nucleoside to form the corresponding nucleotide. Use the letter A to represent the adenine portion.

24-5 STRUCTURE AND PROPERTIES OF NUCLEOTIDE POLYMERS

In proteins, amino acids are joined together into long chains by peptide bonds formed between amino and carboxyl functional groups. How are nucleotides joined to form DNA and RNA? The bonds that hold these polymers together are ester linkages formed between the phosphate on the number 5' carbon of ribose in one nucleotide and the hydroxyl on the number 3' carbon of ribose in the next nucleotide (deoxyribose in the case of DNA). Thus, nucleic acids are said to have **3',5'-phosphate ester bridges** between their nucleotide components (see Figure 24-3). A nucleic acid molecule has a free phosphate on the ribose (or deoxyribose) located at one end and a free hydroxyl on the number 3' carbon of ribose (or deoxyribose) at the other end. This is similar to proteins, which have a free amino end and a free carboxyl end. When referring to nucleic acids, we use abbreviations to indicate the sequence of nucleotides. By convention, the nu-

Figure 24-3   The structure of a small piece of RNA. This piece of RNA can be represented by the simple notation $^{5'}ACGU^{3'}$.

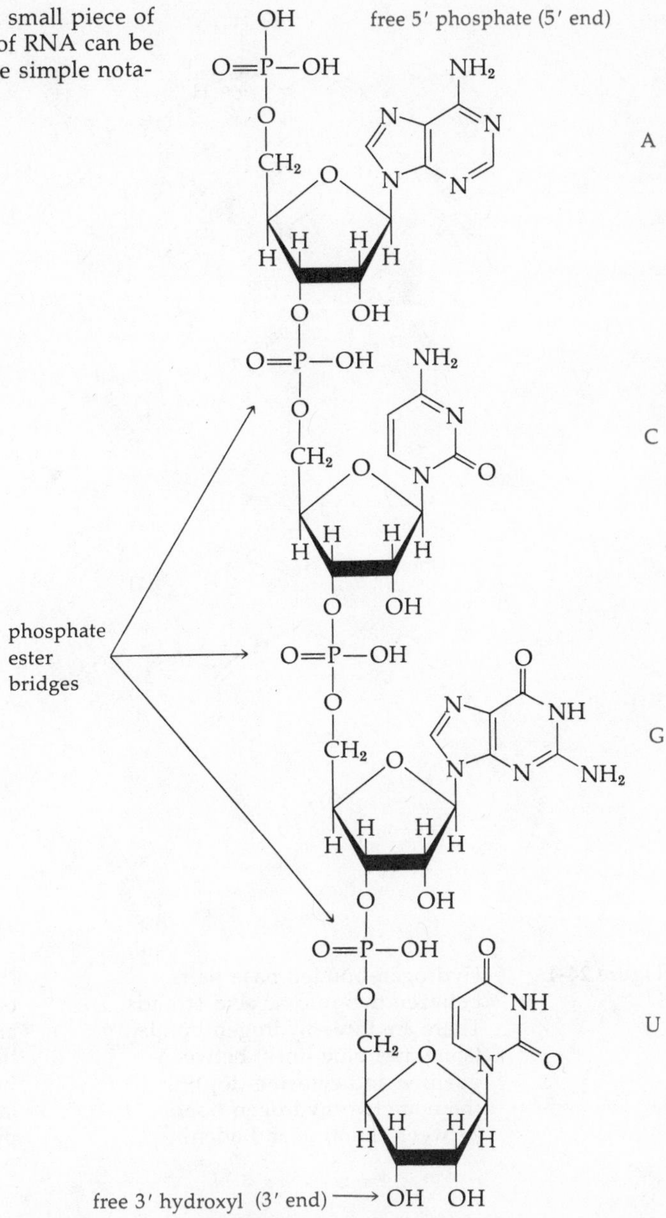

free 5′ phosphate (5′ end)

phosphate ester bridges

free 3′ hydroxyl (3′ end) →

cleotide with the free phosphate group (the 5′ end) is written on the left of the sequence and the nucleotide with the free hydroxyl on the number 3′ carbon of the sugar (the 3′ end) is written on the right. Thus, for example, we can represent the sequence of nucleotides in Figure 24-3 by $^{5'}ACGU^{3'}$.

The structure and function of a nucleic acid is determined by the particular sequence of nucleotides in that nucleic acid. We discussed in Chapters 21 and 22 how the sequence of amino acids in a protein determines the unique structure and function of that protein. What special feature of DNA allows it to function as a blueprint? The unique feature is that DNA consists of two long polynucleotide strands, and the base component of each nucleotide on one strand can form hydrogen bonds with only one specific nucleotide base on the other strand.

For example, a guanine base on one DNA strand can hydrogen bond only to a cytosine on another DNA strand. Two nucleotide units that are bound together by hydrogen bonds are called a **base pair.** Guanine-cytosine constitute one such base pair. We say that cytosine is the **complementary base** to guanine and that the complementary base of guanine is cytosine.

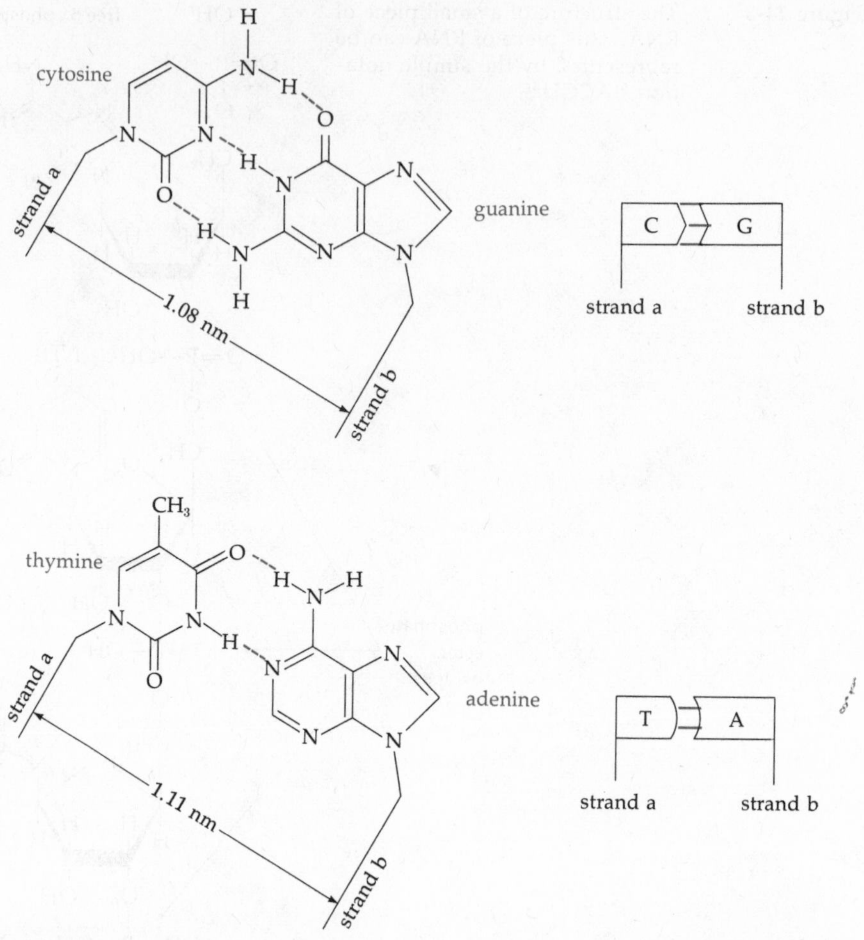

Figure 24-4   Hydrogen-bonded base pairs between two nucleic acid strands. There are three hydrogen bonds (shown as blue lines) between guanine and cytosine (top), and there are two hydrogen bonds between thymine and adenine (bottom). Adenine also forms a base pair with uracil. On the right are schematic representations for these base pairs in which the larger purine bases are represented by larger symbols than those used for pyrimidine bases.

Base pairing is illustrated in Figure 24-4. Note that A is capable of forming two hydrogen bonds with T or U, whereas G can form three hydrogen bonds with C, and G or C cannot form multiple hydrogen bonds with A, T, or U. For DNA, there are thus only two possible base pairs, A-T and G-C. Base pairing also occurs in processes involving RNA. For RNA, only the base pairs A-U and G-C are possible.

Note that the two long polynucleotide strands in DNA are held together by hydrogen-bonded base pairs. Now, hydrogen bonds are fairly weak bonds, so two or three hydrogen bonds could not hold these strands together very tightly. However, in one strand of DNA there are millions of nucleotide bases, so the forces holding the two DNA strands together can be quite strong. In order for this relatively strong bonding to occur, *each* nucleotide base on one strand must be able to hydrogen bond to its complementary base on the opposite strand. The sequences of two DNA strands arranged in this fashion are then said to be **complementary** to each other. For example, in complementary strands, wherever there is a G in one strand, it must be hydrogen bonded to a C in the other strand, and every T must be hydrogen bonded to an A. This is shown in Figure 24-5. Also notice in Figure 24-5 that if the direction of one strand is 3'-5', its

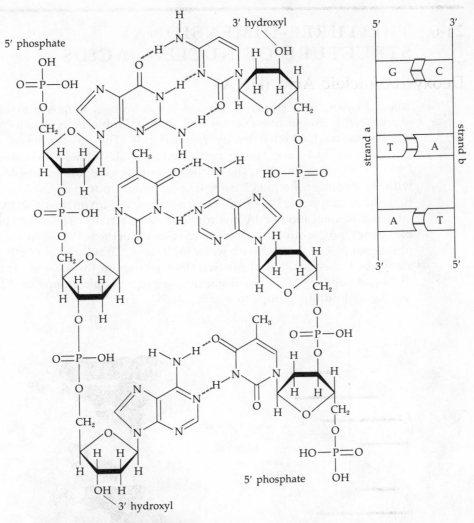

Figure 24-5   Complementary strands of DNA are represented by structural formulas on the left and schematically on the right. Hydrogen bonds are shown as blue lines. Notice that the sequence of bases in one DNA strand specifies the sequence of its complementary strand.

complementary strand must be in the opposite direction (5'-3'). This has been observed to be true for all double-stranded nucleic acids. Thus, <u>the sequence of one strand of DNA specifies the sequence and direction of its complementary strand. This is the crucial structural property of nucleic acids!</u> For example, if the sequence of a portion of one DNA strand is $^{5'}CCA^{3'}$, then the sequence of the corresponding portion of its complementary strand must be $^{5'}TGG^{3'}$, and these hydrogen-bonded strands can be represented by the notation

$^{5'}CCA^{3'}$
$^{3'}GGT^{5'}$

Exercise 24-2
Given the following abbreviated notation for part of one DNA strand,

$^{5'}ATCGGATTCC^{3'}$

show the sequence and direction of its complementary strand.

24-6 THE THREE-DIMENSIONAL STRUCTURE OF NUCLEIC ACIDS

Deoxyribonucleic Acid (DNA)

About 25 years ago, Francis Crick and James Watson proposed that the DNA strands twist around one another and form a **double helix** in which the two strands are held together by hydrogen bonds. The double helical model of DNA, shown in Figure 24-6, is consistent with all of the known properties of DNA. In this double helix, the nitrogenous bases are located inside the helix with the hydrogen-bonded base pairs perpendicular to the main axis. In addition, the hydrophobic base pairs, which are stacked up on top of each other, are attracted to one another by hydrophobic interactions. The sugar-phosphate backbones, on the other hand, are on the outside of the helix, where these hydrophilic groups can interact with water molecules. Thus a double helix structure stabilizes the association of the two DNA strands in three ways: (1) base-pair hydrogen bonding, (2) internal stacking of hydrophobic groups, and (3) exposing the hydrophilic groups to water.

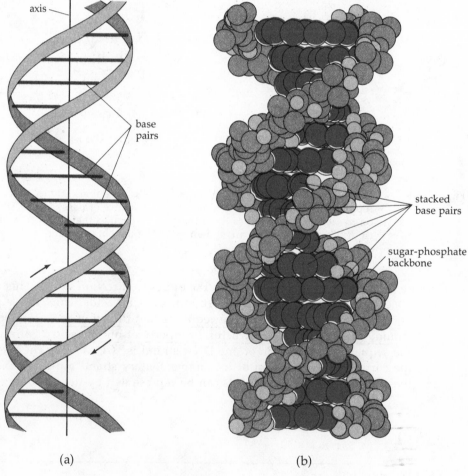

(a)   (b)

Figure 24-6   The double helix model of DNA. (a) The sugar-phosphate backbones of the two complementary strands are indicated as ribbons to which the hydrogen-bonded base pairs are attached. (b) A molecular model, with the sugar-phosphate backbone drawn in gray and the base pairs in blue.

Ribonucleic Acid (RNA)

Each RNA molecule consists of a single strand. There are three main types of RNA molecules: messenger RNA, ribosomal RNA, and transfer RNA.

Messenger RNA (mRNA) is the copy of one strand of the DNA gene that is used to specify the amino acid sequence of the corresponding protein. We do not believe that mRNA has an elaborate three-dimensional structure, but that it exists as a long filament, or a tape (see Figure 24-7a).

Ribosomal RNA (rRNA) molecules are components of ribosomes, the particles on which amino acids are bound together to form proteins. Ribosomes are structurally very complex. More is being learned about their structure and function every year. For our purposes, we can visualize ribosomes as composed of a large and a small subunit. Each subunit is made up of rRNA and several proteins (see Figure 24-7b). There are over 70 proteins in a human ribosome.

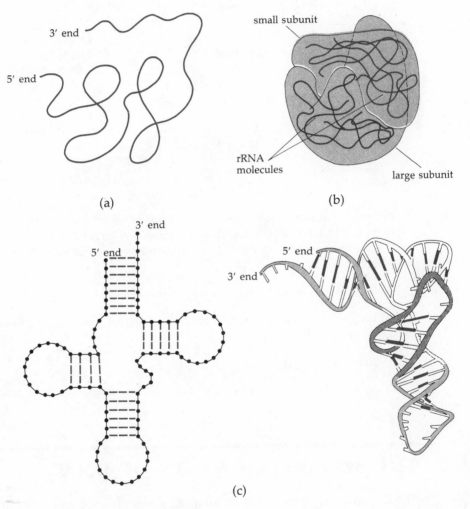

(a)

small subunit

rRNA molecules

large subunit

(b)

3' end

5' end

5' end

3' end

(c)

Figure 24-7   The three-dimensional structure of RNA molecules. (a) Messenger RNA molecules probably exist as long strands. (b) A recently proposed model for the shape of ribosomes in *E. coli*. The presence of rRNA is represented by the blue lines. (c) Representations for a tRNA molecule. A tRNA can be represented as a cloverleaf (left), which twists up to form the three-dimensional shape represented on the right. The scales used in these four drawings are not the same. A tRNA molecule is actually much smaller than a ribosome.

Transfer RNA (tRNA) molecules are the carriers for the amino acids used in protein synthesis. Only one specific amino acid can bind to a particular tRNA carrier. There are more than 20 different tRNAs, at least one for each of the 20 different amino acids. The particular amino acid that can bind to a given tRNA molecule is indicated by a subscript. For example, a tRNA molecule that can bind glycine is represented by $tRNA_{gly}$.

A tRNA molecule is composed of only a single RNA strand, but it does have a characteristic three-dimensional structure. The general overall structure of these molecules is similar to a twisted-up cloverleaf. Base-pair formation between complementary bases on the same strand is responsible for this structure (see Figure 24-7c). Each tRNA has a different sequence of nucleotide components and thus has a slightly different overall shape.

Table 24-1 Size and Structure of Nucleic Acids

Nucleic Acid	Approximate Number of Nucleotides per Molecule	Structure
DNA*	2.5×10^8	Double helix
mRNA	300–9000	Long filament (single strand)
rRNA	100–4000	In complex ribosomes that contain two subunits and dozens of proteins
tRNA	75	Twisted cloverleaf

* There are 46 of these DNA molecules in the nucleus of every human cell except sperm and ova. Each DNA molecule consists of about 200 genes plus many segments that do not serve as blueprints for protein synthesis.

The structures and relative sizes of DNA and RNA molecules are compared in Table 24-1. Notice that DNA polymers are much larger than RNA polymers, since each DNA molecule carries the information for the sequence of several RNA molecules.

Exercise 24-3

Would you expect the two following nucleotide strands to form part of a double helix? Explain your answer.

Strand A: 5'ATCGCCG3'
Strand B: 3'ATCGCCG5'

24-7   THE SELF-REPLICATION OF DNA

The sequence of bases in the component genes of DNA molecules are coded blueprints for the amino acid sequences of proteins. If cells are to divide, and if parents are to pass hereditary information on to their children, there must be a mechanism for producing exact copies of DNA molecules. The process that accomplishes this is called **replication.** Replication is an example of the elegant simplicity of many biochemical processes. Basically, all that needs to be done is to separate the two complementary strands in the "parent" DNA helix and use each of these as a pattern or template for the synthesis of a new complementary strand (see Figure 24-8). The replication process is outlined in Figure 24-9.

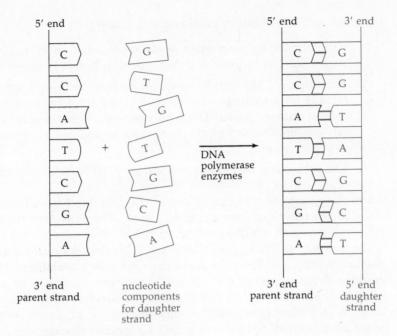

Figure 24-8   A parent DNA strand serves as a template for the synthesis of a daughter strand.

Figure 24-9   Replication of DNA involves (1) unwinding of the double helix, (2) polymerization of nucleotide components to form new "daughter" strands, and (3) formation of two new DNA helices, each composed of one parent and one daughter strand.

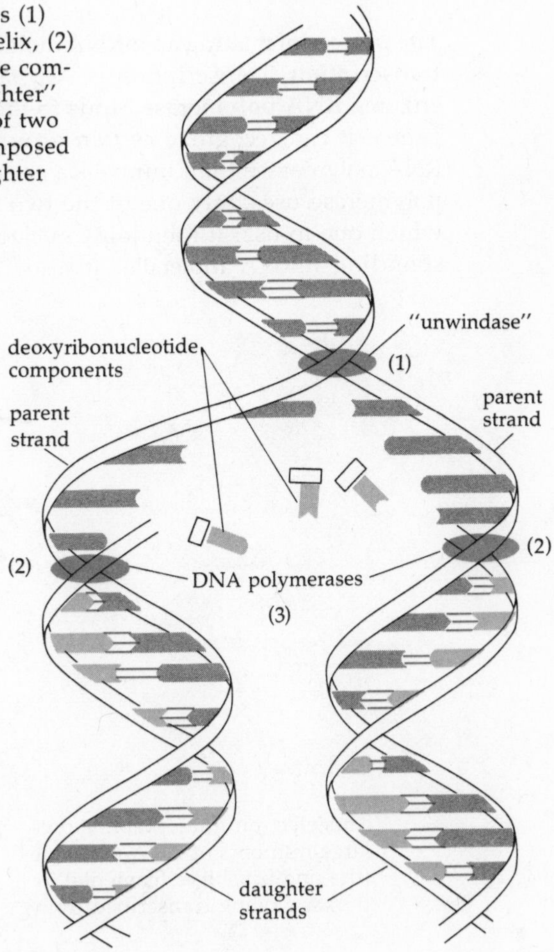

The steps in the replication process are as follows:

1. An enzyme (appropriately nicknamed *unwindase*) unravels the two parent strands by breaking the hydrogen bonds between base pairs.

2. The unpaired nucleotide bases on each of the single unwound parent strands form hydrogen bonds with new complementary nucleotides. A second enzyme, called **DNA polymerase,** takes these new complementary nucleotides and binds them together to form a new complementary daughter strand.

3. The parent and daughter strands then twist back up into double helices.

When the process is complete, there are two double helical DNAs, each composed of one parent and one daughter strand. The parent cell is now able to divide and give each daughter cell an identical set of genes.

The two sets of DNA genes that result from the replication process must be absolutely identical. Errors in replication are called **mutations.** Mutations can produce defective genes, which can direct the synthesis of defective proteins. Rarely, mutations are beneficial to the survival of the organism (Section 22-3). A defective gene often results in disease or death. When defective genes are inherited, the disease is called a **genetic disease.**

24-8   TRANSCRIPTION

The process of making an mRNA molecule corresponding to a DNA gene is called **transcription.** Transcription is very similar to replication. In transcription, the enzyme **RNA polymerase** binds to the DNA double helix at the beginning of a gene—it can recognize certain groups of nucleotides as starting points. The RNA polymerase then unravels a *portion* of the helix to expose the bases. RNA polymerase uses only one of the two DNA strands as a template—it "knows" which one to use. It then joins nucleotide components together into the corresponding mRNA molecule. It also "knows" when to stop the synthesis of

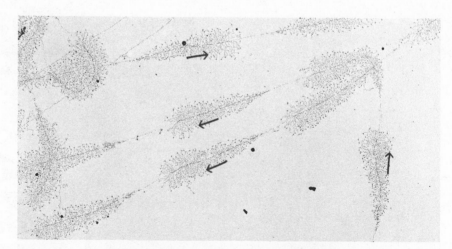

This electron micrograph shows large numbers of RNA molecules (the fine brushlike filaments) as they are being transcribed from DNA strands (the straighter lines). The arrows indicate the direction of movement of RNA polymerase molecules.

RNA—certain groups of nucleotides are recognized as stopping points. The RNA polymerase enzyme and the newly synthesized mRNA then detach from the gene, and the DNA re-forms its helical shape. As you can see, RNA polymerase is a versatile and complex enzyme. The steps in the transcription process are illustrated in Figure 24-10.

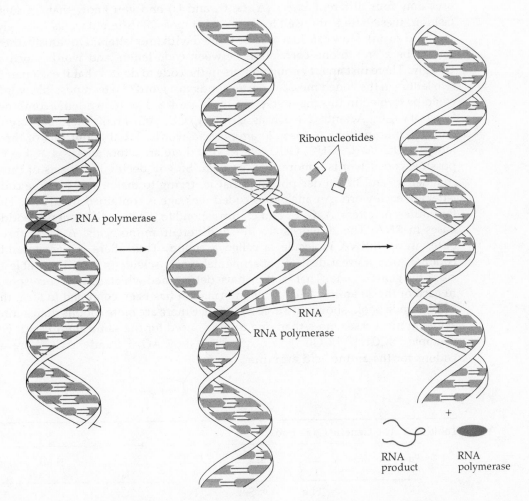

Ribonucleotides

RNA polymerase

RNA

RNA polymerase

+

RNA product

RNA polymerase

Figure 24-10   The process of RNA transcription begins with RNA polymerase binding to DNA at an appropriate point (left) and separating a portion of the helix. Using one DNA strand as a template, RNA polymerase then catalyzes the polymerization of ribonucleotides to form an RNA strand complementary to the DNA template (center). Transcription terminates when a stop signal on the DNA template is reached, and RNA polymerase and the newly synthesized RNA separate from the DNA template (right).

Exercise 24-4
Given the following sequence of nucleotides in part of a DNA gene, what is the sequence of nucleotides in the corresponding portion of the RNA molecule transcribed from this DNA strand?

DNA strand: $\cdots$ $^{5'}$TCATGCA$^{3'}$ $\cdots$

24-9 CRYPTOGRAPHY—ON CODES AND DECIPHERING THEM

The Genetic Code

Imagine yourself as an international spy. You have stolen a coded message that uses only four different *letters* (A, G, C, and U) and you know that, in some fashion, these letters are used to represent at most 20 different *words* (phe, gly, met, and so on). How can 20 words be formed with four letters? Obviously, there cannot be a one-to-one correlation between code letters and words, such as U = gly. There just aren't enough letters in the code to do it. What if each pair of two letters in the coded message corresponds to a word? Sixteen possible words could be formed in this manner, since there are $4 \times 4 = 16$ two-letter combinations of U, C, A, and G (such as UU, GA, CU, AG, etc.). There still aren't enough combinations to form 20 amino acid words. What about sets of three letters (GUG, AAU, GUC, GUU, etc.)? Now there are a total of $4 \times 4 \times 4 = 64$ possible coded words—more than enough. So you decide to try sets of three code letters and begin deciphering, that is, trying to match sets of three code letters to each word. An analogous coded message is contained in mRNA. The four different letters, A, G, C, and U, correspond to the four different nucleotide bases in RNA. The 20 words are the 20 different amino acids. A set of three bases in an mRNA molecule is called a **codon.** Thus there are 64 possible codons. After a great deal of experimental work, scientists have been able to crack the **genetic code.** That is, they have determined which codons correspond to each of the 20 amino acids. This information has been compiled to form the genetic code book, shown in Table 24-2. Since there are more codons than amino acids, in most cases more than one codon is used for the same amino acid. For example, $^{5'}CGU^{3'}$, $^{5'}CGC^{3'}$, $^{5'}CGA^{3'}$, $^{5'}CGG^{3'}$, $^{5'}AGA^{3'}$, and $^{5'}AGG^{3'}$ are all codons for the amino acid arginine.

Table 24-2 The Genetic Code Book

First Position (5' end)	Second Position				Third Position (3' end)
	U	C	A	G	
U	Phe	Ser	Tyr	Cys	U
	Phe	Ser	Tyr	Cys	C
	Leu	Ser	Term*	Term*	A
	Leu	Ser	Term*	Trp	G
C	Leu	Pro	His	Arg	U
	Leu	Pro	His	Arg	C
	Leu	Pro	Gln	Arg	A
	Leu	Pro	Gln	Arg	G
A	Ile	Thr	Asn	Ser	U
	Ile	Thr	Asn	Ser	C
	Ile	Thr	Lys	Arg	A
	Met**	Thr	Lys	Arg	G
G	Val	Ala	Asp	Gly	U
	Val	Ala	Asp	Gly	C
	Val	Ala	Glu	Gly	A
	Val	Ala	Glu	Gly	G

* Term = chain-terminating codon (stop signal)
** Met = chain-initiating codon (start signal)

Some codons are reserved to punctuate the message. The codons $^{5'}UAA^{3'}$, $^{5'}UAG^{3'}$, and $^{5'}UGA^{3'}$ are not codons for any amino acid. These three codons signal the termination of protein synthesis (the period at the end of the message). The codon $^{5'}AUG^{3'}$ codes for the amino acid methionine. This codon is also used to indicate the start of protein synthesis (the beginning of the message). All mRNA messages begin with the codon $^{5'}AUG^{3'}$.

Exercise 24-5
Without reference to Table 24-2, list all of the codons beginning with U.

Exercise 24-6
Using Table 24-2, write the sequence of amino acids corresponding to the mRNA sequence $^{5'}AUGCCCUGUAAUAGGCGAUAUUAG^{3'}$ (Note: Read the coded message from left to right.)

Machinery for Deciphering the Code

How do cells decipher the genetic code? The tRNA molecules do this job. You might consider tRNA molecules as interpreters that help translate an RNA language into a protein language. Every tRNA molecule has a set of three bases called an **anticodon** that can base pair with the three bases of a specific codon in mRNA. For example, the set of bases $^{5'}AGA^{3'}$ is the anticodon for the mRNA codon $^{5'}UCU^{3'}$, which specifies the amino acid serine. When an anticodon and a codon bind, the 3′ end of the anticodon binds to the 5′ end of the codon (as in the case of complementary strands in DNA). Thus, for example, the anticodon $^{5'}GAA^{3'}$ (rather than $^{5'}AAG^{3'}$) binds to the codon $^{5'}UUC^{3'}$ (see Figure 24-11). The codon $^{5'}UUC^{3'}$ corresponds to the amino acid phenylalanine according to the genetic code book.

Figure 24-11   Bases of an mRNA codon hydrogen bond with complementary bases of the anticodon of the appropriate tRNA molecule. Note that the 3′ end of an anticodon binds to the 5′ end of a codon.

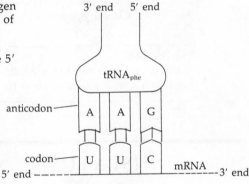

How does the tRNA molecule with the anticodon $^{5'}AGA^{3'}$ match up with the amino acid serine? This is accomplished by an enzyme called serine aminoacyl-tRNA synthetase. It binds serine—and no other amino acid—to the 3′ end of a tRNA$_{ser}$ molecule, which has the anticodon AGA. The hydrolysis of ATP supplies the energy required to join serine to its appropriate tRNA, forming serine-tRNA$_{ser}$. A schematic representation of this reaction is

$$\text{serine} + \text{tRNA}_{ser} + \text{ATP} \xrightarrow[\substack{\text{serine aminoacyl-}\\ \text{tRNA synthetase}}]{} \text{serine-tRNA}_{ser} + \text{AMP} + \text{HO}-\overset{\overset{\displaystyle O}{\|}}{\underset{\underset{\displaystyle OH}{|}}{P}}-O-\overset{\overset{\displaystyle O}{\|}}{\underset{\underset{\displaystyle OH}{|}}{P}}-\text{OH}$$

Similarly, each of the tRNA molecules is bound to the correct amino acid by a specific aminoacyl-tRNA synthetase. An amino acid bound to its tRNA is called an **aminoacyl-tRNA.** The important structural features of one such aminoacyl-tRNA are illustrated in Figure 24-12.

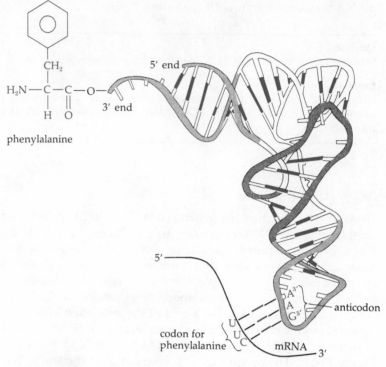

Figure 24-12   Schematic representation of the important features of phenyl-alanine-tRNA$_{phe}$. Note the binding of the anticodon portion of the tRNA to the appropriate codon on mRNA. The representations for the amino acid, for tRNA, and for mRNA are drawn to different scales.

Exercise 24-7

A tRNA molecule has the anticodon $^{5'}GUA^{3'}$. To what mRNA codon does this tRNA bind? What amino acid binds to this tRNA molecule?

24-10 TRANSLATION

With a good supply of aminoacyl-tRNAs and a messenger RNA template, the cell is ready to assemble the amino acids into a protein. This process, appropriately called **translation,** takes place on the ribosomes. Figure 24-13 shows the major steps of this process, which will be described here:

1. The 5' end of the mRNA molecule binds to the smaller subunit of the ribosome. Recall that all mRNA messages begin with the codon $^{5'}AUG^{3'}$, which corresponds to the amino acid methionine.* Methionyl-tRNA$_{met}$ recognizes this codon and binds to the mRNA.

---

* Methionine is always the N-terminal amino acid in newly made proteins. However, specific enzymes can cut an N-terminal peptide away from many newly synthesized proteins, so methionine is not the N-terminal amino acid of all "mature" proteins.

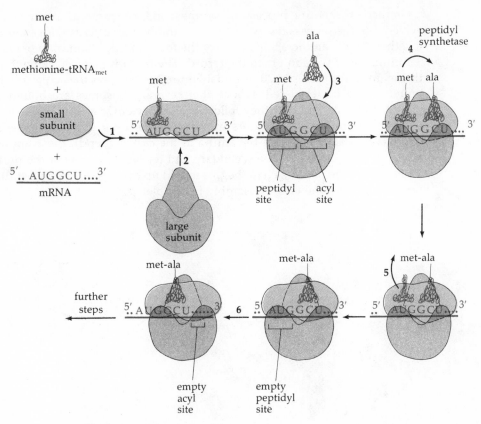

Figure 24-13   A schematic representation of the process of translation.

2. The larger ribosome subunit then binds to form a complete complex. The larger subunit has two active sites into which tRNA molecules can fit (think of them as keyholes), called the **peptidyl site** and the **acyl site.**

3. The next codon on the mRNA binds its aminoacyl-tRNA at the acyl site.

4. Now an enzyme, called **peptidyl synthetase,** located in the ribosome, removes the methionine from tRNA$_{met}$ and forms a peptide bond between methionine and the next amino acid (alanine, in this case), which remains attached to its tRNA at the acyl site.

5. Having lost its amino acid, tRNA$_{met}$ is released from the peptidyl site.

6. The ribosome slides along the mRNA tape by one codon's length, placing the newly formed dipeptide into the peptidyl site. The energy for this movement is supplied by hydrolysis of GTP (which can be represented by GTP + H$_2$O → GDP + P$_i$ + energy; see Chapter 25).

7. The process continues in this fashion, with the next aminoacyl-tRNA binding at the acyl site, followed by peptide bond formation and ribosome movement.

8. When a chain-terminating codon is finally encountered, the completed protein is detached from the last tRNA. The ribosome splits up into two subunits, which can go back and start the translation process over again.

In the past few years, several additional facts have been discovered about the process of translation. It is now known that a number of proteins that are not components of ribosomes are also involved in translation and that a second molecule of GTP must be hydrolyzed for every amino acid that is added to the growing peptide chain. Hence, the steps described here are somewhat simplified.

Granted, the entire process of synthesizing one protein chain is very involved, but the process has been clocked, and human cells can, for example, put together the 141 amino acids in one of the four peptide chains of hemoglobin in 3 minutes. And we don't hold the record. The bacterium *E. coli* can put together that many amino acids in 20 seconds! Figure 24-14 shows the actual synthesis of proteins in a bacterial cell. Notice that several ribosomes are bound to each mRNA molecule and that they follow each other down the message. One final word about protein assembly. As amino acids are being added to a growing protein chain (assembled from the amino to the carboxyl end), these amino acids begin to interact to form the secondary and tertiary structure of the protein. In other words, the peptide chain begins to fold up even as it is being synthesized, rather than after the entire assembly is completed.

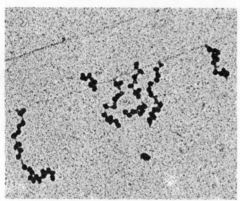
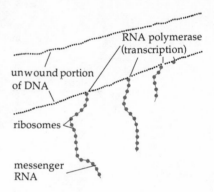

RNA polymerase (transcription)

unwound portion of DNA

ribosomes

messenger RNA

Figure 24-14   An electron micrograph showing transcription and translation in a bacterial cell. Along the lower DNA strand, several RNA polymerase molecules are at various stages in the transcription of mRNA molecules. As each mRNA is transcribed, it is used for the assembly of several identical protein molecules (not visible here) by several ribosomes. The mRNA molecules and the ribosomes are colored blue in the corresponding sketch on the right.

24-11 REGULATION OF PROTEIN SYNTHESIS

We have now seen how proteins are made. But since cells do not constantly need to synthesize all of the proteins coded for in their DNA blueprints, the process must be controlled. If a particular mRNA existed forever, we could never turn off the synthesis of the protein it codes for. The solution to this problem is simple. An enzyme in the cytoplasm, **ribonuclease,** chops up mRNA molecules into their component nucleotides. Thus a single mRNA molecule is usually able to supervise the assembly of only a small number of protein molecules before it is broken down.

Protein synthesis is controlled primarily by regulating the types and amounts of mRNA molecules produced in a cell. Thus the main regulation of protein synthesis is accomplished by controlling the process of transcription. In 1961, F. Jacob and J. Monod proposed a model that describes how transcription is regulated in *E. coli.* Two similar processes, called **induction** and **repression,** are involved in this model. Some of the same general mechanisms are used to control transcription in human cells, but the control process in human cells is very much more complicated than in *E. coli.*

Repression of Transcription

Genes that code for enzymes involved in a series of enzyme-catalyzed reactions are called **related genes.** For example, for the series of reactions

$$A \xrightarrow{E_1} B \xrightarrow{E_2} C \xrightarrow{E_3} D$$

the genes that code for the enzymes E_1, E_2, and E_3 are related genes. In *E. coli*, related genes are generally located next to each other on the chromosome, allowing them to be transcribed sequentially by the enzyme RNA polymerase. However, preceding each group of related genes is a set of nucleotides called an **operator site,** which can bind to a specific protein called a **repressor protein.** When the specific repressor protein in its active form is bound to its corresponding operator site, it prevents RNA polymerase from unwinding that portion of the DNA helix and transcribing the genes associated with that operator site (see Figure 24-15). The group of genes, together with its operator site, form a unit called an **operon.**

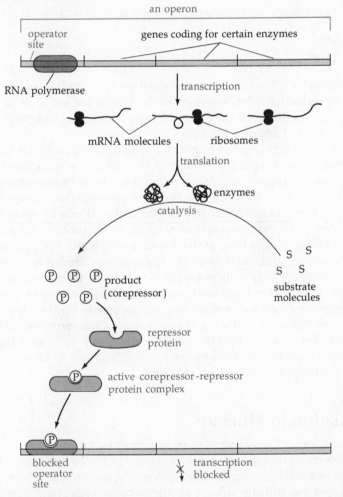

Figure 24-15   A model for the repression of genes. Enzyme molecules that are coded for by the genes in this operon catalyze the formation of a product P. Acting as a corepressor, P binds to a specific repressor protein, altering the conformation of the repressor protein to an active form, which then binds to its operator site on the DNA and prevents further transcription of the genes for the enzymes that produce P.

Now, the repressor protein itself can exist in two states, *active* and *inactive*. In its active state, the repressor binds to its operator site and prevents transcription. In its inactive state, it cannot bind to the operator site and transcription occurs. The repressor protein is normally inactive. However, when a small molecule called a **corepressor** binds to an allosteric site on a repressor protein, the repressor protein changes shape and becomes active. The active repressor protein stops transcription. The corepressor molecule is often the end product of the series of reactions catalyzed by the enzymes that are coded for by the genes in the operon. Do you now see how repression control operates? The more end product corepressor molecules are present, the more active repressor protein is formed. This results in a greater repression of mRNA formation and hence synthesis of less of the enzymes needed to form that end product (Figure 24-15). Thus, this process of repression prevents wasteful synthesis of unneeded enzymes.

The binding of the corepressor to the repressor protein is analogous to the binding and change of shape that occurs for allosteric enzymes. However, end product repression of enzyme synthesis is different from feedback inhibition of enzyme activity. Repression controls the amount of enzyme synthesized, whereas feedback inhibition controls the level of activity of enzymes.

Induction of Transcription

In induction, a different type of repressor protein binds to an operator site. However, in induction, the repressor protein is in its active state when *no* small molecule is bound to it. In its active state it binds to its operator and prevents transcription of appropriate genes.

However, when large amounts of a specific small molecule called an **inducer** are present, binding of the inducer at an allosteric site on the repressor protein inactivates the repressor protein, removing it from its operator site so that transcription and enzyme synthesis can then proceed (see Figure 24-16). Inducers are typically small molecules that can be broken down by enzymes produced by a cell. When the inducer inactivates its repressor protein, the inactive repressor protein can no longer bind at the operator site. Now, RNA polymerase can synthesize mRNA, and, via translation, the enzymes needed to break up the inducer are synthesized. This is illustrated in Figure 24-16. Of course, after a while, when the large supply of inducer has been used up by the newly synthesized enzymes, the repressor protein no longer has any inducer to bind to. It is therefore active again and turns off synthesis of any more of these enzymes. The overall effect of induction, like the overall effect of repression, is to stop the production of particular enzymes when they are not needed, and to allow the production of these enzymes when they are required.

Gene Regulation in Humans

How is transcription controlled in humans? Unfortunately, we do not have all the answers to this question yet. Some control is by induction and repression, but control mechanisms in humans are much more complicated than those in *E. coli* and are still the subject of intensive investigation. Some of these complications are the following: First, related genes are not always grouped together into an operon. For example, the genes for the α and β chains of hemoglobin are on different chromosomes. Second, the vast majority of genes in a given adult human cell are always kept turned off. Third, human cells undergo the process of **differentiation** as they mature from embryonic to specialized

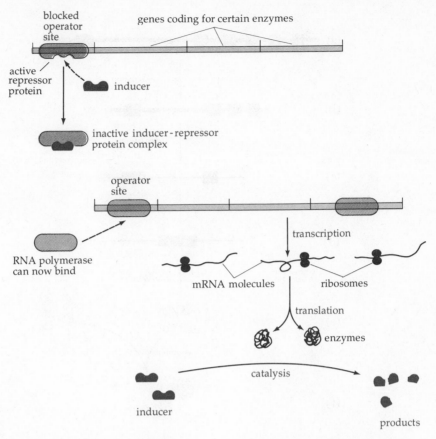

Figure 24-16   A model for the induction of genes. When the inducer is present, it binds to an allosteric site on its repressor protein. The repressor protein with its bound inducer now has a different conformation, and it "falls off" the operator site. This allows RNA polymerase to transcribe genes for enzymes catalyzing the conversion of inducer molecules to products.

adult cells; several genes must be switched on early in development, then turned off later. Fourth, hormones are used to help control transcription in humans. Fifth, the synthesis of at least some proteins in humans is regulated at the level of translation. For example, recent studies have shown that translation of the mRNA for hemoglobin does not occur if the heme prosthetic group is not available.

In the last few years, scientists have discovered that portions of some genes in higher organisms, including humans, are reorganized during differentiation, and that the messenger RNA molecules for some proteins are cut and spliced prior to translation. This reorganization of DNA segments during differentiation, called **recombination,** and the **splicing** of mRNA molecules are known to occur in the case of the DNA and RNA coding for antibody molecules (see Figure 24-17). The cutting and splicing of mRNA molecules to remove noncoding intravening segments, called **introns,** also occurs prior to the translation of several but not all other proteins. Introns have also been found in some newly synthesized tRNA molecules, and they must be removed in order for these tRNAs to function. The possible function of introns in the expression and regulation of human genes is currently being studied by a large number of scientists.

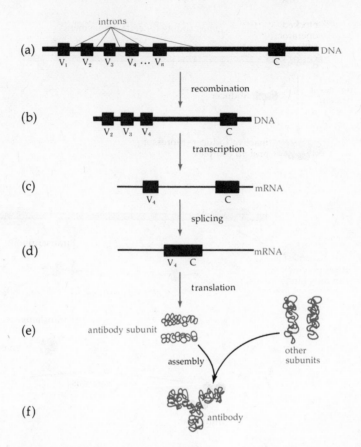

Figure 24-17   Steps involved in the expression of genes coding for antibody proteins. Part (a) represents a segment of DNA that contains genes coding for a variety of antibody subunits, as they are organized in embryonic cells. The solid rectangles represent individual genes coding for the antigen recognition portions, $V_1, \ldots, V_n$, of various antibody subunits, and a gene coding for C, a constant portion of these antibody subunits. The V and C genes are separated by introns. During differentiation, these DNA segments undergo recombination in some cells to allow for the various combinations of V and C genes found in the antibody-producing cells of adults. The recombinant DNA shown in (b), for example, can be transcribed to yield (c), a messenger RNA with a single V gene (V_4 in this case) separated from a C gene by an intron. This intron is spliced out to give (d), the mRNA that codes for the specific antibody subunit in (e). Assembly of this subunit with other subunits is followed by secretion from the cell of a functional antibody molecule (f).

Exercise 24-8

Consider the following sequence of enzyme-catalyzed reactions in which a molecule A is sequentially broken down to smaller molecules B, C, and D:

$$A \xrightarrow{E_1} B \xrightarrow{E_2} C \xrightarrow{E_3} D$$

Which of the molecules, A, B, C, D, might act as a corepressor molecule and which as an inducer?

24-12 VIRUSES

The ability to replicate DNA, transcribe DNA, and make proteins are major criteria that distinguish living organisms from nonliving things. Viruses do not contain DNA polymerase, RNA polymerase, or ribosomes. By themselves viruses are incapable of reproduction or growth. In fact, the simplest **viruses** are composed of just a small DNA molecule containing a few genes, and a protein coat to house the DNA (see Figure 24-18). However, when viruses infect (invade) living cells, they can use the machinery of the host cell to transcribe and replicate their own viral DNA. The mRNA molecules produced in this process are translated and viral proteins are assembled. Some of these proteins prevent normal function of the host cell, whereas the remainder are those needed for new envelopes to contain newly replicated viral DNA. Eventually, many new viruses are made and the host cell bursts, releasing the new viruses (see Figure 24-18).

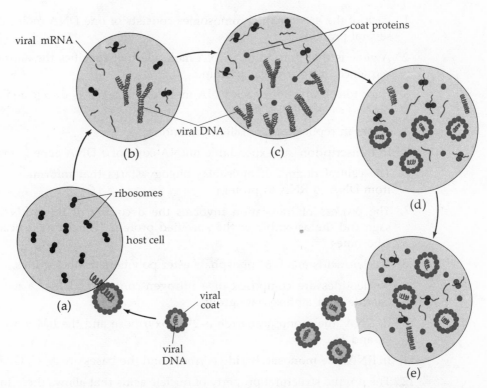

Figure 24-18   The life cycle of a DNA virus begins with a virus particle attaching to a host cell and injecting its viral DNA molecule inside (a). The viral DNA is then replicated and transcribed using the host cell's enzymes (b). Translation of viral mRNA on host ribosomes (c) yields viral coat proteins, which assemble spontaneously with new viral DNA molecules to form complete virus particles (d). Eventually the virus particles cause the host cell to burst (e), releasing the virus particles.

Other viruses contain RNA instead of DNA. They are called **RNA viruses.** The first thing an RNA virus must do when it infects a host cell is to make a master DNA blueprint from its RNA. This is accomplished with the aid of an enzyme called **reverse transcriptase:**

$$\text{Host deoxyribonucleotides} \xrightarrow[\text{reverse transcriptase}]{\text{viral RNA template}} \text{viral DNA}$$

Once viral DNA blueprints are made, they are used to make more viral RNA and protein coats for the daughter viruses.

Occasionally, a viral DNA will not replicate immediately upon infecting a host cell. Instead the viral DNA is inserted into the host cell's chromosome. The host cell's master blueprints now contain blueprints for this virus as well as for the normal host cell. Thus, the viral genes can be replicated along with the host's chromosome and given to daughter cells. At some later time, the viral DNA may be transcribed and new viruses made.

Very rarely a different process can occur. Viral DNA may cause rapid replication of the host DNA. This rare occurrence has been proposed to be one cause of human cancer. Although this has been shown to be a cause of cancer in many animals, it is unclear what role viruses play in human cancer.

24-13 SUMMARY

1. Each of the 46 human chromosomes consists of one DNA molecule and several proteins.

2. A gene is that portion of a DNA molecule that specifies the amino acid sequence in a particular protein.

3. Prior to cell division, each DNA molecule in a cell is duplicated in the process called replication.

4. Errors in replication are called mutations.

5. In transcription, an expendable mRNA copy of a DNA gene is made.

6. The central dogma of molecular biology states that information flows from DNA to RNA to protein.

7. The process of translation involves the decoding of the mRNA message and the assembly of the specified protein. This process occurs on ribosomes.

8. Nucleic acids are 3',5'-phosphate ester polymers of nucleotides.

9. Nucleotides are composed of a nitrogen-containing base, a monosaccharide, and a phosphate group.

10. In DNA, the monosaccharide is 2'-deoxyribose and the bases are A, C, G, and T.

11. In RNA, the monosaccharide is ribose and the bases are A, C, G, and U.

12. The unique structural property of nucleic acids that allows them to function as blueprints is the ability of the bases to form specific base pairs by hydrogen bonding.

13. The allowable base pairs are A and T, C and G, and A and U.

14. Base pairing is responsible for the complementary relationship of the two DNA strands in the double helix.

15. There are three classes of RNA: mRNA, which are the copies of DNA genes; rRNA, which are components of ribosomes; and tRNA, which are used as carriers of amino acids in protein synthesis.

16. Codons on an mRNA molecule are sets of three nucleotides that specify one amino acid. They are recognized by the three complementary nucleotides of the anticodon on the tRNA for that amino acid.

17. In *E. coli*, protein synthesis is controlled by the induction or repression of mRNA transcription. Induction and repression involve the reversible binding of allosteric repressor proteins to DNA operator sites.

18. In humans, embryonic cells undergo the process of differentiation.

19. Viruses, which are particles containing a nucleic acid and proteins, cannot reproduce by themselves. Rather, they use their host cell's machinery for replication.

PROBLEMS

1. Draw the structural formula for one purine nucleotide triphosphate and the structural formula for one pyrimidine deoxyribonucleoside.

2. Draw simple diagrams representing the structures of DNA and the different types of RNA.

3. Translation is often divided into three operations, *initiation, elongation,* and *termination.* Which of the steps of protein synthesis shown in Figure 24-13 correspond to these operations?

4. Several enzymes in *E. coli* are required for the synthesis of the amino acid histidine from smaller molecules. Is the synthesis of these enzymes controlled by induction, or is it controlled by repression?

5. What differences in their substrates must be recognized by the enzymes DNA polymerase and RNA polymerase?

SOLUTIONS TO EXERCISES

24-1

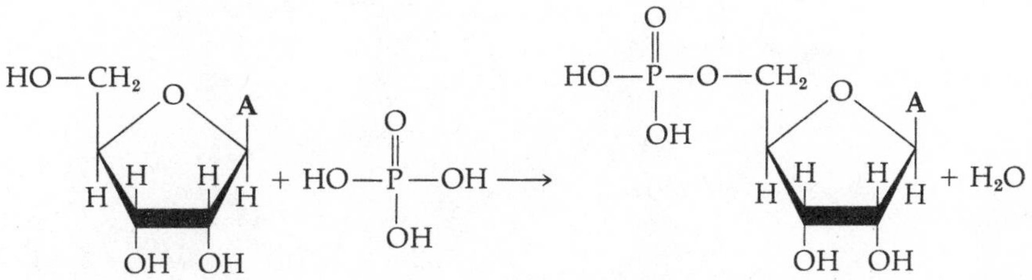

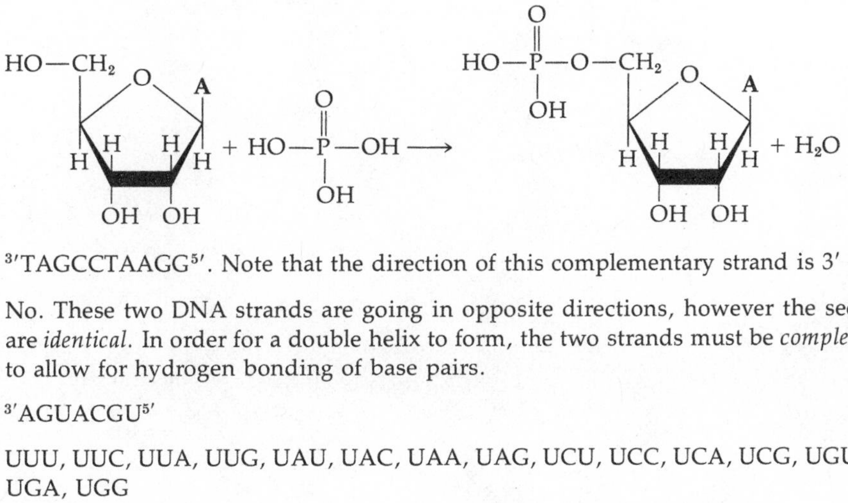

24-2 $^{3'}$TAGCCTAAGG$^{5'}$. Note that the direction of this complementary strand is $3' \rightarrow 5'$.

24-3 No. These two DNA strands are going in opposite directions, however the sequences are *identical*. In order for a double helix to form, the two strands must be *complementary* to allow for hydrogen bonding of base pairs.

24-4 $^{3'}$AGUACGU$^{5'}$

24-5 UUU, UUC, UUA, UUG, UAU, UAC, UAA, UAG, UCU, UCC, UCA, UCG, UGU, UGC, UGA, UGG

24-6 N-met-pro-cys-asn-arg-arg-tyr-C. The last codon, $^{5'}$UAG$^{3'}$, is a termination signal.

24-7 This anticodon will bind to the codon $^{5'}$UAC$^{3'}$, which specifies the amino acid tyrosine. Thus, this tRNA molecule binds to tyrosine.

24-8 The final product of this series of reactions, D, might serve as a corepressor, whereas the starting material, A, might be an inducer.

METABOLISM

Overview

In the last four chapters we looked at the structure and function of nucleic acids and proteins, the management and assembly-line workers in the cellular factories of the human body. We also looked at the complex machinery involved in the synthesis of these polymers from their monomeric components. In the following chapters we shall turn our attention to the cellular assembly lines themselves—we shall study the chemical reactions cells use to synthesize and break down molecules. The synthesis and breakdown of molecules occur in **metabolic pathways,** the collective name for cellular assembly lines. **Anabolic pathways** lead to the synthesis of molecules; **catabolic pathways** break apart molecules.

Most of these cellular assembly lines, or metabolic pathways, are used to assemble or take apart small molecules that are components of larger biomolecules. For example, some of these pathways make amino acids and nucleotides. Since these small molecules are generally used as *intermediates* in the manufacture of large biomolecules, pathways involving smaller molecules are frequently lumped together under the term **intermediary metabolism.**

Obviously, assembly lines need a source of energy to keep them moving. Cells need energy to fuel anabolic pathways and for other processes as well. This energy is usually supplied by the exergonic hydrolysis of a molecule called ATP. In Chapter 25 we shall discuss how the hydrolysis of ATP drives anabolic cellular assembly lines.

If cells hydrolyze ATP to get energy, then they need an adequate supply of energy to synthesize ATP. Where do they get it? We shall see that ATP is formed in conjunction with the catabolism of food molecules such as fats and sugars. The exergonic catabolism of these food molecules provides cells with the energy needed to form ATP. In order to understand how cells derive energy from the breakdown of sugars (carbohydrates) and fats (lipids), we shall need to understand some of the chemistry of these molecules. Carbohydrates have been discussed in Chapter 20. The chemistry of lipids will be discussed in Chapter 26, where we shall see that, in addition to serving as a source of energy, some fats are components of cell membranes and others are hormones and vitamins. We shall also see how carbohydrates and lipids are synthesized.

There are a great many cellular assembly lines, and we shall not be able to study all of them. In addition to pathways involving fats and sugars, we shall look at the pathways used to make and degrade amino acids (Chapter 28). However, we shall not study each and every step in all of these pathways. Rather, we shall concentrate on enzyme-catalyzed reactions that are similar in the synthesis of many amino acids and on the common features of amino acid catabolism. We shall see that amino acids are the body's major source of nitrogen. We shall also look at another pathway called the urea cycle. Urea is used by the human body as a vehicle for the excretion of excess nitrogen.

When we study a particular cellular assembly line, it is important to bear in mind that in a living cell many processes take place at the same time, and that all of the cellular assembly lines are interconnected. In fact, all of the metabolic pathways of intermediary metabolism are interrelated; they operate in an integrated manner, as we shall discuss in Chapter 29. When something goes wrong in one metabolic pathway, problems can arise in other connected pathways. Hence, cells use elaborate mechanisms to control their metabolic pathways. These control mechanisms prevent excess buildup or breakdown of any one type of molecule. We have discussed the basic features of these mechanisms: allosterism, hormones, chemical modification of enzymes, and competitive inhibition. In the following chapters we shall discuss a few important metabolic control points in some detail in order to gain an understanding of the principles of cellular control mechanisms.

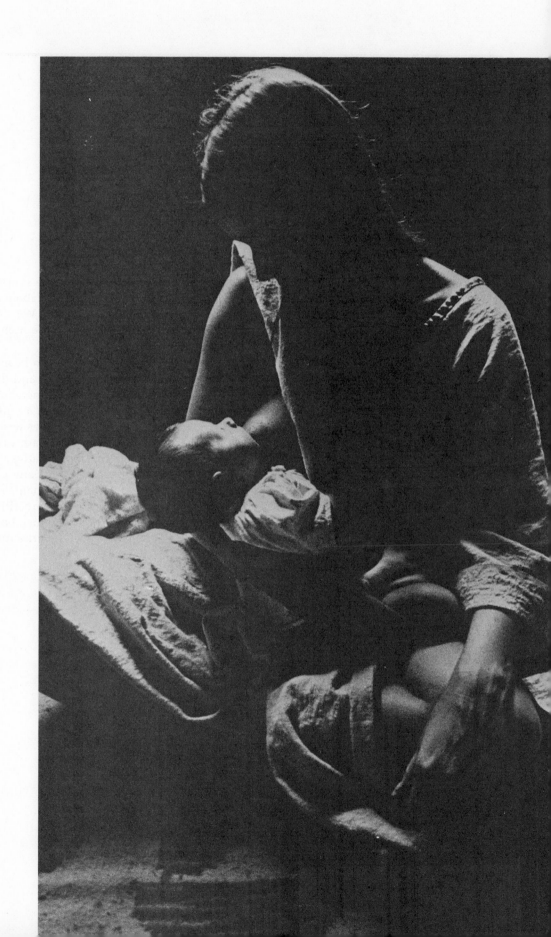

CHAPTER 25

Carbohydrate Metabolism and Bioenergetics

25-1 INTRODUCTION

Many people have a "sweet tooth." They crave sweets, even though an excess of sugar in the diet can lead to tooth decay, obesity, or worse. No one is more aware of the need for carefully controlling sugar intake than those persons who suffer from diabetes. More than 10 million people in the United States alone are afflicted with this disease. *Diabetes* is a disorder in which the metabolism of carbohydrates is not properly controlled. Understanding the pathology of diabetes requires an understanding of carbohydrate metabolism and how it is normally controlled in the human body.

Did you ever stop to wonder why people like sweet foods so much, especially when excess carbohydrate intake can cause medical problems? Our collective desire for sweets is probably due to the fact that our bodies actually need a somewhat generous (but not excessive) supply of carbohydrates—simple sugars or polymeric starch and glycogen—in order to supply the energy and carbon atoms needed to make nucleic acids, proteins, and membranes. We also need energy for movement, breathing, vision, active transport, maintaining body temperature, and so on. Carbohydrates are the major energy source for the human body. Even if we eat a lot of meat and our diet is therefore very rich in protein, the animals we eat lived off carbohydrates found in plants. Thus, directly or indirectly, the source of our biochemical energy is carbohydrates. In addition to energy, we need food molecules to supply the carbon, hydrogen, oxygen, and nitrogen that we need to make other biomolecules.

We do not use all of the atoms in our food to make biomolecules. The majority of the carbon atoms in the food we eat ends up in the carbon dioxide we exhale. Plants, in turn, require a supply of carbon dioxide so that they can make carbohydrates. Since green plants use solar energy in the process of photosynthesis to make carbohydrates from carbon dioxide and water, the ultimate source of energy for all plants and animals is the sun. Thus, all living things are dependent on one another for the atoms that are essential for life, and all rely on the sun

---

Nursing infants hydrolyze lactose to glucose and galactose. Galactosemic infants cannot properly metabolize galactose, and some of the galactose is converted to harmful byproducts. The current therapy for galactosemia involves a lactose-free diet.

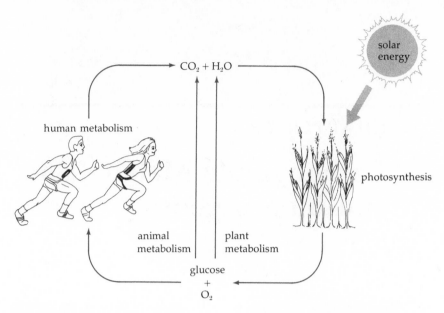

Figure 25-1   The flow of carbon atoms and energy in the biosphere. Photosynthetic organisms, such as green plants, use carbon dioxide, water, and solar energy to synthesize glucose, which is usually polymerized to cellulose and starch. All organisms rely on these carbohydrates for their energy needs, breaking glucose down to carbon dioxide and water, and using the energy released in this process.

to supply the necessary energy for life. This interdependence of the organisms that live on the earth for food and energy forms a network called the *biosphere*. The flow of energy and carbon in the biosphere is outlined in Figure 25-1. Other elements that are essential for life, such as nitrogen, are also circulated through the biosphere.

The overall reaction for the formation of 1 mole of glucose from carbon dioxide and water by the process of photosynthesis requires an input of at least 686 kcal of energy supplied by the sun:

(25-1) $$6CO_2 + 6H_2O + 686 \text{ kcal} \longrightarrow \underset{\text{glucose}}{C_6H_{12}O_6} + 6O_2$$

Glucose therefore contains stored chemical energy. In this chapter we shall see how glucose is broken down in the human body in order to convert some of this stored chemical energy into more usable forms. We shall also see how some other carbohydrates—sucrose (table sugar) and the lactose in milk—are metabolized. Finally, we shall look at the control of carbohydrate metabolism and the imbalance in this control mechanism that is present in diabetics.

25-2 STUDY OBJECTIVES

After careful study of the material in this chapter, you should be able to:

1. Define the terms catabolism, anabolism, metabolism, and biosphere.
2. Differentiate endergonic from exergonic biochemical reactions.
3. Describe why ATP is called an energy carrier.
4. Recognize biochemical reactions that are energy coupled.

5. Show the steps in glycolysis that produce ATP from ADP, as well as those that produce NADH + H$^+$ from NAD$^+$.

6. List the different end products obtained from the glycolysis of glucose under aerobic and anaerobic conditions.

7. Explain what happens to the six carbon atoms in glucose as a result of either aerobic or anaerobic glycolysis.

8. Explain why the TCA cycle is called a cycle, where acetyl-S-CoA is fed into it, where carbon dioxide is released, and where ATP and reduced coenzymes are produced.

9. Recognize similarities in the enzymatic reactions of glycolysis and those of the TCA cycle.

10. Show the steps in the electron transport system and describe how energy can be made available for ATP synthesis via this system.

11. **(Optional)** Explain how the pentose phosphate pathway is used as a source of ribose 5-phosphate and the reduced coenzyme NADPH.

12. Write the overall reaction for gluconeogenesis, and describe the conditions that favor use of this pathway.

13. Explain the functions of the following three enzymes: lactase, sucrase, and hexose 1-phosphate uridylyltransferase.

14. Describe the disease galactosemia and explain its cause.

15. Describe the role of UDP-glucose in the synthesis of hexoses from glucose.

16. Describe in detail the glycogen synthesis/glycogen phosphorylase control point for glucose metabolism, including the roles of the hormones epinephrine (adrenaline), glucagon, and insulin.

17. Describe the major control point for glycolysis/gluconeogenesis.

18. Describe the various ways glucose is used in the human body and the general conditions under which each of these metabolic pathways is used.

19. Define the terms hyperglycemia, hypoglycemia, and hyperinsulinism.

20. Explain the role that insulin plays in diabetes and hypoglycemia.

21. Describe how a glucose tolerance test can be used to diagnose diabetes mellitus and hypoglycemia.

25-3 BIOENERGETICS

We burn food, and not just in the kitchen! In some areas crops are being grown for use as fuel to supplement our dwindling petroleum resources. This is an indirect use of solar energy. We can let plants capture the solar energy, then burn them to generate heat. In the body we burn food in a different way.

We use some energy to keep our bodies warm. But we need large amounts of energy in forms other than heat: We need a great deal of energy to synthesize complex biomolecules.

Energy-Coupled Reactions

Energy for biosynthesis is provided by the coupling of chemical reactions in the body. When a chemical reaction that releases energy, an **exergonic** reaction, supplies the energy to drive another reaction for which energy is required, an

endergonic reaction, we say that these two reactions are **energy coupled.** Recall that an exergonic reaction is one in which the free energy (G) of the products is less than the free energy of the reactants. Thus, for an exergonic reaction, the change in the free energy (ΔG) of the chemical system is negative. For an endergonic reaction, ΔG is positive. The concept of energy-coupled reactions is illustrated schematically in Figure 25-2.

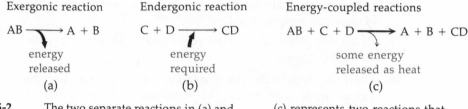

Exergonic reaction

$$AB \longrightarrow A + B$$

energy
released

(a)

Endergonic reaction

$$C + D \longrightarrow CD$$

energy
required

(b)

Energy-coupled reactions

$$AB + C + D \longrightarrow A + B + CD$$

some energy
released as heat

(c)

Figure 25-2   The two separate reactions in (a) and (b) are not energy coupled, whereas (c) represents two reactions that are energy coupled.

The term **bioenergetics** is used to refer to energy changes and energy coupling in biochemical reactions. This coupling of two reactions is never perfect. Some of the energy released by the exergonic reaction is not used to drive the endergonic reaction, but rather is released in the form of heat.

Energy-coupled reactions occur together and are interdependent. A coupled reaction is somewhat analogous to the motion of two people on a seesaw (see Figure 25-3). When one person goes down, the other person is forced up. Similarly, in a coupled biochemical reaction, the decrease in free energy in the exergonic reaction supplies the energy needed for the endergonic reaction. Of course, the exergonic reaction must supply *at least* as much energy as is required to drive the endergonic reaction. The coupling of endergonic reactions to exergonic reactions occurs at the active sites of enzymes. Thus, enzymes are like the seesaw of our analogy.

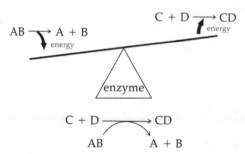

$$C + D \longrightarrow CD$$
energy

$$AB \longrightarrow A + B$$
energy

enzyme

$$C + D \longrightarrow CD$$
$$AB \qquad A + B$$

Figure 25-3   The interdependence of two energy-coupled reactions is somewhat like the motion of two people on a seesaw. The energy coupling of two reactions is often represented by using a straight reaction arrow for one and a curved arrow for the other.

In the human body, carbohydrates, lipids, and proteins are broken down in a stepwise fashion. These reactions are exergonic, and the energy released when they occur is captured very efficiently by coupled endergonic reactions in which bonds are formed. Breaking foods down in a stepwise manner does not change the total amount of energy that is released. It does, however, permit us to capture some of the released energy in a usable form. The overall energy change in any chemical reaction still depends only on the reactants and products.

ATP

The energy required for essential endergonic reactions in the human body is not obtained directly from the stepwise breakdown of foods, but rather involves an energy-carrier intermediary as follows:

1. Many of the different exergonic reactions that occur in the stepwise breakdown of foods are coupled to the *same* endergonic reaction, in which a substance called adenosine diphosphate (ADP) is converted to one called adenosine triphosphate (ATP).

2. Hydrolysis of ATP to ADP releases energy.

3. This exergonic hydrolysis of ATP is the reaction that is directly coupled to most of the endergonic reactions in the human body.

Therefore, ATP acts as an energy carrier for the vast majority of endergonic biochemical reactions. The energy required for a few endergonic biochemical reactions is supplied directly by the hydrolysis of other substances. The hydrolysis reactions for these substances are similar to those for ATP.

Let us investigate some of the properties of ATP. Adenosine triphosphate, ATP, is very similar in structure to the RNA nucleotide containing the base adenine (Chapter 24). Recall that this RNA nucleotide has three component parts: the base adenine, the sugar ribose, and a phosphate group. ATP also has the base adenine and the sugar ribose. However, it has three phosphate groups bonded together by phosphoanhydride bonds (Chapter 17). The structural formula for ATP in its neutral form is shown in Figure 25-4. Notice that ATP has four acidic hydrogens and one basic amine group.

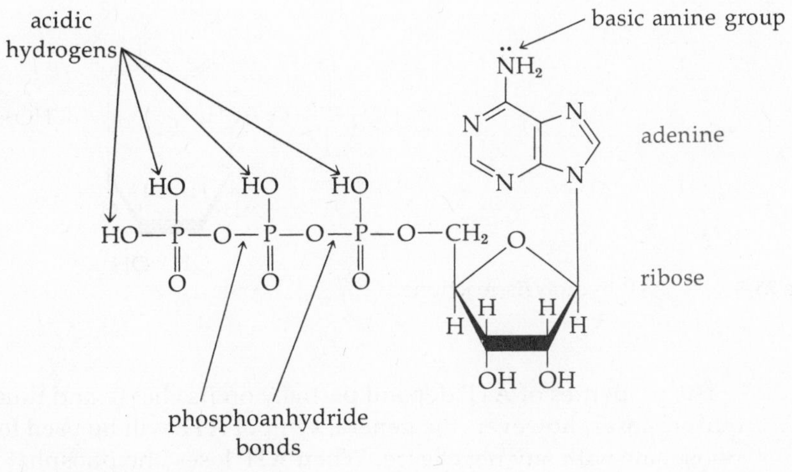

Figure 25-4   The structural formula for adenosine triphosphate, ATP.

We have seen that the actual charge on a molecule with many acidic and basic sites depends on the pH. The higher the pH, the more negatively charged the molecule is; and the lower the pH, the more positively charged the molecule is. At pH 7, which is approximately the pH of cells in the human body, about half of the ATP molecules have a net charge of −3 and about half have a net charge of −4. The structural formula for ATP with a net charge of −3 is shown on the left of the reactions in Figure 25-5.

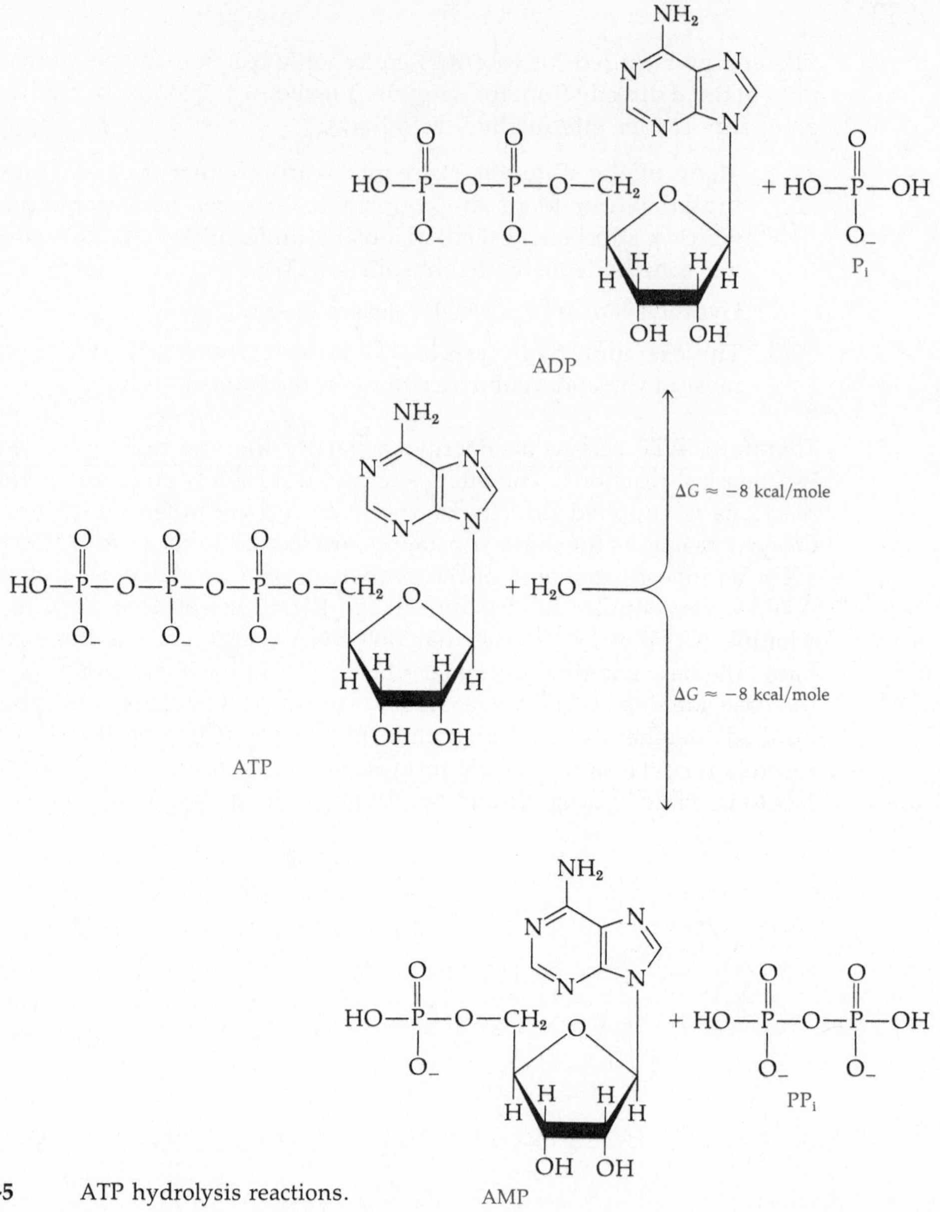

Figure 25-5   ATP hydrolysis reactions.

The properties of ATP depend partially on its charge and thus on the pH. For our purposes, however, the general symbol **ATP** will be used for adenosine triphosphate with any net charge. When ATP loses one phosphate group, a molecule of adenosine diphosphate, **ADP**, is produced. When ATP loses two phosphate groups, the product is adenosine monophosphate, **AMP**. We shall also use the general symbols P_i for inorganic phosphate (whether in the form H_3PO_4, $H_2PO_4^-$, HPO_4^{2-}, or PO_4^{3-}) and similarly, PP_i for pyrophosphate.

ATP can undergo two hydrolysis reactions (see Figure 25-5). Using our general symbols for phosphates, we can represent these reactions as

(25-2) $ATP + H_2O \longrightarrow ADP + P_i + energy$

and

(25-3) $ATP + H_2O \longrightarrow AMP + PP_i + energy$

In each of these reactions, a phosphoanhydride bond is broken and energy is released. Both of these reactions are therefore exergonic. Each reaction has a ΔG of about -8 kcal/mole. The precise value of ΔG depends on several factors, including the pH, because the charges on ATP, ADP, AMP, P_i, and PP_i depend on pH.

Catabolism

Chemical reactions in the body that break larger molecules (such as glucose or amino acids) into smaller ones (such as CO_2, H_2O, and NH_3) are called catabolic reactions, and the overall process is called **catabolism.** As we have noted, food molecules are broken down in a stepwise fashion in order to capture some of the free energy released in catabolism. Some catabolic reactions are coupled directly to the formation of ATP, but others are not. In the next section we shall consider the catabolism of glucose and the coupling of glucose breakdown to ATP formation. In later chapters we shall discuss how lipids (Chapter 27) and proteins (Chapter 28) can also be broken down in reactions that are coupled to ATP synthesis.

Exercise 25-1

Indicate which of the following reactions are endergonic and which are exergonic:

(a) 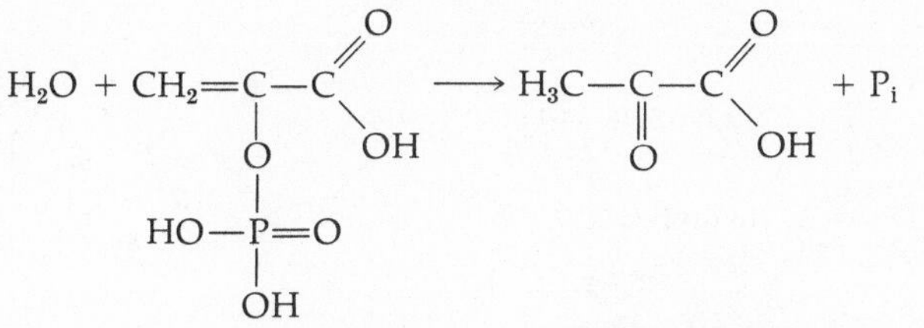 $H_2O + CH_2{=}C{-}C$ ⟶ $H_3C{-}C{-}C$ + P_i $\Delta G = -12$ kcal/mole

(b) 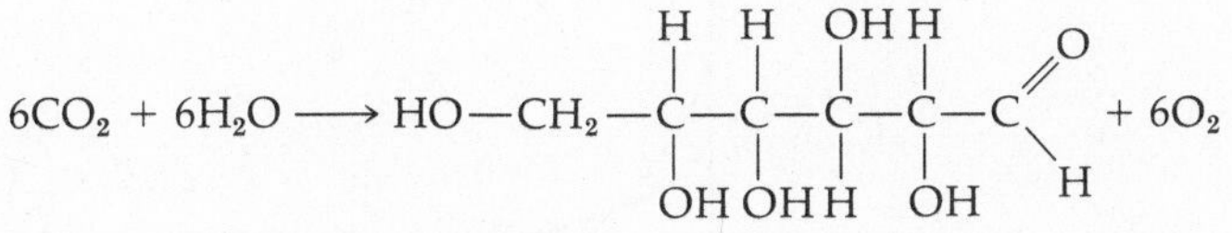 $6CO_2 + 6H_2O$ ⟶ $HO{-}CH_2{-}C{-}C{-}C{-}C{-}C$ + $6O_2$

(c) $ADP + P_i$ ⟶ $ATP + H_2O$

25-4 CARBOHYDRATE CATABOLISM AS A SOURCE OF ENERGY

The first catabolic pathway we shall study is the group of reactions that break down glucose to pyruvic acid. This series of reactions is called the glycolytic (sugar-breaking) pathway, or simply **glycolysis.** We shall look first at glucose because of its primary importance in our diet and its use in the body. We eat a lot of plant starch, animal starch (glycogen), and just plain sugar (sucrose). In the human body the concentration of glucose in the blood is carefully controlled so that enough will be available when needed. In fact, cells in the brain use only glucose to supply their energy needs and are extremely sensitive to changes in glucose concentration in the blood. We also store additional glucose in the polymeric form of glycogen.

We shall see that pyruvic acid, $H_3C-\overset{\overset{\textstyle O}{\|}}{C}-\overset{\overset{\textstyle O}{\|}}{C}-OH$, which is the end product of glycolysis, can suffer two different fates in the human body. If oxygen is not available, pyruvic acid is converted to lactic acid. On the other hand, if oxygen is available, pyruvic acid is further broken down to produce CO_2 and a substance called **acetyl-S-CoA,** $H_3C-\overset{\overset{\textstyle O}{\|}}{C}-S-$Coenzyme A.

Acetyl-S-CoA plays a central role in metabolism. It is produced during the catabolism of lipids and proteins as well as carbohydrates. Acetyl-S-CoA is the input for the second catabolic pathway we shall study (Section 25-5), the *tricarboxylic acid pathway* or TCA cycle.

In Section 25-6 we shall look at another pathway linked to the TCA cycle and glycolysis. In this pathway, called the *electron transport system* (ETS), the reduced coenzymes produced during reactions in glycolysis and the TCA cycle are reoxidized.

The oxidation of coenzymes by the ETS is coupled to ATP synthesis from ADP and P_i. For these coenzymes to be oxidized, oxygen must be available. When cells need energy, and if enough oxygen is present, glucose is oxidized completely to CO_2 and water. Under these conditions, ATP is produced by the combined action of glycolysis, the TCA cycle, and the electron transport system (see Figure 25-6).

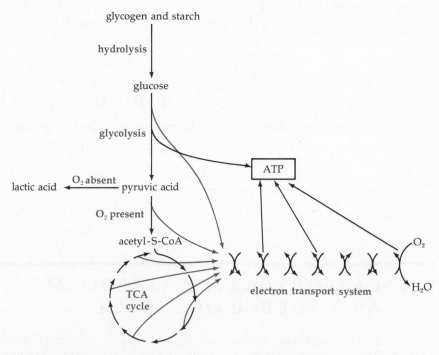

Figure 25-6 The relationship of glycolysis to the TCA cycle and the electron transport system. Hydrolysis of dietary polysaccharides in the gastrointestinal tract produces glucose, which is absorbed and subsequently catabolized in cells (via glycolysis) to pyruvic acid. When oxygen is not available, pyruvic acid is converted to lactic acid. When oxygen is available (it is needed for the electron transport system), pyruvic acid is further broken down via the TCA cycle. Some ATP is produced directly in glycolysis and the TCA cycle. Much more ATP is produced by the electron transport system, which reoxidizes coenzymes (indicated by blue arrows) that have been reduced in glycolysis and the TCA cycle.

Glucose in Polymeric Form

Before we begin our discussion of the catabolism of glucose, let us see what happens to food that contains glucose in polymeric form. The bonds that hold the glucose units together in both starch and glycogen are α-1,4- and α-1,6-glycosidic bonds, as we saw in Chapter 20. When we ingest starch or glycogen, these bonds are quickly broken by enzymes produced in the salivary glands and pancreas. In humans the major enzyme involved in this digestive process is α-amylase, which hydrolyzes α-1,4-glycosidic bonds (see Figure 25-7). Recall that α-amylase cannot break the β-1,4-glycosidic bonds in cellulose. Thus human beings cannot digest wood or vegetable cellulose, but a few animals can. For instance, organisms that can catalyze the hydrolysis of β-1,4-glycosidic bonds in cellulose live in the digestive systems of termites and ruminant animals.

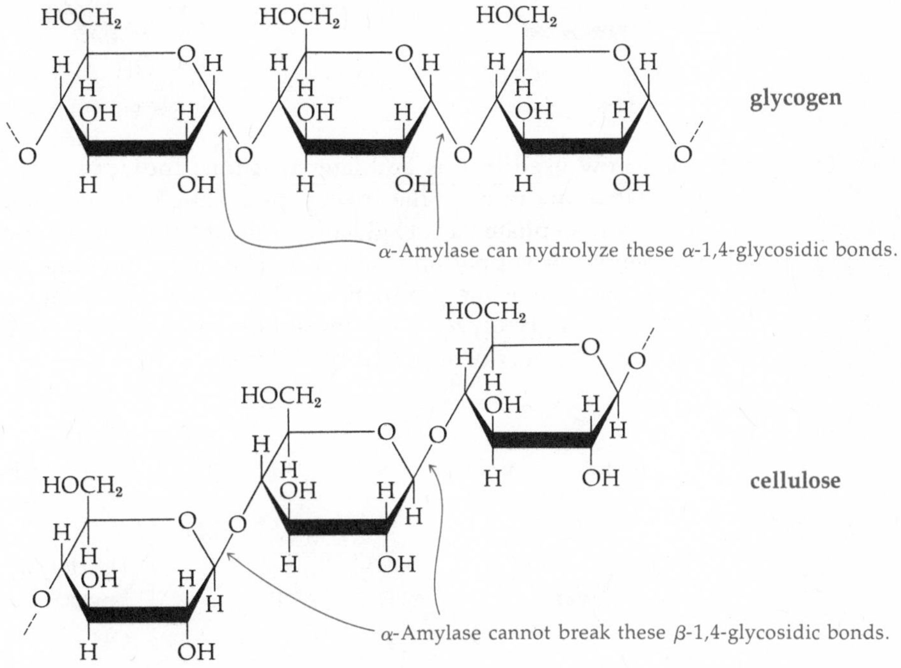

α-Amylase can hydrolyze these α-1,4-glycosidic bonds.

glycogen

cellulose

α-Amylase cannot break these β-1,4-glycosidic bonds.

Figure 25-7 α-Amylase is specific for α-1,4-glycosidic bonds.

Glycolysis

Glycolysis involves the breakdown of a glucose molecule to two molecules of pyruvic acid in a series of reactions coupled to the synthesis of two ATP molecules from ADP. Since the major purpose of glycolysis is the capture of energy, we shall keep close track of the energy-coupled reactions.

In the first step of glycolysis, energy is actually used to add a phosphate group to glucose, a process called **phosphorylation.** The energy for this reaction comes from the conversion of ATP to ADP:

Step 1:

(25-4)

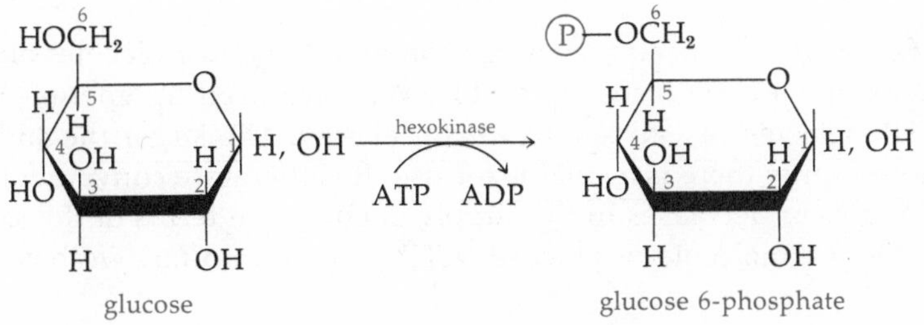

glucose   glucose 6-phosphate

Remember that the curved arrow indicates that these two reactions are energy coupled. The endergonic phosphorylation of glucose is coupled to the exergonic conversion of ATP to ADP. ATP actually donates the phosphate group in this reaction as well. The symbol Ⓟ is a shorthand way of indicating a phosphate group in an organic molecule.

In the next step the aldose, glucose 6-phosphate, is converted to its structural isomer, the ketose, fructose 6-phosphate:

Step 2:

(25-5)

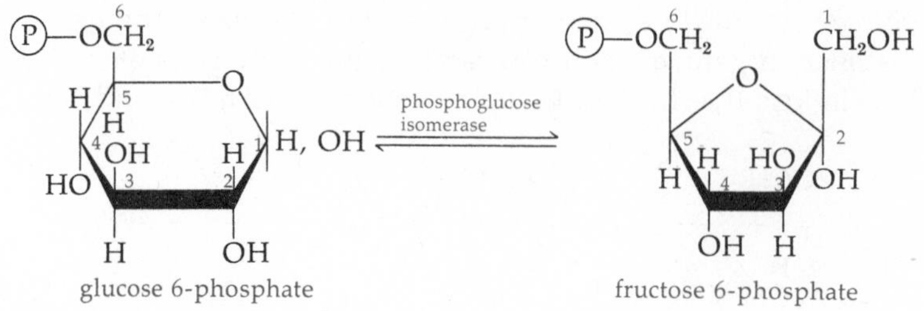

glucose 6-phosphate

fructose 6-phosphate

The double arrow used in this and later reactions indicates a reaction that is readily reversible. In this case, the enzyme phosphoglucose isomerase also converts fructose 6-phosphate back to glucose 6-phosphate. The net direction of the reaction depends on supply and demand. The larger the amount of glucose 6-phosphate, the greater the production of fructose 6-phosphate.

The third step in glycolysis is the phosphorylation of fructose 6-phosphate, which also requires energy supplied by ATP:

Step 3:

(25-6)

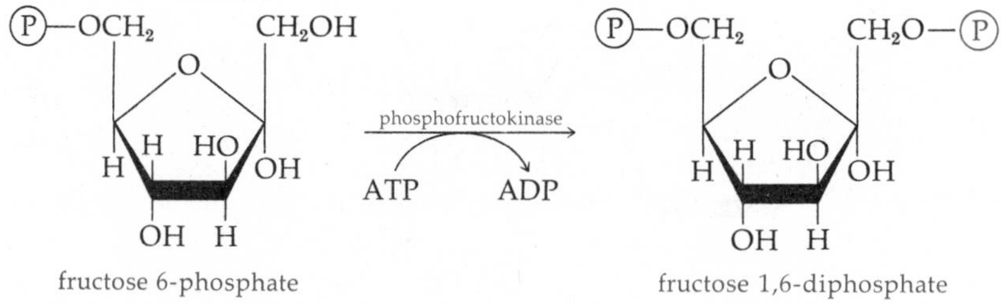

fructose 6-phosphate

fructose 1,6-diphosphate

In the fourth step of glycolysis, fructose 1,6-diphosphate is split into two three-carbon units by the enzyme aldolase:

Step 4:

(25-7)

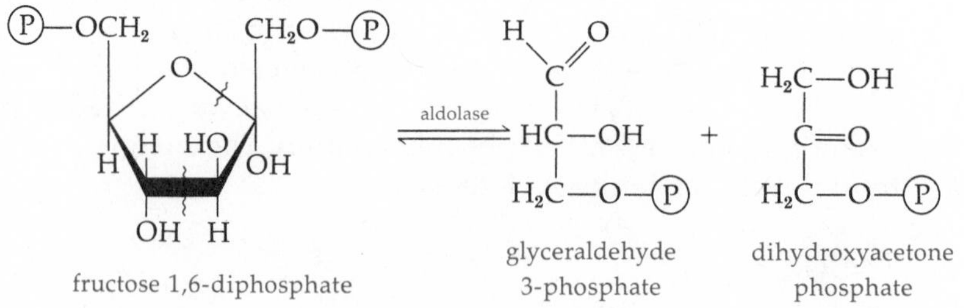

fructose 1,6-diphosphate

glyceraldehyde 3-phosphate

dihydroxyacetone phosphate

So far we have used the energy contained in two molecules of ATP to partially break down one glucose molecule. We have used up energy. We can think of ATP molecules as storage places for energy. The larger the supply of ATP, the more energy there is available for use. It is therefore convenient to keep track of increases or decreases in the supply of energy in terms of the ratio of ATP molecules per molecule of glucose (ATP/glucose). So far, we have lost two molecules of ATP per molecule of glucose (one in step 1 and one in step 3), so our ATP/glucose ratio is −2.

The fifth step of glycolysis is a reaction that allows the dihydroxyacetone phosphate produced in step 4 to be converted to glyceraldehyde 3-phosphate, the other product of step 4. Because of step 5, only one subsequent pathway is required. Separate pathways for breaking down glyceraldehyde 3-phosphate and dihydroxyacetone phosphate are not needed.

Step 5:

(25-8)

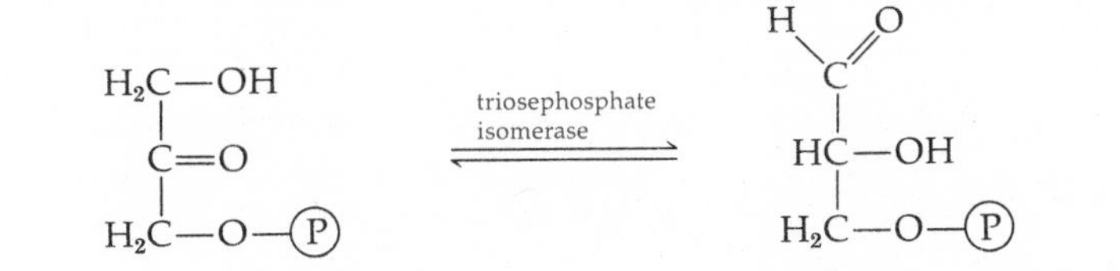

$$\text{dihydroxyacetone phosphate} \xrightarrow[\text{isomerase}]{\text{triosephosphate}} \text{glyceraldehyde 3-phosphate}$$

dihydroxyacetone phosphate   glyceraldehyde 3-phosphate

There is no change in the amount of ATP in step 5, and—like steps 2 and 4—this step is rapidly reversible. Keep in mind that after step 5, we have *two* molecules of glyceraldehyde 3-phosphate for every glucose molecule with which we started.

Step 6:

(25-9)

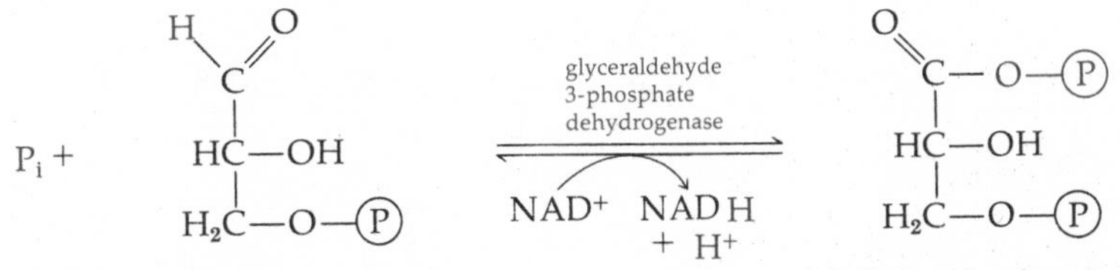

$$P_i + \text{glyceraldehyde 3-phosphate} \xrightarrow[\substack{NAD^+ \quad NADH \\ + \; H^+}]{\substack{\text{glyceraldehyde} \\ \text{3-phosphate} \\ \text{dehydrogenase}}} \text{1,3-diphosphoglyceric acid}$$

glyceraldehyde 3-phosphate   1,3-diphosphoglyceric acid

In step 6 glyceraldehyde 3-phosphate is oxidized. This oxidation involves the conversion of an aldehyde to its corresponding acid, and the subsequent reaction of this acid to form an anhydride with phosphoric acid. The oxidation of glyceraldehyde 3-phosphate is coupled to the reduction of the coenzyme NAD^+ to NADH and H^+ (see Chapter 23). Notice that step 6 involves a combination oxidation-reduction reaction and the binding of a phosphate group to the acid product. The energy needed to form the phosphate anhydride comes from the energy released by the oxidation-reduction reaction.

Up to now in the glycolytic pathway, the energy resources of the cell have been drained. In the remaining steps of the pathway, energy reserves of the cell are replenished and increased.

Step 7:

(25-10)

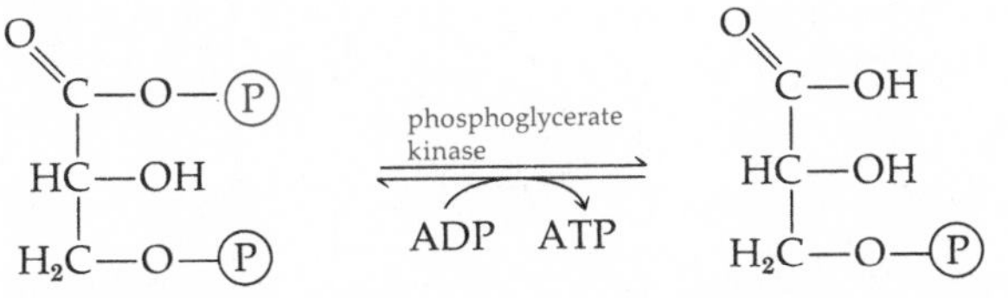

$$\text{1,3-diphosphoglyceric acid} \xrightarrow[\substack{ADP \quad ATP}]{\substack{\text{phosphoglycerate} \\ \text{kinase}}} \text{3-phosphoglyceric acid}$$

1,3-diphosphoglyceric acid   3-phosphoglyceric acid

ATP is produced in step 7. For every molecule of glucose with which we started, there are now two molecules of 1,3-diphosphoglyceric acid. So step 7 gives us an ATP/glucose ratio of +2. Before step 7, the ATP/glucose ratio was −2. So now our *net* ATP/glucose ratio is zero.

We are now approaching the end of glycolysis. In step 8 another isomerization reaction takes place in which the phosphate group is moved to the next carbon on the chain:

Step 8:

(25-11)

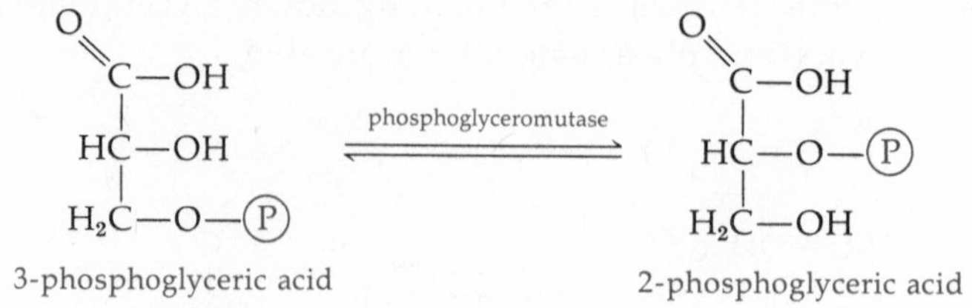

3-phosphoglyceric acid   2-phosphoglyceric acid

In step 9, the 2-phosphoglyceric acid formed in step 8 is dehydrated:

Step 9:

(25-12)

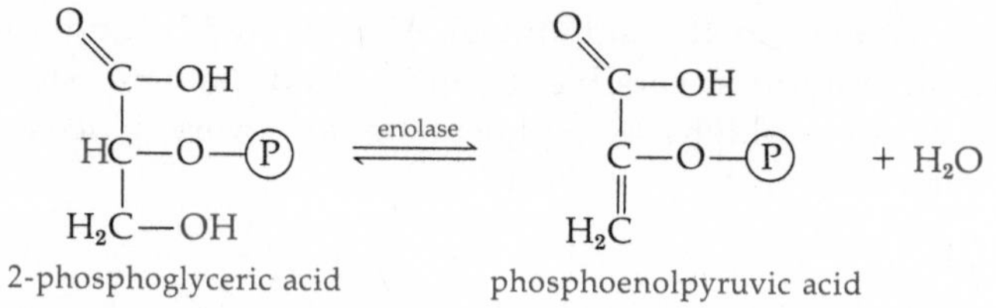

2-phosphoglyceric acid   phosphoenolpyruvic acid

In the tenth and last reaction of glycolysis, the phosphate group in the phosphoenolpyruvic acid is donated to ADP, and ATP is produced.

Step 10:

(25-13)

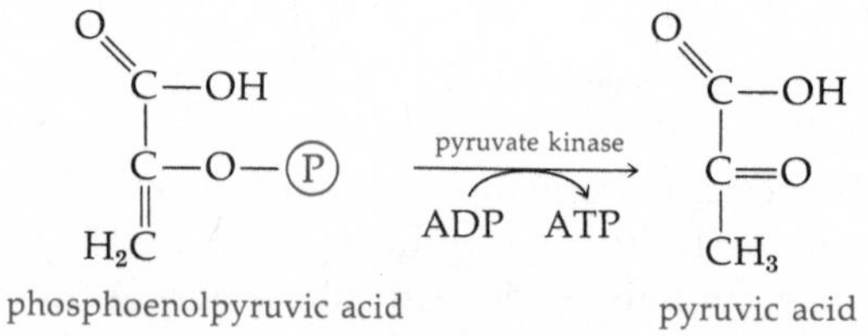

phosphoenolpyruvic acid   pyruvic acid

For the ten steps of glycolysis, the net ATP/glucose ratio is +2. The *overall* process of glycolysis can be symbolized as

(25-14)

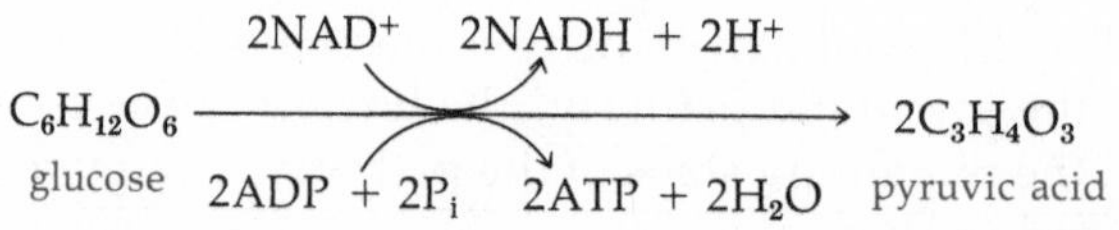

$$C_6H_{12}O_6 \xrightarrow[\text{2ADP + 2P}_i \quad \text{2ATP + 2H}_2O]{\text{2NAD}^+ \quad \text{2NADH + 2H}^+} 2C_3H_4O_3$$

glucose   pyruvic acid

We can consider the overall glycolytic process to be the sum of two separate reactions: (1) an oxidation-reduction reaction in which one molecule of glucose is oxidized—forming two molecules of pyruvic acid—and two molecules of the coenzyme NAD^+ are reduced,

(25-15) $C_6H_{12}O_6 + 2NAD^+ \longrightarrow 2C_3H_4O_3 + 2NADH + 2H^+$

and (2) a reaction in which two molecules of ATP are synthesized,

(25-16) $2ADP + 2P_i \longrightarrow 2ATP + 2H_2O$

As we shall see, the NADH formed in glycolysis can be very useful. It can be used in the electron transport system to generate more ATP (see Section 25-6). The overall pathway of glycolysis is illustrated in Figure 25-8. What happens to the end product, pyruvic acid, depends on the availability of oxygen.

Figure 25-8   Glycolysis. The blue numbers indicate the number of moles of ATP per mole of glucose produced or consumed at the indicated points in the pathway.

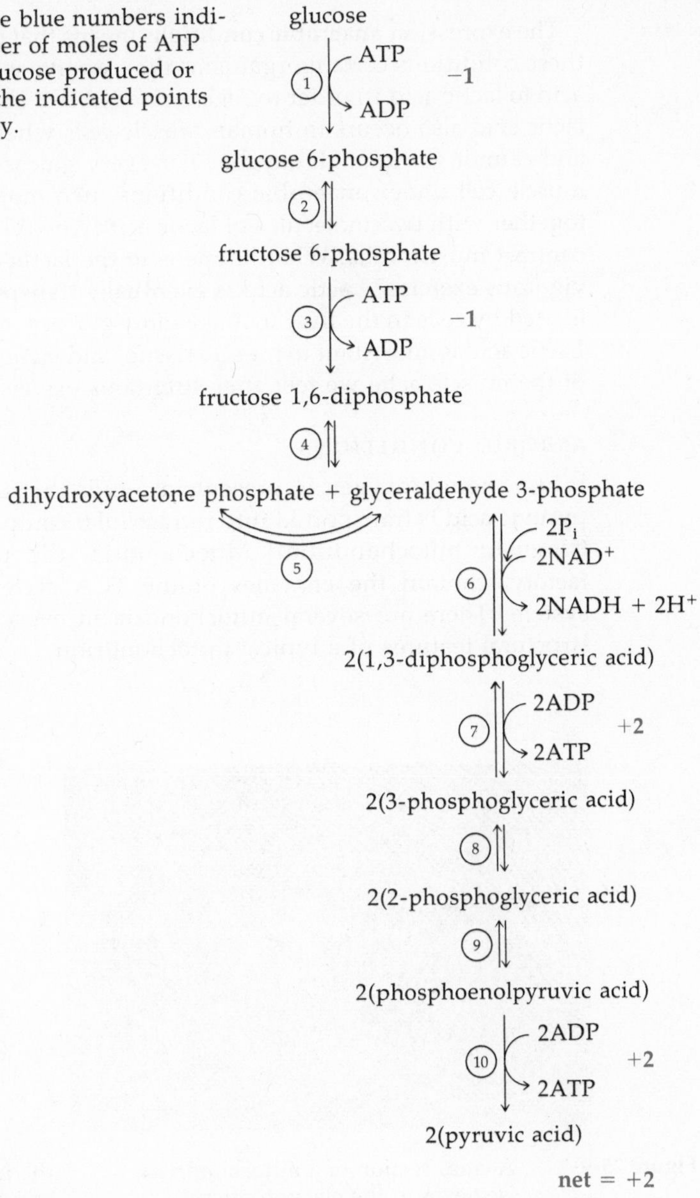

glucose

①   ⌐ ATP
     └→ ADP   −1

glucose 6-phosphate

②

fructose 6-phosphate

③   ⌐ ATP
     └→ ADP   −1

fructose 1,6-diphosphate

④

dihydroxyacetone phosphate + glyceraldehyde 3-phosphate

⑤

⑥   ⌐ 2P$_i$
     ⌐ 2NAD$^+$
     └→ 2NADH + 2H$^+$

2(1,3-diphosphoglyceric acid)

⑦   ⌐ 2ADP
     └→ 2ATP   +2

2(3-phosphoglyceric acid)

⑧

2(2-phosphoglyceric acid)

⑨

2(phosphoenolpyruvic acid)

⑩   ⌐ 2ADP
     └→ 2ATP   +2

2(pyruvic acid)

net = +2

The Fate of Pyruvic Acid

Glycolysis and the production of pyruvic acid cannot continue for very long unless the coenzyme NAD$^+$, which was reduced to NADH and H$^+$ in step 6 of glycolysis, is oxidized back to NAD$^+$. There is a simple reason for this. The human body contains only a limited amount of NAD$^+$. Therefore, when NAD$^+$ is reduced to NADH, it must be reoxidized and reused.

ANAEROBIC CONDITIONS

In the absence of oxygen, NAD$^+$ is regenerated from NADH and H$^+$ in a reaction that is coupled to the reduction of pyruvic acid to lactic acid:

(25-17)

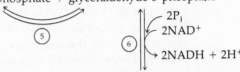

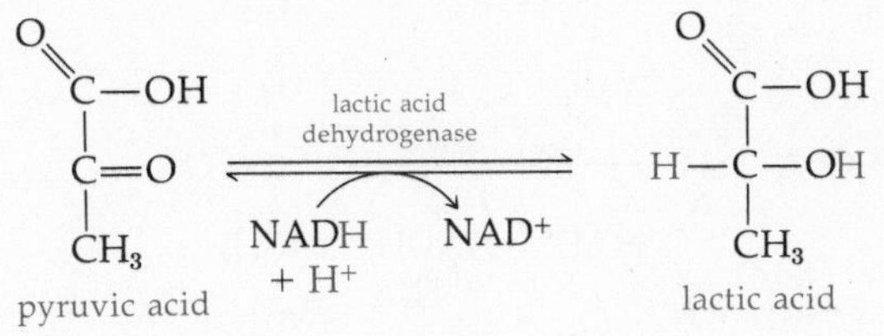

pyruvic acid   lactic acid

The expression **anaerobic conditions** means that oxygen is not available. Under these conditions certain organisms that can live without oxygen reduce pyruvic acid to lactic acid in order to regenerate NAD^+. The reduction of pyruvic acid to lactic acid also occurs in human muscle cells when they are working very hard and cannot get enough oxygen. For every glucose molecule metabolized by a muscle cell under anaerobic conditions, two molecules of ATP are produced, together with two molecules of lactic acid. The ATP is used to supply energy to contract muscle cells. What happens to the lactic acid that accumulates during vigorous exercise? Lactic acid is eventually transported out of muscle cells, and is used by cells in the liver to make more glucose, as we shall see in Section 25-9. Lactic acid is an irritant to muscle tissue, and lactic acid buildup is a major cause of the muscle ache we feel after strenuous exercise.

AEROBIC CONDITIONS

When enough oxygen is present in a cell, that is, under **aerobic conditions,** pyruvic acid is transported into intracellular compartments called **mitochondria** (singular: mitochondrion). Mitochondria, the powerhouses of the cellular factory, contain the enzymes of the TCA cycle and the electron transport system. There are several mitochondria in every cell. Figure 25-9 shows the structural features of a typical mitochondrion.

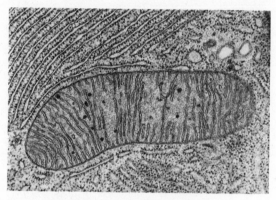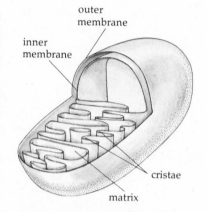

Figure 25-9   A cross section of a mitochondrion is shown in the electron micrograph on the left. The drawing on the right shows structural details of a mitochondrion, which has two membranes. The infoldings of the inner membrane, called cristae, contain the proteins of the electron transport system. The enzymes of the TCA cycle are located in the innermost matrix spaces of mitochondria.

Once inside the mitochondria, pyruvic acid is oxidized by a very complex enzyme, pyruvate dehydrogenase. This oxidation is coupled to the reduction of more NAD^+. Although this reaction produces even more NADH, under aerobic conditions NADH is readily reoxidized to NAD^+, as we shall see in Section 25-6. Pyruvate dehydrogenase catalyzes the reaction

(25-18)

$$\underset{\text{pyruvic acid}}{\overset{\displaystyle O \atop \displaystyle \underset{\displaystyle \underset{CH_3}{|}}{\overset{\displaystyle \underset{C}{\parallel}}{\underset{C=O}{|}}}C-OH} + \underset{\text{coenzyme A}}{CoA-SH} \xrightarrow[\underset{NAD^+ \quad NADH}{\quad}]{\text{pyruvate}\atop\text{dehydrogenase}} \underset{\text{acetyl-S-CoA}}{\overset{\displaystyle CoA \atop \displaystyle \underset{\displaystyle \underset{CH_3}{|}}{\overset{\displaystyle \underset{S}{|}}{\underset{C=O}{|}}}} + CO_2}$$

The structural formula for coenzyme A was given in Chapter 18. We shall use the abbreviation CoA—SH for coenzyme A. Recall that —SH is the sulfhydryl functional group. Other coenzymes are also involved in the pyruvate dehydrogenase reaction, but they are reduced and then reoxidized during the reaction, so we need only consider the overall reaction (25-18).

Recall that two molecules of pyruvic acid are produced for every glucose with

which we started. The molecule $CH_3-\overset{\overset{\displaystyle O}{\|}}{C}-S-CoA$ is known as acetyl-S-CoA

because $CH_3-\overset{\overset{\displaystyle O}{\|}}{C}-$ is the **acetyl group.** Of the six carbon atoms in glucose, two are lost as CO_2 in reaction 25-18. The other four carbon atoms end up as part of two acetyl-S-CoA molecules. Acetyl-S-CoA is the entry point into the TCA cycle.

ANAEROBIC FERMENTATION

In some organisms, such as yeast, pyruvic acid can be converted to ethanol in a series of two reactions:

(25-19)

$$
\underset{\substack{\text{pyruvic acid}}}{\overset{\displaystyle O}{\underset{\displaystyle CH_3}{\overset{\displaystyle \|}{\underset{\displaystyle |}{C=O}}}}{\overset{\displaystyle C-OH}{|}}} \xrightarrow[\text{decarboxylase}]{\text{pyruvic acid}} \underset{\substack{\text{acetaldehyde}}}{\overset{\displaystyle H}{\underset{\displaystyle CH_3}{\overset{\displaystyle |}{\underset{\displaystyle |}{C=O}}}}} + CO_2
$$

(25-20)

$$
\underset{\substack{\text{acetaldehyde}}}{\overset{\displaystyle H}{\underset{\displaystyle CH_3}{\overset{\displaystyle |}{\underset{\displaystyle |}{C=O}}}}} \xrightarrow[\substack{\text{NADH} \\ + H^+ \quad NAD^+}]{\text{alcohol} \\ \text{dehydrogenase}} \underset{\substack{\text{ethanol}}}{\overset{\displaystyle H}{\underset{\displaystyle CH_3}{\overset{\displaystyle |}{\underset{\displaystyle |}{H-C-OH}}}}}
$$

These reactions also reoxidize the NADH produced in step 6 of glycolysis. The fate of pyruvic acid under different conditions is summarized in Figure 25-10.

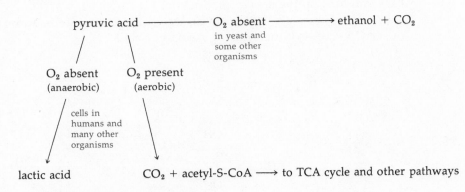

Figure 25-10   Possible fates of pyruvic acid.

Exercise 25-2
Write the overall reaction for glucose catabolism under anaerobic conditions.

Exercise 25-3
Why must active muscle cells convert pyruvic acid to lactic acid?

25-5   THE TRICARBOXYLIC ACID CYCLE

The tricarboxylic acid (TCA) cycle, also called the Krebs cycle or citric acid cycle, occurs in mitochondria under aerobic conditions. It is used to oxidize further the acetyl-S-CoA obtained from glycolysis and from other metabolic pathways in order to extract more energy (i.e., ATP). It is not a self-sufficient cycle. It is more like a gasoline engine: Acetyl-S-CoA is the fuel, and CO_2 and water are exhaust products. The molecules involved in the intermediate steps of the TCA cycle must be present in order for the cycle to operate. They are used in the cycle but are regenerated so that they can be used again. See Figure 25-11 for a comparison of the TCA cycle and the operation of a gasoline engine.

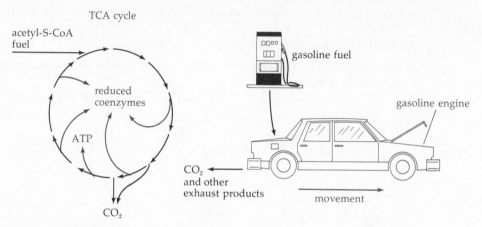

Figure 25-11   In the TCA cycle (left), breakdown of acetyl-S-CoA fuel is coupled to the synthesis of ATP and the reduction of coenzymes. The gasoline en-gine in a car (right) couples the breakdown of gasoline molecules to the movement of the car.

As we study the reactions of the TCA cycle (there are 10 steps in all), look for places where ATP or NADH are produced. In one step $FADH_2$, another reduced coenzyme, is produced from FAD. As we shall see, $FADH_2$ can also be used to make more ATP. After we have studied the individual reactions, we shall go back and total up the energy production.

In the first step of the TCA cycle, an acetyl group is transferred from a molecule of acetyl-S-CoA to a molecule of oxaloacetic acid:

Step 1:

(25-21)

$$H_3C-\overset{\overset{O}{\|}}{C}-S-CoA \;+\; \begin{matrix} O=C-\overset{\overset{O}{\|}}{C}-OH \\ | \\ H_2C-\overset{\overset{O}{\|}}{C}-OH \end{matrix} \;+\; H_2O \;\xrightarrow{\text{citrate synthetase}}\; \begin{matrix} H_2C-\overset{\overset{O}{\|}}{C}-OH \\ | \\ HO-C-\overset{\overset{O}{\|}}{C}-OH \\ | \\ H_2C-\overset{\overset{O}{\|}}{C}-OH \end{matrix} \;+\; CoA-SH$$

acetyl-S-CoA   oxaloacetic acid   citric acid   coenzyme A

You can now see why the cycle is called the tricarboxylic acid or citric acid cycle. In this first step, two carbon atoms enter the TCA cycle via the acetyl group. Also notice that CoA—SH is regenerated in this step.

In the next two reactions of the TCA cycle, citric acid is converted to its structural isomer, isocitric acid. Both reactions are catalyzed by the same enzyme.

Step 2:

(25-22)

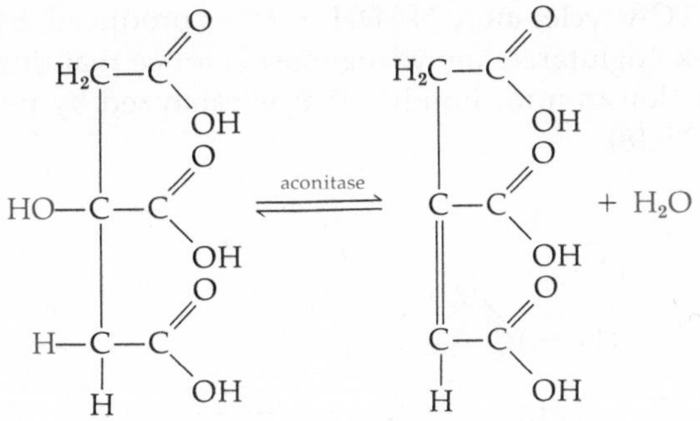

citric acid   cis-aconitic acid

Step 3:

(25-23)

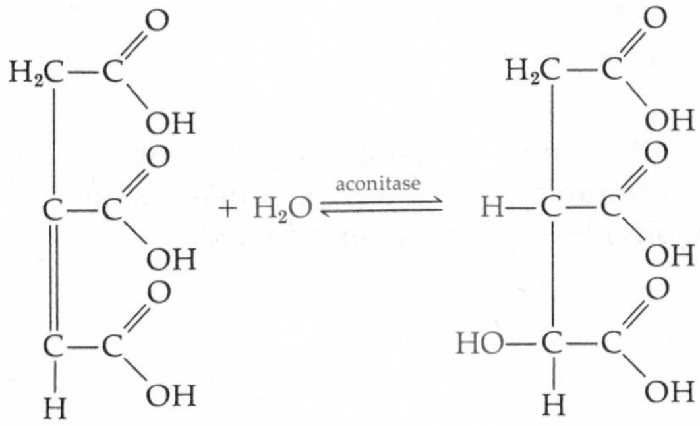

cis-aconitic acid   isocitric acid

In the next two steps of the TCA cycle, isocitric acid is converted to α-ketoglutaric acid. Both steps 4 and 5 are catalyzed by the enzyme isocitrate dehydrogenase. In step 4, NAD^+ is reduced to $NADH + H^+$ in a reaction coupled to the oxidation of the secondary alcohol group of isocitric acid to form a carbonyl group. The product of the reaction is oxalosuccinic acid.

Step 4:

(25-24)

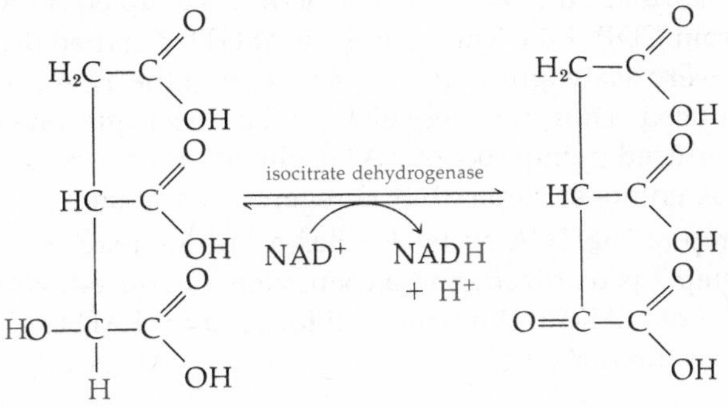

isocitric acid   oxalosuccinic acid

In step 5, a carboxyl group is removed in a decarboxylation reaction:

Step 5:

(25-25)

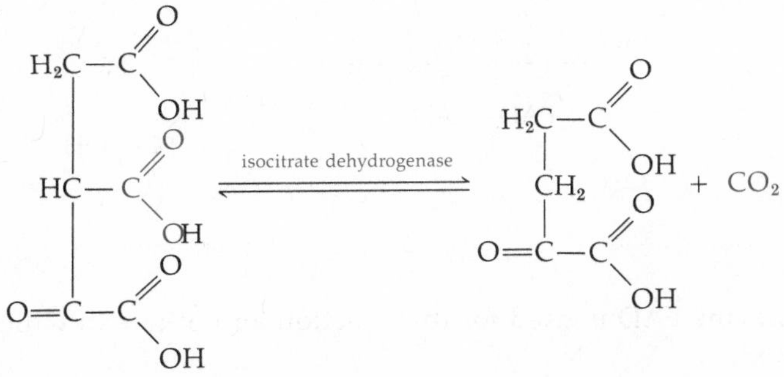

oxalosuccinic acid α-ketoglutaric acid

In step 6 of the TCA cycle, more NADH + H$^+$ is produced. Step 6 is catalyzed by the enzyme α-ketoglutarate dehydrogenase. Notice that this reaction is very similar to the reaction in mitochondria that is catalyzed by pyruvate dehydrogenase (reaction 25-18).

Step 6:

(25-26)

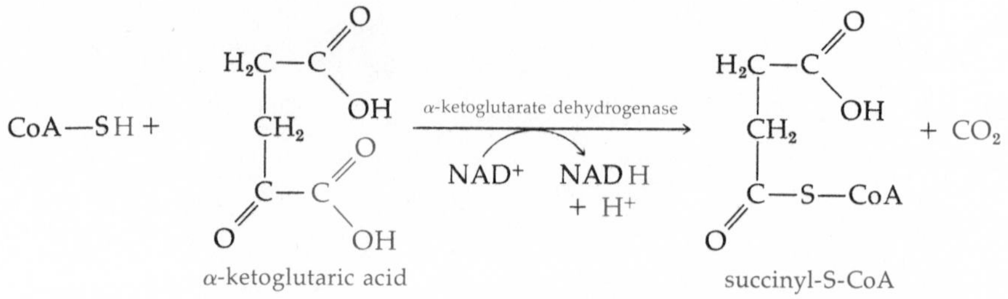

$$CoA-SH +$$

α-ketoglutaric acid

α-ketoglutarate dehydrogenase

NAD$^+$ NADH + H$^+$

succinyl-S-CoA

$+ CO_2$

In step 6, another carbon atom is lost as CO_2. This is the last CO_2 lost in one round of the cycle. A round of the cycle is only fueled with two carbon atoms, so *two and only two* molecules of CO_2 are eliminated.

Step 7:

(25-27)

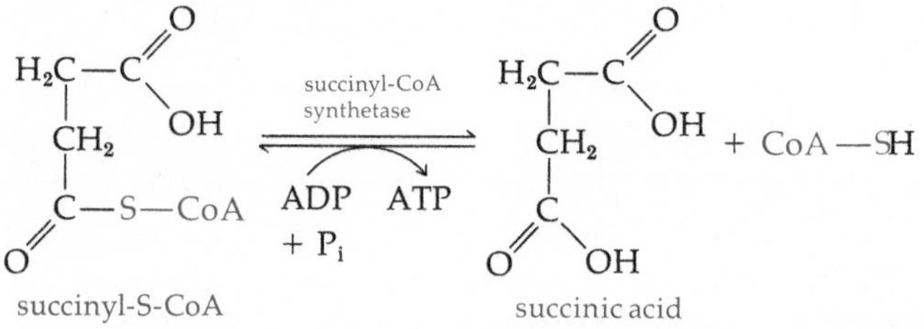

succinyl-S-CoA

succinyl-CoA synthetase

ADP   ATP + P$_i$

succinic acid

$+ CoA-SH$

In step 7 the highly exergonic hydrolysis of the succinyl-S-CoA thioester is coupled to the formation of ATP. (This reaction is coupled directly to the formation of GTP from GDP + P$_i$. The hydrolysis of GTP can then drive the formation of ATP.) For every acetyl group that is used to fuel the TCA cycle, one molecule of ATP is produced. Thus, for every glucose that enters glycolysis, *two* molecules of ATP are produced (remember our ATP/glucose units). Notice that a molecule of coenzyme A enters at step 6, but is regenerated at step 7.

The next step of the TCA cycle also has a unique feature. Succinic acid obtained from step 7 is oxidized, and a coenzyme is reduced. However, the coenzyme is FAD, not NAD$^+$. (The structural formula for FAD is given in Appendix 3 at the back of the book.)

Step 8:

(25-28)

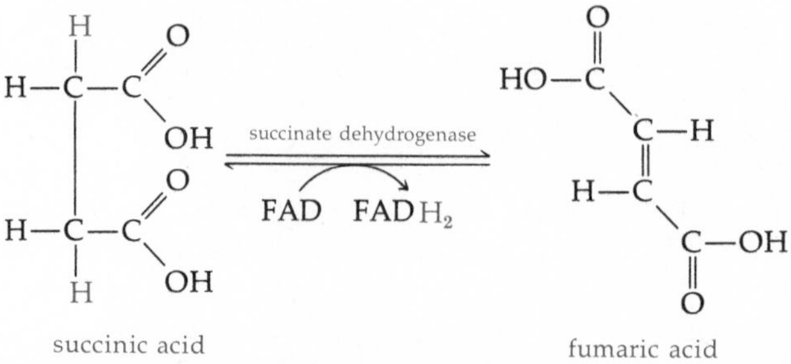

succinic acid

succinate dehydrogenase

FAD   FADH$_2$

fumaric acid

The coenzyme FAD is used for this reaction and others in which double bonds are formed.

In the final two steps of the TCA cycle, water is added to the double bond in fumaric acid, and the resulting alcohol group is oxidized to a carbonyl group. This regenerates oxaloacetic acid for use in the next turn of the cycle.

Step 9:

(25-29)

fumaric acid $+ H_2O \rightleftharpoons$ (fumarase)   L-malic acid

Step 10:

(25-30)

L-malic acid   L-malate dehydrogenase NAD^+ $NADH + H^+$ oxaloacetic acid   to step 1

This last step of the TCA cycle also produces more NADH. The TCA cycle is now complete. Oxaloacetic acid can pick up more acetyl group fuel and start the cycle all over again. A summary of the TCA cycle is shown in Figure 25-12.

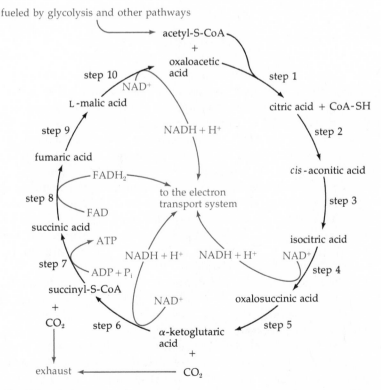

Figure 25-12   The TCA cycle.

What does our acetyl-group-burning TCA cycle engine give us? For every acetyl group we get one high-energy ATP molecule. We also get a lot of reduced coenzymes, $1FADH_2 + 3NADH + 3H^+$. We shall see that reoxidation of these coenzymes in the electron transport system can be used to generate more ATP. Reoxidation of $FADH_2$ and NADH is also necessary in order to supply the TCA cycle with the FAD and NAD^+ coenzymes that it needs.

Exercise 25-4
Write a balanced overall reaction for the TCA cycle.

Exercise 25-5
What is the net ATP/glucose ratio generated in the TCA cycle? How many molecules of NADH and $FADH_2$ are obtained in the TCA cycle for every glucose molecule that entered glycolysis?

25-6   THE ELECTRON TRANSPORT SYSTEM; PHOSPHORYLATION OF ADP TO ATP

Some ATP is produced in a combination of glycolysis and the TCA cycle. But an ATP/glucose of $+4$ (or 32 kcal/mole of glucose) is all that is obtained. This is a fairly small amount compared to the 686 kcal that is available from the complete oxidation of 1 mole of glucose to CO_2 and H_2O. Also, several reduced coenzymes are produced in the TCA cycle and in glycolysis. They need to be reoxidized so that they can be used again. Reduced coenzymes such as NADH and $FADH_2$ are very strong reducing agents. For example, the oxidation of NADH back to NAD^+ is highly exergonic. It would be wasteful to release all of this energy without putting it to work. The **electron transport system**, ETS, is used to pass electrons and hydrogen from NADH (and $FADH_2$) to oxygen. The net overall reaction for the oxidation of NADH in the ETS is

(25-31)
$$NADH + H^+ + \tfrac{1}{2}O_2 \longrightarrow NAD^+ + H_2O$$

We can think of this net overall reaction in the following manner: A hydrogen atom with two electrons, called a **hydride ion** ($H:^-$), leaves NADH, forming NAD^+. The hydride ion, $H:^-$, and the proton, H^+, combine with an atom of oxygen, $\tfrac{1}{2}O_2$, to form a molecule of water. It is in this sense that we speak of NADH passing electrons and hydrogen to oxygen in the ETS. To do this in one step would be a waste of energy. So, just as in glycolysis and the TCA cycle, this exergonic process is broken down into a series of steps with smaller energy changes. To make this point clear, consider the following: The overall free-energy change (ΔG) for reaction 25-31 is -53 kcal/mole. If we coupled this to the formation of one molecule of ATP, which requires 8 kcal/mole, we would be wasting a lot of energy. However, when the overall reaction 25-31 is broken down into a number of steps in the ETS, three molecules of ATP are formed for each molecule of NADH available.

In the ETS, electrons and hydrogen are passed along through a series of intermediates, or carriers. Each transfer to the next carrier is an exergonic oxidation-reduction reaction. Three of these transfers are sufficiently exergonic to be coupled to ATP formation.

All of the carriers in the ETS are found clustered together in the infolded inner membrane (cristae) of the mitochondria (see Figure 25-9). Each of the carriers in the ETS can exist in two states: oxidized and reduced. The major reactions, as we currently understand them, are as follows: The first carrier contains riboflavin (vitamin B_2) and is therefore a **flavoprotein.** It is represented by the symbol FP_1. This protein accepts the hydride ion from NADH along with a proton, H^+ (see Figure 25-13). Flavoprotein$_1$ is sometimes called NADH dehydrogenase. The next major electron carrier is **coenzyme Q,** abbreviated CoQ. CoQ accepts the hydrogen atoms from FP_1H_2 and is reduced to $CoQH_2$. The structure of CoQ is given in Appendix 3 at the back of the book. The remaining carriers in the ETS are called cytochromes. **Cytochromes** are proteins that contain a heme prosthetic group, as does the protein hemoglobin (Chapter 22). The major differences between individual cytochromes are in the protein portion of these molecules. All of the cytochromes are capable of undergoing the following oxidation-reduction reaction:

(25-32) $Fe^{2+} \rightleftharpoons Fe^{3+} + 1e^-$

Figure 25-13   Schematic representation of the major components in the transport of electrons from NADH to oxygen. The blue brackets represent oxidative phosphorylation sites, which are discussed on the following pages. These sites are the only ones sufficiently exergonic to drive ATP synthesis.

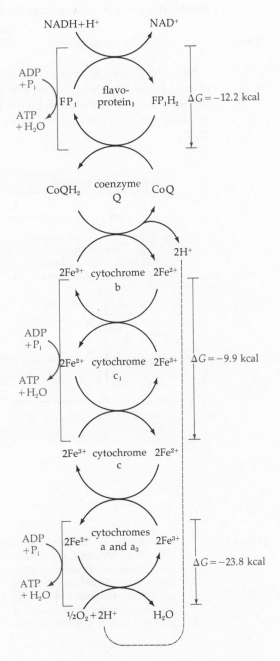

Notice that only one electron is involved in reaction 25-32. Thus, in order to pass the two electrons from $CoQH_2$ along to the cytochromes, two molecules of each cytochrome must be reduced at each step. Also notice that the hydrogen atoms from $CoQH_2$ are not passed along to the cytochromes. They are released as protons, $2H^+$, and are later used to form one molecule of water.

Since most of the ETS carriers are bound to the inner mitochondrial membrane, they have been difficult to study. Our information about the ETS has therefore been hard to acquire and is not complete. The important features of the ETS are represented in Figure 25-13, where the notation

$$\begin{array}{cc} A & B \\ & \diagup\!\!\!\!\diagdown \\ C & D \end{array}$$

is used as a symbolic representation for the reaction $A + C \rightarrow B + D$. Note that the arrows go from reactants (arrow tails) to products (arrow heads). For example, the transfer of electrons from $CoQH_2$ to cytochrome b can be written as

(25-33) $CoQH_2 + 2(\text{cytochrome b} \cdot Fe^{3+}) \longrightarrow CoQ + 2(\text{cytochrome b} \cdot Fe^{2+}) + 2H^+$

Note the following important features of the ETS. *First*, NADH (and $FADH_2$) is reoxidized. The NAD^+ (and FAD) produced can be reused in the TCA cycle. We now see why there must be oxygen available to the cell (aerobic conditions) in order for the TCA cycle to work. The TCA cycle needs the ETS to supply it

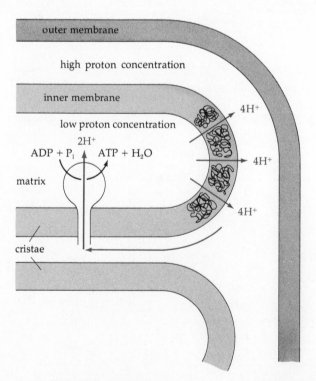

Figure 25-14   A model for oxidative phosphorylation. Peter Mitchell, who received the 1978 Nobel Prize in chemistry, has proposed that the coupling of electron transport to ATP synthesis occurs by protons being pumped out of the matrix space at three sites in the electron transport system. Recent experiments indicate that four protons are pumped out at each of these sites. According to this model, the diffusion of these protons back into the matrix drives oxidative phosphorylation.

with NAD^+ (and FAD), and the ETS cannot work without O_2. *Second,* you should see that each of the carriers in the ETS is first reduced and then reoxidized so that it can be used again. *Third,* notice that three portions of the ETS are exergonic enough to be coupled to ATP synthesis. The coupling of these three portions to ATP formation is called **oxidative phosphorylation,** and each portion of the ETS at which such coupling takes place is called an oxidative phosphorylation site. It has been proposed that the actual coupling mechanism involves the pumping of protons through the mitochondrial inner membrane (see Figure 25-14).

The electron transport system is also used to reoxidize $FADH_2$. However, $FADH_2$ cannot reduce FP_1. It enters the ETS at a different point. $FADH_2$ first reduces a different flavoprotein, FP_2, to FP_2H_2, and then FP_2H_2 reduces CoQ to $CoQH_2$ (see Figure 25-15). These steps ($FADH_2$ to CoQ) are not exergonic enough to allow for ATP synthesis, so that from the oxidation of one molecule of $FADH_2$ via the ETS, we get only two molecules of ATP.

Figure 25-15   The transfer of electrons from $FADH_2$ to a flavoprotein, FP_2, and then to coenzyme Q, is not sufficiently exergonic to drive ATP synthesis.

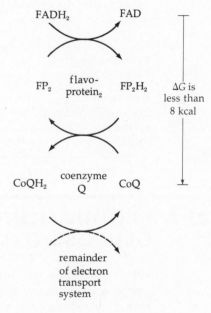

Hence, using the ETS, we can get three ATP molecules for every $NADH + H^+$ and two ATP molecules for every $FADH_2$ that were generated in the TCA cycle from the oxidation of acetyl-S-CoA. The NADH produced by the pyruvate dehydrogenase reaction (see reaction 25-18) in mitochondria also feeds the ETS and provides three ATPs per molecule of pyruvic acid.

NADH was also produced in step 6 of glycolysis. But step 6 of glycolysis takes place in cytoplasm, not in mitochondria. Can we generate any ATP in this case? In order for the NADH from glycolysis to be used in the ETS, it needs to enter the mitochondria. The NADH, however, cannot cross the mitochondrial membrane, so additional steps, involving additional molecules, are required by which NADH shuttles its electrons and hydrogen across to another molecule inside the mitochondria. In liver cells, the hydrogen atoms are shuttled to NAD^+ inside of the mitochondria. In muscle cells they are shuttled to a molecule of FAD. We do not need to examine this shuttle system in detail, but the overall reaction, shown in Figure 25-16, is important. Because $FADH_2$, not NADH, is the input to the ETS in muscle cells, only *two* ATP molecules can be made via the ETS for each NADH generated in glycolysis.

Figure 25-16   The shuttle of electrons across the mitochondrial membrane in a muscle cell is represented schematically. The process involves intermediates, X and XH_2, which are able to pass through the mitochondrial membrane.

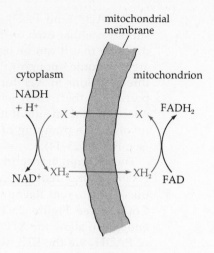

Several substances strongly inhibit, or block, the electron transport system. For example, the cyanide ion, CN^-, is poisonous because it blocks the transfer of electrons to oxygen.

Exercise 25-6

Write the reaction for the last step of the ETS.

Exercise 25-7

The overall change in free energy for the oxidation of NADH in the electron transport system is -53 kcal/mole. Why aren't six ATP molecules produced for every molecule of NADH oxidized in the ETS?

25-7 PRODUCTION OF ATP BY AEROBIC GLUCOSE OXIDATION

Let us determine the number of molecules of ATP that can be made by the complete oxidation of glucose. To do this, we shall take 1 mole of glucose and look at the net reactions of glycolysis, pyruvate dehydrogenase, the tricarboxylic acid cycle, and the electron transport system. We must keep track of the reduced coenzymes, NADH and $FADH_2$, because they are the inputs into the ETS.

A. Overall Reactions

Glycolysis (in the cytoplasm):

(25-34)   glucose + $2NAD^+$ + 2ADP + $2P_i \longrightarrow$
$$2 \text{ pyruvic acid} + 2ATP + 2H_2O + 2NADH + 2H^+$$

Pyruvate dehydrogenase (in mitochondria):

(25-35)   2(pyruvic acid + NAD^+ + CoA-SH $\longrightarrow$ acetyl-S-CoA + CO_2 + NADH + H^+)

The TCA cycle (in mitochondria):

(25-36)   2(acetyl-S-CoA + $3NAD^+$ + FAD + $2H_2O$ + ADP + $P_i \longrightarrow$
$$2CO_2 + 3NADH + 3H^+ + FADH_2 + ATP + CoA\text{-}SH)$$

B. Yield per Mole of Glucose

Yield

Glycolysis plus shuttle system
(in muscle cells): $2ATP + 2FADH_2$
Pyruvate dehydrogenase: $2NADH + 2H^+$
The TCA cycle: $2ATP + 2FADH_2 + 6NADH + 6H^+$

Subtotal: $\overline{4ATP + 4FADH_2 + 8NADH + 8H^+}$

Recall that in liver cells the shuttle system yields $2NADH^+$ in the mitochondria rather than $2FADH_2$ as in muscle cells. Thus, for liver cells, this subtotal would be $4ATP + 2FADH_2 + 10NADH + 10H^+$.

C. The Electron Transport System

Overall reactions:

(25-37) $8(NADH + H^+ + \tfrac{1}{2}O_2 + 3ADP + 3P_i \longrightarrow NAD^+ + 4H_2O + 3ATP)$

(25-38) $4(FADH_2 + \tfrac{1}{2}O_2 + 2ADP + 2P_i \longrightarrow FAD + 3H_2O + 2ATP)$

ATP production per mole of glucose:

$8(NADH + H^+) \longrightarrow 24ATP$
$4(FADH_2) \longrightarrow \underline{8ATP}$
Subtotal: $\overline{32ATP}$

D. ATP per Mole of Glucose

1. From part B: $4ATP + \underline{4FADH_2 + 8NADH + 8H^+}$

2. From part C: $\underline{32ATP} \longleftarrow$

3. Total $36ATP$ in muscle cells

Because of its different shuttle system, a total of 38 ATP/glucose are produced in liver cells.

The overall reaction for the complete oxidation of one molecule of glucose via glycolysis, the TCA cycle, and the ETS can now be derived.
First, take the sum of reaction 25-34 (glycolysis) + reaction 25-35 (pyruvate dehydrogenase) + reaction 25-36 (the TCA cycle), then (for muscle cells) adjust for the NADH shuttle into mitochondria (Figure 25-16) to obtain

(25-39) $Glucose + 8NAD^+ + 4FAD + 4ADP + 4P_i + 2H_2O \longrightarrow$
$6CO_2 + 4FADH_2 + 8NADH + 8H^+ + 4ATP$

Now, add to reaction 25-39 the ETS reactions 25-37 and 25-38 to get

(25-40) $Glucose + 36ADP + 36P_i + 6O_2 \longrightarrow 6CO_2 + 36ATP + 42H_2O$

Notice that in this overall reaction 25-40 the coenzymes are neither reactants nor products. That is because they are reoxidized and reused. The overall reaction 25-40 can be written as the sum of two reactions:

(25-41a) $\text{Glucose} + 6O_2 \longrightarrow 6CO_2 + 6H_2O$

and

(25-41b) $36\text{ADP} + 36P_i \longrightarrow 36\text{ATP} + 36H_2O$

Thus, in muscle cells under aerobic conditions, complete oxidation of glucose produces 36 moles of ATP per mole of glucose. If we use the value 8 kcal/mole for the endergonic reaction $\text{ADP} + P_i \to \text{ATP} + H_2O$, we have captured $8 \times 36 = 288$ kcal of energy in the form of ATP. The free-energy change for reaction 25-41a is about -686 kcal/mole. Thus we have captured 288/686 or about 42% of the energy released when 1 mole of glucose is oxidized. The rest of the energy is released in the form of heat. Hence the catabolism of glucose is not 100% efficient—no process is. However, 42% efficiency is not bad; it is much better than the energy efficiency of any machine ever made. For example, the average automobile has an operating efficiency of only about 3% under standard driving conditions.

Don't just memorize the number 36. It is important to know the reactions in which the ATP and reduced coenzymes are formed. We shall see that lactic acid and pyruvic acid can be made from molecules other than glucose, and we shall want to know how much ATP can be formed by the oxidation of these other molecules via the TCA cycle and the electron transport system.

Optional

25-8 THE PENTOSE PHOSPHATE PATHWAY

Glycolysis is the major pathway used by cells in the body to catabolize glucose. It is not, however, the only pathway. Another catabolic pathway, the **pentose phosphate pathway,** is used by cells under certain conditions. Skeletal and heart muscle cells hardly use it at all, but about 30% of the glucose in the liver is broken down by this pathway. Mammary gland cells use it even more.

Why should another pathway for glucose oxidation be needed? Glycolysis, operating in conjunction with the TCA cycle and the electron transport system, is highly efficient for capturing energy in the form of ATP. The pentose phosphate pathway is used for two other reasons: (1) production of the sugar ribose 5-phosphate, which is needed for the synthesis of nucleotides and nucleic acids; and (2) reduction of the coenzyme nicotinamide adenine dinucleotide phosphate, NADP^+, to form NADPH, which is structurally similar to NADH (see Appendix 3). NADPH is used to reduce molecules, regenerating NADP^+ in several anabolic biochemical pathways, as, for example, in the synthesis of lipids (see Chapter 27).

The starting material for the pentose phosphate pathway is glucose 6-phosphate, which is generated in step 1 of glycolysis (reaction 25-4). Glucose 6-phosphate can be converted into ribose 5-phosphate by several reaction steps. The overall reaction for this process is

(25-42) $\text{Glucose 6-phosphate} + 2\text{NADP}^+ + H_2O \longrightarrow$
$$\text{ribose 5-phosphate} + CO_2 + 2\text{NADPH} + 2H^+$$

By some additional steps, one molecule of glucose 6-phosphate can be completely degraded to CO_2. In this process, 12 molecules of NADPH are formed per molecule of glucose 6-phosphate. The overall reaction in this case is

(25-43) $\text{Glucose 6-phosphate} + 12\text{NADP}^+ + 7H_2O \longrightarrow 6CO_2 + 12\text{NADPH} + 12H^+ + P_i$

Thus, in the pentose phosphate pathway, glucose 6-phosphate can suffer two fates. The extent to which reactions 25-42 and 25-43 take place in a given cell depends on the requirements for ribose 5-phosphate and NADPH, respectively. If the need for NADPH is much larger than the need for ribose 5-phosphate, then reaction 25-43 will predominate. If neither NADPH nor ribose 5-phosphate is in great demand, almost all of the glucose 6-phosphate will be catabolized via glycolysis.

Exercise 25-8

During strenuous exercise, what would be the pathway used to metabolize glucose 6-phosphate? Explain your answer.

25-9   GLUCONEOGENESIS

Our major source of the glucose that we catabolize is dietary—coming primarily from plants, which make glucose in the process of photosynthesis. However, human beings are also capable of making glucose by a process (pathway) called **gluconeogenesis,** which means "making *new* glucose."

Gluconeogenesis occurs primarily in the liver, and is significant in the following situations (see Figure 25-17): (1) The lactic acid produced in active muscle cells as the product of anaerobic glycolysis can be used as a starting material for the synthesis of glucose, which is stored in the liver as glycogen for future use. (2) When the dietary sources of glucose are not sufficient, amino acids from proteins in the body can also be used as starting materials for gluconeogenesis.

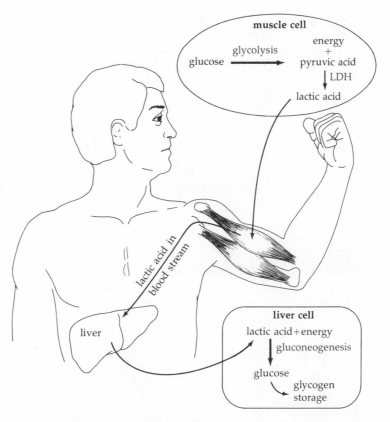

Figure 25-17   Lactic acid, which is produced in active muscle cells, can be converted to glucose via gluconeogenesis in liver cells.

As we have mentioned, lactic acid is produced from glucose in the process of anaerobic glycolysis. However, every step in gluconeogenesis cannot be the exact reverse of a step in the glycolysis pathway. At least some steps in the two pathways must be different. There is a very simple reason for this. The overall process of glycolysis is exergonic. Thus, gluconeogenesis must be an endergonic process. Now, since no biochemical process is 100% efficient, more ATP must be used to convert lactic acid into glucose than is produced when glucose is broken down to lactic acid in glycolysis. Some of the energy required for gluconeogenesis is supplied by the hydrolysis of ATP, and some by the hydrolysis of guanosine triphosphate, GTP, which is similar in structure to ATP, but contains the nitrogenous base guanine instead of the adenine found in ATP. The hydrolysis of GTP is similar to that of ATP:

(25-44) $GTP + H_2O \longrightarrow GDP + P_i$

The ΔG for this reaction is also about the same as for the hydrolysis of ATP, $-8\ kcal/mole$. Thus, the energy supplied by hydrolysis of one GTP is equivalent to that supplied by one ATP.

Gluconeogenesis starts with the oxidation of lactic acid to pyruvic acid.

Step 1:

(25-45)

$$
\underset{\text{lactic acid}}{\begin{array}{c} O \\ \parallel \\ C-OH \\ | \\ H-C-OH \\ | \\ CH_3 \end{array}}
\quad\underset{NAD^+ \quad NADH + H^+}{\overset{\text{lactic acid dehydrogenase}}{\rightleftharpoons}}\quad
\underset{\text{pyruvic acid}}{\begin{array}{c} O \\ \parallel \\ C-OH \\ | \\ C=O \\ | \\ CH_3 \end{array}}
$$

Step 1 is simply the reverse of the last step of anaerobic glucose metabolism (reaction 25-17). However, three of the steps in glycolysis (steps 1, 3, and 10) are not reversible. The next two steps in gluconeogenesis are designed to get around the irreversible step 10 of glycolysis (reaction 25-13).

Step 2:

(25-46)

$$
\underset{\text{pyruvic acid}}{\begin{array}{c} O \\ \parallel \\ C-OH \\ | \\ C=O \\ | \\ CH_3 \end{array}} + CO_2
\quad\underset{H_2O + ATP \quad ADP + P_i}{\overset{\text{pyruvate carboxylase}}{\rightleftharpoons}}\quad
\underset{\text{oxaloacetic acid}}{\begin{array}{c} O \\ \parallel \\ C-OH \\ | \\ C=O \\ | \\ CH_2 \\ | \\ C-OH \\ \parallel \\ O \end{array}}
$$

Step 3:

(25-47)

$$
\underset{\text{oxaloacetic acid}}{\begin{array}{c} O \\ \parallel \\ C-OH \\ | \\ C=O \\ | \\ CH_2 \\ | \\ C-OH \\ \parallel \\ O \end{array}}
\quad\underset{GTP \quad GDP}{\overset{\text{phosphoenol pyruvate carboxykinase}}{\rightleftharpoons}}\quad
\underset{\begin{array}{c}\text{phosphoenol-}\\\text{pyruvic acid}\end{array}}{\begin{array}{c} O \\ \parallel \\ C-OH \\ | \\ C-O-\textcircled{P} \\ \parallel \\ CH_2 \end{array}} + CO_2
$$

In these two steps, ATP and GTP are used to make the phosphoenolpyruvic acid molecule. The next several steps in gluconeogenesis use the reversible steps (9, 8, 7, 6, 5, and 4) of glycolysis (Figure 25-8) to produce fructose 1,6-diphosphate. The next irreversible step of glycolysis (step 3, reaction 25-6) is bypassed by the enzyme fructose 1,6-diphosphatase, which serves to catalyze the following reaction:

(25-48)

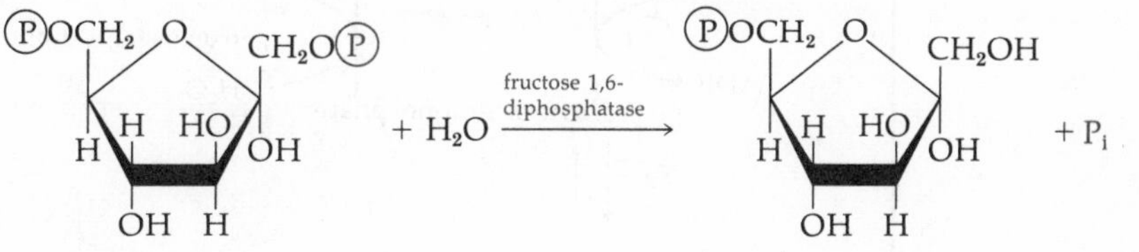

fructose 1,6-diphosphate   fructose 6-phosphate

The remaining irreversible step in glycolysis (step 1) is bypassed by the enzyme glucose 6-phosphatase in the reaction

(25-49)

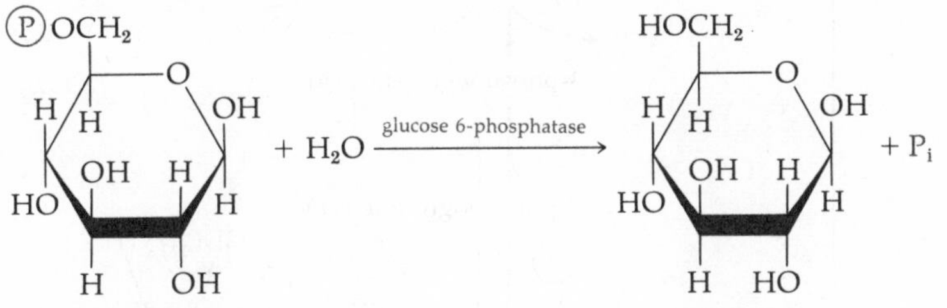

glucose 6-phosphate   glucose

Glycolysis and gluconeogenesis are compared in Figure 25-18 on page 636. The net overall reaction in gluconeogenesis is

(25-50) $2 \text{ Lactic acid} + 4\text{ATP} + 2\text{GTP} + 6\text{H}_2\text{O} \longrightarrow \text{glucose} + 4\text{ADP} + 2\text{GDP} + 6\text{P}_i$

As you can see from reaction 25-50, the formation of one molecule of glucose from two molecules of lactic acid requires the hydrolysis of four molecules of ATP and two molecules of GTP, which is equivalent to a total of six ATP molecules.

We might ask the questions: Why isn't lactic acid just eliminated? Why does the human body utilize energy to convert lactic acid back to glucose? Isn't this a waste of energy? Actually, a lot of energy can be obtained from lactic acid, and eliminating lactic acid would be the wasteful procedure. Energy is used to convert lactic acid to glucose, but once formed, a molecule of glucose represents a potential source of an even larger amount of energy. Recall that a very large amount of ATP is produced when a molecule of glucose is completely degraded to CO_2 and H_2O.

Exercise 25-9
Compare the amount of energy (in terms of ATP) produced when 1 mole of glucose is completely catabolized to CO_2 and H_2O to the amount of ATP needed to form 1 mole of glucose from lactic acid.

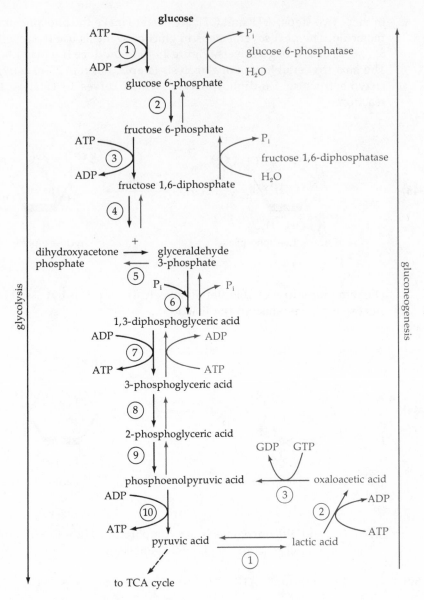

Figure 25-18 Reactions in glycolysis versus gluconeogenesis. Reactions of gluconeogenesis are shown in blue.

25-10 METABOLISM OF OTHER CARBOHYDRATES

Gluconeogenesis in humans and photosynthesis in plants lead to the formation of glucose. By now you are aware of the tremendous importance of glucose and glucose metabolism in the human body. However, in Chapter 20 we saw that glucose is only one of several six-carbon sugars. Fructose, which is part of the disaccharide sucrose, is another. Humans ingest about 100 g of sucrose each day. Are other hexoses important to the human body? If so, how are they metabolized? The answer to the first question is yes. Hexoses and their derivatives play several roles in the human body.

Functions of Hexoses in Humans

Large concentrations of free hexoses, other than glucose, are not found in human body cells. However, several hexoses are rather abundant as components of disaccharides and polysaccharides. For example, the sugar galactose is a component of lactose, the major carbohydrate in milk. Galactose is also a vital component of the carbohydrate prosthetic groups of several glycoproteins (see Chapter 22 for a review of prosthetic groups). Galactose, together with other hexoses and their chemical derivatives, are also components of polysaccharides found on cell surfaces. For example, the different ABO blood types of humans arise because of the presence of different sequences of sugars in certain polysaccharides on cell membranes.

Hexose Metabolism

In order for lactose and sucrose to be metabolized, these disaccharides must first be split into their monosaccharide components. Hydrolysis of lactose and sucrose is accomplished by two enzymes found in intestinal cells. These enzymes, lactase and sucrase, catalyze the following reactions:

(25-51)

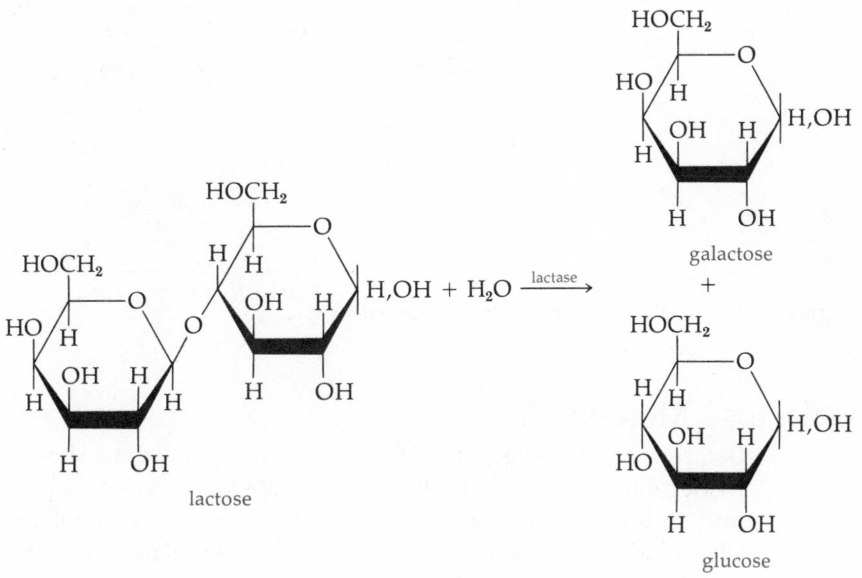

(25-52)

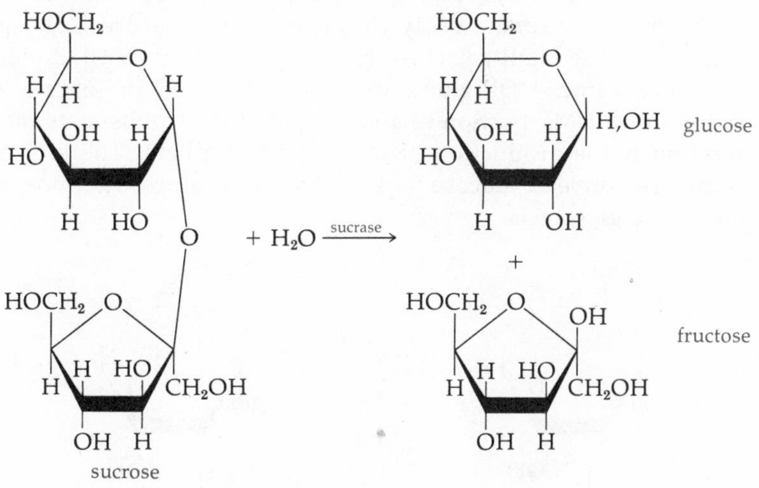

We have already discussed how the glucose portions of these two disaccharides are metabolized via glycolysis. What about galactose and fructose? Let us look at fructose first (because it is easier). Fructose can be fed into glycolysis via the fructose 1-phosphate pathway shown in Figure 25-19. Notice in this pathway that the fructose is first phosphorylated. The product of this reaction, fructose 1-phosphate, is then cleaved by a specific enzyme to give glyceraldehyde and dihydroxyacetone phosphate. Dihydroxyacetone phosphate can immediately enter the glycolytic pathway (step 5, Figure 25-8). The glyceraldehyde molecule is phosphorylated, so that it too can enter into glycolysis, instead of requiring another pathway.

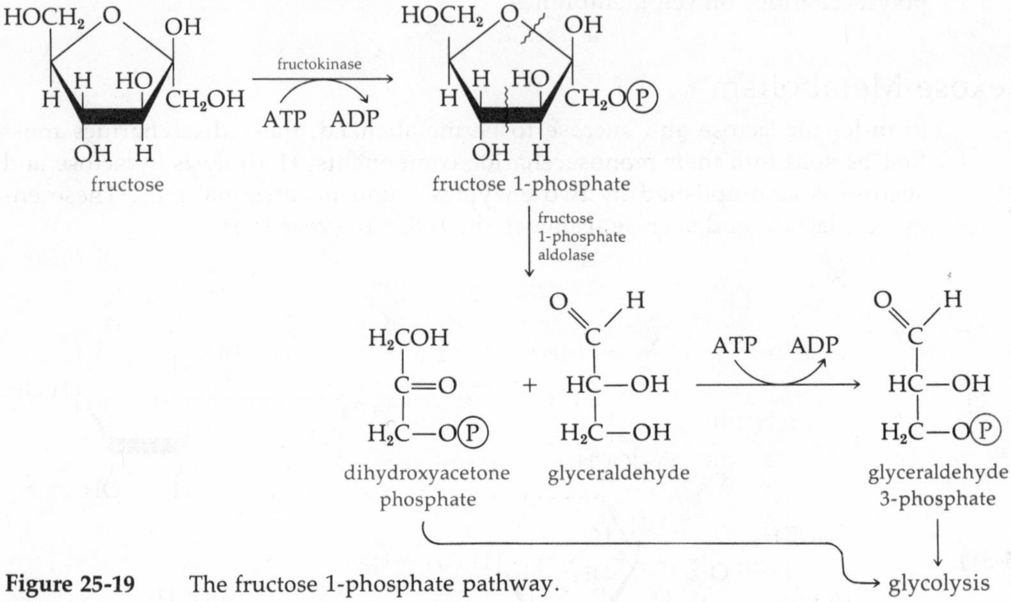

Figure 25-19   The fructose 1-phosphate pathway.

Galactose Metabolism

Galactose is converted to glucose in a series of reactions shown in Figure 25-20. Notice that galactose is first phosphorylated to galactose 1-phosphate. The second step of this pathway involves the transfer of a molecule of UMP from UDP-glucose to galactose 1-phosphate. UMP stands for uridine monophosphate, which is similar in structure to AMP but contains the base uracil instead of the base adenine (see Figure 24-2). UDP is likewise similar to ADP. (We shall see shortly where UDP-glucose comes from.) The products of the reaction catalyzed by hexose 1-phosphate uridylyltransferase are glucose 1-phosphate and UDP-galactose. (The UDP-galactose can be isomerized to UDP-glucose by an epimerase enzyme. UDP-glucose can then be used in several ways, including donating its UMP portion to another galactose 1-phosphate molecule.) The net reaction is the production of glucose 1-phosphate. The enzyme phosphoglucomutase converts glucose 1-phosphate into glucose 6-phosphate, which can enter into glycolysis.

(25-53)

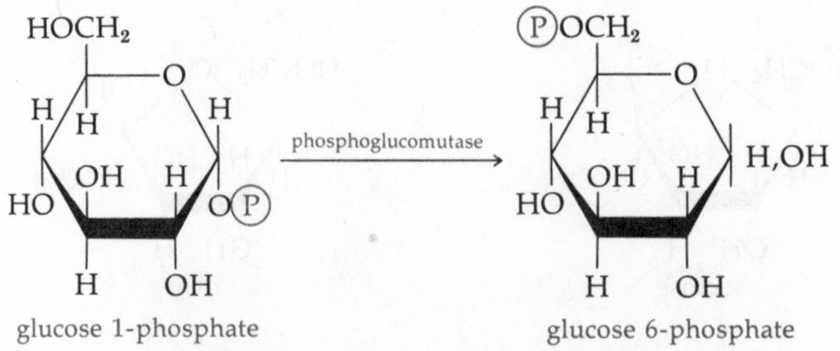

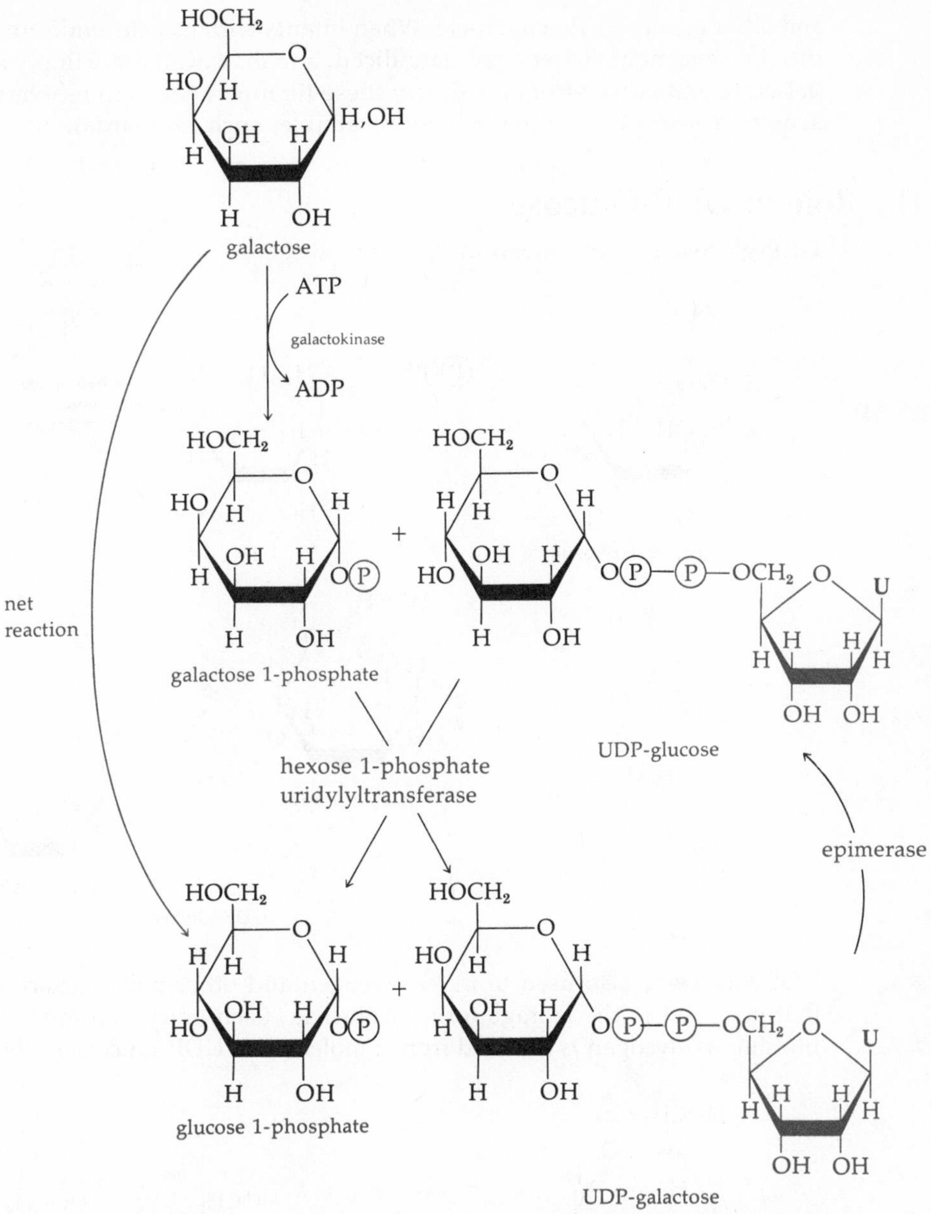

Figure 25-20   The conversion of galactose to glucose 1-phosphate. The UMP group that is transferred in the reaction catalyzed by hexose 1-phosphate uridylyltransferase is shown in blue.

A crucial step in the conversion of galactose to glucose is the transfer of UMP. The enzyme hexose 1-phosphate uridylyltransferase, which catalyzes the transfer of UMP, is not present in some people because of a hereditary defect. The lack of this enzyme results in a disease called **galactosemia.** Infants suffering from galactosemia cannot metabolize milk properly, and large amounts of galactose 1-phosphate build up since it cannot be metabolized further. Some of the excess galactose 1-phosphate is converted to galactitol,

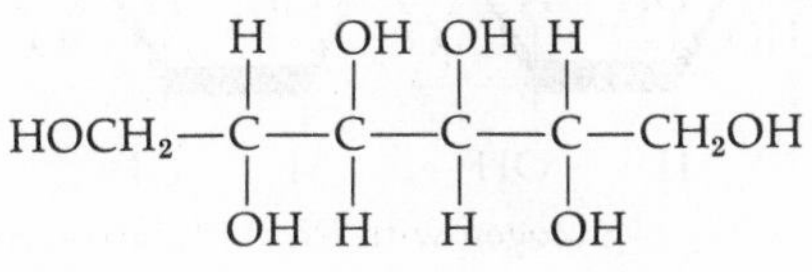

$$HOCH_2\!-\!\overset{\displaystyle H}{\underset{\displaystyle OH}{C}}\!-\!\overset{\displaystyle OH}{\underset{\displaystyle H}{C}}\!-\!\overset{\displaystyle OH}{\underset{\displaystyle H}{C}}\!-\!\overset{\displaystyle H}{\underset{\displaystyle OH}{C}}\!-\!CH_2OH$$

and other molecules that are toxic. When infants with galactosemia are fed milk, they become mentally retarded, jaundiced, and may even die. Simply removing galactose and lactose from the diet of these infants can lead to recovery from all symptoms except those that are nonreversible, such as retardation.

The Role of UDP-Glucose

UDP-glucose is synthesized in the reaction

(25-54)

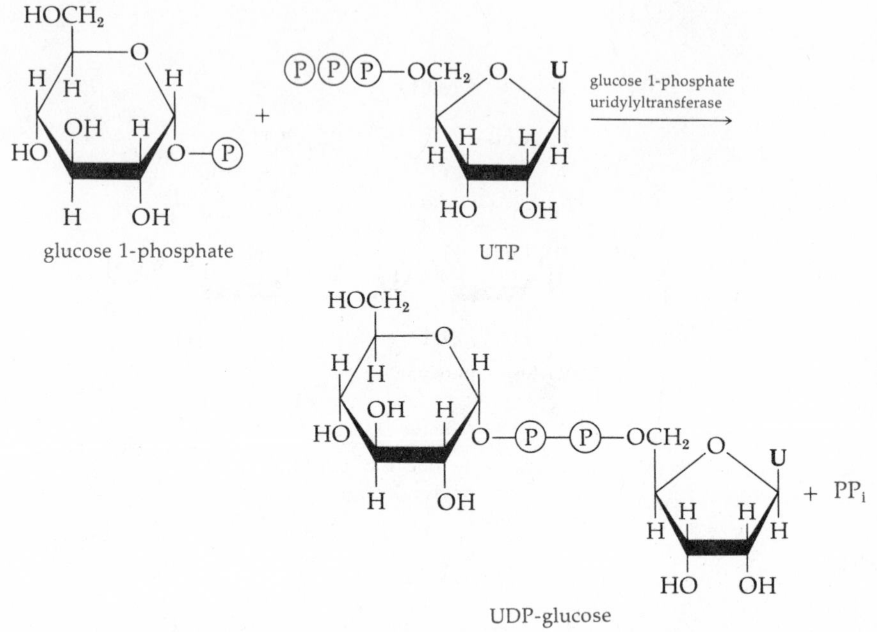

glucose 1-phosphate   UTP

glucose 1-phosphate
uridylyltransferase

UDP-glucose

+ PP$_i$

UDP-glucose is also used to make glycogen and other polysaccharides. Recall that glycogen is the storage form of glucose. Each glucose component in the buildup of glycogen is donated from a molecule of UDP-glucose in the reaction

(25-55)

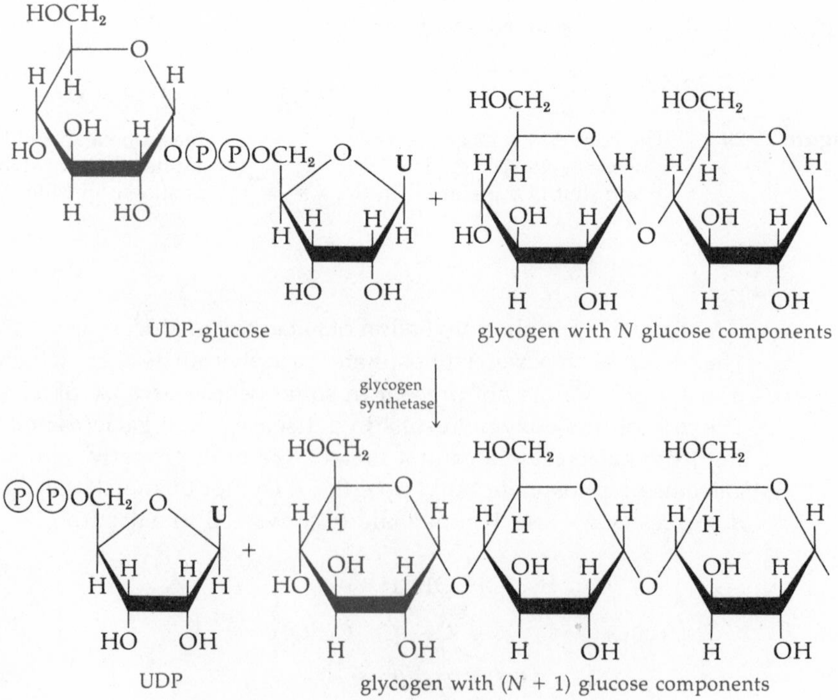

UDP-glucose   glycogen with N glucose components

glycogen
synthetase

UDP   glycogen with (N + 1) glucose components

The glycogen synthetase reaction (25-55), as we shall see shortly, is a key control point in the metabolism of carbohydrates.

UDP-glucose and other UDP sugars serve as the precursors for all disaccharides and polysaccharides. For example, the disaccharide lactose is made from UDP-galactose and glucose by the enzyme lactose synthetase. Lactose synthetase is found only in mammary glands and catalyzes the reaction

(25-56)   Glucose + UDP-galactose $\xrightarrow{\text{lactose synthetase}}$ lactose + UDP

Several other hexoses and their derivatives are also produced in the human body. Several of these, such as mannose, mannosamine, glucosamine, and N-acetylgalactosamine, are important components of glycoprotein prosthetic groups and blood group substances.

Exercise 25-10
How many UDP-glucose molecules are needed to convert 100 galactose 1-phosphate molecules to 100 glucose 1-phosphate molecules? Why?

Exercise 25-11
How many UDP-glucose molecules are required in order to synthesize a glycogen molecule containing 100 glucose components? Why?

25-11 REGULATION OF CARBOHYDRATE METABOLISM

Human cells, as we have seen, have several assembly lines for carbohydrate metabolism. Glucose and its UDP and phosphate derivatives serve as the common connections for these pathways. Thus cells can make glucose via gluconeogenesis, break it down via glycolysis, polymerize it into glycogen, or convert it to other hexoses (see Figure 25-21). Operation of all of these assembly lines at

Figure 25-21   Pathways for glucose metabolism.

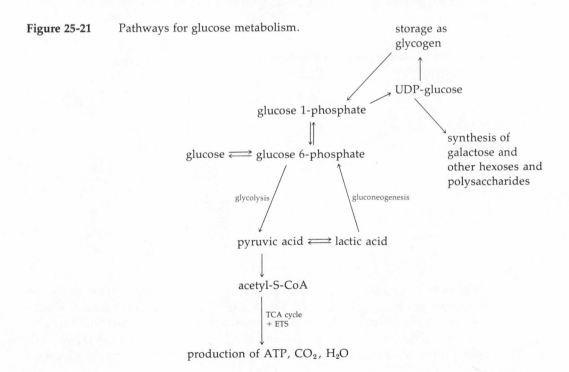

once would be counterproductive. For example, it is a waste of energy to run both the glycolytic and gluconeogenic pathways at the same time. Therefore cells have mechanisms to turn these assembly lines on and off as needed.

There is another important reason for carefully regulating carbohydrate metabolism. Certain cells in the human body, especially those in the brain, use the glycolytic pathway as their sole source of energy. These cells, whose operation is critically dependent on glucose, need a continuous and relatively constant supply of glucose.

A measure of the available glucose supply in the human body is the concentration of glucose in blood. The units usually used to express this concentration are milligrams of glucose/100 cc of blood, or mg/deciliter, or "milligrams %." The glucose concentration goes up after eating, but in normal healthy humans it quickly (within a few hours) returns to a level between 80 and 120 mg %. This range (80 to 120 mg %) is the normal "fasting" glucose level. (The normal range is somewhat dependent on the method used to analyze for glucose in blood. Subjects are required to fast 8 to 10 hours prior to testing.)

Too much or too little glucose in the blood is detrimental to cells, and can be fatal. **Hyperglycemia** is the term used for excessive amounts of glucose in the blood (above 120 mg %). **Hypoglycemia** refers to blood glucose concentrations less than 80 mg %. The glucose concentration in the blood is a sensitive measure of carbohydrate metabolism, and is used to diagnose and monitor certain diseases, most notably, diabetes.

Regulation of Glycogen Synthesis/Breakdown

You are already familiar with the types of mechanisms employed to control carbohydrate metabolism. They include allosteric and chemical modification of enzymes, and hormonal control. In referring back to Chapter 23 to review enzyme control mechanisms, you will notice that one of the processes discussed (Figure 23-14) was the control of glucose production by hydrolysis of glycogen. This process, together with the synthesis of glycogen from glucose, is a very important *control point* in the metabolism of carbohydrates because stored glycogen serves as a readily available source of glucose. Conversely, excess glucose can be stored as glycogen. Figure 25-22 shows a simplified view of this control point.

Figure 25-22   Reactions of glycogen synthesis and breakdown.

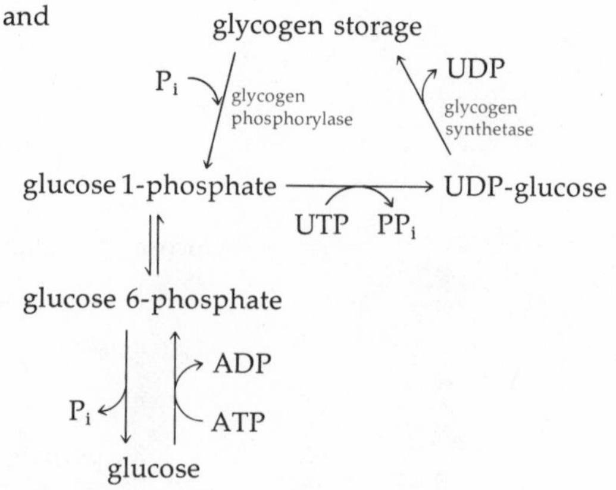

The entry of glucose into cells and its polymerization into glycogen, as well as hydrolysis of glycogen to glucose, are controlled in a fairly complicated way by hormones. The polypeptide hormone **insulin** increases polymerization of glucose into glycogen, thereby decreasing the blood glucose concentration. Both **epinephrine** (also called adrenaline) and the polypeptide hormone **glucagon**

increase hydrolysis of glycogen to glucose 1-phosphate, thereby increasing the blood glucose concentration. Epinephrine,

is a derivative of the amino acid tyrosine.

How do these hormones work? Epinephrine and glucagon directly influence the production of the compound **cyclic AMP** inside the cell. Cyclic AMP is frequently called "the second messenger" because it transmits hormonal messages to the proteins in the cell. Cyclic AMP is produced by the reaction

(25-56)

The enzyme adenyl cyclase is located in cell membranes, right next to receptor proteins that can bind the hormones epinephrine or glucagon (Figure 25-23). The hormone receptor proteins on some cell membranes bind epinephrine, whereas the receptors on other cell membranes bind glucagon. These hormones induce a change in the shape of their respective receptor proteins when they bind to them. This, in turn, causes a change in the shape of the adenyl cyclase enzyme, converting it to a much more active form (see Figure 25-23). The result, for either epinephrine or glucagon binding, is increased production of cyclic AMP.

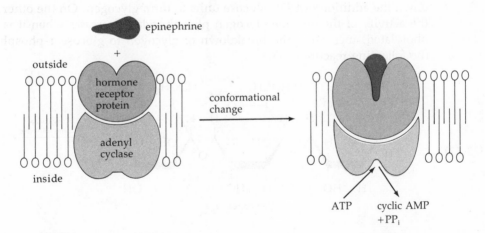

Figure 25-23   The binding of epinephrine to its receptor protein causes the allosteric activation of adenyl cyclase, which results in the increased production of cyclic AMP within the target cell. A number of other hormones function in a similar manner.

Cyclic AMP then binds to the allosteric site of an enzyme called a protein kinase, making it active. The active protein kinase is then capable of catalyzing reactions that result in the covalent bonding of phosphate groups to the enzymes glycogen synthetase and glycogen phosphorylase (see Figure 25-24). Attachment of phosphate groups to one of these enzymes, glycogen phosphorylase, involves an intermediate step that we need not consider.

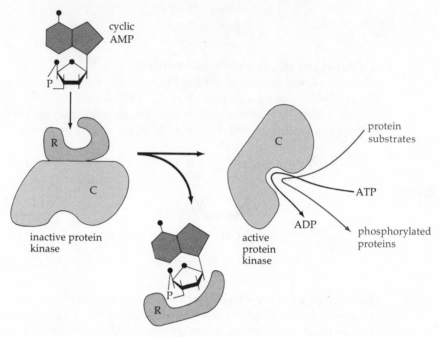

Figure 25-24   The enzyme protein kinase has two subunits, called the catalytic (C) and regulatory (R) subunits. When cyclic AMP binds to the regulatory subunit, R dissociates from C. The catalytic subunit then catalyzes the phosphorylation of certain proteins.

Glycogen synthetase catalyzes the synthesis of glycogen (reaction 25-55). Phosphorylation *lowers* the activity of glycogen synthetase, and thus slows down the addition of UDP-glucose units to form glycogen. On the other hand, the activity of the enzyme glycogen phosphorylase *increases* when it is phosphorylated, increasing the breakdown of glycogen to glucose 1-phosphate in the following reaction:

(25-57)

glycogen $+ P_i \xrightarrow{\text{glycogen phosphorylase}}$

glucose 1-phosphate $+$ remainder of glycogen molecule

Thus, the net result of epinephrine or glucagon hormonal stimulation is an increased supply of glucose, accomplished by the allosteric activation and chemical modification of enzymes, and it can be quite rapid (see Figure 25-25). When you are frightened, there is a rapid rise in your blood glucose level caused by epinephrine. This extra glucose is then available for use by your muscle cells in whatever response (to "flee or fight") you may choose to make to the frightening stimulus.

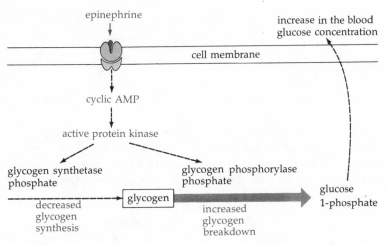

Figure 25-25   Steps leading to the increase in blood glucose concentration upon stimulation of a liver cell by the hormone epinephrine. The phosphorylated form of glycogen synthetase is less active than its unphosphorylated form, whereas phosphorylated glycogen phosphorylase is more active than its unphosphorylated form.

How does insulin influence the control of the concentration of glucose? Insulin has the opposite net effect from epinephrine and glucagon when it binds to its receptor protein on cell membranes. That is, it turns off glycogen phosphorylase, thus decreasing the breakdown of glycogen, while it turns on glycogen synthetase, leading to polymerization of glucose into glycogen. The precise molecular action of insulin is not yet understood. Insulin binding is known to increase glucose uptake by cells, thereby lowering the glucose concentration in the blood. In addition, glucose itself is known to bind to phosphorylated glycogen phosphorylase and promote its dephosphorylation by the enzyme phosphoprotein phosphatase, making it inactive. Phosphoprotein phosphatase also catalyzes the removal of the phosphate group from the phosphorylated form of glycogen synthetase, returning it to its more active, nonphosphorylated form. Research is actively being conducted in an attempt to arrive at a complete picture of the action of insulin.

Regulation of Glycolysis/Gluconeogenesis

The remaining control points in carbohydrate metabolism are less complicated than the control of glycogen metabolism. Let us look at the major control point of glycolysis and gluconeogenesis. Recall that in glycolysis, glucose is degraded to pyruvic acid and ATP is produced, whereas in gluconeogenesis, energy in the form of ATP is used to synthesize glucose.

You may remember that there are a few reactions in glycolysis that are irreversible (see Figure 25-18). In order for gluconeogenesis to occur, these steps must be bypassed. One of these irreversible reactions (step 3 of glycolysis) is fructose 6-phosphate $\xrightarrow[\text{ATP \quad ADP}]{\text{phosphofructokinase}}$ fructose 1,6-diphosphate

In gluconeogenesis, this irreversible step is bypassed by reaction 25-48:

fructose 1,6-diphosphate $+ H_2O \xrightarrow{\text{fructose 1,6-diphosphatase}}$ fructose 6-phosphate $+ P_i$

This pair of reactions is the major control point for glycolysis/gluconeogenesis. Phosphofructokinase and fructose 1,6-diphosphatase are allosteric enzymes that can be activated or inhibited by appropriate molecules (see Figure 25-26). ATP is an allosteric inhibitor of phosphofructokinase, as are citric

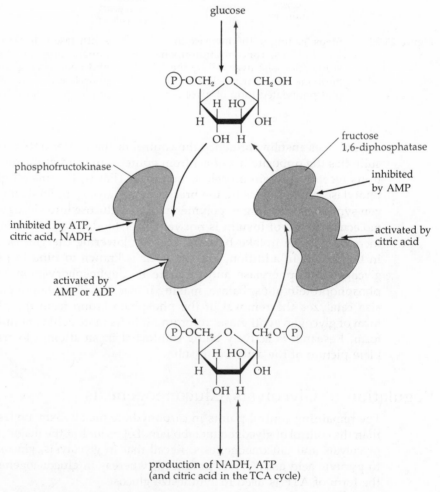

Figure 25-26   The allosteric control of glycolysis (black arrows) and gluconeogenesis (blue arrows).

acid and NADH. On the other hand, citric acid enhances the activity of fructose 1,6-diphosphatase. Conversely, AMP allosterically inhibits the phosphatase enzyme, whereas AMP and ADP enhance the activity of phosphofructokinase. These controls are logical. If the cell already has a lot of ATP, citric acid, and NADH, it does not need to produce any more. But if the cell has used up all of its ATP, then it will have a lot of AMP and ADP, which will signal the need for an increase in the rate of glycolysis and a shutdown of energy-using gluconeogenesis (see Figure 25-27)

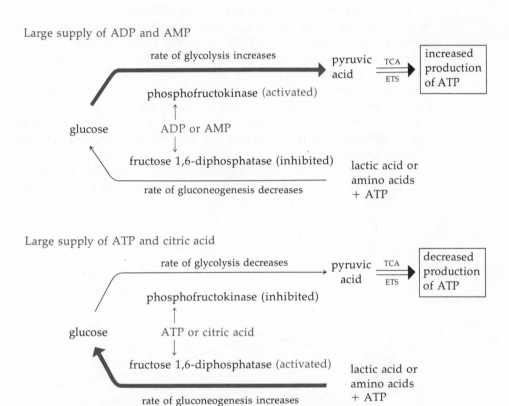

Figure 25-27   A high concentration of AMP or ADP signals the need for increased ATP production, thereby activating glycolysis and inhibiting gluco- neogenesis (top). A high concentration of ATP or critic acid, on the other hand, promotes gluco- neogensis (bottom).

There are several other control points in carbohydrate metabolism, but the two we have looked at appear to be the most important. Also, these two control points serve to illustrate the logical application of control mechanisms involved in cellular metabolism.

Exercise 25-12
When you are frightened, cells in your adrenal medulla produce adrenaline. What effect does this have on blood glucose levels? Does this help your body? How?

Exercise 25-13
Why does a large amount of ADP and AMP in a cell indicate that the cell has a decreased supply of ATP?

25-12 DIABETES

Diabetes mellitus is the disease that results from insufficient production of insulin and is commonly referred to simply as diabetes. The term comes from the Greek and means "sweet-tasting urine." Since an important symptom of this disease is the presence of glucose in the patient's urine, the taste of the urine was formerly used to diagnose the disease. Modern doctors and nurses are fortunate that more quantitative methods have been developed to determine the amount of glucose in the urine. We have seen that the polypeptide hormone insulin facilitates the transport of glucose into cells from the blood and promotes its polymerization into glycogen. Insulin is made by certain cells (called β-cells) in the pancreas, and is needed to keep the concentration of blood glucose within the critical range of 80 to 120 mg %. Diabetics make little or no insulin (or the insulin they do make is not active). Thus diabetics have difficulty keeping the blood glucose concentration under 120 mg %. Patients with mild cases of diabetes may produce insulin, but not always enough to do the job.

Currently the best diagnostic indication of mild or severe diabetes is the **glucose tolerance test** (GTT). In this test, after 8 to 10 hours of fasting, a person is given a doctored soft drink, for example, a sickeningly sweet syrup called glucola, which contains 100 g of glucose. This person's blood glucose level is then measured over a period of 2, 5, or more hours (see Figure 25-28).

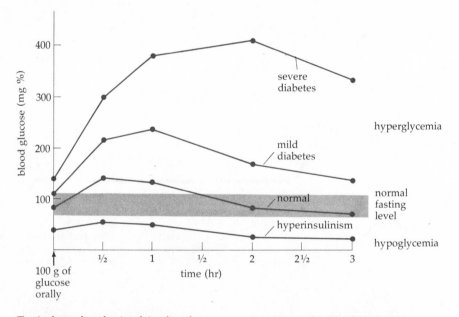

Figure 25-28   Typical results obtained in the glucose tolerance test. Each individual drinks 100 g of glucose at time zero, and the blood glucose concentration is then determined at given time intervals. The blood glucose concentration in normal fasting subjects is indicated by the blue shaded area.

Since glucose is rapidly absorbed into the bloodstream, a normal person will show a rapid increase in blood glucose concentration, peaking at about 150 mg % within 1 hr. This is followed by a rapid decrease in blood glucose concentration as insulin is produced. The blood glucose level of a healthy individual will return to the normal fasting range within 2 hr. People with mild diabetes will need a few hours longer to reduce the glucose concentration. The blood glucose level of severe diabetics may reach very high levels (see Figure

The incidence of diabetes is increasing in our population. Current therapy for severe diabetes involves daily injections of insulin obtained from animals—which is not exactly identical to human insulin. Researchers in California, however, have recently succeeded in splicing the human insulin gene into a bacterial cell, thus opening the way to large-scale production of human insulin by bacterial "slaves."

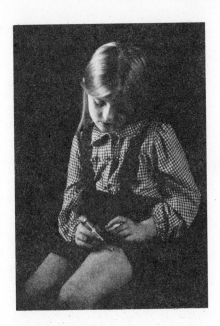

25-28) after ingestion of 100 g of glucose, and the blood glucose level will remain elevated for an extended period. When the blood glucose level reaches values of more than about 180 mg %, a decidedly hyperglycemic condition, some glucose is excreted by the kidneys and appears in the urine. Determination of glucose in urine is sometimes used to screen for diabetes. Testing for glucose in urine is easily done at home, and is routinely done by diabetics to monitor their condition.

Mild cases of diabetes can be controlled by diets that include only small amounts of carbohydrates. Severe diabetics require injections of insulin. There are more than 1 million insulin-dependent diabetics in the United States alone. Commercial insulin is usually obtained from the pancreas of animals. Synthetic insulin, or natural human insulin produced by genetic engineering, may soon be available. Uncontrolled diabetes can lead to a decrease in the pH of the blood, a condition called acidosis (Chapter 29), and can result in coma or even death. One major problem that contributes to these symptoms is the formation of large amounts of "ketone bodies" as a result of defective lipid metabolism (see Chapter 27).

Another clinical condition that results from the improper control of carbohydrate metabolism is hypoglycemia. In certain individuals, an increase in blood glucose level stimulates the pancreas to produce too much insulin. A condition in which the insulin level is above normal is called **hyperinsulinism.** What happens to an individual when the insulin level is too high? In this person, too much glucose is extracted from the blood, and the blood glucose level drops below normal, a hypoglycemic condition. Hypoglycemia is an extremely dangerous condition. When the blood glucose level drops below normal, brain cells quickly starve, which can lead to coma and even death. Hypoglycemic individuals can control their condition by a low-carbohydrate diet. Hypoglycemia can be detected by the glucose tolerance test. Diabetics who inject insulin must also be very careful not to inject too much insulin, which can lead to a hypoglycemic coma.

Exercise 25-14
Can you explain why diabetics frequently carry sugar cubes in addition to insulin?

25-13 SUMMARY

1. Carbohydrates are the primary source of usable energy for humans.

2. Catabolic pathways break down food molecules in a stepwise fashion, with energy-coupled steps leading to ATP synthesis.

3. The hydrolysis of ATP serves as the immediate source of energy used to drive most endergonic biochemical reactions.

4. In the glycolytic pathway, glucose is hydrolyzed to pyruvic acid, coupled to the net synthesis of 2 moles of ATP from ADP and 2 moles of reduced coenzyme NADH per mole of glucose.

5. Pyruvic acid can be reduced to lactic acid under anaerobic conditions, but under aerobic conditions it can be used to generate more ATP and reduced coenzymes via the TCA cycle.

6. The TCA cycle is fueled by acetyl-S-CoA and generates ATP and reduced coenzymes. CO_2 is the exhaust product of the cycle.

7. Reduced coenzymes from glycolysis, the pyruvate dehydrogenase reaction, and the TCA cycle are reoxidized and coupled to ATP synthesis in the electron transport system, with reduction of O_2 to H_2O.

8. In mitochondria, oxidation of 1 mole of $NADH + H^+$ generates 3 moles of ATP via the electron transport system, whereas oxidation of 1 mole of $FADH_2$ yields only 2 moles of ATP.

9. In the overall aerobic catabolism of glucose, about 42% of the liberated free energy is captured in ATP formation. The remainder is lost as heat. A net of 36 moles of ATP are formed in the complete aerobic catabolism of 1 mole of glucose in a muscle cell.

10. **(Optional)** The pentose phosphate pathway is an alternative pathway for glucose oxidation. It is a source of ribose 5-phosphate and the coenzyme NADPH.

11. The pathway for gluconeogenesis is not simply a reversal of glycolysis, but requires four different steps to circumvent the three irreversible steps of glycolysis.

12. Glucose can be converted to other hexoses, or incorporated into glycogen, via the formation of UDP-glucose.

13. Galactosemia is a disease resulting from the absence of the enzyme hexose 1-phosphate uridylyltransferase, which is required for the conversion of galactose to glucose.

14. The concentration of glucose in the blood is normally kept in the critical range of 80 to 120 mg %. This regulation is accomplished primarily by the complex hormonal control of glycogen formation and glycogen breakdown.

15. The key intermediate in controlling glycogen metabolism is cyclic AMP, which acts by allosteric activation of the enzyme protein kinase.

16. Glycolysis and gluconeogenesis are controlled by the allosteric activation and inhibition of the enzymes phosphofructokinase and fructose 1,6-diphosphatase.

17. Diabetes mellitus is a disease that results from the inability to synthesize sufficient amounts of active insulin. People with uncontrolled diabetes have high blood glucose levels, whereas those with hypoglycemia have inadequate blood glucose levels. Both of these conditions can result in death, but both can be detected by the glucose tolerance test.

PROBLEMS

1. Only two steps in glycolysis are sufficiently exergonic to drive the formation of ATP. Which steps are these?

2. Calculate the moles of ATP produced in muscle cells by the oxidation of the following:
   (a) 1 mole of glyceraldehyde 3-phosphate under aerobic conditions
   (b) 2 moles of fructose 1,6-diphosphate under anaerobic conditions
   (c) 1 mole of pyruvic acid under aerobic conditions
   (d) 1 mole of succinic acid under aerobic conditions

3. A glucose molecule contains six carbon atoms, but each acetyl group entering the TCA cycle contains only two. Account for the fate of all six carbon atoms of glucose under aerobic conditions.

4. Describe the similarities and differences between the pyruvate dehydrogenase reaction, the α-ketoglutarate dehydrogenase reaction, and the glyceraldehyde 3-phosphate dehydrogenase reaction.

5. Write a balanced equation for the aerobic oxidation of 1 mole of lactic acid in a liver cell. How many moles of ATP are produced in this process?

6. The fourth step of the TCA cycle involves the oxidation of a hydroxyl group on isocitric acid to form a carbonyl group. Explain why citric acid cannot serve as a substrate for such a reaction.

7. What effect does the ingestion of cyanide have on the TCA cycle?

8. NADH and $FADH_2$ donate electrons and protons to the ETS, but not with equal results. (a) Why are only two ATP molecules formed for each $FADH_2$ molecule in the ETS? (b) Is there a difference between these two coenzymes in the amount of water formed in the ETS (with oxidative phosphorylation)?

9. Explain the difference between the amount of ATP produced by the complete aerobic oxidation of 1 mole of glucose in muscle cells versus liver cells.

10. Draw structural formulas for the products of the following reactions:

(a)

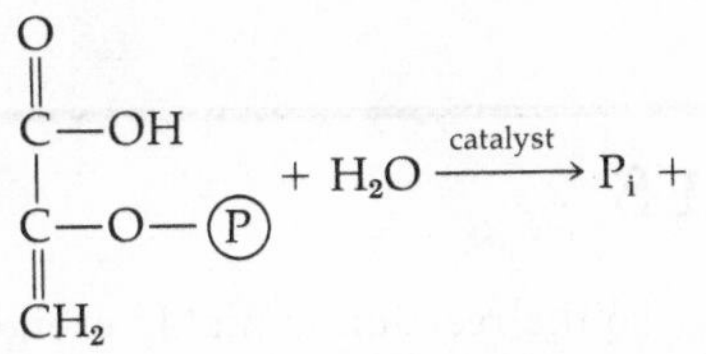

$$\begin{array}{c} O \\ \parallel \\ C-OH \\ | \\ C-O-\textcircled{P} \\ \parallel \\ CH_2 \end{array} \quad + H_2O \xrightarrow{\text{catalyst}} P_i \; +$$

(b)

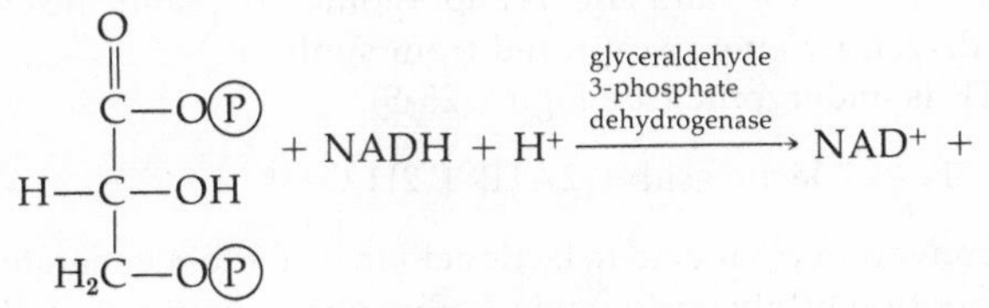

$$\begin{array}{c} O \\ \parallel \\ C-O\textcircled{P} \\ | \\ H-C-OH \\ | \\ H_2C-O\textcircled{P} \end{array} \quad + NADH + H^+ \xrightarrow[\text{dehydrogenase}]{\substack{\text{glyceraldehyde} \\ \text{3-phosphate}}} NAD^+ \; +$$

(c)

$$\begin{array}{c} O \\ \parallel \\ C-OH \\ | \\ C-O-\textcircled{P} \\ \parallel \\ CH_2 \end{array} \quad + H_2O \xrightarrow{\text{enolase}}$$

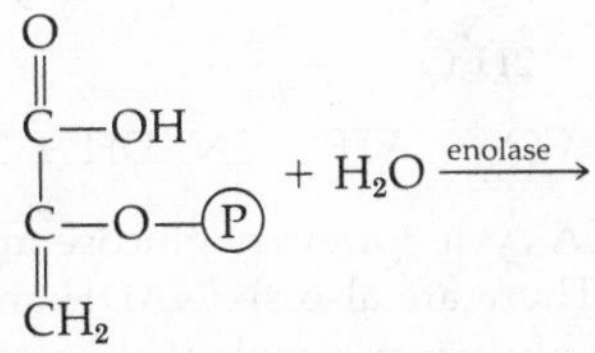

11. The total oxidation of 1 mole of glucose releases 686 kcal. Calculate the amount released per gram of glucose and compare this value to the number of kilocalories per gram of glucose that are captured in ATP when glucose is oxidized in muscle cells. (Use a value of 8.0 kcal/mole for ATP formation.)

12. **(Optional)** Is the operation of the pentose phosphate pathway required for the production of ribosomes? Explain your answer.

13. List any common intermediates in gluconeogenesis and the TCA cycle.

14. Do galactosemic children outgrow their disease? Explain.

15. List two metabolic pathways that involve UDP-galactose and two metabolic pathways that involve UDP-glucose.

16. Outline the reactions involved in the regulation of glycogen metabolism by epinephrine and show which reactions are controlled by allosteric modifiers and which are controlled by chemical modification.

17. In order to add one glucose unit to glycogen, how much energy (in ATP → ADP + P_i units) must be used (a) starting with glucose, and (b) starting with two molecules of lactic acid?

18. "Sugarholics," such as people who constantly drink soda pop and/or eat candy, should not stop their habit "cold turkey." Why?

19. Can you explain why many people feel very tired within a few hours after taking a morning break for coffee and donuts?

20. Free and phosphorylated "forms" of glucose are involved in several metabolic pathways. Outline the reactions that convert these different "forms" of glucose one to another.

21. Draw structural formulas for the products of the following reactions (assume that necessary coenzymes, ATP, etc., are present).

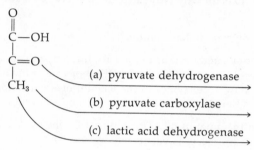

$$
\begin{array}{l}
\text{O} \\
\parallel \\
\text{C—OH} \\
\mid \\
\text{C=O} \\
\mid \\
\text{CH}_3
\end{array}
$$

(a) pyruvate dehydrogenase

(b) pyruvate carboxylase

(c) lactic acid dehydrogenase

SOLUTIONS TO EXERCISES

25-1 (a) ΔG for this reaction is -12 kcal/mole, so the reaction is highly exergonic.
(b) Synthesis of glucose from CO_2 and H_2O is endergonic. (In plants the endergonic synthesis of glucose is driven by energy captured from sunlight.)
(c) Formation of ATP is endergonic (see Figure 25-5).

25-2 Glucose + 2ADP + 2P_i → 2 lactic acid + 2ATP + 2H_2O

25-3 Active muscle cells convert pyruvic acid to lactic acid in order to regenerate NAD^+, which is needed for the oxidation of glyceraldehyde 3-phosphate in step 6 of glycolysis.

25-4 Acetyl-S-CoA + ADP + P_i + 3NAD^+ + FAD + 2H_2O $\xrightarrow{\text{TCA cycle}}$
CoA-SH + 2CO_2 + ATP + 3NADH + 3H^+ + $FADH_2$

25-5 Two ATP molecules are synthesized in the TCA cycle for every glucose molecule that is broken down to two pyruvic acid molecules. There are also six NADH and two $FADH_2$ molecules generated in the TCA cycle for each glucose molecule that entered glycolysis.

25-6 2Fe^{2+} + 2H^+ + $\frac{1}{2}O_2$ → 2Fe^{3+} + H_2O

25-7 Only three steps are sufficiently exergonic to be coupled to ATP synthesis.

25-8 Glycolysis. To provide the energy for muscle contractions during strenuous exercise, cells in the body need to generate ATP, not NADPH or ribose.

25-9   Catabolism of 1 mole of glucose to CO_2 and H_2O in muscle yields 36 moles of ATP. Formation of 1 mole of glucose from lactic acid requires the hydrolysis of 6 moles of ATP. Thus the investment of 6 moles of ATP ensures the production of 36 moles of ATP at a later time.

25-10 Only one. The UMP group is donated to form UDP-galactose, which is then converted to UDP-glucose, which can donate its UMP group to the next galactose 1-phosphate molecule, and so on.

25-11 A hundred. Addition of each glucose unit to the growing glycogen chain requires the hydrolysis of one UDP. The UDP group is recycled.

25-12 Adrenaline (epinephrine) stimulates the breakdown of glycogen, resulting in a higher blood glucose level, which is then available for ATP production via glycolysis. The extra ATP may be needed to flee or fight.

25-13 The hydrolysis of ATP is the major exergonic reaction used to drive endergonic biochemical processes. Hydrolysis of ATP yields AMP + PP_i or ADP + P_i. Thus increased concentrations of ADP, AMP, or P_i indicate that a lot of ATP has been used and that more is needed.

25-14 If a diabetic accidentally injects too much insulin, he or she will begin to feel faint and may lapse into a hypoglycemic coma unless some glucose is quickly administered. Sugar cubes are a convenient source of glucose.

CHAPTER 26

Lipid Structure and Function

26-1 INTRODUCTION

Have you ever heard of "fat chemists"? They are not necessarily obese. Fat chemists are people who study fat molecules. Fat molecules aren't necessarily obese either. Fat molecules or, more generally, lipids, are biomolecules that are insoluble in water and other polar liquids, but soluble in nonpolar organic solvents such as benzene, chloroform, and ether. Many foods, including milk, cream sauces, and salad dressings, involve mixtures of lipids suspended in an aqueous medium. To remove these foods from dishes and cooking utensils, we use detergents to form emulsions of the lipids in water.

Several types of lipid molecules have important functions in the human body. Some, such as the phospholipids, are components of cell membranes. Others, called steroids, can serve as hormones. Still other lipids are vitamins. Some lipids are also a very energy-rich source of nourishment for humans. We shall see that catabolism of some simple lipids, called fatty acids, can be coupled to the production of large amounts of ATP. Because of this, the human body stores fatty acids and other lipids in adipose (fatty) tissue as a reserve supply of energy. Some humans have excessive reserves of lipids—they are obese. Obesity and other disorders, such as atherosclerosis, can result from improper nutrition and/or genetic defects in the metabolism of lipids.

Many lipids play very interesting roles in the human body, but it is not completely known how they function. For example, estrogen and testosterone, the major female and male sex hormones in humans, belong to a class of lipids called steroids. They carry very specific chemical messages to their target cells, but we are still not sure precisely how they work. In living cells, lipid molecules are often combined with other kinds of biomolecules. We saw in Chapter 22, for example, that they can combine with proteins to form lipoproteins and proteolipids. Lipids can also combine with carbohydrates to form molecules called glycolipids. These hybrid molecules also have unique functions, which in many cases are not very well understood.

In this chapter we shall study the structures and properties of lipid molecules. We shall also see how different classes of lipid molecules function in the human body. In the next chapter we shall study the metabolism of lipid molecules.

This experimental chemist is investigating new methods for producing synthetic rubber. Since rubber is a lipid, one could call such a scientist a "fat chemist."

26-2 STUDY OBJECTIVES

After careful study of the material in this chapter, you should be able to:

1. Define the terms fatty acid and essential fatty acid.

2. Describe the difference between a saturated fatty acid and an unsaturated fatty acid.

3. Describe the basic components of the 10 lipid classes.

4. Describe the major biological functions of the 10 classes of lipids.

5. Draw the structural formulas for glycerol and glycerol 3-phosphate.

6. Draw the structural formula for a monoglyceride, a diglyceride, or a triglyceride, given the structural formulas of its component parts.

7. Define the term saponification, and predict the products formed when a glyceride is saponified.

8. Describe a micelle and an emulsion, and show how a soap works.

9. Identify the hydrophilic and hydrophobic parts of a lipid, given its structural formula.

10. Draw a structural formula for an isoprene unit and the characteristic four-ring structure found in steroids.

11. Describe the relationship between fat-soluble vitamins and terpenes.

12. Describe a lipid bilayer and the mosaic model of cell membranes.

26-3 CLASSIFICATION OF LIPIDS

Lipids are a large group of naturally occurring organic molecules that are insoluble in water but soluble in nonpolar solvents. This is a very broad definition, based only on the relatively nonpolar nature of these molecules. Therefore you should not be surprised to discover that there are a variety of lipid molecules, each with a unique type of structure.

We have stressed the very close relationship between the structure of a biomolecule and its function in the human body. Lipids are divided into 10 classes, based on their structural features, and different classes of lipids have different functions in the cells of the human body. It is useful to consider the various roles of lipids in cells, using our analogy of cells as factories. For example, one class of lipid, the phosphoglycerides, are present in cell membranes and play a role comparable to the walls of a factory. We shall see that the structures of phosphoglycerides allow them to serve as excellent walls for cellular factories, separating the aqueous cytoplasm of the cell from its aqueous external surroundings. Some lipids serve as insulation against heat or cold for cellular factories, and for the entire body as well. Other lipids provide a source of energy to run cellular assembly lines.

The basic features of the 10 major classes of lipids are outlined in Table 26-1. Fatty acids comprise the first class. We shall define a **fatty acid** as a carboxylic acid with a long chain of carbon atoms (generally 14 to 22) extending from the carboxyl group. Fatty acids are components of glycerides, phosphoglycerides, sphingolipids, and waxes. Prostaglandins are also derivatives of fatty acids with a 20-carbon atom chain. Terpenes and steroids are classes of lipids that do not contain fatty acids.

Table 26-1 Classes of Lipids

Lipid Class	Fundamental Components
1. Fatty acids	Fatty acids
2. Glycerides	Fatty acids + glycerol
3. Phosphoglycerides	Fatty acids + glycerol + phosphoric acid
4. Sphingolipids	Fatty acids + sphingosine
5. Waxes	Fatty acids + long-chain alcohols
6. Prostaglandins	Derivatives of certain fatty acids
7. Terpenes	Contain isoprene units
8. Steroids	Characteristic four-ring structural unit
Hybrid Lipids	
9. Glycolipids	Lipids + carbohydrates
10. Lipoproteins	Lipids + proteins

Exercise 26-1

Is acetic acid, $CH_3—\overset{\overset{\displaystyle O}{\|}}{C}—OH$, a fatty acid? Is it a lipid? Explain your answer.

26-4 FATTY ACIDS

A fatty acid contains a long chain of carbon atoms. Most of the fatty acids found in the human body consist of a total of 14 to 22 carbon atoms. Most of them contain an even number of carbon atoms. The reason for an even number of carbon atoms will become evident in the next chapter, when we study the synthesis of fatty acids. Table 26-2 shows the condensed structural formulas for several fatty acids. The long carbon chains of these fatty acids may be **saturated** (contain only C—C single bonds), or they may be **unsaturated** (contain C═C double bonds).

Unsaturated fatty acids have lower melting points than saturated fatty acids of the same chain length. For example, compare the melting points of the two 16-carbon fatty acids in Table 26-2. Saturated palmitic acid melts at 63°C, but the double bond in palmitoleic acid lowers its melting point to −0.5°C.

Table 26-2 Some Common Fatty Acids

Name	Number of C Atoms	Condensed Structural Formula	Melting Point (°C)
Saturated			
Lauric	12	$CH_3(CH_2)_{10}COOH$	44
Myristic	14	$CH_3(CH_2)_{12}COOH$	54
Palmitic	16	$CH_3(CH_2)_{14}COOH$	63
Stearic	18	$CH_3(CH_2)_{16}COOH$	70
Arachidic	20	$CH_3(CH_2)_{18}COOH$	77
Lignoceric	22	$CH_3(CH_2)_{20}COOH$	86
Unsaturated			
Palmitoleic	16	$CH_3(CH_2)_5CH{=}CH(CH_2)_7COOH$	−0.5
Oleic	18	$CH_3(CH_2)_7CH{=}CH(CH_2)_7COOH$	13
Linoleic	18	$CH_3(CH_2)_4CH{=}CHCH_2CH{=}CH(CH_2)_7COOH$	−5
Linolenic	18	$CH_3CH_2CH{=}CHCH_2CH{=}CHCH_2CH{=}CH(CH_2)_7COOH$	−11
Arachidonic	20	$CH_3(CH_2)_4(CH{=}CHCH_2)_3CH{=}CH(CH_2)_3COOH$	−50

We shall see that cell membranes contain a mixture of saturated and unsaturated fatty acids and exhibit properties in between those of a liquid and a solid. Humans cannot synthesize some of the required unsaturated fatty acids containing more than one double bond. We require rather large amounts of linoleic and linolenic acids (Table 26-2) for use in making glycerides, phosphoglycerides, and prostaglandins, and we must get these fatty acids in our diet. They are quite plentiful in plants, which serve as our source for these **essential fatty acids** (linoleic acid and linolenic acid). In countries such as the United States, the typical human diet contains a larger ratio of saturated to unsaturated fatty acids than it does in underdeveloped countries. This high dietary ratio favoring saturated fatty acids also favors the development of atherosclerosis, for reasons not well understood (see Section 26-6).

The long hydrocarbon chain of either saturated or unsaturated fatty acids is hydrophobic (Chapter 14), whereas the polar carboxyl group is hydrophilic (Chapter 17). Thus a fatty acid contains both hydrophobic and hydrophilic parts.

What happens when a fatty acid, say, oleic acid, is mixed with water? A solution does not form, since the lipid oleic acid is not soluble in water. However, there is an attraction between the hydrophilic carboxyl groups on the oleic acid molecules and water molecules. Droplets of oleic acid form. Each of these droplets consists of a large number of oleic acid molecules, with their hydrophobic carbon chains close together and pointed toward the center of the droplet, and their hydrophilic carboxyl groups on the exterior of the droplet (see Figure 26-1). Droplets of this type are called **micelles.**

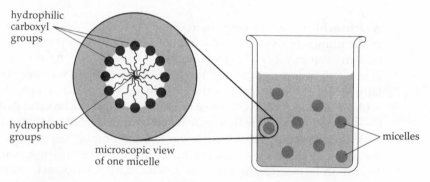

Figure 26-1 An emulsion of fatty acids in water. Emulsions are made up of microscopic droplets, called micelles, each consisting of many fatty acid molecules oriented so that their hydrophobic portions are tightly packed at the center of the droplet and their hydrophilic parts are exposed to the aqueous environment.

A mixture of micelles in water is called an **emulsion.** It is also possible to layer oleic acid and other lipids on top of water, like an oil slick. In this case (see Figure 26-2), the hydrophobic "tails" of the oleic acid molecules stick up into the air, whereas the hydrophilic carboxyl groups stick into the water. We shall see

Figure 26-2 A layer of oleic acid molecules on a water surface. Note that the hydrophilic "heads" of the fatty acid molecules interact with the water, whereas the hydrophobic "tails" stick up into the air.

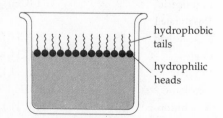

that such a layer of oleic acid on water is actually very similar in structure and properties to one-half of a cell membrane. For centuries people have made extensive use of the K^+ and Na^+ salts of fatty acids as soaps (see Figure 26-3). You should be able to explain how a fatty acid salt can function as a soap and emulsify hydrophobic greases and oils.

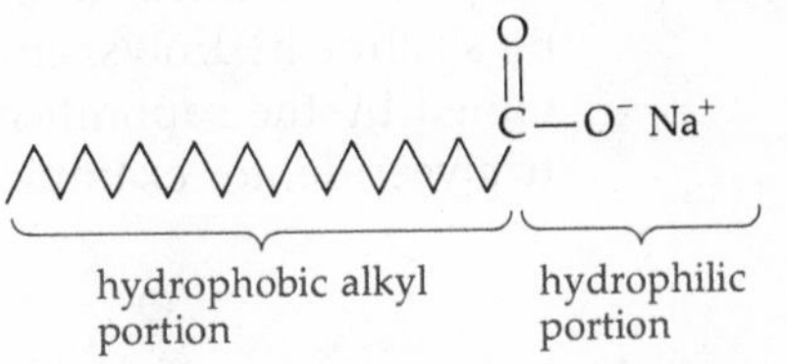

Figure 26-3   A soap molecule contains hydrophilic and hydrophobic parts.

hydrophobic alkyl portion   hydrophilic portion

Exercise 26-2
What is the difference between a 20-carbon saturated fatty acid and a 20-carbon unsaturated fatty acid? Which has the lower melting point?

26-5 LIPIDS DERIVED FROM FATTY ACIDS

In the human body, fatty acids are not generally found as such, but almost always as parts of esters. Recall from Chapter 17 that an ester functional group is formed by the reaction of a carboxyl group on one molecule and an alcohol functional group on another molecule.

Glycerides

Glycerides, esters formed from fatty acids and glycerol, are one of the most common types of esters formed by fatty acids (see Figure 26-4). Since a molecule of glycerol has three alcohol groups, it is possible for one glycerol molecule to form one, two, or three fatty acid esters, called monoglycerides, diglycerides, and triglycerides, respectively.

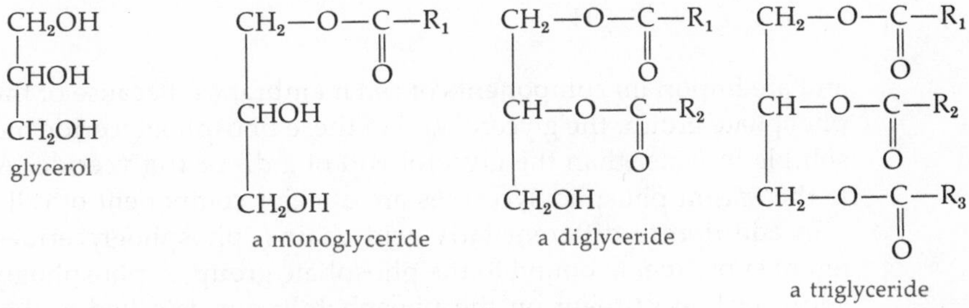

Figure 26-4   The structural formula for glycerol and general structural formulas for glycerides.

Obviously, the exact structure of a glyceride and its properties will depend on which fatty acids are in fact esterified to the glycerol molecule. Triglycerides are used as a means of food storage in animals. Much more reserve energy can be stored in triglyceride molecules than in glycogen. Later, we shall see that 1 g of triglycerides is capable of yielding more than double the energy of 1 g of carbohydrates. It is more convenient to store energy in the form of fat, since it requires less body weight to store a given amount of energy. Our triglyceride reserves, called **fat depots,** increase when we eat more food than is needed to supply us with the energy we use. We are all aware that it is very difficult to get rid of such fat depots once they have accumulated.

Since free fatty acids are not generally found in nature, you may wonder how soaps are obtained. In Chapter 17 we mentioned that ester bonds can be broken by alkaline hydrolysis in a process called **saponification.** Soaps are generally obtained by the saponification of triglycerides in animal fat. Saponification of a triglyceride, for example, will give a mixture of soaps plus a glycerol molecule:

(26-1)

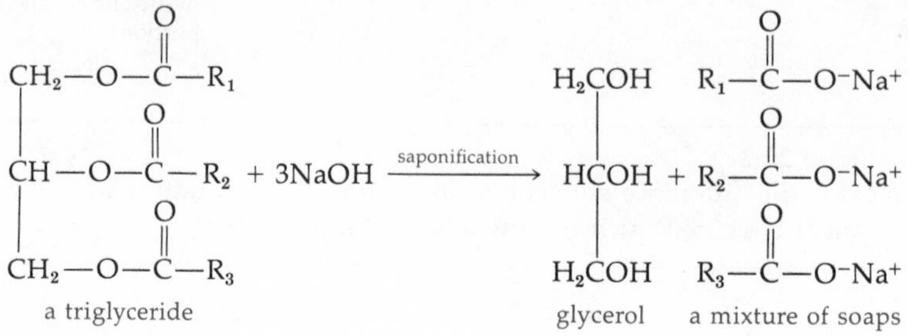

$$
\begin{array}{c}
\underset{\text{a triglyceride}}{
\begin{array}{l}
CH_2-O-\overset{\displaystyle O}{\overset{\|}{C}}-R_1 \\
CH-O-\overset{\displaystyle O}{\overset{\|}{C}}-R_2 \\
CH_2-O-\overset{\displaystyle O}{\overset{\|}{C}}-R_3
\end{array}}
+ 3NaOH
\xrightarrow{\text{saponification}}
\underset{\text{glycerol}}{
\begin{array}{l}
H_2COH \\
HCOH \\
H_2COH
\end{array}}
+
\underset{\text{a mixture of soaps}}{
\begin{array}{l}
R_1-\overset{\displaystyle O}{\overset{\|}{C}}-O^-Na^+ \\
R_2-\overset{\displaystyle O}{\overset{\|}{C}}-O^-Na^+ \\
R_3-\overset{\displaystyle O}{\overset{\|}{C}}-O^-Na^+
\end{array}}
\end{array}
$$

Exercise 26-3
Draw the structural formula for a monoglyceride containing a 16-carbon saturated fatty acid.

Phosphoglycerides

Phosphoglycerides are esters that are formed from one or two fatty acids and glycerol 3-phosphate,

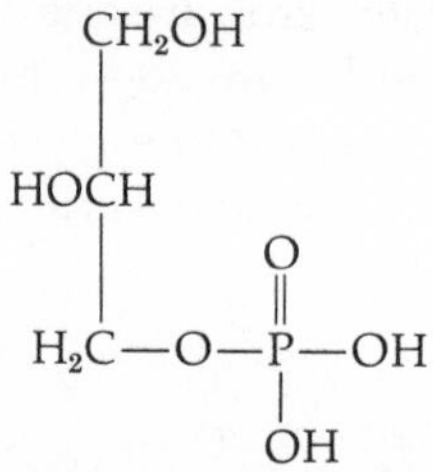

$$
\begin{array}{l}
CH_2OH \\
| \\
HOCH \\
| \quad\quad\quad O \\
| \quad\quad\quad \| \\
H_2C-O-P-OH \\
\quad\quad\quad\quad | \\
\quad\quad\quad\quad OH
\end{array}
$$

and are important components of cell membranes. Because of the polarity of the phosphate group, the glycerol end of these phosphoglyceride molecules is more soluble in water than the glycerol end of a di- or triglyceride. Apparently, this is the reason phosphoglycerides are a major component of cell membranes.

In addition to different fatty acid chains, phosphoglycerides can have different substituents bound to the phosphate group. A phosphoglyceride with no additional substituent on the phosphate group is called a phosphatidic acid. Phosphatidic acids are not abundant in the body. They serve as the precursors for synthesis of other phosphoglycerides. The names and general structural formulas for the major phosphoglycerides are given in Table 26-3. Of these, phosphatidyl ethanolamine (or cephalin) and phosphatidyl choline (or lecithin) are those most commonly found in cell membranes.

Exercise 26-4
What is the difference between a phosphatidic acid and a cephalin? Identify the polar and nonpolar portions of these molecules.

Table 26-3 Phosphoglycerides

Name	General Structural Formula

Phosphatidic acid

$$\begin{array}{c} O \\ \parallel \\ H_2C-O-C-R_1 \\ \\ O \\ \parallel \\ R_2-C-O-CH \\ \\ O \\ \parallel \\ H_2C-O-P-OH \\ \mid \\ OH \end{array}$$

Phosphatidyl ethanolamine (cephalin)

$$\begin{array}{c} O \\ \parallel \\ H_2C-O-C-R_1 \\ \\ O \\ \parallel \\ R_2-C-O-CH \\ \\ O \\ \parallel \\ H_2C-O-P-O-CH_2-CH_2-NH_2 \\ \mid \\ OH \end{array}$$

phosphorylethanolamine

Phosphatidyl serine

$$\begin{array}{c} O \\ \parallel \\ H_2C-O-C-R_1 \\ \\ O \\ \parallel \\ R_2-C-O-CH \qquad\qquad O \\ \qquad\qquad\qquad \parallel \\ \qquad\qquad\qquad\quad C-OH \\ O \qquad\qquad\qquad\quad \mid \\ \parallel \\ H_2C-O-P-O-CH_2-C-NH_2 \\ \mid \qquad\qquad\qquad \mid \\ OH \qquad\qquad\qquad H \end{array}$$

phosphorylserine

Phosphatidyl choline (lecithin)

$$\begin{array}{c} O \\ \parallel \\ H_2C-O-C-R_1 \\ \\ O \\ \parallel \\ R_2-C-O-CH \qquad\qquad\qquad\qquad CH_3 \\ O \qquad\qquad\qquad\qquad\quad \mid \\ \parallel \qquad\qquad\qquad\qquad + \\ H_2C-O-P-O-CH_2-CH_2-N-CH_3 \\ \mid \qquad\qquad\qquad\qquad \mid \\ OH \qquad\qquad\qquad\qquad CH_3 \end{array}$$

phosphorylcholine

Sphingolipids

Sphingolipids contain a fatty acid joined by an amide bond to a molecule called sphingosine:

$$H-\overset{\overset{\displaystyle H}{|}}{\underset{\underset{\displaystyle H}{|}}{C}}-\overset{\overset{\displaystyle H}{|}}{\underset{\underset{\displaystyle H}{|}}{C}}-\overset{\overset{\displaystyle H}{|}}{\underset{\underset{\displaystyle H}{|}}{C}}-\overset{\overset{\displaystyle H}{|}}{\underset{\underset{\displaystyle H}{|}}{C}}-\overset{\overset{\displaystyle H}{|}}{\underset{\underset{\displaystyle H}{|}}{C}}-\overset{\overset{\displaystyle H}{|}}{\underset{\underset{\displaystyle H}{|}}{C}}-\overset{\overset{\displaystyle H}{|}}{\underset{\underset{\displaystyle H}{|}}{C}}-\overset{\overset{\displaystyle H}{|}}{\underset{\underset{\displaystyle H}{|}}{C}}-\overset{\overset{\displaystyle H}{|}}{\underset{\underset{\displaystyle H}{|}}{C}}-\overset{\overset{\displaystyle H}{|}}{\underset{\underset{\displaystyle H}{|}}{C}}-\overset{\overset{\displaystyle H}{|}}{\underset{\underset{\displaystyle H}{|}}{C}}-\overset{\overset{\displaystyle H}{|}}{\underset{\underset{\displaystyle H}{|}}{C}}-\overset{\overset{\displaystyle H}{|}}{\underset{\underset{\displaystyle H}{|}}{C}}-\overset{\overset{\displaystyle H}{|}}{\underset{\underset{\displaystyle H}{|}}{C}}=\overset{\overset{\displaystyle H}{|}}{\underset{\underset{\displaystyle HO}{|}}{C}}-\overset{\overset{\displaystyle H}{|}}{\underset{\underset{\displaystyle NH_2}{|}}{C}}-\overset{\overset{\displaystyle H}{|}}{\underset{\underset{\displaystyle H}{|}}{C}}-OH$$

sphingosine

Sphingolipids are important components of membranes, especially in brain and nerve cells. The sphingolipid compound formed from sphingosine and a fatty acid is sometimes called a **ceramide.** The most abundant sphingolipids in human brain and nerve cells are called **sphingomyelins.** A sphingomyelin consists of a ceramide to which a polar molecule such as phosphorylcholine or phosphorylethanolamine (Table 26-3) is attached. The structure of one sphingomyelin is given in Figure 26-5.

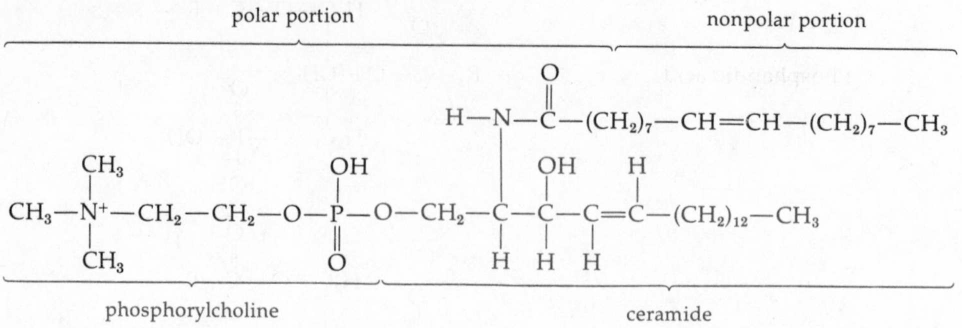

Figure 26-5   A sphingomyelin in which the ceramide consists of sphingosine (in blue) and oleic acid, joined by an amide bond.

Waxes

Waxes are esters of fatty acids and long-chain alcohols or steroid alcohols (see Figure 26-6 for an example). You should be familiar with some of the physical properties of waxes—for example, the flammability of candle wax. Waxes are soft and pliable when they are reasonably warm, but harden upon cooling. Waxes are used as protective coats by certain insects and by some plants. They are also found in the skin and fur of animals. The wax in your ears serves to trap potentially harmful entities that might enter the ear channel.

Figure 26-6   A wax.

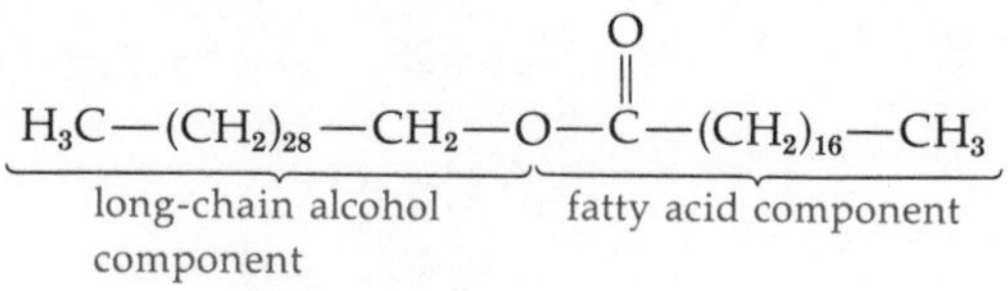

Prostaglandins

Another class of lipid molecules, the **prostaglandins,** are formed from polyunsaturated fatty acids containing 20 carbon atoms such as arachidonic acid (Table 26-2). These polyunsaturated fatty acids in turn must be synthesized from the essential fatty acids linoleic acid and linolenic acid. Two typical prostaglandins are prostaglandin E_1,

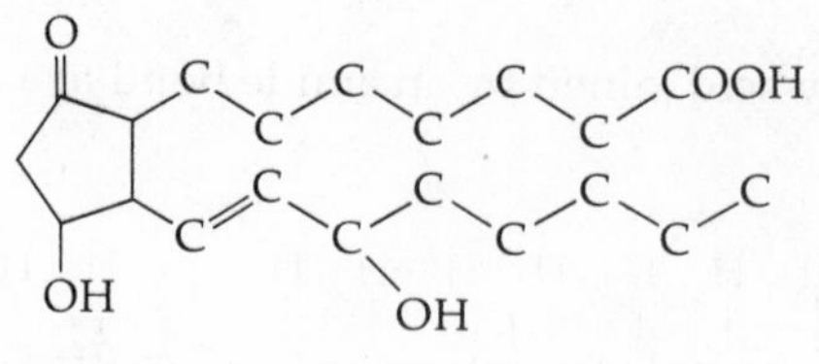

and prostaglandin $F_{1\alpha}$,

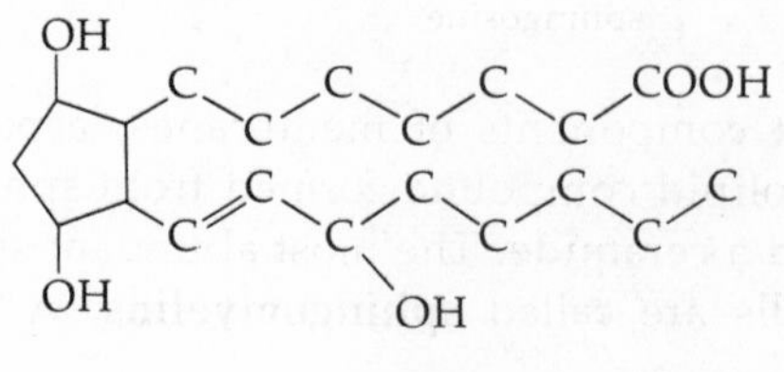

Prostaglandins have very interesting biological properties. They are generally thought to function as hormones and are currently the subject of intense investigations, which may have important medical applications. Some prostaglandins, for example, are known to be involved in the aggregation of platelets during blood clotting, whereas others may be useful for inducing relatively safe abortions.

26-6 TERPENES AND STEROIDS

Terpenes and steroids are quite different in structure from most of the lipids we have discussed so far. **Terpenes** are lipid molecules that can be considered to contain two or more isoprene units. An **isoprene unit** contains five carbon atoms. Its structure is given in Figure 26-7. Therefore, terpenes, by definition, must contain groups of carbon atoms in multiples of five, that is, 10, 15, 20, 25, and so on, carbon atoms.

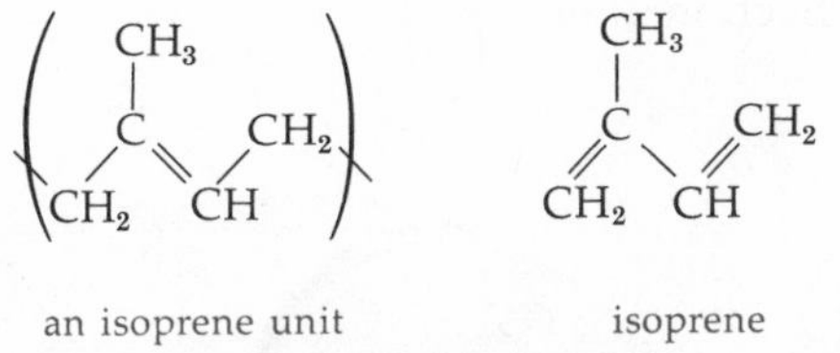

an isoprene unit   isoprene

Figure 26-7   The isoprene units found in terpenes are named after the structurally similar compound isoprene.

Many of the terpenes found in plants have odors and flavors that are familiar to you. For example, menthol and camphor are small terpenes, made from only two isoprene units. On the other hand, natural rubber is a polyterpene made from a very large number of isoprene units. In humans, the most important terpenes are vitamins A, E, and K, which we shall study in Section 26-7, and coenzyme Q, which is a component of the electron transport system (Chapter 25). The structural formula for coenzyme Q is given in Appendix 3. One terpene molecule with 30 carbon atoms, called **squalene,** has the following skeleton structural formula:

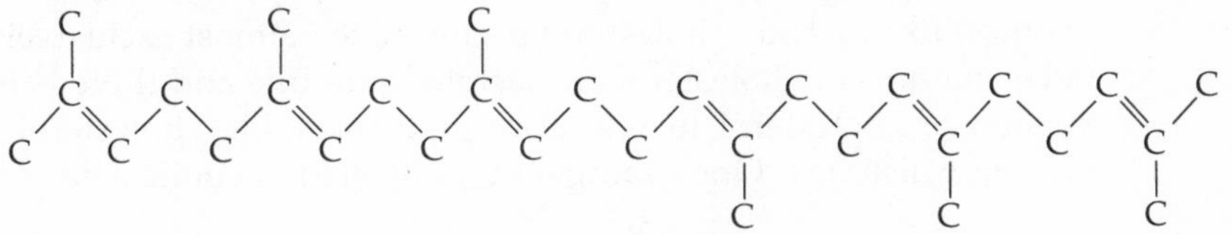

Squalene is used by the body to make steroid molecules.

Steroids are lipids containing a common structural element, which consists of 3 six-membered and 1 five-membered ring of carbon atoms, with the following skeleton structural formula:

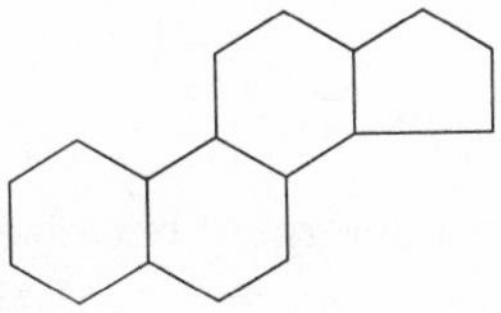

Steroids are all synthesized from acetyl-S-CoA, with squalene as an intermediate. The most abundant steroid molecule in humans is **cholesterol,**

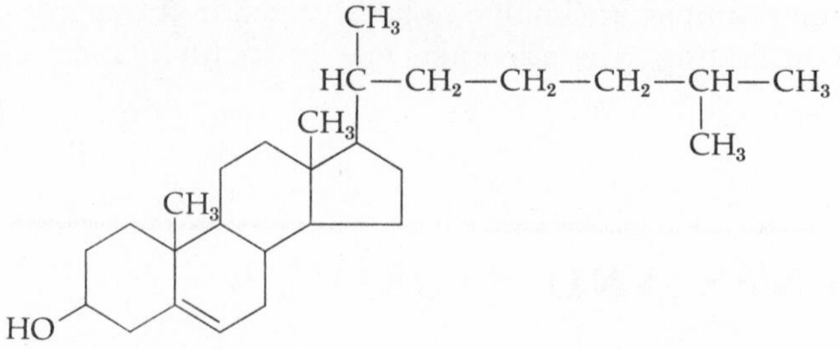

$$\begin{array}{c}CH_3\\|\\HC-CH_2-CH_2-CH_2-CH-CH_3\\|\\CH_3\end{array}$$

Cholesterol is a component of cell membranes and the starting material from which the human body manufactures various steroid hormones. Human beings synthesize cholesterol rapidly and do not need to obtain cholesterol in the diet. In fact, excess cholesterol can be deposited by overworked carrier proteins onto the walls of blood vessels. The disease that results from clogged vessels is called **atherosclerosis.** Restriction of dietary cholesterol is recommended to prevent atherosclerosis, but the solution is not that simple. Saturated fatty acids in the diet are also involved (see Section 26-4).

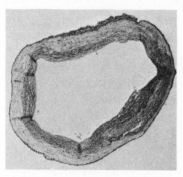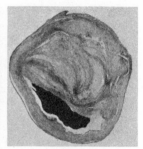

A cross-section of a normal artery is shown on the left. In atherosclerosis, cholesterol is deposited on artery walls and restricts blood flow (center). In severe cases, restricted blood flow may lead to the formation of deadly blood clots (right).

Other steroids function as vitamins and hormones. The structural formulas of several steroid hormones, as well as their functions, are given in Table 26-4.

Another group of steroids are the **bile acids,** which act as detergents in the intestines. Excess body cholesterol is eliminated almost exclusively in bile. When large amounts of cholesterol are secreted into bile and there is insufficient conversion of cholesterol to bile acids, the insoluble cholesterol may crystallize to form gallstones. One example of a bile acid is cholic acid,

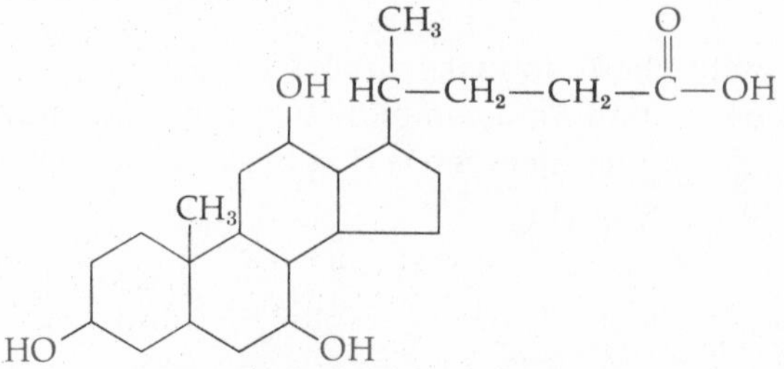

$$\begin{array}{c}CH_3\qquad\qquad O\\|\qquad\qquad\|\\OH\ HC-CH_2-CH_2-C-OH\end{array}$$

Can you see how cholic acid can function as a detergent because of its polar and nonpolar components?

Table 26-4 Steroid Hormones

Name	Structural Formula*	Produced by	Target Tissue	Function
Corticosterone	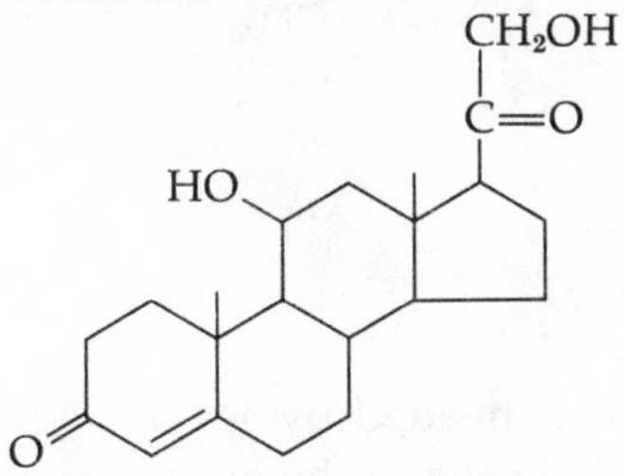	Adrenal gland	Liver and others	Increases glycogen synthesis and gluconeo-genesis
Aldosterone	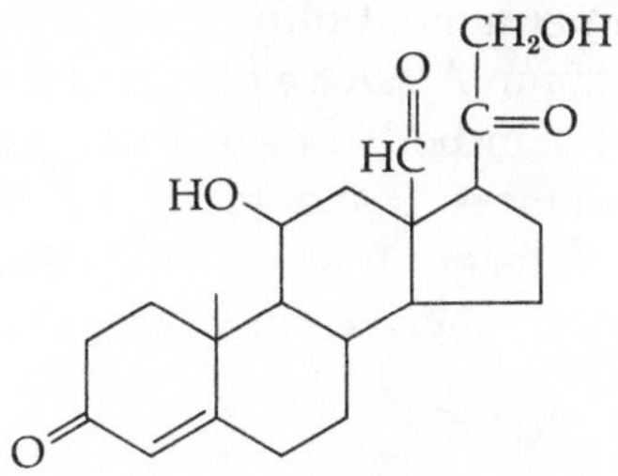	Adrenal gland	Primarily kidney	Na+ and water retention
Estradiol	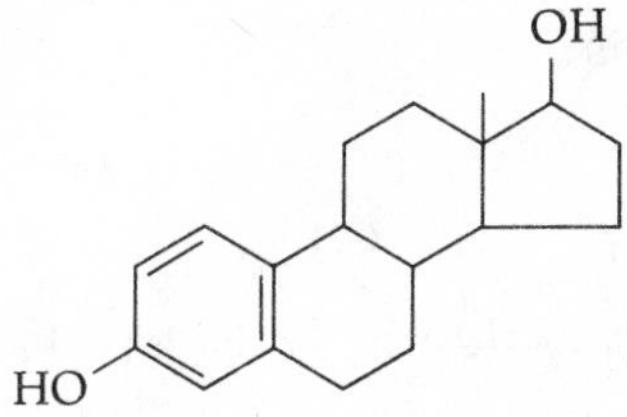	Ovary	Vagina, breast	Maturation and maintenance of reproductive capacity
Testosterone	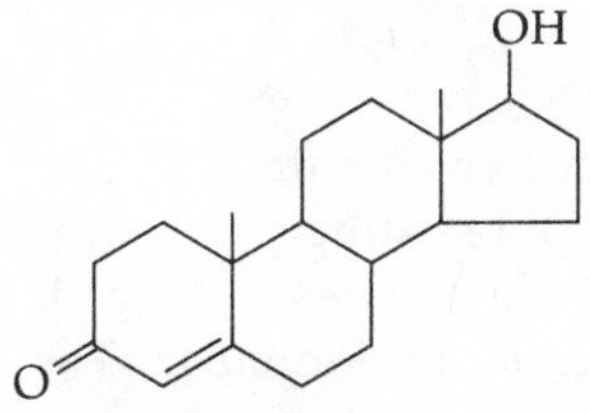	Testis and others	Seminal vesicles	Increased protein synthesis
Progesterone	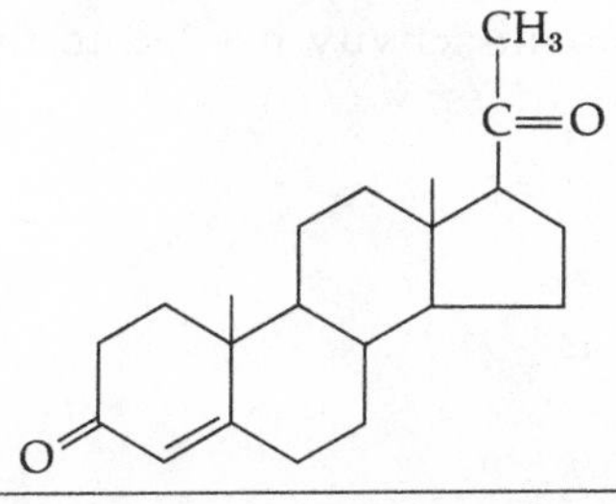	Ovary and others	Uterus	Keeps uterine muscle relaxed during pregnancy

* The short lines protruding from the steroid rings represent methyl groups.

26-7 FAT-SOLUBLE VITAMINS

Some steroids are vitamins. Recall that vitamins are compounds that are required in the human diet in small quantities. In Chapter 23 we mentioned that several coenzymes are derivatives of water-soluble vitamins. **Fat-soluble vitamins** are vitamins that are lipids and hence insoluble in water. The precise functions of a number of these vitamins are not clearly understood. All of the fat-soluble vitamins contain isoprene units.

Vitamin A

The structural formula of **vitamin A** is

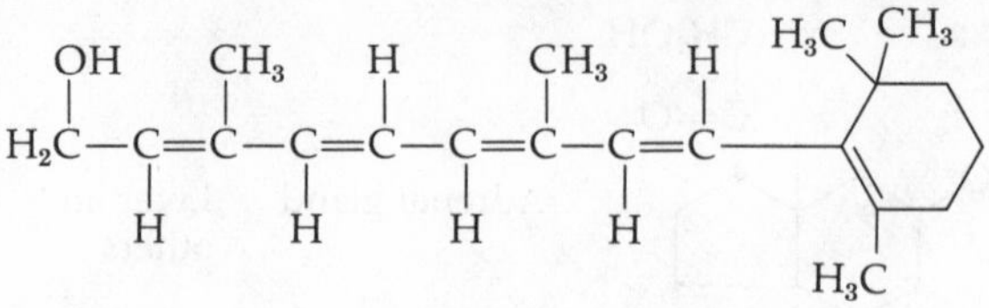

Vitamin A is also called **retinol,** and is needed by several types of cells in the human body, but most of all by retinal cells in the eyes. Absence of vitamin A causes "night blindness" and even permanent blindness in people whose nutrition is especially poor. Dietary vitamin A can be obtained from several vegetables and from fish liver oil. The human body needs about 1 mg of vitamin A per day and can store rather large amounts. It should be noted that excess vitamin A is toxic and can cause liver damage, fragile bones, and deformities in infants. One should keep this in mind when using vitamin supplements.

Vitamin E

The structural formula of **vitamin E** is

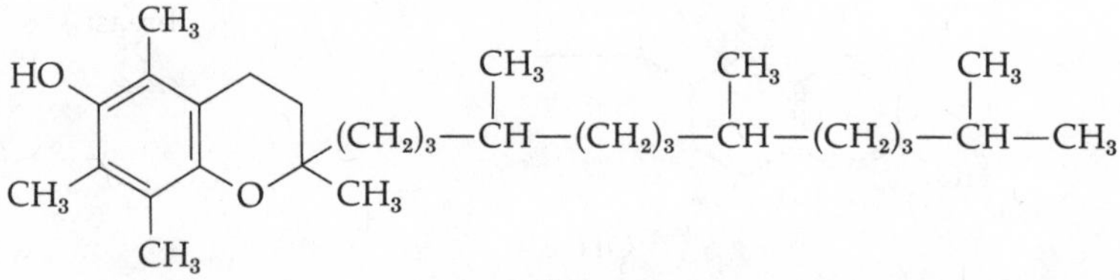

Vitamin E is also called **tocopherol.** Lack of vitamin E results in several symptoms in animals, including infertility. Interestingly, vitamin E has been suggested to be a natural human aphrodisiac. However, there is no concrete evidence that vitamin E affects human sexual performance. The dietary need for vitamin E may be due to its antioxidant activity, as it prevents the oxidation of unsaturated fatty acids. This antioxidant activity has led to the proposal that vitamin E retards aging.

Vitamin K

The structural formula of **vitamin K** is

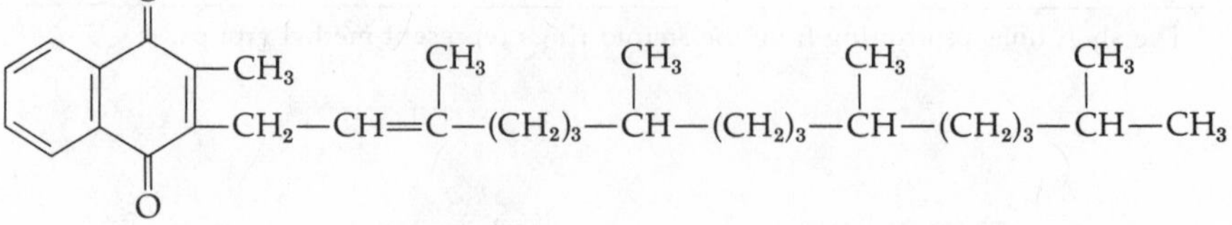

The only known effect of vitamin K deficiency is the inability of liver cells to synthesize an enzyme called **proconvertin.** Proconvertin is one of several enzymes required for blood clotting. A lack of vitamin K, therefore, can lead to defective blood clotting. We don't generally have to worry about getting enough vitamin K, however. Bacteria living in our intestines produce vitamin K which we absorb. It was recently discovered that vitamin K functions as a coenzyme in a number of reactions involving the formation of dicarboxylic acid side chains on certain proteins, including proconvertin.

The drug **Dicumarol,**

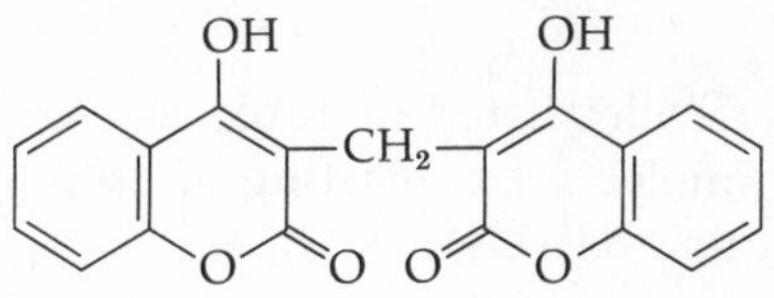

can act as a competitive inhibitor to the formation of proconvertin. Therefore, Dicumarol is used clinically to prevent the formation of blood clots in the blood vessels of patients suffering from phlebitis and other clotting disorders. A closely related compound, called warfarin, is the active ingredient in some rat poisons. Warfarin causes death by uncontrolled internal bleeding.

Vitamin D

There are several related lipids that are collectively called the **D vitamins.** The most important of these are vitamin D_2 (ergocalciferol),

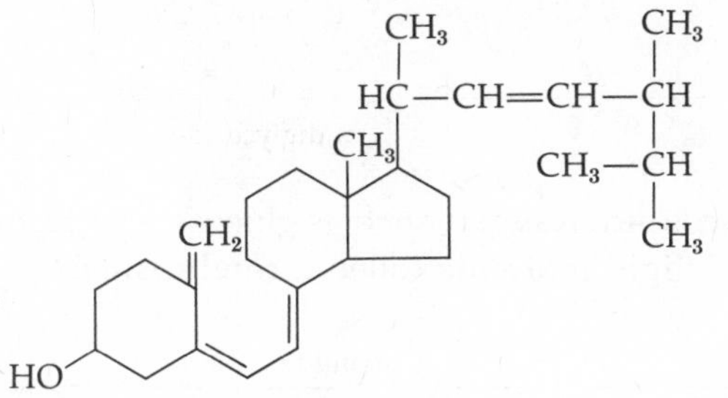

and vitamin D_3 (cholecalciferol),

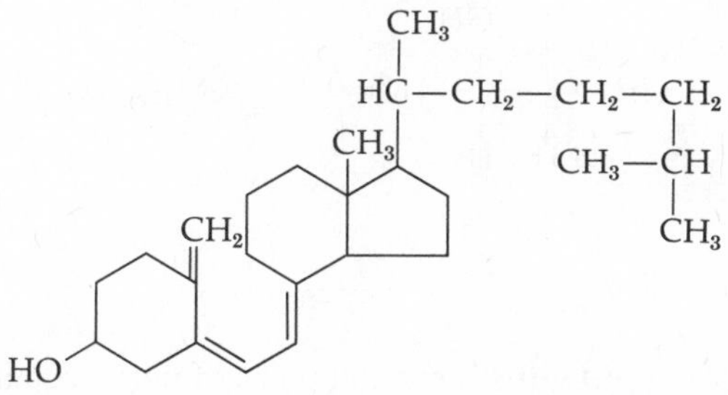

Vitamin D_3 is formed from a derivative of cholesterol by the action of sunlight on the skin. Thus, exposure to sufficient sunlight will assure an adequate supply of vitamin D_3 for humans. In fact, vitamin D-enriched milk is usually obtained merely by exposing milk to light.

What does vitamin D do in the body? Recent studies by Professor Hector De-Luca at the University of Wisconsin, and others, are providing an answer to this question. It appears that vitamin D is converted by kidney cells to a compound called 1,25-dihydroxycholecalciferol. This molecule appears to act as a hormone necessary for proper bone growth. When there is insufficient vitamin D, the kidney cannot make enough 1,25-dihydroxycholecalciferol, which in children can result in **rickets,** a disease in which bone growth is severely impaired.

Exercise 26-5
Dicumarol is used clinically as an anticoagulant. Why is it useful for this?

Exercise 26-6
Are fat-soluble vitamins terpenes or steroids? Explain.

26-8   HYBRID LIPIDS

The last two classes of lipid molecules listed in Table 26-1 are glycolipids and lipoproteins, which are hybrid molecules consisting of lipid molecules bound to carbohydrates and proteins, respectively.

Glycolipids

A **glycolipid** consists of a lipid component that is attached to a carbohydrate component by a glycosidic bond. For example, galactose attached to a diglyceride forms a galactosyldiacylglycerol:

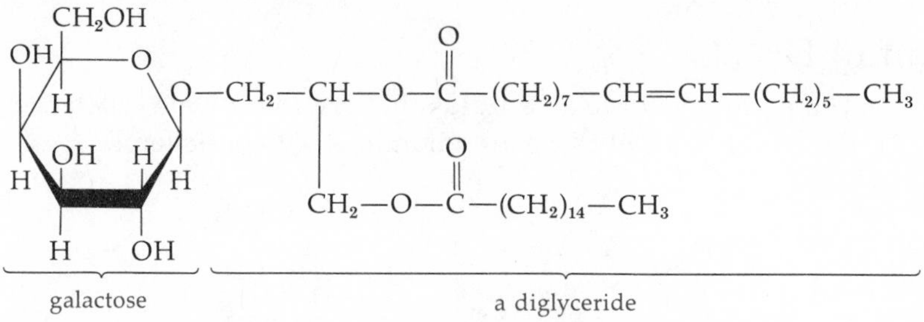

galactose   a diglyceride

In another type of glycolipid, a sugar, such as glucose, is attached to a ceramide to form a glycosphingolipid molecule called a **cerebroside:**

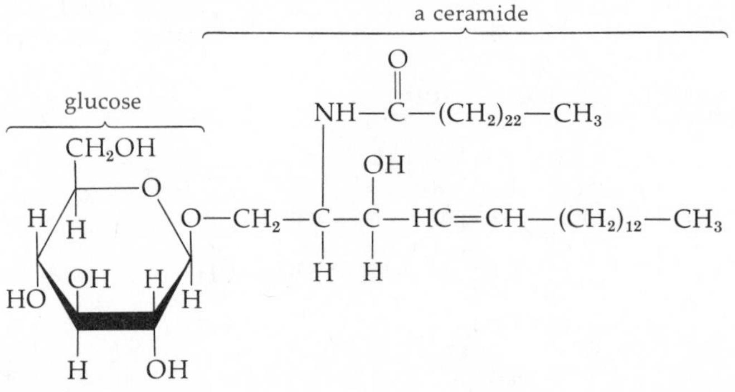

Cerebrosides, like sphingomyelins, are vital parts of nerve cell membranes. The functions of the sphingolipids and glycolipids in nerve cell membranes is being studied intensively. Several human diseases are now known to result from abnormal accumulation of these complex lipids (see Table 26-5). Recent research has shown that each of these diseases is due to the inherited absence of an enzyme needed to break down a sphingolipid or a glycolipid.

Table 26-5 Diseases Resulting from Abnormal Metabolism of Glycolipids and Sphingomyelins

Disease	Organ Affected	Lipid Accumulated
Tay-Sachs	Brain	A glycolipid called a GM_2 ganglioside
Fabry's	Several	A glycolipid consisting of the trisaccharide galactose-galactose-glucose and ceramide
Gaucher's	Spleen, liver	Cerebrosides containing glucose
Niemann-Pick	Several, particularly liver and spleen	Sphingomyelins

Lipoproteins

Lipoproteins, you may recall, are proteins that contain lipid prosthetic groups. Lipoproteins are held together mainly by hydrophobic interactions between nonpolar amino acid side chains and the lipid molecules. Lipoproteins are used as a means of transporting nonpolar lipids through the bloodstream. Thus, as shown in Figure 26-8, the protein component of a lipoprotein is a carrier protein for the lipid component.

Figure 26-8   A schematic representation of a lipoprotein. The carrier protein binds its lipid passenger by means of hydrophobic interactions.

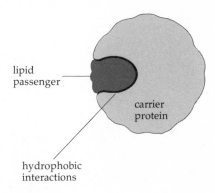

lipid passenger

carrier protein

hydrophobic interactions

Another close association between lipids and proteins is found in cell membranes. We have already seen that proteins are needed in cell membranes to transport polar molecules across the nonpolar lipid membrane, and to serve as receptors for hormone molecules.

Exercise 26-7

Match each of the lipid classes listed on the left with the most appropriate function on the right.

Lipid	Function
(a) Prostaglandins	(1) Hormones and vitamins
(b) Glycerides	(2) Carry lipids in the blood
(c) Steroids	(3) Hormonelike activity
(d) Lipoproteins	(4) Food storage
(e) Phosphoglycerides	(5) Cell membranes

The biochemical cause of Tay-Sachs disease involves the absence of an enzyme that breaks down a lipid found in the brain. The lack of this enzyme leads to the harmful accumulation of fatty deposits within brain cells. Such fatty deposits appear at the bottom of this micrograph. A part of the nucleus of the brain cell is shown at the top.

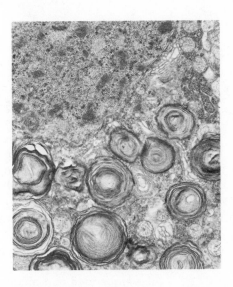

26-9   BIOLOGICAL MEMBRANES

We can consider cell membranes as the walls of cellular factories, and think of phosphoglycerides, cholesterol, and some other lipids as the bricks used to make the walls. A phosphoglyceride molecule, for example, is well suited for this purpose, since it has a polar "head" and two nonpolar, hydrophobic "tails" (see Figure 26-9).

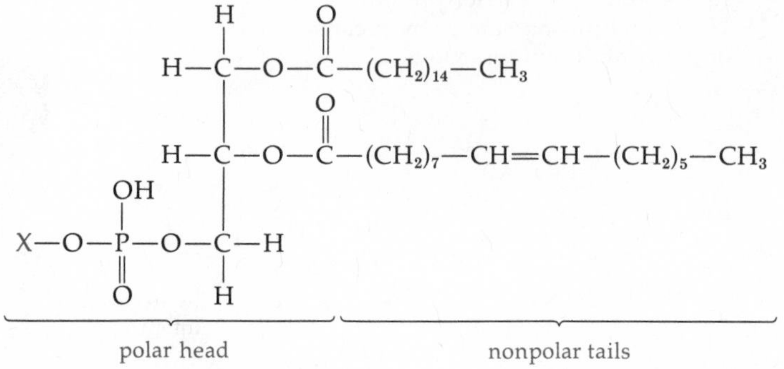

$$
\begin{array}{c}
\text{H} \qquad\quad \text{O} \\
| \qquad\qquad || \\
\text{H}-\text{C}-\text{O}-\text{C}-(\text{CH}_2)_{14}-\text{CH}_3 \\
| \qquad\qquad \text{O} \\
\qquad\qquad\quad || \\
\text{H}-\text{C}-\text{O}-\text{C}-(\text{CH}_2)_7-\text{CH}=\text{CH}-(\text{CH}_2)_5-\text{CH}_3 \\
\text{OH} \quad | \\
| \qquad\quad \\
\text{X}-\text{O}-\text{P}-\text{O}-\text{C}-\text{H} \\
|| \qquad | \\
\text{O} \qquad \text{H}
\end{array}
$$

polar head   nonpolar tails

Figure 26-9   The polar and nonpolar regions of a phosphoglyceride, where X can represent a serine, ethanolamine, or choline component.

Cell membranes are composed primarily of a lipid bilayer, with several protein components added in. The structure of a **phospholipid bilayer** is shown in Figure 26-10. Notice that the polar heads of the phosphoglycerides are on the aqueous outsides of the bilayer, and that the hydrophobic tails of the phosphoglycerides form the nonpolar interior of the bilayer. Because of this nonpolar interior, cell membranes can act as barriers to polar molecules and ions.

Figure 26-10   A phospholipid bilayer. The hydrophobic tails of the phosphoglycerides form a hydrophobic region separating two aqueous regions.

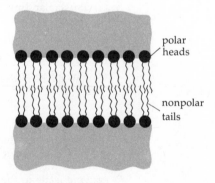

polar heads

nonpolar tails

Remember that the melting points of saturated fatty acids are above human body temperature, whereas those of unsaturated fatty acids are below (see Table 26-2). Cell membranes usually have just the right mixture of saturated and unsaturated fatty acid components to allow them to be semisolid. This gives cell membranes structural strength, but does not make them rigid or brittle.

We have seen that membrane transport proteins and hormone receptor proteins (Chapter 22) are also components of cell membranes. If we stretch our analogy a bit, we might consider them the windows and doors in the walls of the cellular factory. Thus, transport proteins, for example, help specific polar molecules to get into, and out of, cells. The complete structure of a cell membrane is currently viewed as consisting of a large number of these protein molecules embedded in a pliable wall of phosphoglycerides (see Figure 26-11). This is called the **mosaic model** of membrane structure. Notice in this illustration that the protein components are "floating" in a "sea" of phosphoglycerides. Since the phosphoglyceride bilayer is semisolid, the protein molecules can move around in the membrane. The portions of these proteins that are in the interior of the phosphoglyceride bilayer must be nonpolar. The proteins are actually held in the membrane by hydrophobic interactions between the fatty acid tails of the phosphoglycerides and the hydrophobic amino acid side chains of the proteins.

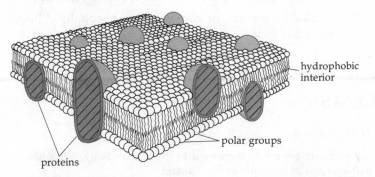

hydrophobic interior

polar groups

proteins

Figure 26-11   The mosaic model of membrane structure. The membrane consists of a phospholipid bilayer with protein molecules penetrating into either side of, or extending through, the membrane.

Exercise 26-8
Why can't phosphoglyceride monolayers act as cell membranes?

26-10 SUMMARY

1. Lipids are insoluble in water but soluble in nonpolar organic solvents.

2. A fatty acid consists of a long nonpolar chain of carbon atoms, with a carboxyl group on one end.

3. Unsaturated fatty acids have lower melting points than saturated fatty acids.

4. Sodium and potassium salts of fatty acids are soaps.

5. The human body cannot synthesize some of the unsaturated fatty acids it requires. These fatty acids are called essential fatty acids because they must be obtained in the diet.

6. Glycerides are esters of glycerol and one, two, or three fatty acids. They serve as food reserves in the body.

7. Phosphoglycerides consist of two fatty acids that are esterified to glycerol 3-phosphate. Other polar groups may also be bonded to the phosphate group of phosphoglycerides.

8. Sphingolipids are esters of fatty acids and sphingosine.

9. Waxes are esters of fatty acids and long-chain alcohols.

10. Terpenes and steroids are lipid molecules made from isoprene units.

11. Cholesterol is the most abundant steroid in humans. It is a component of cell membranes. Other steroid molecules, such as estrogen and testosterone, are hormones.

12. Prostaglandins are lipids made from 20-carbon fatty acids and have hormonelike functions.

13. Glycolipids are lipid molecules that contain a carbohydrate portion attached to a lipid portion by a glycosidic bond.

14. Lipoproteins consist of lipids bound to their carrier proteins.

15. Vitamins A, E, K, and D are fat-soluble vitamins.

16. Dicumarol prevents clotting by acting as a competitive inhibitor to the formation of proconvertin, which requires vitamin K.

17. A biological membrane consists of a phospholipid bilayer with proteins embedded in it.

PROBLEMS

1. List five major functions of lipid molecules.

2. Fatty acids are extremely important to the human body, yet a normal human has only small amounts of free fatty acids. Explain.

3. Draw the structural formula for a triglyceride composed of glycerol and three lauric acid components.

4. Draw the structural formulas for the products obtained upon saponification of the triglyceride in Problem 3.

5. How does a soap consisting primarily of sodium palmitate enable you to remove grease, that is, hydrocarbons, from your hands and clothes?

6. Name two potential problems that could result from a lack of essential fatty acids in the diet.

7. Draw the characteristic structural component of steroids.

8. Why are molecules similar to vitamin K used in rat poison?

9. Siberian children have historically suffered form a high incidence of rickets. Explain a possible reason for this and suggest a solution that would not require dietary vitamin supplements.

10. The presence of phosphoglycerides containing unsaturated fatty acid components is an important part of the "fluid" mosaic model of cell membranes. Why wouldn't a membrane composed of phosphoglycerides with only saturated fatty acids function properly?

11. Describe the type of bonding found in lipoproteins.

SOLUTIONS TO EXERCISES

26-1 We do not consider acetic acid a fatty acid or a lipid because it does not have a long chain of carbon atoms and it is soluble in water and aqueous solutions.

26-2 The unsaturated fatty acid has at least one double bond, which results in a lower melting point than the saturated fatty acid.

26-3
$$
\begin{array}{l}
\mathrm{H_2C-O-\overset{\displaystyle O}{\overset{\|}{C}}-(CH_2)_{14}-CH_3} \\
\mathrm{H-C-OH} \\
\mathrm{H_2C-OH}
\end{array}
$$

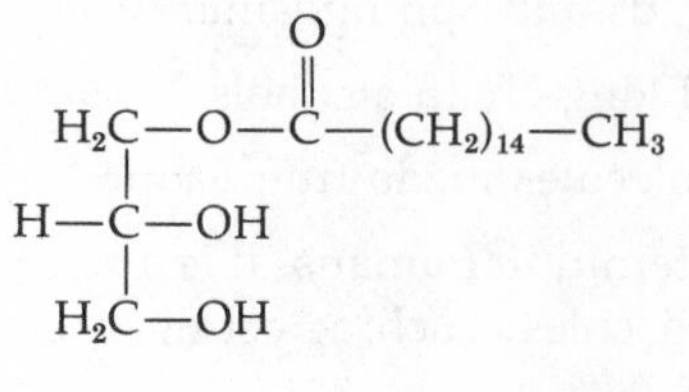

26-4 A phosphatidic acid can be considered to be the phosphate ester of a diglyceride, whereas a cephalin is a phosphatidic acid esterified to a molecule of ethanolamine. The phosphate and phosphorylethanolamine portions of these two types of molecules are polar, whereas the long fatty acid chains (indicated in Table 26-3 as R_1 and R_2) are not.

26-5 It has been found that Dicumarol can act as a competitive inhibitor of enzymes that require vitamin K. One such enzyme is needed to make proconvertin (it adds an additional carboxyl group to proconvertin to make it active). By reducing the rate of proconvertin production, the tendency to form blood clots is also reduced.

26-6 Some fat-soluble vitamins are terpenes (K, A, and E), whereas others (the D vitamins) are steroids.

26-7 (a) Prostaglandins: (3) Hormonelike activity
      (b) Glycerides: (4) Food storage
      (c) Steroids: (1) Hormones and vitamins
      (d) Lipoproteins: (2) Carry lipids in the blood
      (e) Phosphoglycerides: (5) Cell membranes

26-8 A phosphoglyceride monolayer has one polar side and one nonpolar side. The nonpolar side of such a monolayer would not be compatible with the aqueous environment inside or outside of a cell.

CHAPTER 27

Lipid Metabolism

27-1 INTRODUCTION

We have seen that lipids are a diverse group of compounds with correspondingly diverse biological functions. In this chapter we shall discuss the major pathways for the biosynthesis and catabolism of lipids in the human body. Since fatty acids are components of most classes of lipids, we shall pay particular attention to their metabolism.

We shall not study the detailed synthesis of each and every type of lipid molecule. Rather, we shall emphasize the basic features of the synthesis of fatty acids, glycerides, phosphoglycerides, and steroids. The starting material for the biosynthesis of all of these molecules is acetyl-S-CoA. Because they have this common feature, you should find the pathways for lipid biosynthesis fairly easy to follow.

What about pathways for the catabolism of lipids? In Chapter 26 we mentioned that triglycerides are the major food storage depot in the human body and are the source of a great deal of the energy needed by our cellular factories. We shall therefore study the major catabolic pathways for lipids and determine the amount of ATP that is synthesized in the process.

We shall see that the catabolism of a triglyceride involves its hydrolysis to glycerol and three fatty acids. The fatty acids are then degraded by a series of enzyme-catalyzed reactions, called the β-oxidation pathway, to give acetyl-S-CoA, whereas the glycerol is converted in a few steps to dihydroxyacetone phosphate. Since we have previously studied the catabolism of dihydroxyacetone phosphate via glycolysis and the catabolism of acetyl-S-CoA in the TCA cycle, we need only study β-oxidation and the steps in the conversion of glycerol to dihydroxyacetone phosphate in order to have a complete understanding of the catabolism of triglycerides.

Notice that acetyl-S-CoA is both the starting material for lipid biosynthesis and the common intermediate for lipid catabolism. Acetyl-S-CoA is also an intermediate in the catabolism of carbohydrates (Chapter 25) and some amino acids (Chapter 28). Since acetyl-S-CoA is involved in so many metabolic pathways, the human body must carefully regulate which of these pathways is used under a given set of conditions.

The flesh of many species of fish is rich in lipids. These fats contribute to the flavor and texture of such delicacies as smoked whitefish and salmon.

27-2   STUDY OBJECTIVES

After careful study of the material in this chapter, you should be able to:

1. Outline the relationship of lipid metabolism to carbohydrate metabolism and show the central role of acetyl-S-CoA in the metabolism of lipids and carbohydrates.

2. Write a balanced equation showing the action of a lipase enzyme on a triglyceride.

3. Write equations for the reactions involved in the conversion of glycerol to dihydroxyacetone phosphate.

4. Outline the steps of the β-oxidation pathway for fatty acids.

5. Calculate the amount of ATP produced by the complete catabolism of a fatty acid to CO_2 and H_2O, given its chemical formula.

6. Describe the need for the molecule carnitine in fatty acid catabolism.

7. Explain why storage of energy in the form of triglycerides is preferred to storage in the form of glycogen.

8. Describe the role of the acyl carrier portion of fatty acid synthetase, and the relationship between this prosthetic group and coenzyme A.

9. Describe how malonyl-S-CoA is synthesized and used for fatty acid synthesis.

10. Outline the fatty acid biosynthesis pathway.

11. Show how triglycerides are synthesized from fatty acids and glycerol 3-phosphate.

12. Show how phosphatidyl ethanolamine is synthesized in the human body.

13. List several controls for the synthesis and catabolism of fatty acids and triglycerides, and discuss how each works.

14. Define the term ketone bodies, explain their production in diabetics, and tell why they are harmful.

27-3   THE CENTRAL ROLE OF Acetyl-S-CoA

Before we proceed to study some of the individual pathways for lipid metabolism, it will be helpful to have an overall view of lipid metabolism and its interrelationship to carbohydrate metabolism (see Figure 27-1).

When the human body's need for energy cannot be met by the catabolism of glucose, stored triglycerides are catabolized. In this process, a triglyceride is first hydrolyzed to give fatty acids, plus glycerol. The glycerol is converted into dihydroxyacetone phosphate, which is an intermediate in the glycolytic pathway. The fatty acids are broken up into two-carbon-atom fragments which are used to form molecules of acetyl-S-CoA, the fuel for the TCA cycle. Since we have already studied glycolysis and the TCA cycle, we shall only need to study a few new enzyme-catalyzed reactions to understand glyceride catabolism. Phosphoglycerides are catabolized in an essentially identical manner.

When the human body has more than enough glucose to meet its energy requirements, glucose is converted to glycerides, which are stored for later use. The biosynthesis of the fatty acid components of triglycerides starts with acetyl-S-CoA produced, for example, from glucose by glycolysis. Fatty acids are made on a different assembly line from the one used for their catabolism.

Figure 27-1   An overview of lipid metabolism.

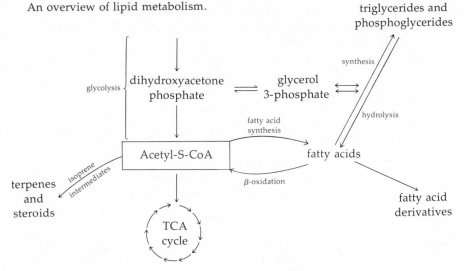

The glycerol or glycerol 3-phosphate portions of glycerides and phosphoglycerides are made from the dihydroxyacetone phosphate intermediate in the catabolism of glucose, in exactly the reverse of the process used to feed them into the glycolysis pathway.

How are needed steroids and terpenes synthesized by the human body? These molecules are built from isoprene units, which in turn are synthesized from acetyl-S-CoA molecules (see Figure 27-1). Therefore, acetyl-S-CoA is the common precursor for the synthesis of all lipids.

27-4   LIPID CATABOLISM

Triglycerides and phosphoglycerides are very good sources of energy for the human body. Triglycerides are stored in fat depots in the body in order to meet future nutritional needs. In fact, the body stores only enough glycogen to provide about one day's supply of energy, but enough triglycerides to supply energy for about one month! In a normal individual the stored triglycerides amount to about 10% of the total body weight. For an average 150-lb adult this amounts to about 15 lb. Storage of an equivalent amount of energy in the form of glycogen would require about 35 lb of glycogen.

We have already seen that both triglycerides and phosphoglycerides contain fatty acid molecules esterified to glycerol. The catabolism of these molecules begins with their hydrolysis to fatty acids and glycerol. This hydrolysis is accomplished by enzymes called lipases in the reaction

(27-1)

$$
3H_2O \; + \;
\begin{matrix}
R_1-\overset{\overset{\displaystyle O}{\|}}{C}-O-CH_2 \\[4pt]
R_2-\overset{\overset{\displaystyle O}{\|}}{C}-O-CH \\[4pt]
R_3-\overset{\overset{\displaystyle O}{\|}}{C}-O-CH_2
\end{matrix}
\;\xrightarrow{\text{lipases}}\;
\begin{matrix}
HO-CH_2 \\[4pt]
HO-CH \\[4pt]
HO-CH_2
\end{matrix}
\; + \;
\begin{matrix}
R_1-\overset{\overset{\displaystyle O}{\|}}{C}-OH \\[4pt]
R_2-\overset{\overset{\displaystyle O}{\|}}{C}-OH \\[4pt]
R_3-\overset{\overset{\displaystyle O}{\|}}{C}-OH
\end{matrix}
$$

a triglyceride   glycerol   3 fatty acids

One lipase is produced by the pancreas and released into the intestines, where it hydrolyzes the triglycerides in our diet. Another lipase is used to hydrolyze the triglycerides in our fat cells when this is required. Hormones, such as epinephrine and glucagon, activate this lipase when energy-rich fatty acids are needed (see Figure 27-2).

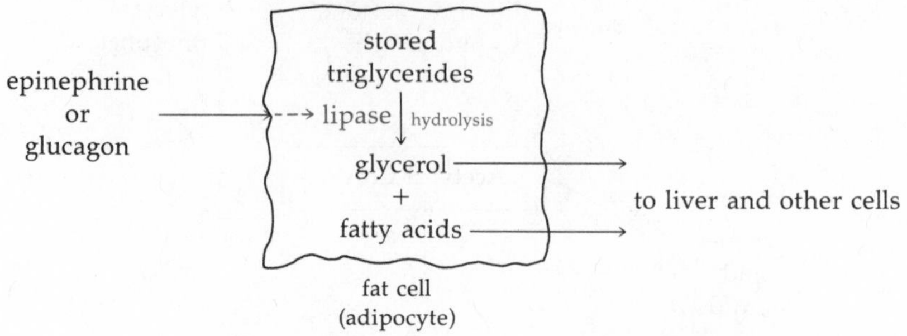

Figure 27-2   The hormones epinephrine and glucagon activate the lipase enzyme in fat cells. The lipase, in turn, hydrolyzes triglycerides to give glycerol and fatty acids.

Glycerol

First, let us see what happens to the glycerol. Glycerol produced by the hydrolysis of one lipid can be used, if needed, as a component in the synthesis of other lipids. Glycerol can also be converted to dihydroxyacetone phosphate, which is then further catabolized via the glycolytic pathway. This conversion is accomplished in two steps: phosphorylation to glycerol 3-phosphate,

(27-2)

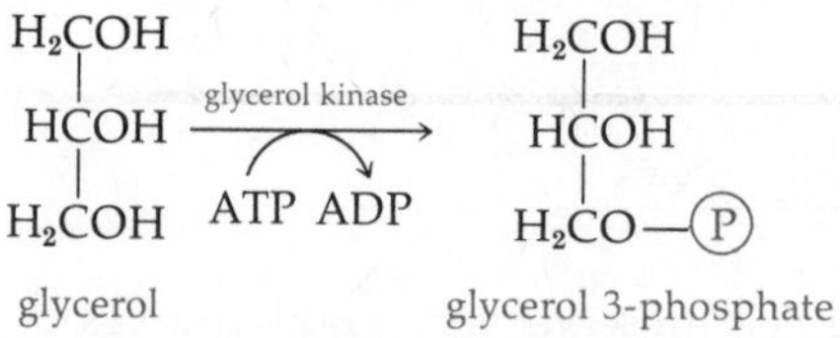

and oxidation to dihydroxyacetone phosphate,

(27-3)

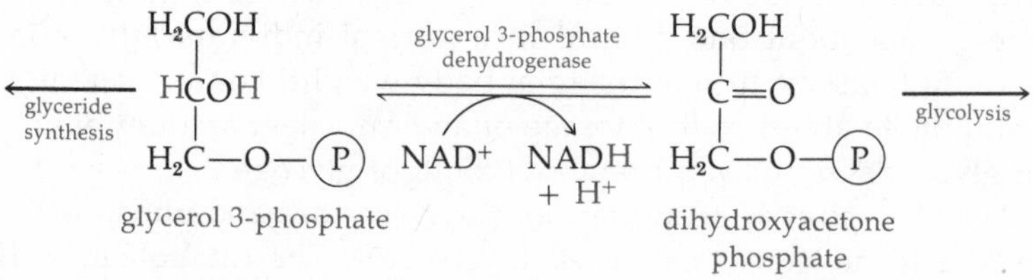

The reverse of reaction 27-3 is used to make glycerol 3-phosphate from dihydroxyacetone phosphate for use in lipid synthesis.

Fatty Acid Catabolism

The majority of fatty acids contain an even number of carbon atoms, and the pathway for their degradation involves their cleavage into two-carbon-atom fragments. This pathway is called the **β-oxidation pathway,** and the two-carbon-atom fragments produced are coupled to coenzyme A molecules.

The first step in the catabolism of fatty acids is called **activation.** In this step, the esterification of free fatty acids to coenzyme A molecules is catalyzed by an enzyme called acyl-S-CoA synthetase, thus producing acyl-S-CoA molecules—activated fatty acids. Recall from Chapter 17 that an acyl group is derived from a carboxylic acid by removing the —OH group.

Step 1:

(27-4)

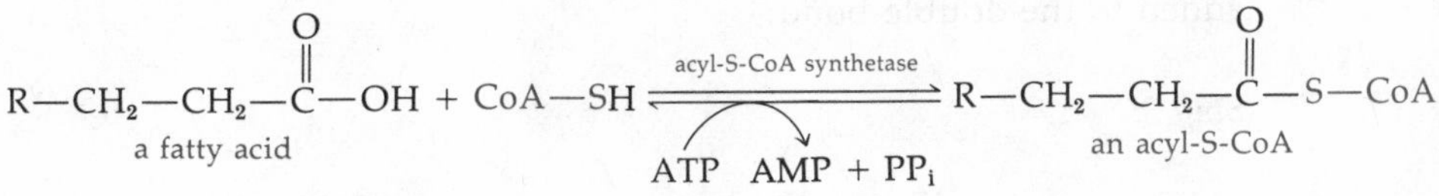

$$R-CH_2-CH_2-\overset{\overset{\displaystyle O}{\|}}{C}-OH + CoA-SH \underset{\text{ATP} \quad \text{AMP} + PP_i}{\overset{\text{acyl-S-CoA synthetase}}{\rightleftharpoons}} R-CH_2-CH_2-\overset{\overset{\displaystyle O}{\|}}{C}-S-CoA$$

a fatty acid   an acyl-S-CoA

This activation step takes place in the cytoplasm of cells, and requires energy in the form of hydrolysis of 1 mole of ATP to AMP per mole of fatty acid. An additional mole of ATP is required to convert AMP to ADP:

(27-5) $AMP + ATP \longrightarrow 2ADP$

Thus, this first step requires an energy input equivalent to the hydrolysis of 2 moles of ATP to ADP per mole of fatty acid.

Further catabolism of fatty acids occurs inside the cell's mitochondria, and produces energy. Since an activated fatty acid does not easily permeate the mitochondrial membrane, it is bonded to a molecule of **carnitine** for transport across the membrane (see Figure 27-3). Once inside the mitochondria, an enzyme breaks the acyl carnitine molecule apart, producing an activated fatty acid and carnitine. The carnitine can then return to the cytoplasm for another fatty acid. Stepwise degradation of the activated fatty acid can now begin.

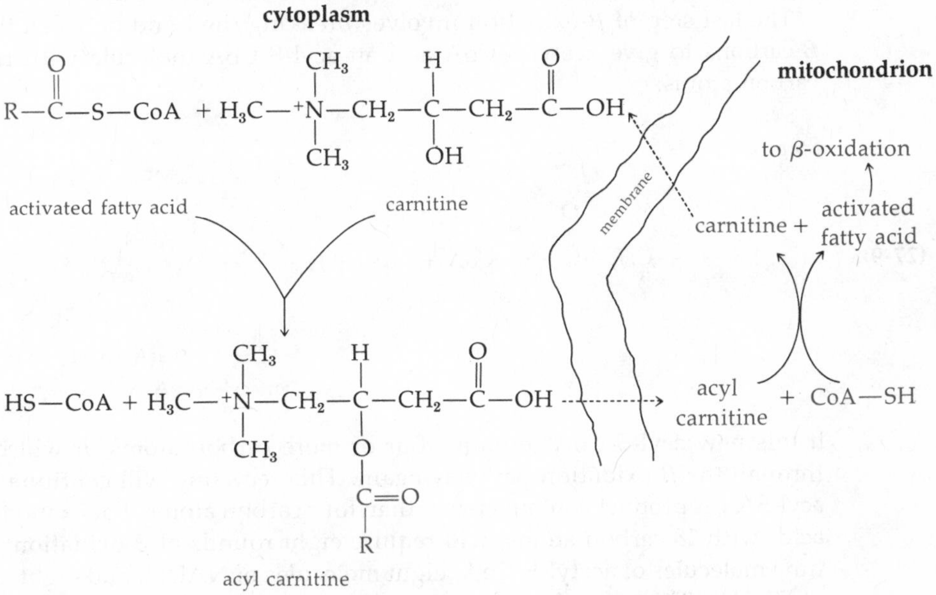

Figure 27-3   Carnitine serves as a carrier for the transport of activated fatty acids   from the cytoplasm into mitochondria.

The next step of the β-oxidation pathway involves the oxidation of the single bond between the α and β carbons of the activated fatty acid:

Step 2:

(27-6)

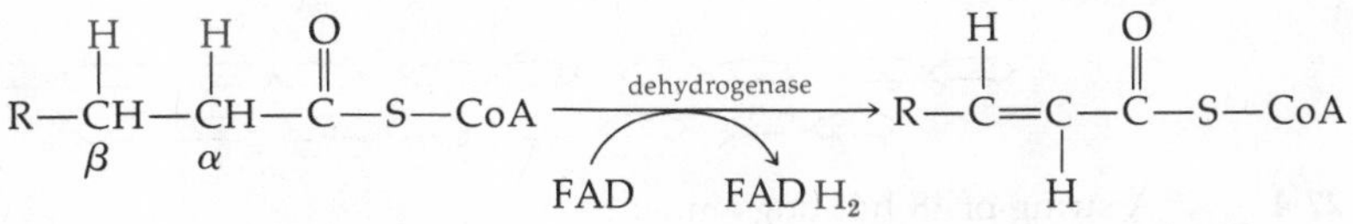

$$R-\overset{\overset{\displaystyle H}{|}}{\underset{\beta}{C}H}-\overset{\overset{\displaystyle H}{|}}{\underset{\alpha}{C}H}-\overset{\overset{\displaystyle O}{\|}}{C}-S-CoA \underset{\text{FAD} \quad \text{FAD}H_2}{\overset{\text{dehydrogenase}}{\longrightarrow}} R-\overset{\overset{\displaystyle H}{|}}{C}=\overset{\overset{\displaystyle O}{\|}}{\underset{\underset{\displaystyle H}{|}}{C}}-C-S-CoA$$

Notice that the oxidation of the saturated fatty acid is coupled to the reduction of an FAD molecule. The FADH$_2$ produced is used to generate ATP in the ETS.

In the third enzyme-catalyzed step of β-oxidation, a molecule of water is added to the double bond:

Step 3:

(27-7)

$$H_2O + R-\underset{H}{\overset{H}{C}}=\underset{}{\overset{}{C}}-\overset{O}{\overset{\|}{C}}-S-CoA \xrightarrow{\text{hydratase}} R-\underset{H}{\overset{OH}{C}}-\underset{H}{\overset{H}{C}}-\overset{O}{\overset{\|}{C}}-S-CoA$$

In this step, the hydroxyl group is always added to the β-carbon.

In the fourth step of this pathway, the alcohol group involving the β-carbon is oxidized to form a ketone, hence the name β-oxidation for this pathway. This step is coupled to the reduction of a NAD⁺ coenzyme molecule.

Step 4:

(27-8)

$$R-\underset{H}{\overset{OH}{\underset{\beta}{C}}}-\underset{\alpha}{CH_2}-\overset{O}{\overset{\|}{C}}-S-CoA \xrightarrow[\substack{NAD^+ \quad NADH \\ + H^+}]{\text{dehydrogenase}} R-\overset{O}{\overset{\|}{C}}-CH_2-\overset{O}{\overset{\|}{C}}-S-CoA$$

The NADH produced is also used to generate ATP in the ETS.

The last step of β-oxidation involves breaking the bond between the α- and β-carbons to give acetyl-S-CoA and an acyl-S-CoA molecule with two fewer carbon atoms:

Step 5:

(27-9)

$$R-\underset{\beta}{\overset{O}{\overset{\|}{C}}}-\underset{\alpha}{CH_2}-\overset{O}{\overset{\|}{C}}-S-CoA + CoA-SH \xrightarrow{\text{thiolase}}$$

$$R-\overset{O}{\overset{\|}{C}}-S-CoA + CH_3-\overset{O}{\overset{\|}{C}}-S-CoA$$

an acyl-S-CoA   acetyl-S-CoA

If this new acyl-S-CoA contains four or more carbon atoms, it will be cycled through the β-oxidation pathway again. This recycling will continue until the acyl-S-CoA product contains fewer than four carbon atoms. For example, stearic acid, with 18 carbon atoms, will require eight rounds of β-oxidation, yielding nine molecules of acetyl-S-CoA, eight molecules of NADH, and eight molecules of FADH₂. This is analogous to breaking a string of 18 hot dogs eight times in order to get nine strings of two hot dogs each (see Figure 27-4).

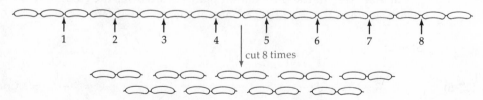

Figure 27-4   A string of 18 hot dogs must be broken eight times to get nine strings of two hot dogs each.

Therefore, for a fatty acid with 18 carbon atoms, the overall reaction for the β-oxidation of the initial activated fatty acid within the mitochondrion is

$$CH_3(CH_2)_{16}\text{—}\underset{\displaystyle O}{\overset{\displaystyle O}{\|}}\text{C—S—CoA} + 8CoA\text{—SH} \longrightarrow 9CH_3\text{—}\overset{\displaystyle O}{\|}\text{C—S—CoA}$$
$$+ 8H_2O + 8FAD + 8NAD^+ \qquad + 8FADH_2 + 8NADH + 8H^+$$

The molecules of acetyl-S-CoA produced by the β-oxidation of fatty acids can be further degraded by the TCA cycle (see Chapter 25). Remember that for each molecule of acetyl-S-CoA that enters the TCA cycle, 12 ATP molecules are formed. Thus, starting with the 18-carbon-atom stearic acid, eight rounds of β-oxidation produce nine molecules of acetyl-S-CoA. The oxidation of these acetyl-S-CoA molecules via the TCA cycle and the ETS results in the formation of 108 ATP molecules:

$$(9 \text{ acetyl-S-CoA}) \times \left(\frac{12 \text{ ATP}}{\text{acetyl-S-CoA}}\right) = 108 \text{ ATP}$$

In addition, eight rounds of β-oxidation produce eight molecules of $FADH_2$ and eight molecules of NADH. When the ETS is used to reoxidize these coenzymes, another 40 ATP molecules are formed:

$$\left(8 \text{ NADH} \times \frac{3 \text{ ATP}}{\text{NADH}}\right) + \left(8 \text{ FADH}_2 \times \frac{2 \text{ ATP}}{\text{FADH}_2}\right) = 40 \text{ ATP}$$

Therefore, the complete oxidation of an 18-carbon acyl-S-CoA molecule yields 40 + 108 = 148 ATP molecules. Of course, the *net* ATP production from the complete oxidation of an 18-carbon-atom fatty acid is 146, since two ATP molecules are hydrolyzed to ADP in the activation step (reactions 27-4 and 27-5). A similar set of calculations are shown in Figure 27-5 for the 16-carbon fatty acid, palmitic acid.

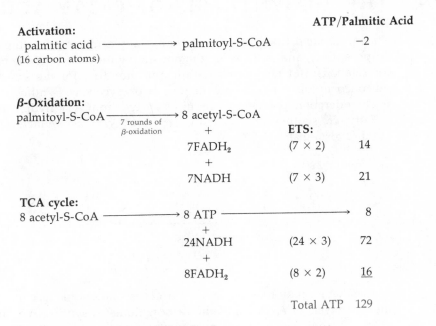

Figure 27-5   The complete oxidation of 1 mole of palmitic acid generates 129 moles of ATP via the β-oxidation pathway.

Complete oxidation of palmitic acid to CO_2 and H_2O involves β-oxidation, the TCA cycle, and the ETS, with the overall reaction

(27-10)
$$CH_3(CH_2)_{14}\overset{\overset{\displaystyle O}{\|}}{C}-OH + 23O_2 \longrightarrow 16CO_2 + 16H_2O + energy$$

How efficient is this process? The experimentally measured ΔG value for reaction 27-10 is -2340 kcal/mole. This is the maximum possible amount of usable energy that can be obtained from this reaction.

Now, in Figure 27-5 we calculated that 129 moles of ATP are produced by the complete oxidation of 1 mole of palmitic acid in the human body. Recall that for the hydrolysis of ATP, $\Delta G = -8$ kcal/mole (Chapter 25). The ΔG for the hydrolysis of the 129 moles of ATP produced by the oxidation of 1 mole of palmitic acid is thus -1032 kcal. Thus this biological oxidation of fatty acids is about 44% efficient, $(-1032 \text{ kcal}/-2340 \text{ kcal}) \times 100\% = 44\%$. This is comparable to the efficiency we determined for the oxidation of glucose in Chapter 25.

Note that in the formation of 129 moles of ATP from ADP + P_i, 129 moles of H_2O are produced in addition to the 16 moles of H_2O from reaction 27-10. Thus, the complete oxidation of 1 mole of palmitic acid, coupled to the formation of 129 moles of ATP, produces 145 moles of H_2O. This is a lot of water. For this reason a camel can store a large supply of water *and* energy in its fat hump.

The catabolism of unsaturated fatty acids and those fatty acids that contain an odd number of carbons require additional steps, which we need not consider.

Exercise 27-1
Compare the amount of ATP that can be formed by the complete oxidation of a six-carbon fatty acid to CO_2 and H_2O with the ATP formed by complete oxidation of a glucose molecule to CO_2 to H_2O.

27-5   THE BIOSYNTHESIS OF FATTY ACIDS

The metabolic pathway used to assemble fatty acids occurs in the cytoplasm of adipose, liver, and other cells. It is not simply the reverse of the pathway used for catabolism of fatty acids within mitochondria, for the same reason that gluconeogenesis is not simply the reverse of glycolysis. Synthesis of fatty acids is an endergonic process and requires energy from ATP hydrolysis.

Fatty acid synthesis begins with acetyl-S-CoA, most of which is first carboxylated to form malonyl-S-CoA:

Step 1:

(27-11)
$$H_3C-\overset{\overset{\displaystyle O}{\|}}{C}-S-CoA + CO_2 \xrightarrow[\substack{ATP \quad ADP \\ + H_2O \quad + P_i}]{\substack{\text{acetyl-S-CoA} \\ \text{carboxylase} \\ + \text{biotin} \\ \text{(a coenzyme)}}} HO-\overset{\overset{\displaystyle O}{\|}}{C}-CH_2-\overset{\overset{\displaystyle O}{\|}}{C}-S-CoA$$

acetyl-S-CoA   malonyl-S-CoA

Notice that formation of malonyl-S-CoA is an endergonic reaction and requires the hydrolysis of ATP. All of the carbon atoms in synthesized fatty acids are obtained from acetyl-S-CoA molecules. However, we shall see that only two carbon atoms are directly supplied by acetyl-S-CoA. The rest of the carbon atoms are supplied by malonyl-S-CoA molecules made from acetyl-S-CoA.

The remaining steps of fatty acid biosynthesis in humans take place on a very complex enzyme, called **fatty acid synthetase.** Many of the chemical reactions that are catalyzed by this enzyme are similar to the reverse of those used for fatty acid catabolism. Fatty acid synthetase has two prosthetic groups, called **acyl carrier portions** (ACP), which are identical to a portion of a coenzyme A molecule (see Figure 27-6).

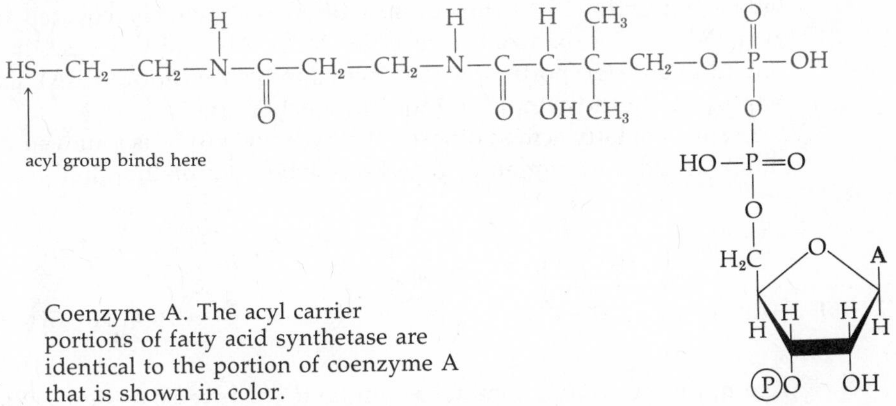

Figure 27-6   Coenzyme A. The acyl carrier portions of fatty acid synthetase are identical to the portion of coenzyme A that is shown in color.

Actual assembly of a fatty acid requires the covalent attachment of one acetyl group and one malonyl group to the ACP portions of fatty acid synthetase, as represented in the following reactions:

(27-12)

$$CoA-S-\overset{O}{\overset{\|}{C}}-CH_3 + ACP-SH \rightleftharpoons ACP-S-\overset{O}{\overset{\|}{C}}-CH_3 + CoA-SH$$

acetyl-S-CoA   acyl carrier portion   acetyl-S-ACP

(27-13)

$$CoA-S-\overset{O}{\overset{\|}{C}}-CH_2-\overset{O}{\overset{\|}{C}}-OH + ACP-SH \rightleftharpoons ACP-S-\overset{O}{\overset{\|}{C}}-CH_2-\overset{O}{\overset{\|}{C}}-OH$$

malonyl-S-CoA   malonyl-S-ACP

$$+ CoA-SH$$

The next step in the assembly of a fatty acid involves the transfer of the acetyl group from its ACP anchor to the malonyl group anchored to the other ACP portion of fatty acid synthetase. In this process, the recently added carboxyl group on malonyl-S-ACP is removed:

Step 2:

(27-14)

$$ACP-S-\overset{O}{\overset{\|}{C}}-CH_2-\overset{O}{\overset{\|}{C}}-OH + ACP-S-\overset{O}{\overset{\|}{C}}-CH_3 \xrightarrow{\text{fatty acid synthetase}}$$

malonyl-S-ACP   acetyl-S-ACP

$$ACP-S-\overset{O}{\overset{\|}{C}}-CH_2-\overset{O}{\overset{\|}{C}}-CH_3 + CO_2 + ACP-SH$$

acetoacetyl-S-ACP

The product of reaction 27-14, acetoacetyl-S-ACP, remains anchored to fatty acid synthetase, which now catalyzes the reduction of the ketone functional group to an alcohol. Note that the carbonyl group bonded to the sulfur atom is not a ketone, but rather is part of a thioester.

Step 3:

(27-15)

$$\text{ACP}-\text{S}-\overset{\overset{\text{O}}{\|}}{\text{C}}-\text{CH}_2-\overset{\overset{\text{O}}{\|}}{\text{C}}-\text{CH}_3 \xrightarrow[\substack{\text{NADPH} \quad \text{NADP}^+ \\ + \text{H}^+}]{\text{fatty acid synthetase}} \text{ACP}-\text{S}-\overset{\overset{\text{O}}{\|}}{\text{C}}-\text{CH}_2-\overset{\overset{\text{OH}}{|}}{\underset{\underset{\text{H}}{|}}{\text{C}}}-\text{CH}_3$$

acetoacetyl-S-ACP

Notice that this step is somewhat similar to the reverse of step 4 of fatty acid catabolism (reaction 27-8). However, a different enzyme is required and the coenzyme NADPH is oxidized in reaction 27-15. Also, in this case the substrate is bound to an ACP portion of fatty acid synthetase (not to CoA), and this reaction occurs in the cytoplasm (not in mitochondria).

In step 4 of fatty acid synthesis, the hydroxyl group is removed in a dehydration reaction, forming an α, β-carbon-carbon double bond:

Step 4:

(27-16)

$$\text{ACP}-\text{S}-\overset{\overset{\text{O}}{\|}}{\text{C}}-\overset{\overset{\text{H}}{|}}{\text{CH}}-\overset{\overset{\text{OH}}{|}}{\text{CH}}-\text{CH}_3 \xrightarrow{\substack{\text{fatty acid} \\ \text{synthetase}}} \text{ACP}-\text{S}-\overset{\overset{\text{O}}{\|}}{\text{C}}-\underset{\alpha}{\text{CH}}=\underset{\beta}{\text{CH}}-\text{CH}_3 + \text{H}_2\text{O}$$

Again, this reaction is somewhat similar to the reverse of step 3 of fatty acid catabolism (reaction 27-7).

In step 5 of fatty acid synthesis, the C=C double bond is reduced:

Step 5:

(27-17)

$$\text{ACP}-\text{S}-\overset{\overset{\text{O}}{\|}}{\text{C}}-\text{CH}=\text{CH}-\text{CH}_3 \xrightarrow[\substack{\text{NADPH} \quad \text{NADP}^+ \\ + \text{H}^+}]{\substack{\text{fatty acid} \\ \text{synthetase}}} \text{ACP}-\text{S}-\overset{\overset{\text{O}}{\|}}{\text{C}}-\overset{\overset{\text{H}}{|}}{\text{CH}}-\overset{\overset{\text{H}}{|}}{\text{CH}}-\text{CH}_3$$

Again, as in step 3, this reduction step is coupled to the oxidation of a molecule of NADPH. Step 5 completes one round of action of the fatty acid synthetase enzyme. Butyric acid, $\text{CH}_3-\text{CH}_2-\text{CH}_2-\text{COOH}$, a four-carbon carboxylic acid, could be obtained by the hydrolysis of this thioester. However, long-chain fatty acids are needed by the human body, so the product of step 5 is used for further rounds of fatty acid synthetase action. Each subsequent round begins with the binding of a malonyl group to the free ACP portion of the fatty acid synthetase (reaction 27-13). That is, additional two-carbon-atom pieces are added from additional malonyl-S-ACP molecules. The synthesis of a 16-carbon fatty acid thus requires seven rounds of fatty acid synthetase action. This process is summarized in Figure 27-7. The acyl group produced by seven rounds of fatty acid synthetase action (the palmitoyl group) is then transferred to a molecule of coenzyme A in the reaction

(27-18)

$$\text{ACP}-\text{S}-\overset{\overset{\text{O}}{\|}}{\text{C}}-(\text{CH}_2)_{14}-\text{CH}_3 + \text{CoA}-\text{SH} \rightleftharpoons \text{CoA}-\text{S}-\overset{\overset{\text{O}}{\|}}{\text{C}}-(\text{CH}_2)_{14}-\text{CH}_3 + \text{ACP}-\text{SH}$$

palmitoyl-S-ACP   palmitoyl-S-CoA

The hydrolysis of this palmitoyl-S-CoA thioester yields palmitic acid, the usual product of fatty acid synthetase action. The net overall reaction for the synthesis of palmitic acid from acetyl-S-CoA is

(27-19)

$$8\text{CH}_3-\overset{\overset{\text{O}}{\|}}{\text{C}}-\text{S}-\text{CoA} + 7\text{ATP} + \text{H}_2\text{O} \longrightarrow \text{CH}_3(\text{CH}_2)_{14}\overset{\overset{\text{O}}{\|}}{\text{C}}-\text{OH} + 8\text{CoA}-\text{SH}$$

$$+ \ 14\text{NADPH} + 14\text{H}^+ \qquad\qquad\qquad\qquad\qquad\qquad + \ 7\text{ADP} + 7\text{P}_i + 14\text{NADP}^+$$

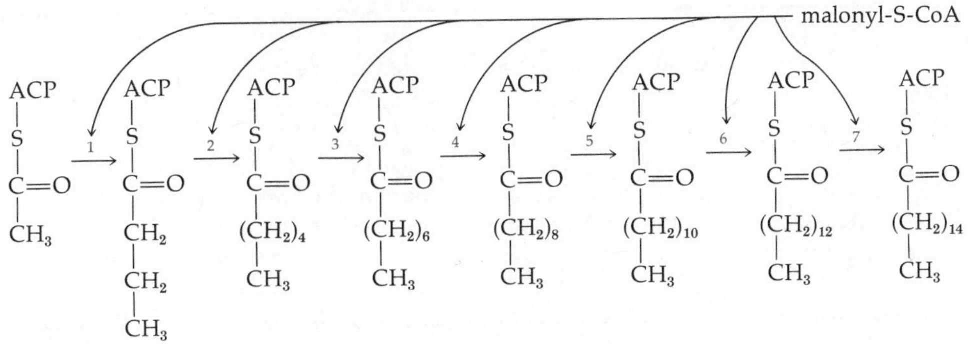

Figure 27-7   Synthesis of the 16-carbon-atom palmitic acid requires seven rounds of fatty acid synthesis. In each round, malonyl-S-CoA donates two carbon atoms.

Each round of the fatty acid synthesis pathway is fed by malonyl-S-CoA, which is immediately decarboxylated. Thus, only two carbon atoms are added per round, and you can now see why the majority of fatty acids found in nature have even-numbered chains of carbon atoms. The synthesis of unsaturated fatty acids begins with the synthesis of the corresponding saturated fatty acids. Double bonds are then introduced by a specific enzyme.

Conversion of Glucose into Lipids

Now let us consider why the human body converts glucose into lipids when the supply of glucose is in excess of the immediate need for energy. Take palmitic acid ($C_{16}H_{32}O_2$) as an example. To synthesize palmitic acid, eight acetyl-S-CoA molecules are needed. In glycolysis, one glucose molecule yields two molecules of acetyl-S-CoA (Chapter 25). Therefore, four glucose molecules are needed to make one molecule of palmitic acid. The overall reaction for this process is

(27-20) $$4C_6H_{12}O_6 + O_2 \longrightarrow C_{16}H_{32}O_2 + 8CO_2 + 8H_2O$$

Synthesis of palmitic acid from acetyl-S-CoA requires ATP hydrolysis, but palmitic acid is a rich potential source for ATP synthesis. Is the process of converting glucose molecules into molecules of palmitic acid worthwhile? In liver cells, complete catabolism of glucose to CO_2 and H_2O yields 38 ATP/glucose, so complete catabolism of four glucose molecules yields $4 \times 38 = 152$ ATP. Complete catabolism of palmitic acid yields 129 ATP, so the potential synthesis of 23 ATP molecules ($152 - 129$) is lost in converting glucose to palmitic acid. However, palmitic acid is a much more compact storage form for energy. One mole of palmitic acid weighs 256 g, whereas the molecular weight of glucose is 180. The potential supply of ATP per gram for each of these molecules is

(27-21) $$\frac{38 \text{ moles of ATP}}{1 \text{ mole of glucose}} \times \frac{1 \text{ mole of glucose}}{180 \text{ g}} = 0.21 \text{ mole of ATP/g of glucose}$$

(27-22) $$\frac{129 \text{ moles of ATP}}{1 \text{ mole of palmitic acid}} \times \frac{1 \text{ mole of palmitic acid}}{256 \text{ g}}$$
$$= 0.50 \text{ mole of ATP/g of palmitic acid}$$

More than twice as much ATP can be produced from the complete catabolism of 1 g of palmitic acid as from 1 g of glucose. If you used only glucose (or glycogen) to store energy, you would need to weigh a good deal more. And if you weighed more, you would need to use more energy to move around.

Exercise 27-2

How many molecules of glucose are required for the synthesis of one molecule of the 20-carbon-atom fatty acid, arachidic acid?

Exercise 27-3

(a) How many molecules of ATP are synthesized from the complete oxidation of arachidic acid? (b) Compare this amount to the amount of ATP produced by the complete oxidation in the liver of the number of glucose molecules you arrived at in Exercise 27-2.

27-6 THE BIOSYNTHESIS OF TRIGLYCERIDES AND PHOSPHOGLYCERIDES

In this section we shall see how newly synthesized fatty acids are used to form two major classes of lipid molecules, triglycerides and phosphoglycerides. The synthesis of both of these classes of lipids requires glycerol 3-phosphate in addition to fatty acyl-S-CoA molecules. Glycerol 3-phosphate can be formed from dihydroxyacetone phosphate by the reaction

(27-23)

$$
\begin{array}{c}
\text{H}_2\text{C}-\text{OH} \\
| \\
\text{C}=\text{O} \\
| \\
\text{H}_2\text{CO}-\text{(P)}
\end{array}
\quad
\xrightleftharpoons[\substack{\text{NADH} \\ + \text{H}^+}]{\substack{\text{glycerol} \\ \text{phosphate} \\ \text{dehydrogenase}}}\quad \text{NAD}^+
\quad
\begin{array}{c}
\text{H}_2\text{COH} \\
| \\
\text{HCOH} \\
| \\
\text{H}_2\text{CO}-\text{(P)}
\end{array}
$$

dihydroxyacetone phosphate   glycerol 3-phosphate

Reaction 27-23 may look familiar, since it is exactly the reverse of reaction 27-3. Glycerol phosphate dehydrogenase is used for both the synthesis and catabolism of glycerol 3-phosphate.

The next step is the formation of a phosphatidic acid molecule by the esterification of two fatty acids to the alcohol groups on the glycerol 3-phosphate. This reaction is catalyzed by a transferase enzyme:

(27-24)

$$
\begin{array}{c}
\text{H}_2\text{COH} \\
| \\
\text{HCOH} \\
| \\
\text{H}_2\text{CO}-\text{(P)}
\end{array}
\;+\;
\begin{array}{c}
\overset{\displaystyle O}{\overset{\|}{\text{CoA}-\text{S}-\text{C}}}-\text{R}_1 \\
\overset{\displaystyle O}{\overset{\|}{\text{CoA}-\text{S}-\text{C}}}-\text{R}_2
\end{array}
\longrightarrow 2\text{CoA}-\text{SH} \;+\;
\begin{array}{c}
\overset{\displaystyle O}{\text{H}_2\text{C}-\text{O}-\overset{\|}{\text{C}}-\text{R}_1} \\
\overset{\displaystyle O}{\text{HC}-\text{O}-\overset{\|}{\text{C}}-\text{R}_2} \\
\text{H}_2\text{C}-\text{O}-\text{(P)}
\end{array}
$$

glycerol 3-phosphate   two fatty acyl-S-CoA's   a phosphatidic acid

The formation of a triglyceride proceeds by removal of the phosphate group from a phosphatidic acid molecule,

(27-25)

$$
\text{H}_2\text{O} \;+\;
\begin{array}{c}
\overset{\displaystyle O}{\text{H}_2\text{C}-\text{O}-\overset{\|}{\text{C}}-\text{R}_1} \\
\overset{\displaystyle O}{\text{HC}-\text{O}-\overset{\|}{\text{C}}-\text{R}_2} \\
\text{H}_2\text{C}-\text{O}-\text{(P)}
\end{array}
\longrightarrow
\begin{array}{c}
\overset{\displaystyle O}{\text{H}_2\text{C}-\text{O}-\overset{\|}{\text{C}}-\text{R}_1} \\
\overset{\displaystyle O}{\text{HC}-\text{O}-\overset{\|}{\text{C}}-\text{R}_2} \\
\text{H}_2\text{C}-\text{OH}
\end{array}
\;+\; \text{P}_i
$$

a phosphatidic acid   a diglyceride

followed by the formation of a third fatty acid ester:

(27-26)

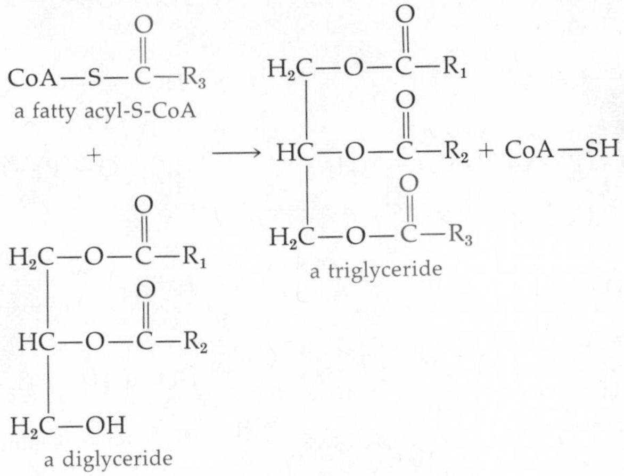

$$CoA—S—\overset{\overset{\displaystyle O}{\|}}{C}—R_3$$

a fatty acyl-S-CoA

$$+$$

$$H_2C—O—\overset{\overset{\displaystyle O}{\|}}{C}—R_1$$
$$HC—O—\overset{\overset{\displaystyle O}{\|}}{C}—R_2$$
$$H_2C—OH$$

a diglyceride

$$\longrightarrow$$

$$H_2C—O—\overset{\overset{\displaystyle O}{\|}}{C}—R_1$$
$$HC—O—\overset{\overset{\displaystyle O}{\|}}{C}—R_2 + CoA—SH$$
$$H_2C—O—\overset{\overset{\displaystyle O}{\|}}{C}—R_3$$

a triglyceride

In the synthesis of phosphoglycerides, human cells also use reaction 27-25 to produce diglyceride intermediates. We shall consider the synthesis of two common phosphoglycerides: phosphatidyl ethanolamine and phosphatidyl serine. The synthesis of phosphatidyl enthanolamine begins with the phosphorylation of ethanolamine to give phosphoethanolamine:

(27-27)

$$H_2N—CH_2—CH_2—OH \xrightarrow[\substack{\\ ATP \quad ADP}]{ethanolamine\ kinase} H_2N—CH_2—CH_2—O—\textcircled{P}$$

ethanolamine $\qquad\qquad\qquad\qquad$ phosphoethanolamine

Phosphoethanolamine is next activated by coupling it to a molecule of CMP in the reaction

(27-28)

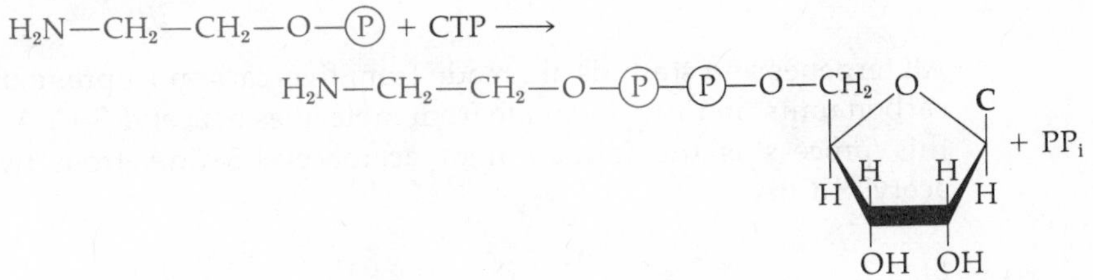

$$H_2N—CH_2—CH_2—O—\textcircled{P} + CTP \longrightarrow$$

$$H_2N—CH_2—CH_2—O—\textcircled{P}—\textcircled{P}—O—CH_2 \quad + PP_i$$

CDP-ethanolamine

CTP is similar in structure to ATP, except that it contains the base cytosine instead of adenine (Chapter 24).

Recall from Chapter 25 that the synthesis of polysaccharides also proceeds via similar nucleotide-activated intermediates. Phosphatidyl ethanolamine is then made in the following reaction:

(27-29)

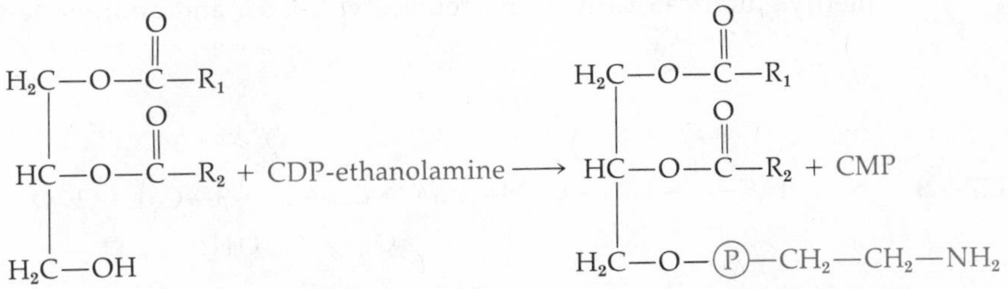

$$H_2C—O—\overset{\overset{\displaystyle O}{\|}}{C}—R_1$$
$$HC—O—\overset{\overset{\displaystyle O}{\|}}{C}—R_2 + CDP\text{-ethanolamine} \longrightarrow$$
$$H_2C—OH$$

a diglyceride

$$H_2C—O—\overset{\overset{\displaystyle O}{\|}}{C}—R_1$$
$$HC—O—\overset{\overset{\displaystyle O}{\|}}{C}—R_2 + CMP$$
$$H_2C—O—\textcircled{P}—CH_2—CH_2—NH_2$$

a phosphatidyl ethanolamine

Synthesis of phosphatidyl serine involves the exchange of the ethanolamine portion of a phosphatidyl ethanolamine molecule for the amino acid serine:

(27-30)

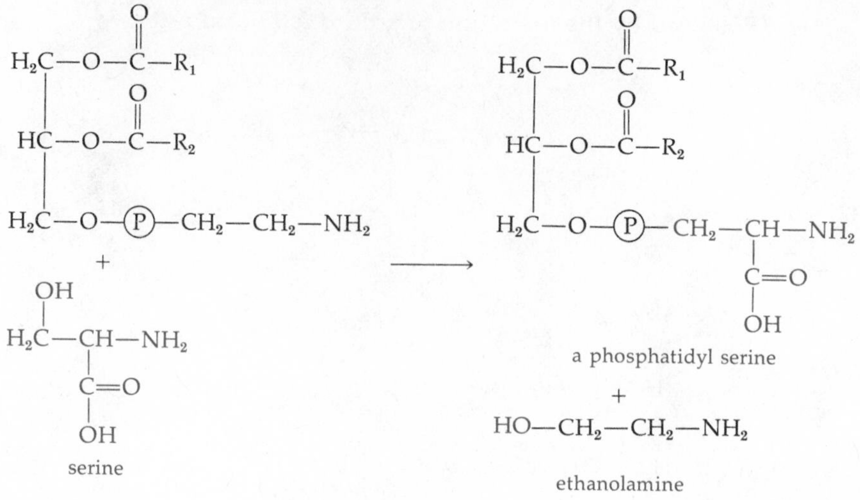

$$H_2C-O-\overset{\overset{\displaystyle O}{\|}}{C}-R_1$$
$$HC-O-\overset{\overset{\displaystyle O}{\|}}{C}-R_2$$
$$H_2C-O-\textcircled{P}-CH_2-CH_2-NH_2$$
+
$$\overset{\overset{\displaystyle OH}{|}}{H_2C-CH-NH_2}$$
$$\overset{|}{C=O}$$
$$\overset{|}{OH}$$
serine

$\longrightarrow$

$$H_2C-O-\overset{\overset{\displaystyle O}{\|}}{C}-R_1$$
$$HC-O-\overset{\overset{\displaystyle O}{\|}}{C}-R_2$$
$$H_2C-O-\textcircled{P}-CH_2-CH-NH_2$$
$$\overset{|}{C=O}$$
$$\overset{|}{OH}$$
a phosphatidyl serine
+
$$HO-CH_2-CH_2-NH_2$$
ethanolamine

Exercise 27-4
What type of molecule is the last common intermediate in the synthesis of triglycerides and phosphoglycerides?

Exercise 27-5
What enzyme-catalyzed step in the synthesis of triglycerides is also used in their degradation?

27-7 BIOSYNTHESIS OF TERPENES AND STEROIDS

All terpenes and steroids are made from five-carbon isoprene units. These five-carbon units, in turn, are made from molecules of acetyl-S-CoA. The first step in this process is the formation of acetoacetyl-S-CoA from two molecules of acetyl-S-CoA:

Step 1:

(27-31)

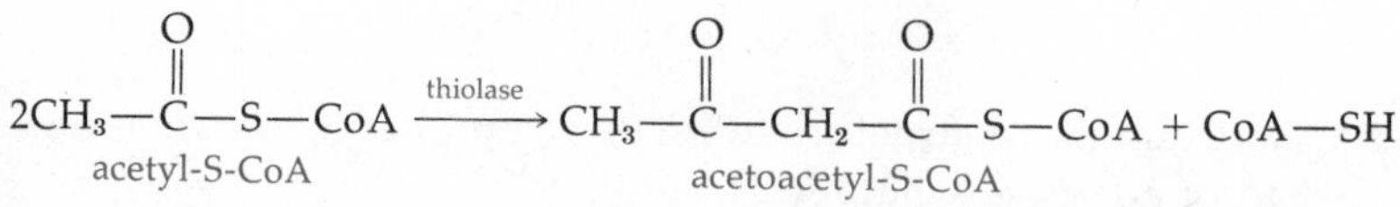

$$2CH_3-\overset{\overset{\displaystyle O}{\|}}{C}-S-CoA \xrightarrow{\text{thiolase}} CH_3-\overset{\overset{\displaystyle O}{\|}}{C}-CH_2-\overset{\overset{\displaystyle O}{\|}}{C}-S-CoA + CoA-SH$$
acetyl-S-CoA   acetoacetyl-S-CoA

Acetoacetyl-S-CoA is sometimes produced in excessively large amounts by people with diabetes, for reasons we shall discuss shortly. The next step in steroid biosynthesis is the formation of a six-carbon-atom unit called β-hydroxy-β-methylglutaryl-S-CoA from acetoacetyl-S-CoA and another acetyl-S-CoA:

Step 2:

(27-32)

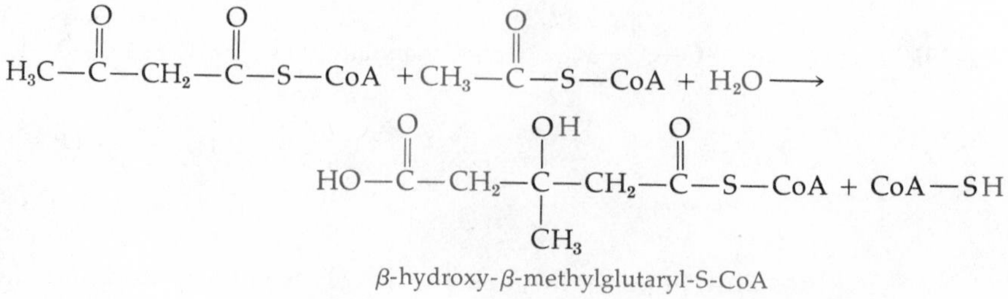

$$H_3C-\overset{\overset{\displaystyle O}{\|}}{C}-CH_2-\overset{\overset{\displaystyle O}{\|}}{C}-S-CoA + CH_3-\overset{\overset{\displaystyle O}{\|}}{C}-S-CoA + H_2O \longrightarrow$$

$$HO-\overset{\overset{\displaystyle O}{\|}}{C}-CH_2-\overset{\overset{\displaystyle OH}{|}}{\underset{\underset{\displaystyle CH_3}{|}}{C}}-CH_2-\overset{\overset{\displaystyle O}{\|}}{C}-S-CoA + CoA-SH$$

β-hydroxy-β-methylglutaryl-S-CoA

In five subsequent reactions of this pathway, β-hydroxy-β-methylglutaryl-S-CoA is converted to 3,3-dimethylallyl pyrophosphate, an isoprene derivative. The overall reaction for these five steps is

(27-33)

$$HO{-}\underset{\underset{O}{\|}}{C}{-}CH_2{-}\underset{\underset{CH_3}{|}}{\overset{\overset{OH}{|}}{C}}{-}CH_2{-}\underset{\underset{O}{\|}}{C}{-}S{-}CoA \longrightarrow CH_3{-}\underset{\underset{CH_3}{|}}{C}{=}CH{-}CH_2{-}\text{P}{-}\text{P}$$

$$+\ 3ATP + 2NADPH + 2H^+ \qquad\qquad +\ 3ADP + P_i + 2NADP^+$$
$$+\ CoA{-}SH + CO_2 + 2H_2O$$

Note that a total of three acetyl-S-CoA molecules are required for the synthesis of one isoprene unit (shown in blue). Subsequent steps in this pathway involve joining isoprene units together to give terpenes with 10 carbon atoms, 15 carbon atoms, and so on. A 30-carbon-atom terpene called squalene is used as the precursor for cholesterol biosynthesis. In humans all other steroids are made from cholesterol.

Exercise 27-6
How many acetyl-S-CoA molecules are needed to synthesize one molecule of cholesterol?

27-8 THE REGULATION OF LIPID METABOLISM

Lipid molecules are all made from, and degraded to, acetyl-S-CoA molecules, which is also the fuel used in the TCA cycle (see Chapter 25). However, acetyl-S-CoA is not used to make fatty acids and other lipids unless there is an adequate supply to fuel the TCA cycle. In fact, when acetyl-S-CoA is needed, fats are broken down to supply it. Conversely, when there is more than enough acetyl-S-CoA present in a cell, it is used to make triglycerides for the fat depots. How are the synthesis and degradation of lipid molecules controlled in cells? There are several control mechanisms, each of which is very logical.

First, the synthesis of fatty acids depends upon the availability of NADPH, which is needed for the reduction steps in this biosynthetic pathway. NADPH can be generated by the pentose phosphate pathway (Chapter 25). Another pathway is also used to generate NADPH for fatty acid biosynthesis. In one of the reactions in this pathway, malic acid is oxidized to pyruvic acid, while $NADP^+$ is reduced to form $NADPH + H^+$:

(27-34)

$$HO{-}\underset{\underset{O}{\|}}{C}{-}CH_2{-}\underset{\underset{OH}{|}}{\overset{\overset{H}{|}}{C}}{-}\underset{\underset{O}{\|}}{C}{-}OH \xrightarrow[\underset{NADP^+\ \ \ NADPH}{\qquad +\ H^+\qquad}]{\text{malic enzyme}} H_3C{-}\underset{\underset{O}{\|}}{C}{-}\underset{\underset{O}{\|}}{C}{-}OH + CO_2$$

malic acid   pyruvic acid

When the cell's supply of acetyl-S-CoA exceeds that which is needed to run the TCA cycle, the excess acetyl-S-CoA activates malic enzyme to produce NADPH by reaction 27-34.

Other controls of fatty acid synthesis take the cell's energy supply into account. If the cell has a lot of ATP, the excess ATP can inactivate the TCA cycle enzyme, isocitrate dehydrogenase, by binding to an allosteric site. With that enzyme turned off, acetyl-S-CoA can be used for fatty acid synthesis instead of serving as fuel for the TCA cycle. A major control of fatty acid biosynthesis involves acetyl-S-CoA carboxylase. This enzyme, which synthesizes the malonyl-S-CoA required for fatty acid biosynthesis, is allosterically activated by the TCA cycle intermediates citric acid and isocitric acid. On the other hand, both acetyl-S-CoA carboxylase and fatty acid synthetase are inhibited by palmitoyl-S-CoA, the major product of the reactions catalyzed by fatty acid synthetase.

Several hormones also aid in the regulation of lipid metabolism. Epinephrine and glucagon stimulate the hydrolysis of triglycerides. Remember that epinephrine and glucagon also stimulate the breakdown of glycogen to glucose (see Chapter 25). Thus, these two hormones tell cells that energy is needed, and stimulate ATP formation from the catabolism of both fats and carbohydrates. Insulin, on the other hand, tells cells to store energy by synthesizing fats and glycogen. Controls for lipid metabolism are summarized in Figure 27-8.

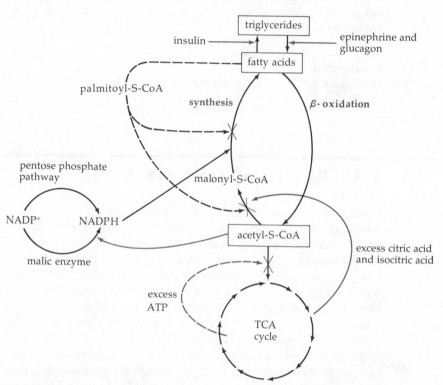

Figure 27-8   Regulation of lipid metabolism. Fatty acid synthesis is stimulated by citric acid, isocitric acid, acetyl-S-CoA, and by excess ATP (which inhibits operation of the TCA cycle). Fatty acid synthesis is inhibited by palmitoyl-S-CoA. Triglyceride synthesis is stimulated by insulin, whereas triglyceride hydrolysis is stimulated by epinephrine and glucagon.

In normal individuals, insulin stimulates cells to synthesize fatty acids and triglycerides when there is an excess supply of acetyl-S-CoA. However, in diabetic individuals, who have insufficient and/or inactive insulin, excess acetyl-S-CoA is not converted into fatty acids, but rather is used to make inordinate amounts of acetoacetyl-S-CoA (reaction 27-31).

Acetoacetyl-S-CoA is used to synthesize β-hydroxy-β-methylglutaryl-S-CoA (reaction 27-32), an intermediary used in terpene and steroid synthesis. In diabetics, however, there is a large excess of β-hydroxy-β-methylglutaryl-S-CoA, and most of it is converted to acetoacetic acid in the reaction

(27-35)

$$CoA—S—\overset{\overset{O}{\|}}{C}—CH_2—\overset{\overset{OH}{|}}{\underset{\underset{CH_3}{|}}{C}}—CH_2—\overset{\overset{O}{\|}}{C}—OH \longrightarrow$$

β-hydroxy-β-methylglutaryl-S-CoA

$$H_3C—\overset{\overset{O}{\|}}{C}—S—CoA + H_3C—\overset{\overset{O}{\|}}{C}—CH_2—\overset{\overset{O}{\|}}{C}—OH$$

acetyl-S-CoA   acetoacetic acid

Acetoacetic acid is also converted to either β-hydroxybutyric acid (reaction 27-36) or to acetone (reaction 27-37).

(27-36)

$$H_3C—\overset{\overset{O}{\|}}{C}—CH_2—\overset{\overset{O}{\|}}{C}—OH \xrightarrow[\substack{NADH \quad NAD^+ \\ + H^+}]{dehydrogenase} H_3C—\overset{\overset{OH}{|}}{\underset{\underset{H}{|}}{C}}—CH_2—\overset{\overset{O}{\|}}{C}—OH$$

acetoacetic acid   β-hydroxybutyric acid

(27-37)

$$H_3C—\overset{\overset{O}{\|}}{C}—CH_2—\overset{\overset{O}{\|}}{C}—OH \xrightarrow{decarboxylase} H_3C—\overset{\overset{O}{\|}}{C}—CH_3 + CO_2$$

acetoacetic acid   acetone

In diabetics, abnormally large amounts of acetoacetic acid, β-hydroxybutyric acid, and acetone, collectively referred to as **ketone bodies,** can accumulate in the blood. Excessive amounts of acetoacetic acid and β-hydroxybutyric acid lower the blood pH and result in a condition called metabolic acidosis, a serious condition that can cause death (see Chapter 29). Acetone can be detected in the urine, and even on the breath, of severely affected diabetics.

Exercise 27-7
How can the supply of NADPH for fatty acid synthesis be increased?

27-9   SUMMARY

1. Acetyl-S-CoA is the starting material used for the synthesis of a large variety of lipids. Catabolism of several lipids also results in acetyl-S-CoA production.

2. The first step in the catabolism of triglycerides and phosphoglycerides is their hydrolysis by lipase enzymes.

3. Glycerol 3-phosphate is converted to dihydroxyacetone phosphate for catabolism in the glycolysis pathway. This compound can also be synthesized from dihydroxyacetone phosphate by the reverse of the same enzymatic reactions.

4. Catabolism of fatty acids begins in the cytoplasm, where they are activated by bonding to coenzyme A molecules. The activated fatty acids are then transported into mitochondria by carnitine molecules.

5. Degradation of fatty acids occurs in mitochondria via the β-oxidation pathway.

6. The β-oxidation pathway cleaves fatty acids into two-carbon-atom fragments, producing acetyl-S-CoA and reduced coenzymes.

7. Fatty acid synthesis occurs in the cytoplasm of human cells, and is not simply a reversal of the catabolic pathway for fatty acids.

8. In fatty acid synthesis, the two-carbon-atom fragments added to growing fatty acid molecules come from malonyl-S-CoA.

9. Fatty acid synthesis requires oxidation of the coenzyme NADPH, which is supplied by the pentose phosphate pathway and the malic enzyme reaction.

10. Triglycerides are made by (1) esterification of two fatty acids to a molecule of glycerol 3-phosphate to give a molecule of phosphatidic acid, (2) removal of the phosphate to give a diglyceride, and (3) esterification with a third fatty acid molecule.

11. Phosphoglycerides are made from diglycerides by attachment of phosphorylated molecules.

12. Fatty acid synthesis is controlled by (1) the available supply of NADPH, (2) the supply of ATP and TCA cycle intermediates, and (3) the concentration of palmitoyl-S-CoA.

13. Hydrolysis of triglycerides is stimulated by glucagon and epinephrine, whereas triglyceride synthesis is stimulated by insulin.

14. Acetoacetic acid, β-hydroxybutyric acid, and acetone, collectively referred to as ketone bodies, are produced in excessive amounts in diabetics. The acidic ketone bodies decrease the blood pH, resulting in metabolic acidosis, which can cause death.

PROBLEMS

1. Suppose that a particular triglyceride contains three hexanoic (six-carbon) acid esters. Outline the route whereby the entire molecule is catabolized to acetyl-S-CoA. Do all of the carbon atoms in this triglyceride become parts of acetyl groups? Explain.

2. Write a balanced equation for the action of pancreatic lipase on the triglyceride in Problem 1.

3. Calculate the total ATP production per mole by the complete aerobic catabolism in liver cells of the triglyceride in Problem 1.

4. Compare the structure and function of coenzyme A and the acyl carrier portion of fatty acid synthetase.

5. What is the role of phosphatidic acids in phosphoglyceride biosynthesis?

6. What is one similarity between the syntheses of phosphatidyl ethanolamine from ethanolamine and glycogen from glucose?

7. Outline the various controls used to regulate lipid metabolism and explain the role of each stimulator or inhibitor.

8. Draw structural formulas for the products of the following reactions:

(a)

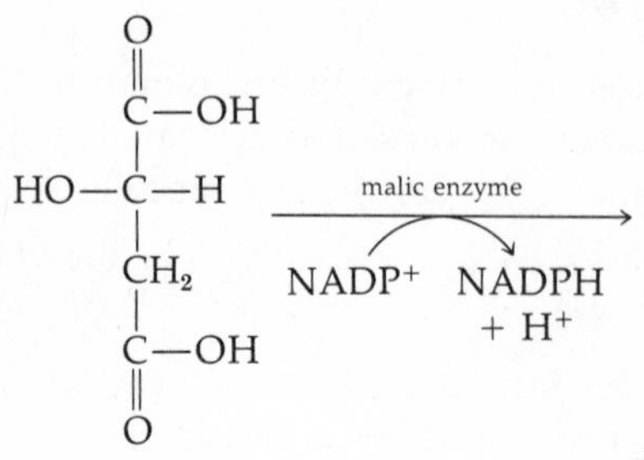

$$\text{HO}-\underset{\underset{\displaystyle C}{\overset{\displaystyle |}{\underset{\displaystyle CH_2}{|}}}}{\overset{\displaystyle \overset{O}{\overset{\|}{C}}-OH}{\underset{\displaystyle |}{C}}}-H \xrightarrow[\text{NADP}^+ \quad \text{NADPH} + \text{H}^+]{\text{malic enzyme}}$$

C—OH at bottom with O double bond

(b) $CH_3-\overset{O}{\overset{\|}{C}}-S-CoA + CO_2 \xrightarrow{\substack{\text{acetyl-S-CoA} \\ \text{carboxylase}}}$

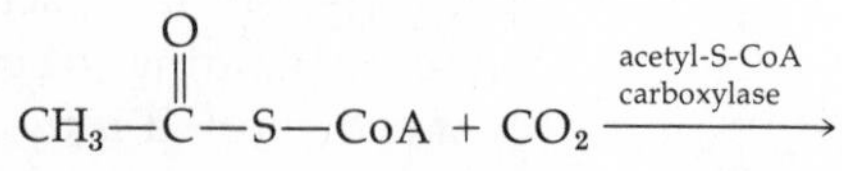

(c) $2[CH_3-\overset{\overset{\displaystyle O}{\|}}{C}-S-CoA] \xrightarrow{\text{thiolase}} HS-CoA +$

(d) $\begin{array}{l} H_2C-O-\overset{\overset{\displaystyle O}{\|}}{C}-R_1 \\ \quad\;\; \overset{\overset{\displaystyle O}{\|}}{} \\ HC-O-\overset{\overset{\displaystyle O}{\|}}{C}-R_2 + 3H_2O \xrightarrow{\text{lipase}} \\ \quad\;\; \overset{\overset{\displaystyle O}{\|}}{} \\ H_2C-O-\overset{\overset{\displaystyle O}{\|}}{C}-R_3 \end{array}$

SOLUTIONS TO EXERCISES

27-1 A six-carbon fatty acid: **ATP/mole**
        activation -2
        $2(\beta\text{-oxidation} + ETS)$ 10
        $3(TCA + ETS)$ $\underline{36}$
                                                          44 ATP/mole

        Glucose:
        glycolysis + ETS $+8$ (in liver)
        2(pyruvate dehydrogenase + ETS)   6
        2(TCA + ETS) $\underline{24}$
                                                       38 ATP/mole

27-2 Since every glucose molecule can yield 2 molecules of acetyl-S-CoA, and 10 molecules of acetyl-S-CoA are required to make 1 molecule of arachidic acid, the total is

$$\frac{1 \text{ glucose}}{2 \text{ acetyl-S-CoA}} \times \frac{10 \text{ acetyl-S-CoA}}{1 \text{ arachidic acid}} = \frac{5 \text{ glucose}}{1 \text{ arachidic acid}}$$

27-3 (a) Arachidic acid: **ATP/mole**
        activation -2
        9 rounds (β-oxidation + ETS)   45
        10 acetyl-S-CoA (TCA + ETS) $\underline{120}$
                                                         163 ATP/mole

        (b) Glucose: 5 moles of glucose $\times$ 38 ATP/mole (in liver) = 190 moles of ATP/5 moles of glucose.

27-4 A diglyceride.

27-5 The glycerol phosphate dehydrogenase reaction,

$$\begin{array}{l} H_2COH \\ \;\;| \\ \;\;C=O \\ \;\;| \\ H_2CO-\textcircled{P} \end{array} + NADH + H^+ \underset{\text{degradation}}{\overset{\text{synthesis}}{\rightleftharpoons}} \begin{array}{l} H_2COH \\ \;\;| \\ HCOH \\ \;\;| \\ H_2CO-\textcircled{P} \end{array} + NAD^+$$

dihydroxyacetone   glycerol 3-phosphate
  phosphate

27-6 Three acetyl-S-CoA molecules are used to make each isoprene unit containing five carbon atoms (the sixth is lost as CO_2). Cholesterol is made from the terpene squalene, which consists of six isoprene units. Therefore, a total of $6 \times 3 = 18$ acetyl-S-CoA molecules are required for each cholesterol molecule.

27-7 The NADPH supply can be increased by activation of malic enzyme by acetyl-S-CoA.

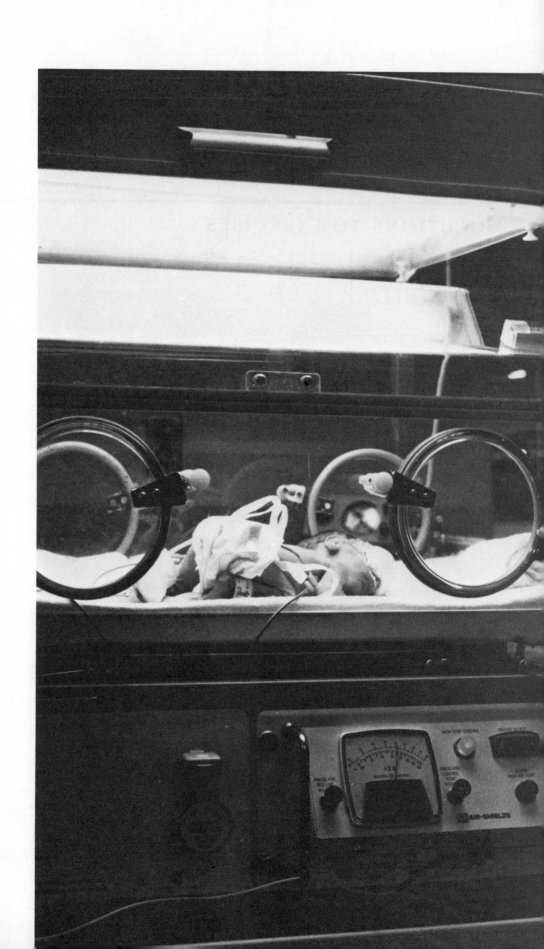

CHAPTER 28

Metabolism of Nitrogenous Compounds

28-1 INTRODUCTION

We have seen the importance of nitrogen-containing functional groups to the structure and function of a large number of biomolecules, including amino acids, proteins, nucleic acids, and some vitamins and lipids. When we discussed the biosynthesis of proteins and nucleic acids, we took for granted the presence in a cell of the requisite amino acids and nucleotides. Since there are 20 amino acids and 4 nucleotides, too much space would be required to discuss the synthesis and degradation of each in detail. However, these nitrogen-containing compounds are vital constituents of the cells in the human body, and we must now consider their metabolism. Fortunately, there are many common features in the metabolic pathways for these compounds. In this chapter we shall discuss the general features of the metabolism of amino acids and some other nitrogen-containing biomolecules. Since amino acids and proteins constitute the bulk of the nitrogen-containing compounds digested and synthesized by the human body, we shall concentrate our efforts on the study of their metabolism.

We shall discuss how the human body digests proteins in order to provide essential amino acids, how proteins are sometimes used as a source of energy, how proteins are constantly being turned over in the body, and how excess nitrogen is incorporated into urea for excretion in urine. In an optional section we shall describe a few of the pathways that lead to the synthesis of some interesting derivatives of amino acids, including some hormones and neurotransmitters. Finally, we shall briefly discuss the degradation of nucleotides. Since most of the reactions discussed in this chapter involve nitrogen-containing functional groups, you may wish to review the physical properties and reactivity of these functional groups, which we discussed in Chapter 18.

The jaundiced (yellow) color of some premature infants is due to a condition known as hyperbilirubinemia, in which the nitrogen-containing compound bilirubin—a degradation product of the heme prosthetic group—cannot be metabolized rapidly enough by the infant's immature liver cells. If not treated, hyperbilirubinemia can cause serious brain damage. This premature infant is being treated for hyperbilirubinemia by phototherapy, in which fluorescent light rays promote the conversion of bilirubin molecules to harmless products. Note that the infant's eyes must be protected from the strong light.

28-2 STUDY OBJECTIVES

After careful study of the material in this chapter, you should be able to:

1. Define the terms protein digestion, protein turnover, nitrogen balance, and amino acid pool.

2. Define the term proteolytic, and predict the products obtained by the action of proteolytic enzymes on peptides.

3. Outline the possible fates of amino acids in the amino acid pool.

4. Draw structural formulas for the products of transamination, oxidative deamination, and direct deamination reactions, given structural formulas for the reactants.

5. Draw structural formulas for the precursors and describe the enzyme-catalyzed reactions that lead to the synthesis of the amino acids alanine, glutamic acid, aspartic acid, glutamine, and asparagine, given structural formulas for these amino acids.

6. Define essential and nonessential amino acids.

7. Write the net overall reaction for the urea cycle, and explain the importance of carbamyl phosphate, glutamic acid, and aspartic acid in this cycle.

8. Describe how the aspartic acid used in the urea cycle is regenerated.

9. Describe the process of nucleic acid digestion.

10. Identify the source of the uric acid produced in the human body and explain its role in gout.

28-3 DIGESTION AND TURNOVER OF PROTEINS

Nitrogen is very abundant on earth. About 80% of air is molecular nitrogen (N_2). Nitrite (NO_2^-) and nitrate (NO_3^-) ions, as well as some ammonia (NH_3), are present in soil. The human body, however, cannot use these forms of nitrogen to synthesize useful nitrogen-containing biomolecules. Our existence depends on a variety of plants and bacteria. Working together, these plants and bacteria produce nitrogen-containing compounds that humans can use. When we ingest these compounds, some of them—including many vitamins—are quickly absorbed by cells in the intestinal tract and distributed by the circulatory system to the cells that need them. The major source of nitrogen in the human diet, however, is proteins, and proteins are too large to be absorbed by cells in the intestines.

Digestion of Dietary Protein

In order for the body to utilize ingested proteins, they must first be broken down into their constituent amino acids. This is accomplished in the digestive tract by a number of enzymes that hydrolyze the peptide bonds of the proteins. This hydrolysis process is aptly called **protein digestion.**

Besides the problem of absorption, there is another reason why the human body bothers to destroy all of the proteins that it eats, instead of immediately putting some of them (e.g., enzymes) to use. The proteins in our diet, unless we are cannibals, are not exactly identical in amino acid sequence to those synthesized in our body. The immune system (Chapter 22) would identify these proteins as foreign invaders and destroy them.

The digestion of proteins begins in the stomach, with the action of the enzyme **pepsin.** Pepsin is secreted into the stomach as a component of gastric juice, which is quite acidic (pH about 1–2). Pepsin is an unusual enzyme in that it is fairly resistant to denaturation by the acidic gastric juice. In fact, pepsin is maximally active at a pH of about 1–2. However, at this pH, ingested proteins are partially denatured and in their unfolded state they are hydrolyzed by pepsin into several smaller peptides. The reaction catalyzed by pepsin is illustrated in Figure 28-1. Pepsin predominately hydrolyzes peptide bonds that involve the aromatic amino acids phenylalanine, tyrosine, or tryptophan.

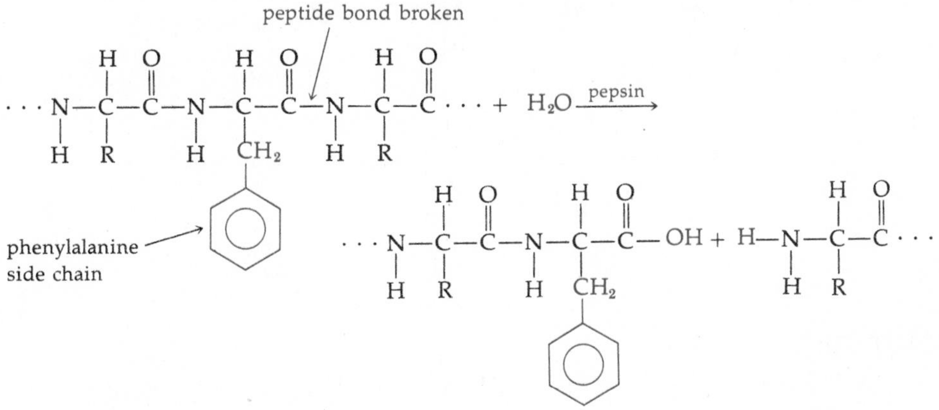

Figure 28-1   The hydrolysis by pepsin of a protein containing phenylalanine. Pepsin also hydrolyzes peptide bonds involving tyrosine, tryptophan, and some other amino acids.

From the stomach, the peptides that have been produced by the action of pepsin travel on to the small intestine, where their digestion continues. Several additional protein-hydrolyzing enzymes, including trypsin, chymotrypsin, carboxypeptidase, and aminopeptidase, are released into the small intestine in pancreatic juice. Protein-hydrolyzing enzymes are usually called **proteolytic** enzymes or proteases. The specificity of these proteolytic enzymes is shown in Table 28-1. Notice that trypsin and chymotrypsin hydrolyze peptide bonds at certain specific points within a peptide, whereas the enzymes carboxypeptidase and aminopeptidase "chew off" one amino acid at a time starting at different ends of a peptide. As a result of the combined action of all of these proteolytic

Table 28-1 Specificity of Intestinal Proteolytic Enzymes

Enzyme	Specificity*
Trypsin	R_2 is arginine or lysine
Chymotrypsin	R_2 is tyrosine, tryptophan, phenylalanine, or leucine
Carboxypeptidase	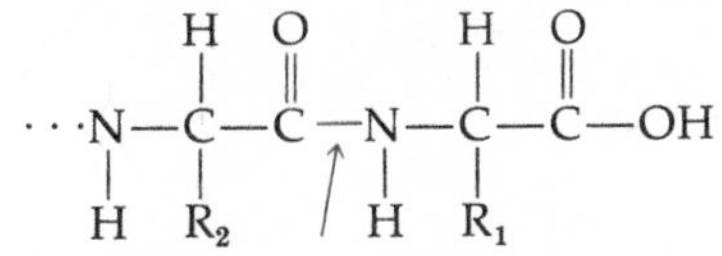 Hydrolyzes C-terminal amino acids
Aminopeptidase	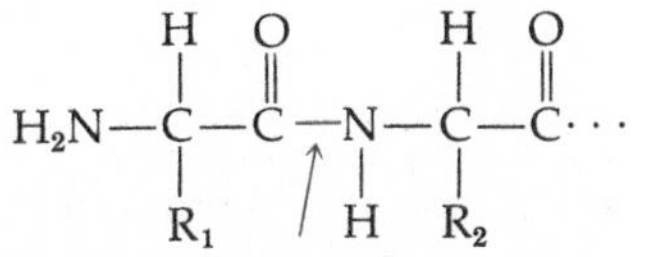 Hydrolyzes N-terminal amino acids

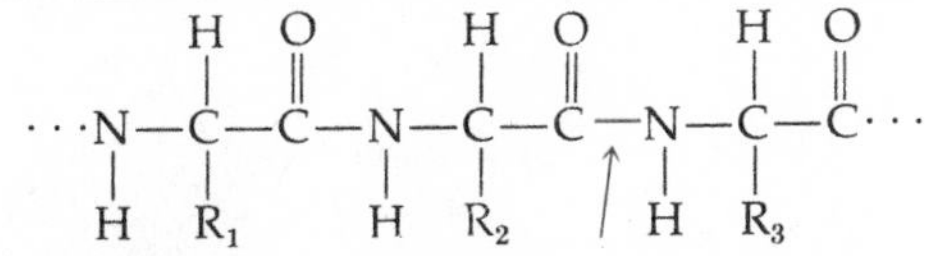

* Arrows indicate the peptide bond hydrolyzed.

enzymes, almost all ingested proteins are finally hydrolyzed into their component amino acids in the small intestine. These free amino acids are then transported across the cells of the intestinal walls into the bloodstream by means of a number of specific active transport proteins (see Figure 28-2).

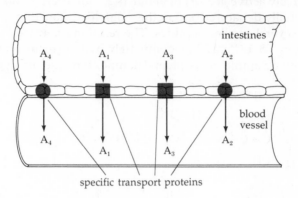

Figure 28-2   The absorption of amino acids from the intestines is accomplished   by several specific active transport proteins.

Nitrogen Balance

Free amino acids in the bloodstream, obtained from ingested proteins and the turnover of body proteins (to be discussed shortly), form what is called the **amino acid pool** (see Figure 28-3). Cells in the body draw upon the amino acid pool for the synthesis of new proteins or, under certain conditions, as a source of energy. Excess amino acids and other small nitrogenous compounds are degraded and the nitrogen atoms excreted from the body in the compound urea (see Section 28-5). Healthy adults normally excrete about as much nitrogen per day as they receive in their diet, and are said to be in **nitrogen balance.** When suffering from certain diseases or when severely undernourished, however, an individual may need to catabolize body proteins for energy. This leads to the excretion of more nitrogen than is received in the diet, a condition referred to as **negative nitrogen balance.** On the other hand, healthy pregnant women and growing children need to ingest more nitrogen than they excrete. They require a state of **positive nitrogen balance.**

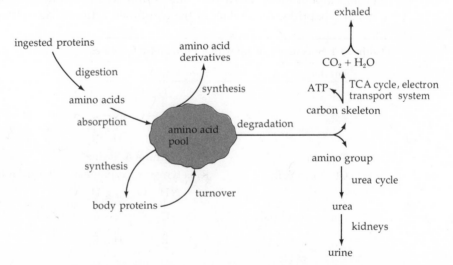

Figure 28-3   The amino acid pool in the bloodstream obtains amino acids from the diet and from protein turnover. The amino acid pool is drawn upon   for the synthesis of proteins and amino acid derivatives, and on occasion to meet energy needs.

Turnover of Body Proteins

All of the various body proteins, including structural proteins, carrier proteins, enzymes, and so forth, are constantly being synthesized and degraded. This continual process of synthesis and breakdown of proteins is called **protein turnover.** Protein turnover provides new active proteins and removes proteins that have been around for a while and that may have become modified or damaged, and may also replace proteins that were not synthesized correctly. Some proteins, such as those in liver cells, are turned over fairly rapidly. About one-half of the proteins in the liver are turned over every 6 days. That is, during a period of 6 days, half of the proteins in the liver are degraded and replaced by newly synthesized ones. Proteins in other parts of the body are degraded and synthesized more slowly. For example, about half of the body's collagen is turned over every 3 years. The normal rate of turnover of various body proteins is generally related to their size, structure, and accessibility to proteolytic enzymes. The free amino acids produced by the degradation of body proteins enter the amino acid pool, where they can be used for synthesis of new proteins and other molecules or, when necessary, for energy production.

Exercise 28-1

List the products that would predominate if the peptide N-lys-phe-met-ser-ala-gly-tyr-C was "digested" by each of the following proteolytic enzymes:

(a) Pepsin   (b) Trypsin   (c) Chymotrypsin   (d) Pepsin, then trypsin

Exercise 28-2

Would you expect a healthy individual on a high-protein diet to be in a state of nitrogen balance? What about an individual on a 2-week fast?

28-4 ASPECTS OF AMINO ACID CATABOLISM

Usually a healthy individual is in nitrogen balance and will catabolize an amount of amino acids approximately equal to the amount ingested. However, when a person fasts or is starving or has diabetes, larger amounts of amino acids are catabolized to provide energy and the total amount of protein in the body becomes depleted. In the catabolism of amino acids, the amino groups of the individual amino acids are removed and then incorporated into the compound urea, which is excreted in urine (see Section 28-5). The remainder of each of the various amino acids, called the **carbon skeletons,** are converted into molecules that are used to produce glucose by the process of gluconeogenesis or are ultimately oxidized via the tricarboxylic acid cycle to produce CO_2 and reduced coenzymes. The reduced coenzymes are then used by the electron transport system to produce ATP from ADP (see Chapter 25). The overall process of amino acid catabolism, when used to supply energy for the body, is summarized in Figure 28-4.

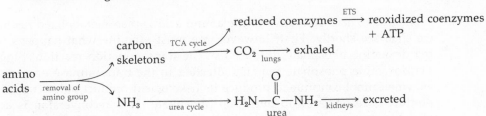

Figure 28-4   An overview of amino acid catabolism for energy production.

The first step in the catabolism of amino acids involves the removal of the amino group. Two general types of enzyme-catalyzed reactions are used for this purpose: transamination and oxidative deamination. Two of the amino acids, serine and threonine, lose their amino groups by a third process called direct deamination.

Transamination

Transamination reactions involve enzymes called **transaminases,** which catalyze the exchange of the amino group located on the α-carbon of particular amino acids for the carbonyl oxygen of the coenzyme **pyridoxal phosphate** (see Figure 28-5), which is bound to the transaminase enzyme.

carbonyl oxygen atom

vitamin B_6

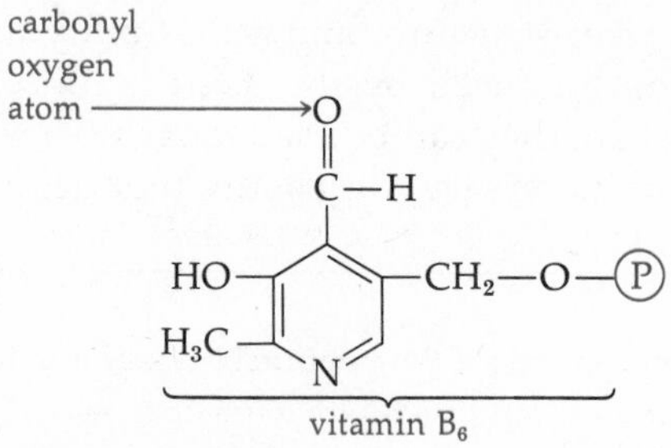

Figure 28-5   The coenzyme pyridoxal phosphate is a derivative of vitamin B_6.

The products of this reaction are the α-keto acid derivative of the reactant amino acid and a molecule of **pyridoxamine phosphate,** which remains attached to the transaminase:

(28-1)

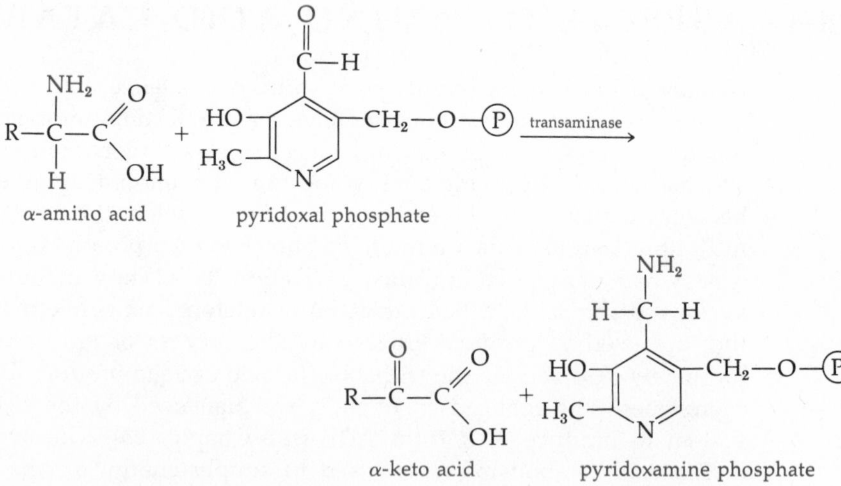

α-amino acid   pyridoxal phosphate   transaminase

α-keto acid   pyridoxamine phosphate

The α-keto acid derivative of the amino acid is then metabolized further, as we shall see shortly. First, however, we must examine what happens to the pyridoxamine phosphate produced in the transamination reaction. Since the pyridoxamine phosphate remains attached to the transaminase enzyme, this enzyme cannot continue to produce its α-keto acid product unless this second product is reconverted to pyridoxal phosphate. This reconversion is accomplished in a second reaction in which the amine group of pyridoxamine phosphate is exchanged for the α-keto oxygen of another α-keto acid, generally, that of α-ketoglutaric acid:

(28-2)

$$\underset{\text{α-ketoglutaric acid}}{\text{HO}-\overset{O}{\overset{\|}{C}}-CH_2-CH_2-\overset{O}{\overset{\|}{C}}-\overset{O}{\overset{\|}{C}}-OH} \quad + \quad \underset{\text{pyridoxamine phosphate}}{\text{pyridoxamine phosphate}} \xrightarrow{\text{transaminase}}$$

$$\underset{\text{glutamic acid}}{\text{HO}-\overset{O}{\overset{\|}{C}}-CH_2-CH_2-\overset{NH_2}{\underset{H}{C}}-\overset{O}{C}\diagdown_{OH}} \quad + \quad \underset{\text{pyridoxal phosphate}}{\text{pyridoxal phosphate}}$$

Notice that reaction 28-2 is basically the reverse of reaction 28-1, except that a *particular* α-keto acid, α-ketoglutaric acid, is used as the donor of the carbonyl oxygen atom.

The combination of reactions 28-1 and 28-2 results in the net transfer of an amino group from a specific amino acid to α-ketoglutaric acid:

(28-3)

$$\text{α-amino acid} + \text{α-ketoglutaric acid} \xrightarrow[\text{pyridoxal phosphate}]{\text{transaminase}} \text{α-keto acid} + \text{glutamic acid}$$

Notice in reaction 28-3 that the amino group becomes part of glutamic acid regardless of the particular reactant amino acid. For example, the enzyme alanine transaminase catalyzes the net conversion of the amino acid alanine to pyruvic acid:

(28-4)

$$\underset{\text{alanine}}{H_3C-\overset{NH_2}{\underset{H}{C}}-\overset{O}{C}\diagdown_{OH}} \quad + \quad \underset{\text{α-ketoglutaric acid}}{HO-\overset{O}{\overset{\|}{C}}-CH_2-CH_2-\overset{O}{\overset{\|}{C}}-\overset{O}{C}\diagdown_{OH}} \xrightarrow[\text{pyridoxal phosphate}]{\text{alanine transaminase}}$$

$$\underset{\text{pyruvic acid}}{H_3C-\overset{O}{\overset{\|}{C}}-\overset{O}{C}\diagdown_{OH}} \quad + \quad \underset{\text{glutamic acid}}{HO-\overset{O}{\overset{\|}{C}}-CH_2-CH_2-\overset{NH_2}{\underset{H}{C}}-\overset{O}{C}\diagdown_{OH}}$$

Notice that reactions 28-3 and 28-4 each depict the sum of two reactions. Since the coenzyme pyridoxal phosphate used in the first reaction (28-1) is regenerated in the second reaction (28-2), we only need to indicate that this coenzyme is used when writing the overall reaction catalyzed by a transaminase.

Oxidative Deamination

The second general type of enzyme-catalyzed reaction that is used to remove amino groups from amino acids is called **oxidative deamination.** A single enzyme, amino acid oxidase, catalyzes a complex reaction that produces ammonia and hydrogen peroxide as products, together with the α-keto acid derivative of the reactant amino acid:

(28-5)

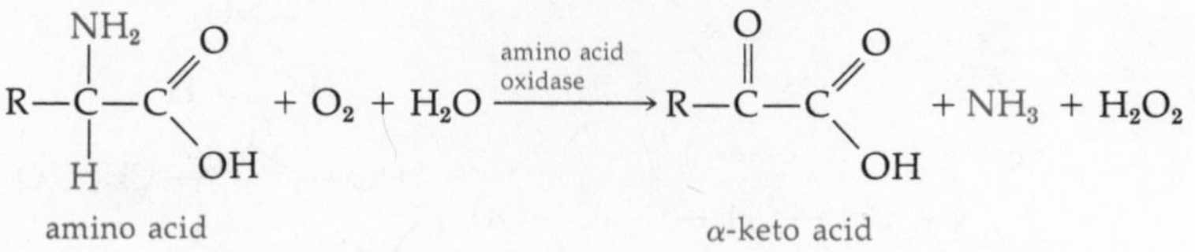

$$R-\underset{\underset{H}{|}}{\overset{\overset{NH_2}{|}}{C}}-C\overset{O}{\underset{OH}{\diagup}} + O_2 + H_2O \xrightarrow{\text{amino acid oxidase}} R-\overset{O}{\underset{}{C}}-C\overset{O}{\underset{OH}{\diagup}} + NH_3 + H_2O_2$$

amino acid   α-keto acid

The hydrogen peroxide produced in this reaction is rapidly destroyed by another enzyme, catalase, which catalyzes the reaction

(28-6)
$$2H_2O_2 \xrightarrow{\text{catalase}} 2H_2O + O_2$$

Amino acid oxidase is used far less often in amino acid catabolism than are transaminases. However, another type of oxidative deamination reaction that is very important in the metabolism of amino acids involves the oxidative deamination of glutamic acid. Recall that glutamic acid is the common product produced from α-ketoglutaric acid upon the transamination of other amino acids (reaction 28-3). The amino group of glutamic acid is eventually incorporated into urea for excretion, but it must first be removed from glutamic acid. The enzyme **glutamic acid dehydrogenase** catalyzes the reaction

(28-7)

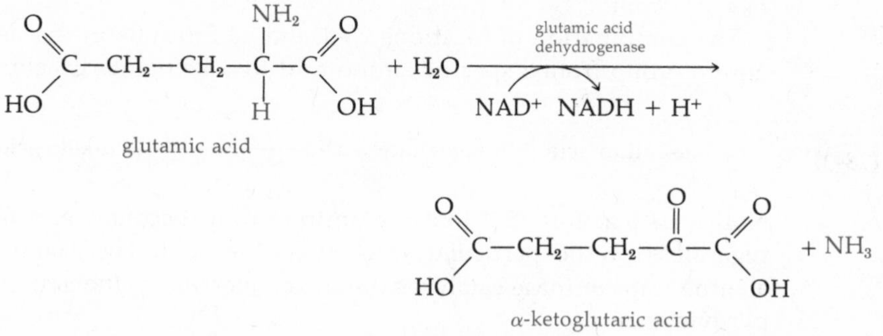

$$\overset{O}{\underset{HO}{\diagdown}}C-CH_2-CH_2-\underset{\underset{H}{|}}{\overset{\overset{NH_2}{|}}{C}}-C\overset{O}{\underset{OH}{\diagup}} + H_2O \xrightarrow[\overset{}{NAD^+ \ NADH + H^+}]{\text{glutamic acid dehydrogenase}}$$

glutamic acid

$$\overset{O}{\underset{HO}{\diagdown}}C-CH_2-CH_2-\overset{O}{\underset{}{C}}-C\overset{O}{\underset{OH}{\diagup}} + NH_3$$

α-ketoglutaric acid

Thus, by the *joint* action of transaminases and glutamic acid dehydrogenase, the amino groups from many amino acids can be converted to ammonia in a similar manner:

α-amino acid   α-ketoglutaric acid   NADH + H⁺ + NH₃

(28-8)   transaminase   glutamic acid
                                          dehydrogenase

α-keto acid   glutamic acid   NAD⁺ + H₂O

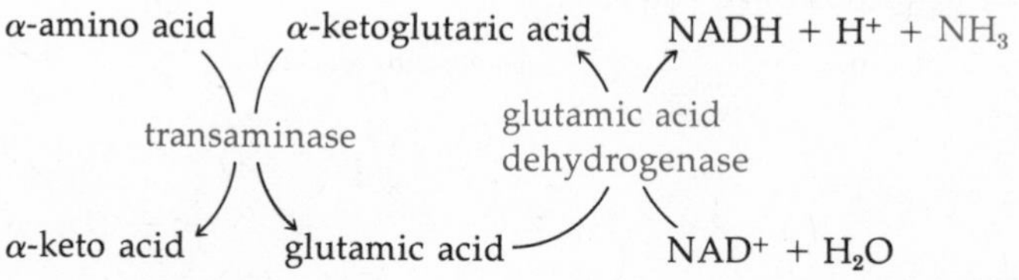

The ammonia thus produced is used to synthesize urea (Section 28-5).

Direct Deamination

The amino acids serine and threonine contain alcohol functional groups in their side chains. These amino acids can be directly deaminated by the enzyme **serine-threonine dehydratase.** This enzyme catalyzes the following reaction:

(28-9)

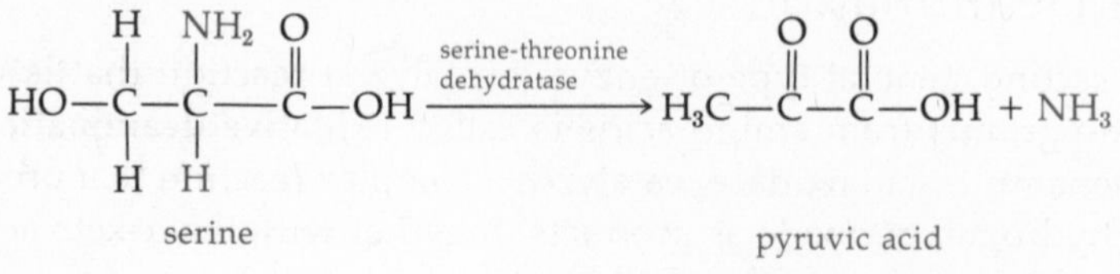

$$HO-\underset{\underset{H}{|}}{\overset{\overset{H}{|}}{C}}-\underset{\underset{H}{|}}{\overset{\overset{NH_2}{|}}{C}}-\overset{O}{\underset{}{C}}-OH \xrightarrow{\substack{\text{serine-threonine} \\ \text{dehydratase}}} H_3C-\overset{O}{\underset{}{C}}-\overset{O}{\underset{}{C}}-OH + NH_3$$

serine   pyruvic acid

Notice that serine-threonine dehydratase also removes the hydroxyl group from the side chain of serine, producing pyruvic acid. This enzyme also catalyzes the direct deamination of threonine. However, two other pathways for threonine catabolism also exist.

Fate of the α-Keto Acid Derivatives

After removal of the α-amino group, the α-keto acid derivatives of the various amino acids, containing the carbon skeletons of the original amino acids, are metabolized further. They may enter the gluconeogenic pathway or, when energy is required, be catabolized via the TCA cycle. We are already familiar with the metabolism of some of these α-keto acids. Pyruvic acid produced by the transamination of alanine (reaction 28-4) or by serine-threonine dehydratase (reaction 28-9) is converted to acetyl-S-CoA by pyruvic acid dehydrogenase (Chapter 25). α-Ketoglutaric acid, produced by glutamic acid dehydrogenase, is one of the intermediates in the TCA cycle (Chapter 25). The transamination of aspartic acid produces oxaloacetic acid, which is also an intermediate in the TCA cycle. The α-keto acids produced from the other amino acids are also funneled into the TCA cycle. For some of them, however, this requires a number of additional reaction steps. We shall not discuss these special pathways. Figure 28-6 summarizes the routes whereby the carbon skeletons of the various amino acids can be shuttled into the TCA cycle. The energy yield from the catabolism of 1 g of a typical protein is comparable to that obtained from the aerobic catabolism of 1 g of glucose.

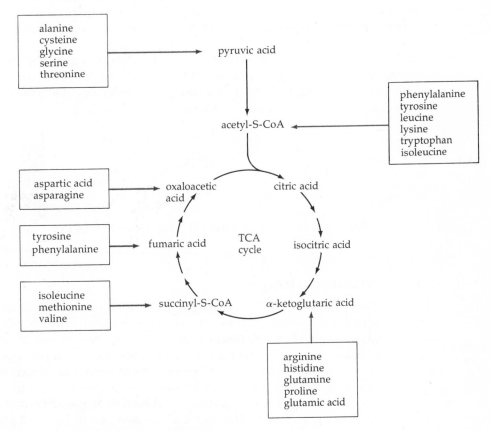

Figure 28-6   Pathways of entry of the carbon skeletons of amino acids into the TCA cycle.

Exercise 28-3

Draw the structural formulas for the products of the following reactions:

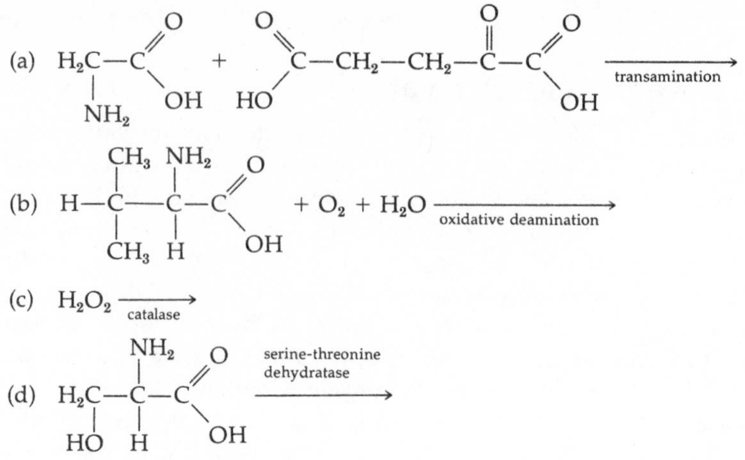

(a)

$$H_2C-C\begin{smallmatrix}O\\\\OH\end{smallmatrix} \quad + \quad \begin{smallmatrix}O\\\\HO\end{smallmatrix}C-CH_2-CH_2-C\begin{smallmatrix}O\\\\\end{smallmatrix}-C\begin{smallmatrix}O\\\\OH\end{smallmatrix}$$

with NH_2 on the first carbon $\xrightarrow{\text{transamination}}$

(b)

$$H-\underset{\underset{CH_3}{|}}{\overset{\overset{CH_3}{|}}{C}}-\underset{\underset{H}{|}}{\overset{\overset{NH_2}{|}}{C}}-C\begin{smallmatrix}O\\\\OH\end{smallmatrix} \quad + O_2 + H_2O \xrightarrow{\text{oxidative deamination}}$$

(c) $H_2O_2 \xrightarrow{\text{catalase}}$

(d)

$$H_2C-\underset{\underset{H}{|}}{\overset{\overset{NH_2}{|}}{C}}-C\begin{smallmatrix}O\\\\OH\end{smallmatrix}$$

with HO on first carbon $\xrightarrow[\text{dehydratase}]{\text{serine-threonine}}$

28-5 THE UREA CYCLE

In order for a normal individual to be in a state of nitrogen balance, excess nitrogen generated in the catabolism of amino acids and other nitrogen-containing compounds must be excreted from the body. The excess nitrogen is incorporated into the nontoxic compound urea within liver cells. The urea is then excreted in urine.

Urea synthesis is accomplished in a cyclic series of reactions that is called the **urea cycle.** The overall net reaction of the urea cycle involves two molecules of ammonia, which originate from two amino acid molecules, and a molecule of carbon dioxide:

(28-10)
$$2NH_3 + CO_2 \longrightarrow H_2N-\overset{\overset{O}{\|}}{C}-NH_2 + H_2O$$

Note that urea is the diamide of carbonic acid, $HO-\overset{\overset{O}{\|}}{C}-OH$.

Source of the Amino Groups

There are three sources for the amino groups used for the synthesis of urea: glutamic acid produced from α-ketoglutaric acid in the transamination of other amino acids (reaction 28-3), the direct deamination of serine or threonine, and the oxidative deamination of other amino acids. These two amino groups enter the urea cycle by different routes. One enters as free ammonia, which is produced either by direct deamination or by oxidative deamination reactions or from glutamic acid by the enzyme glutamic acid dehydrogenase (reaction 28-7). The second amino group enters the urea cycle as part of a molecule of aspartic acid, which is produced from oxaloacetic acid in a reaction catalyzed by the enzyme **aspartic acid transaminase.** Glutamic acid donates the amino group in this reaction.

(28-11)

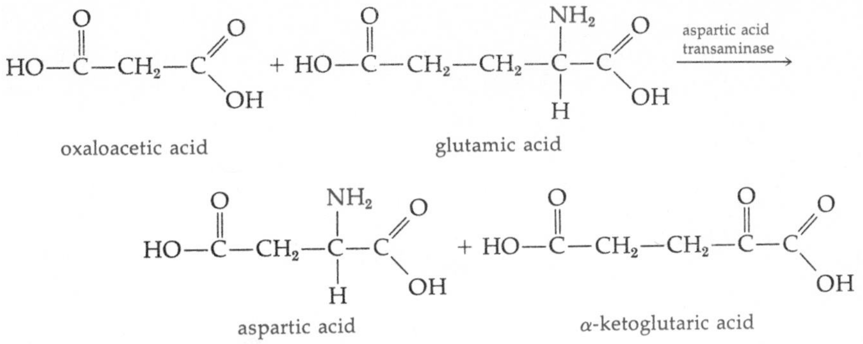

oxaloacetic acid   glutamic acid

aspartic acid
transaminase

aspartic acid α-ketoglutaric acid

Exercise 28-4

The transamination of ingested amino acids leads to the formation of glutamic acid from α-ketoglutaric acid. What happens to the glutamic acid? What is the source of the α-ketoglutaric acid?

Carnivores obtain protein from both plant and animal tissue, whereas

herbivores obtain protein mainly from plant tissue.

Synthesis of Urea

The first step of the urea cycle is the synthesis of **carbamoyl phosphate,** which requires the hydrolysis of two molecules of ATP. The ammonia that is used in this step comes from one of the amino acids, and the CO_2 is a product of the TCA cycle. The carbamoyl phosphate is then used, together with ornithine, for the synthesis of **citrulline** in the second step of the urea cycle. In the third step of the cycle, the citrulline is joined to a molecule of aspartic acid (synthesized in reaction 28-11), which carries the second amino group that will end up in urea. This step requires the hydrolysis of a molecule of ATP to AMP and pyrophosphate, and results in the production of **argininosuccinic acid.** In the fourth step, argininosuccinic acid is cleaved to produce **arginine** and **fumaric acid.** If arginine is needed for protein synthesis, the urea cycle stops at this point. The fumaric acid produced in step 4 is used in the TCA cycle to produce more oxaloacetic acid, which can then be converted to aspartic acid by reaction 28-11.

The last step of the urea cycle (step 5) is the hydrolysis of arginine to give **urea** and **ornithine.** The urea is excreted and the ornithine is used in the next round of the cycle (in step 2). Figure 28-7 summarizes this cyclic process.

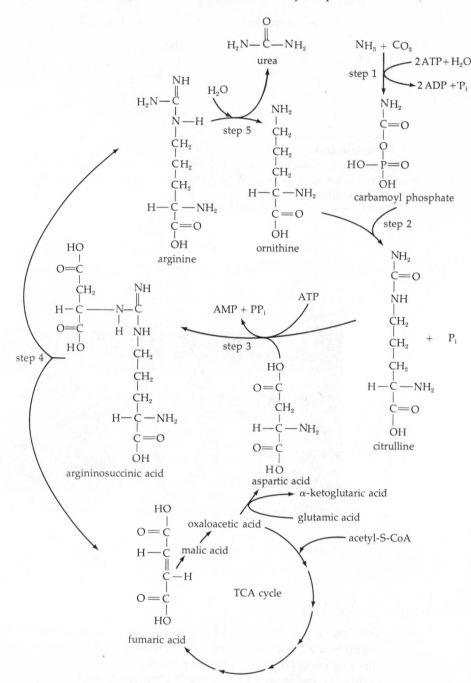

Figure 28-7   The urea cycle. Notice that the argininosuccinic acid is cleaved in step 4 to give arginine and fumaric acid. The arginine is then hydrolyzed to give urea in step 5, whereas the fumaric acid is converted to oxaloacetic acid in the TCA cycle.

Notice that there is no net consumption of aspartic acid nor production of fumaric acid in the urea cycle, since fumaric acid can be used to synthesize another molecule of aspartic acid. When we consider the energy (from ATP hydrolysis) that is required, the net overall reaction of the urea cycle can be written as:

$$\text{(28-12)} \quad 2NH_3 + CO_2 \xrightarrow[\text{urea cycle}]{\overset{\displaystyle 3ATP + 2H_2O \quad\quad 2ADP + AMP + 2P_i + PP_i}{\curvearrowright}} H_2N-\overset{\overset{\displaystyle O}{\|}}{C}-NH_2 \atop \text{urea}$$

Recall that an additional molecule of ATP is required to convert the AMP produced in the urea cycle to ADP. Thus, the energy required for the operation of the entire urea cycle is equivalent to the hydrolysis of 4ATP to $4ADP + 4P_i$.

Exercise 28-5

Two molecules of ATP are hydrolyzed in the first step of the urea cycle. What happens to the two phosphate groups that are removed from these two ATP molecules?

Exercise 28-6

Both of the amino groups in a molecule of urea may originate from glutamic acid molecules, but they are not transferred directly from glutamic acid to urea. How do these amino groups enter the urea cycle?

28-6 BIOSYNTHESIS OF AMINO ACIDS

Of the 20 α-amino acids required for the synthesis of proteins, the adult human body is able to synthesize only 12. Since it is not necessary for these 12 amino acids to be included in our diet, they are called **nonessential amino acids.** The 8 amino acids that we cannot synthesize are called **essential amino acids.** Table 28-2 lists the essential and nonessential amino acids for human adults. Nutritional requirements for amino acids and other compounds will be discussed further in Chapter 29.

Table 28-2 Essential and Nonessential Amino Acids for Human Adults

Nonessential	Essential
Glycine	Isoleucine
Alanine	Leucine
Serine	Lysine
Cysteine	Methionine
Proline	Phenylalanine
Glutamic acid	Threonine
Glutamine	Tryptophan
Aspartic acid	Valine
Asparagine	
Tyrosine	
Histidine*	
Arginine*	

* Although histidine and arginine do not appear to be essential for adults, they apparently are essential for the normal growth of children.

We have already discussed one mechanism used to synthesize amino acids, namely, transamination. (Remember that enzyme-catalyzed reactions are generally reversible.) Alanine, glutamic acid, and aspartic acid are produced directly by the respective transamination of pyruvic acid, α-ketoglutaric acid, and oxaloacetic acid.

(28-13)

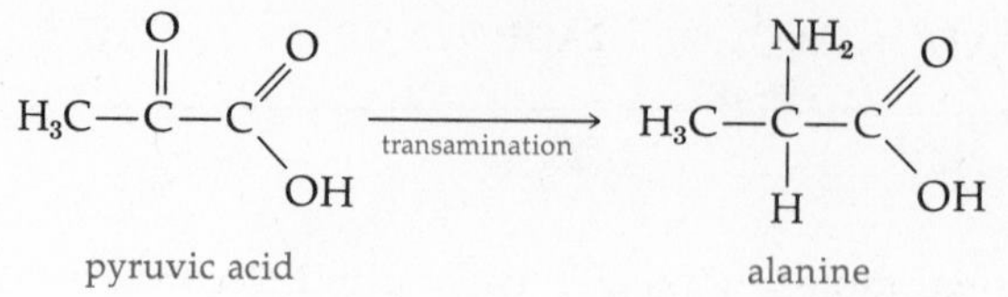

$$H_3C-\underset{\underset{O}{\parallel}}{C}-\underset{\underset{OH}{}}{C}\diagdown^O \xrightarrow{\text{transamination}} H_3C-\underset{\underset{H}{|}}{\overset{NH_2}{C}}-C\diagdown^O_{OH}$$

pyruvic acid   alanine

(28-14)

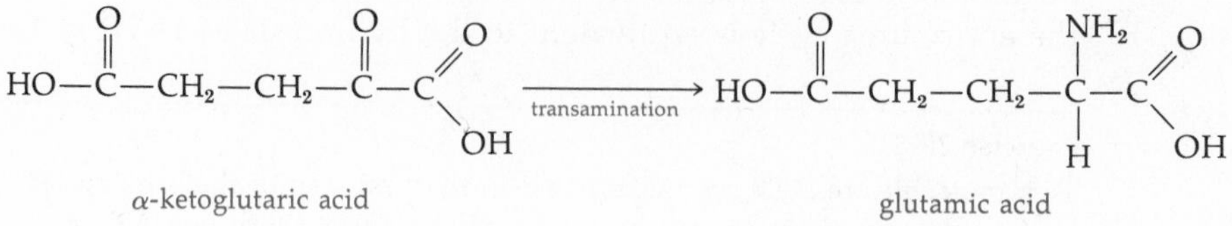

$$HO-\underset{\underset{O}{\parallel}}{C}-CH_2-CH_2-\underset{\underset{O}{\parallel}}{C}-C\diagdown^O_{OH} \xrightarrow{\text{transamination}} HO-\underset{\underset{O}{\parallel}}{C}-CH_2-CH_2-\underset{\underset{H}{|}}{\overset{NH_2}{C}}-C\diagdown^O_{OH}$$

α-ketoglutaric acid   glutamic acid

(28-15)

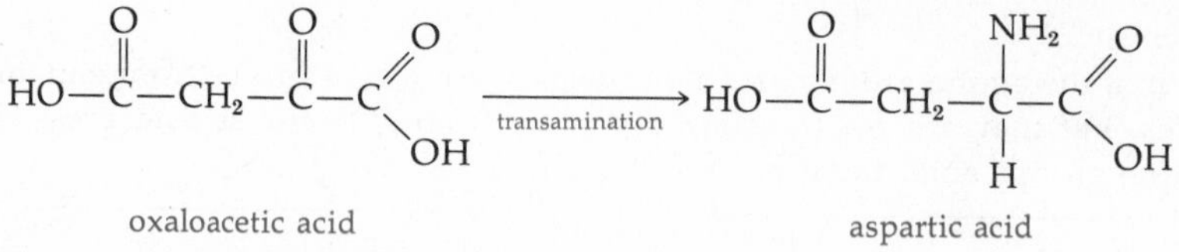

$$HO-\underset{\underset{O}{\parallel}}{C}-CH_2-\underset{\underset{O}{\parallel}}{C}-C\diagdown^O_{OH} \xrightarrow{\text{transamination}} HO-\underset{\underset{O}{\parallel}}{C}-CH_2-\underset{\underset{H}{|}}{\overset{NH_2}{C}}-C\diagdown^O_{OH}$$

oxaloacetic acid   aspartic acid

The amino acid glutamine is produced from glutamic acid and ammonia by the enzyme **glutamine synthetase,** in the following reaction:

(28-16)

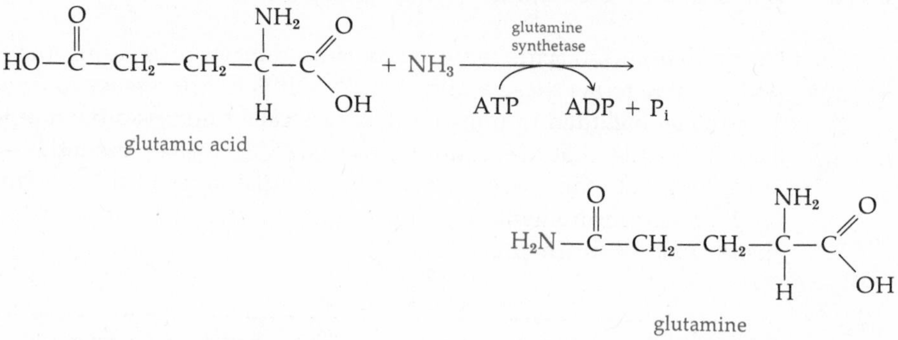

$$HO-\underset{\underset{O}{\parallel}}{C}-CH_2-CH_2-\underset{\underset{H}{|}}{\overset{NH_2}{C}}-C\diagdown^O_{OH} + NH_3 \xrightarrow[\text{ATP} \quad \text{ADP} + P_i]{\text{glutamine synthetase}}$$

glutamic acid

$$H_2N-\underset{\underset{O}{\parallel}}{C}-CH_2-CH_2-\underset{\underset{H}{|}}{\overset{NH_2}{C}}-C\diagdown^O_{OH}$$

glutamine

Asparagine is produced from aspartic acid in an identical manner by the enzyme **asparagine synthetase.**

The human body synthesizes tyrosine by the hydroxylation of phenylalanine, in a reaction catalyzed by the enzyme **phenylalanine hydroxylase:**

(28-17)

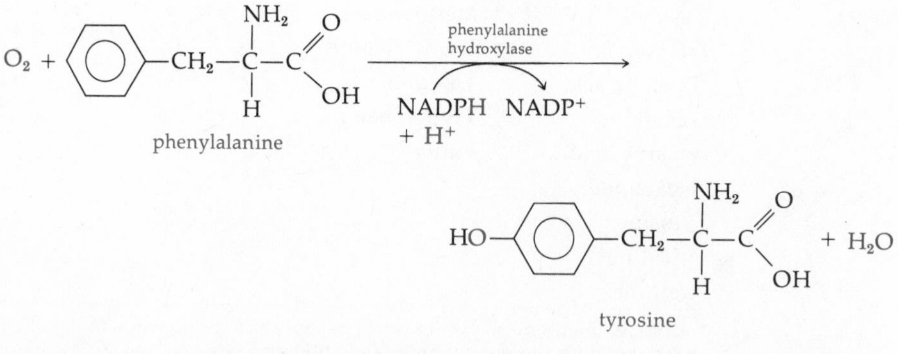

$$O_2 + \underset{\text{phenylalanine}}{\bigcirc-CH_2-\underset{\underset{H}{|}}{\overset{NH_2}{C}}-C\diagdown^O_{OH}} \xrightarrow[\substack{\text{NADPH} \quad \text{NADP}^+ \\ + H^+}]{\text{phenylalanine hydroxylase}}$$

$$HO-\bigcirc-CH_2-\underset{\underset{H}{|}}{\overset{NH_2}{C}}-C\diagdown^O_{OH} + H_2O$$

tyrosine

This reaction is also the first step in the synthesis of several other important compounds, including epinephrine and the neurotransmitter DOPA (see Section 28-7). In the disease **phenylketonuria** (PKU), the absence of this enzyme results in the buildup of phenylalanine, which causes mental retardation. The synthesis of the amino acids glycine, serine, cysteine, proline, and histidine require several specific steps (which we shall not discuss). Recall that the synthesis of arginine occurs as part of the urea cycle.

Exercise 28-7
Tyrosine is a nonessential amino acid. Can an adequate supply of tyrosine be synthesized by the human body without a source of protein for digestion or turnover?

Optional

28-7 BIOSYNTHESIS OF SOME IMPORTANT AMINO ACID DERIVATIVES

In addition to their use as components of proteins and occasionally as sources of energy for the human body, some amino acids are used as starting materials for the synthesis of several compounds involved in the regulation of metabolism. These amino acid derivatives include epinephrine (adrenaline), the thyroid hormones, and a variety of other small molecules with very specific functions. We shall discuss the synthesis and biological effects of a few of the more well-known amino acid derivatives.

Derivatives of Tyrosine

In the last section we saw that cells in the human body synthesize tyrosine by the hydroxylation of the essential amino acid phenylalanine. Tyrosine then serves as the starting material for the synthesis of a number of hormones and neurotransmitters. The biosynthetic pathway used for the synthesis of several of these compounds is shown in Figure 28-8. The first step in this pathway is the hydroxylation of tyrosine to form dihydroxyphenylalanine, or DOPA.

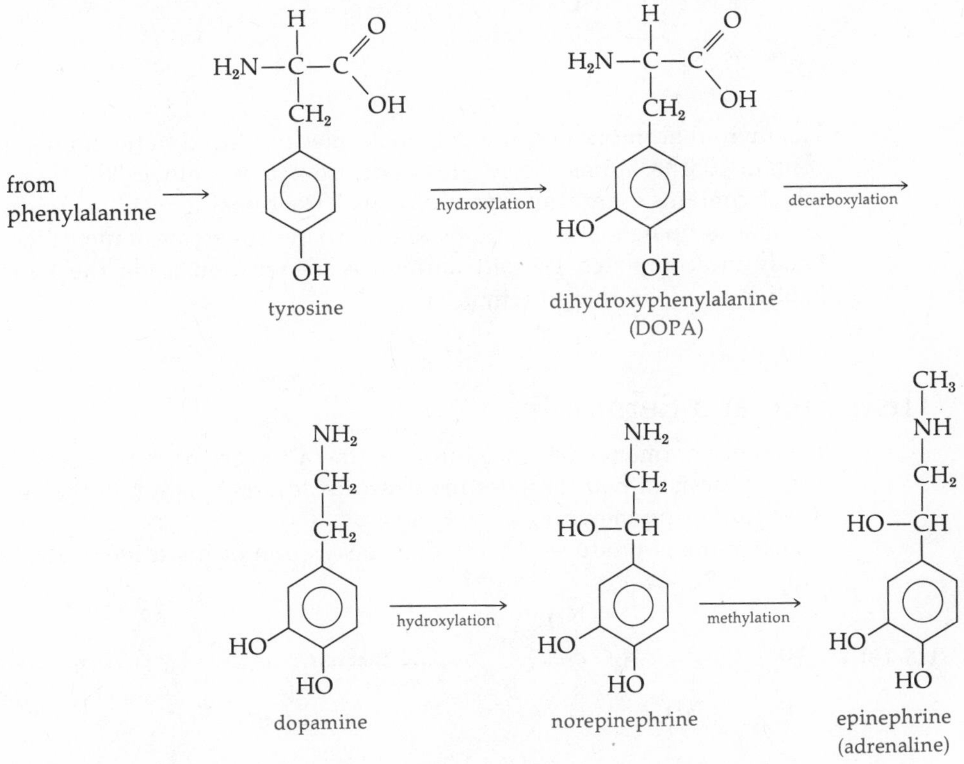

Figure 28-8 The synthesis of hormones and neurotransmitters from tyrosine.

DOPA is then decarboxylated to give the compound **dopamine,** which acts as a **neurotransmitter,** a substance that stimulates nerve cells. Dopamine works specifically at the junctions between nerve cells and muscle cells (also called sympathetic junctions). Dopamine is also the starting material for the further synthesis of norepinephrine (also called noradrenaline) and epinephrine (see Figure 28-8). We have already seen in Chapters 23 and 25 that epinephrine is a hormone used to regulate the synthesis and breakdown of glycogen. Both epinephrine and norepinephrine are also used as neurotransmitters in certain types of nerve junctions. In addition, norepinephrine causes constriction of blood vessels in peripheral areas of the body, helping to regulate blood pressure. Thus these tyrosine derivatives perform a wide range of extremely important regulatory functions in the human body.

Additional derivatives of tyrosine are made in the thyroid gland and are called **thyroid hormones.** Thyroid hormones are essential for maintenance of proper growth and metabolism of cells in the body. However, the exact mechanisms whereby this is accomplished are complex and not totally understood. The two primary thyroid hormones are **triiodothyronine,**

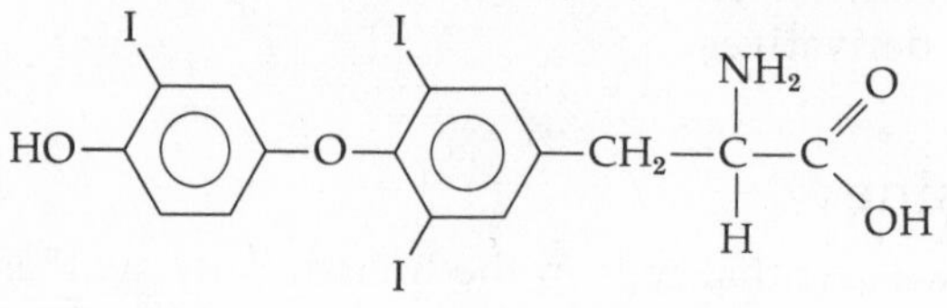

and tetraiodothyronine, or **thyroxine,**

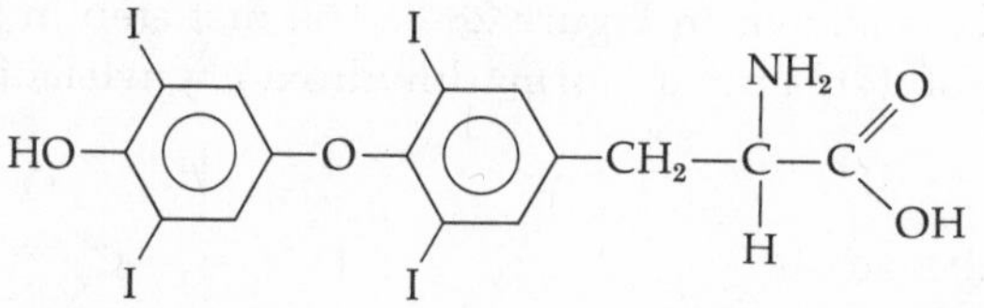

The thyroid hormones are the only molecules produced in the human body that contain iodine atoms. They are made from a protein called thyroglobulin, which contains a large amount of tyrosine. The tyrosine residues on this protein are first iodinated; then pairs of iodinated tyrosines are joined together; and finally the completed thyroid hormones are released upon the hydrolysis of peptide bonds in the protein.

Histamine and Serotonin

Two other hormones produced from amino acids are histamine and serotonin. The synthesis of both of these involves the decarboxylation of the amino acids used as starting materials.

Histamine is produced by the decarboxylation of histidine in the reaction

(28-18)

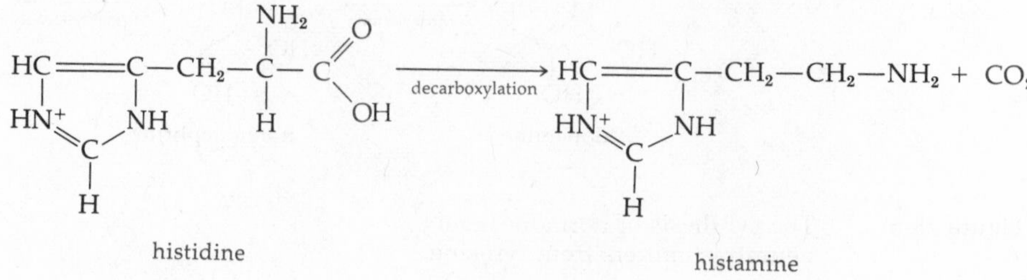

histidine   histamine

Histamine is synthesized predominately in skin and lung tissue, but is also made and stored in mast cells. This hormone causes the dilation of capillary blood vessels, which increases their permeability. Relatively large amounts of histamine are released into the bloodstream in traumatic shock and in allergic reactions. Several drugs called **antihistamines** are used to inhibit the adverse effects caused by an excessive release of histamine.

The synthesis of **serotonin** involves the hydroxylation of tryptophan followed by decarboxylation (see Figure 28-9). Serotonin has a variety of effects, including the constriction of blood vessels and the promotion of peristalsis in the intestinal system. It also serves as a neurotransmitter for certain types of nerve cells.

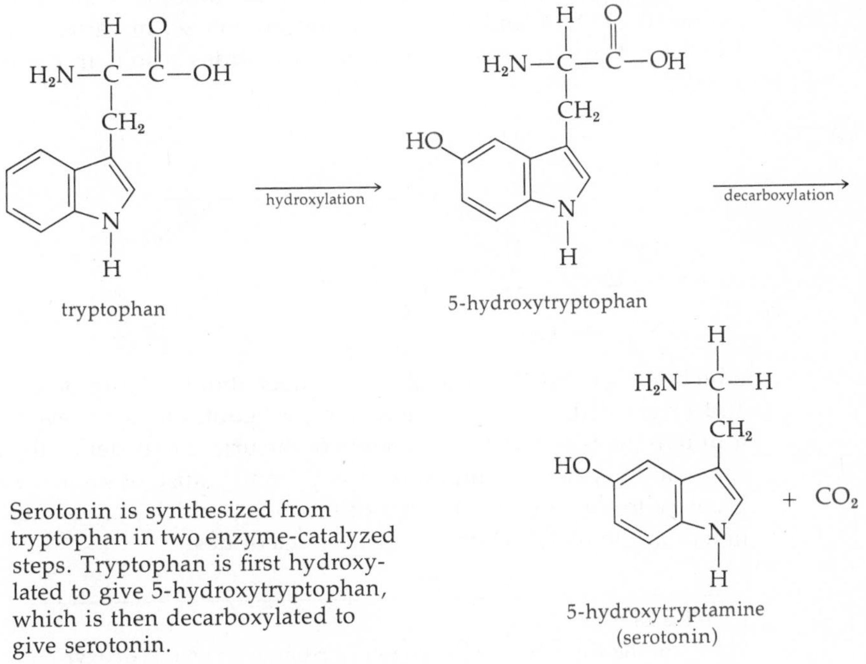

Figure 28-9   Serotonin is synthesized from tryptophan in two enzyme-catalyzed steps. Tryptophan is first hydroxylated to give 5-hydroxytryptophan, which is then decarboxylated to give serotonin.

Derivatives of several other amino acids also have important functions in the human body. The examples we have presented illustrate the vast differences in the functions of biomolecules that are caused by rather small molecular changes, such as the alteration or removal of only one functional group. These amino acid derivatives also illustrate the ability of cells in the human body to produce a large number of molecules with widely differing functions from only a relatively small number of precursors.

You will recall that various nucleotides also have a variety of functions. In addition to serving as component parts of nucleic acids, some nucleotides are used as sources of chemical energy, such as ATP. Others, such as cyclic AMP, are used as messengers, whereas still other nucleotide derivatives function as coenzymes.

Exercise 28-8
What type of reaction is common to the synthesis of serotonin, epinephrine, and histamine from their amino acid precursors? What other type of reaction is common to the synthesis of serotonin, norepinephrine, and DOPA?

28-8   METABOLISM OF NUCLEOTIDES

Cells in the human body are capable of synthesizing *all* of the nucleotides necessary for the assembly of DNA and RNA. These molecules are therefore not essential components of the human diet. However, nucleotides that are present in the human diet can be used for the synthesis of nucleic acids, which is more energy efficient than synthesizing these molecules from scratch. Nucleic acids in the diet are hydrolyzed to their component nucleotides during the course of digestion. This is accomplished by **ribonuclease** and **deoxyribonuclease,** two enzymes that are produced in the pancreas. The free nucleotides are then transported through the cells of the intestinal wall and distributed for use.

When larger amounts of nucleotides are ingested than are needed for biosynthesis of DNA and RNA, they are catabolized in various tissues, especially the liver. The catabolism of purine nucleotides results in the formation of uric acid,

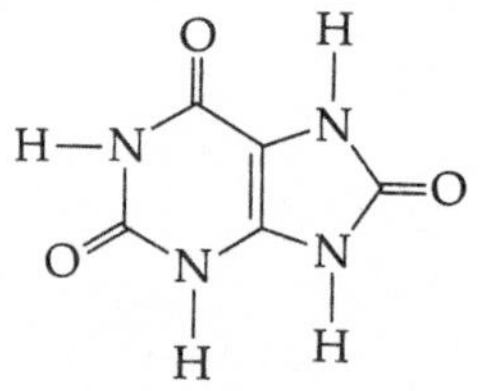

uric acid

which is excreted. In normal individuals about 500 mg of uric acid is formed and excreted daily. In the disease called **gout,** elevated levels of uric acid are found in the blood and urine. Some of the uric acid is deposited in joints, which can lead to arthritic symptoms. The primary cause of gout is believed to be an increase in the rate of synthesis of uric acid, which may be due to an inherited defect in the metabolism of purine nucleotides.

Exercise 28-9

What are the functions of the enzymes ribonuclease and deoxyribonuclease?

Exercise 28-10

Cells in the human body turn over their proteins. Would you also predict substantial DNA turnover in human cells?

28-9   SUMMARY

1. The human body can only utilize nitrogen atoms that are already incorporated into biomolecules.

2. Digestion of proteins involves their hydrolysis into component amino acids by a number of proteolytic enzymes in the digestive system. These enzymes include pepsin, trypsin, chymotrypsin, carboxypeptidase, and aminopeptidase.

3. After hydrolysis of ingested proteins, the free amino acid products are transported into the bloodstream for distribution and use by the various cells in the body.

4. The amino acid pool is the collection of amino acids in the bloodstream that are available for synthesis or energy production by cells in the body.

5. Excess amino acids are catabolized and the nitrogen atoms are incorporated into urea for excretion.

6. Normal human beings, with the exception of pregnant women and growing children, ingest as much nitrogen per day as they excrete, and are said to be in nitrogen balance.

7. All of the proteins in the human body are constantly being synthesized and degraded. This process is called protein turnover.

8. When amino acids are catabolized as a source of energy, the nitrogen atoms are first removed for excretion in the form of urea, whereas the remainder of the molecules, called the carbon skeletons, are ultimately oxidized to CO_2.

9. The removal of amino groups from amino acids is accomplished primarily by transamination, and to a lesser extent by oxidative deamination and direct deamination.

10. Transamination, which is catalyzed by transaminase enzymes, involves the exchange of an amino group for a carbonyl oxygen atom, resulting in the production of an α-keto acid from an amino acid. Transaminase enzymes use the coenzyme pyridoxal phosphate, and generally they transfer amino groups to α-ketoglutaric acid to form glutamic acid.

11. Amino groups on glutamic acid are liberated as ammonia in a reaction that is catalyzed by the enzyme glutamic acid dehydrogenase.

12. The human body is capable of synthesizing 12 of the amino acids found in proteins. These are called the nonessential amino acids. The other 8 amino acids found in proteins must be included in the human diet and are called essential amino acids.

13. Some of the nonessential amino acids are synthesized by the transamination of α-keto acids. The amino acids glutamine and asparagine are synthesized from ammonia and glutamic acid and from ammonia and aspartic acid, respectively.

14. Urea, NH_2—$\overset{\overset{\textstyle O}{\|}}{C}$—$NH_2$, is synthesized in the urea cycle from CO_2 and the α-amino groups from two amino acid molecules. One of the amino groups is used to synthesize carbamoyl phosphate. The second amino group enters the urea cycle via aspartic acid.

15. The net overall reaction of the urea cycle is

$$2NH_3 + CO_2 + 3ATP + 2H_2O \longrightarrow H_2N-\overset{\overset{\textstyle O}{\|}}{C}-NH_2 + 2ADP + 2P_i + AMP + PP_i$$

16. The human body is capable of synthesizing all of the nucleotide components of DNA and RNA.

17. The digestion of ingested nucleic acids involves their hydrolysis into free nucleotides by the enzymes ribonuclease and deoxyribonuclease. These free nucleotides can then be used by the body for the synthesis of DNA and RNA.

18. Excess nucleotides from the diet are catabolized and their products excreted. The product of catabolism of purine nucleotides is uric acid. The disease gout involves the synthesis of abnormally large amounts of uric acid.

PROBLEMS

1. What is the difference between protein digestion and protein turnover?

2. The structural formulas for a few amino acids are shown below. Draw structural formulas for their precursors and briefly describe their enzymatic synthesis.

(a) 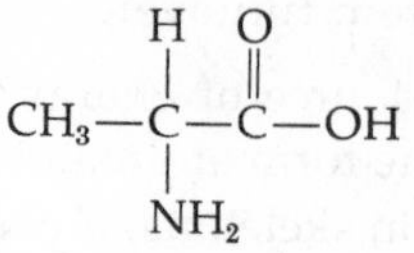 CH_3—C—C—OH
        H O
        | ||
        NH$_2$

   alanine

(b) 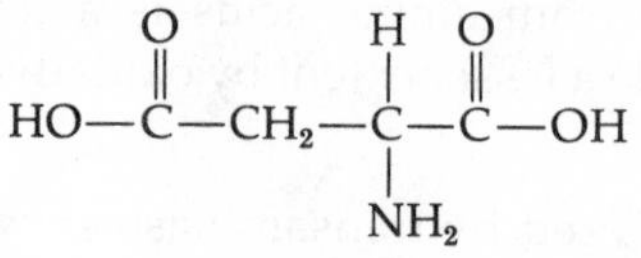 HO—C—CH$_2$—C—C—OH
        O   H O
        || | ||
                   NH$_2$

   aspartic acid

(c) 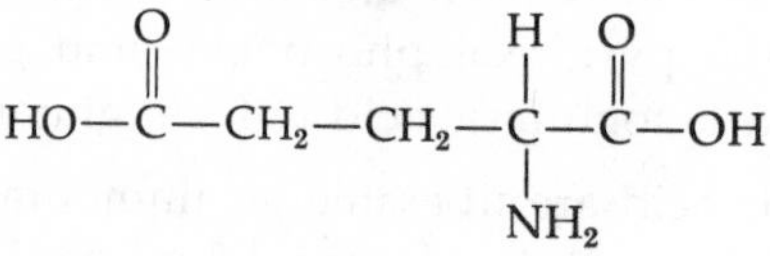 HO—C—CH$_2$—CH$_2$—C—C—OH
        O   H O
        || | ||
                           NH$_2$

   glutamic acid

(d) 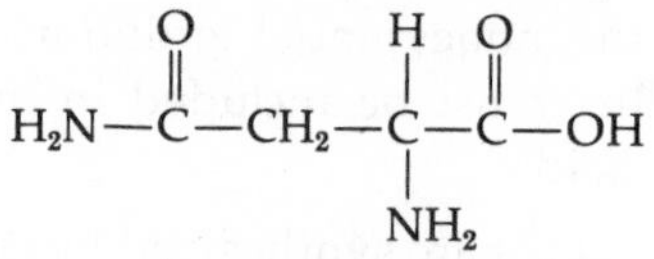 H$_2$N—C—CH$_2$—C—C—OH
          O   H O
          || | ||
                     NH$_2$

   asparagine

3. What is the function of aspartic acid in the urea cycle? How is this aspartic acid regenerated?

4. What is common in the intestinal fates of protein and nucleic acid polymers?

5. Draw the structural formulas for the products of the following reactions.

(a) 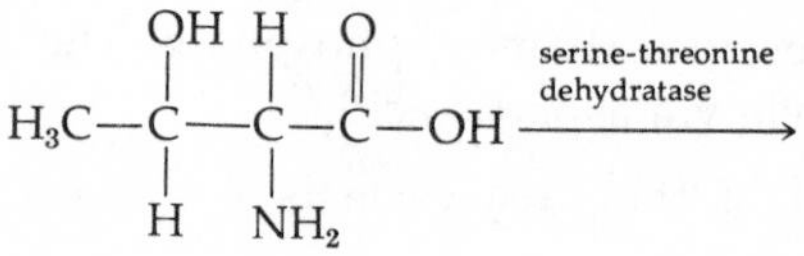 H$_3$C—C—C—C—OH $\xrightarrow{\text{serine-threonine dehydratase}}$
         OH H O
         | | ||
         H NH$_2$

(b) 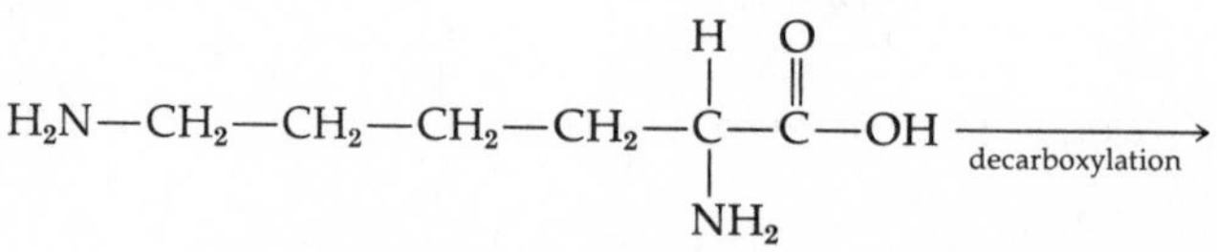 H$_2$N—CH$_2$—CH$_2$—CH$_2$—CH$_2$—C—C—OH $\xrightarrow{\text{decarboxylation}}$
                                    H O
                                    | ||
                                    NH$_2$

(c) $\begin{matrix} H_3C \\ \quad \diagdown \\ \qquad CH \\ \quad \diagup \\ H_3C \end{matrix}$ —C—C—OH + HO—C—CH$_2$—CH$_2$—C—C—OH $\xrightarrow{\text{transamination}}$
          H O   O   O O
          | || || || ||
          NH$_2$

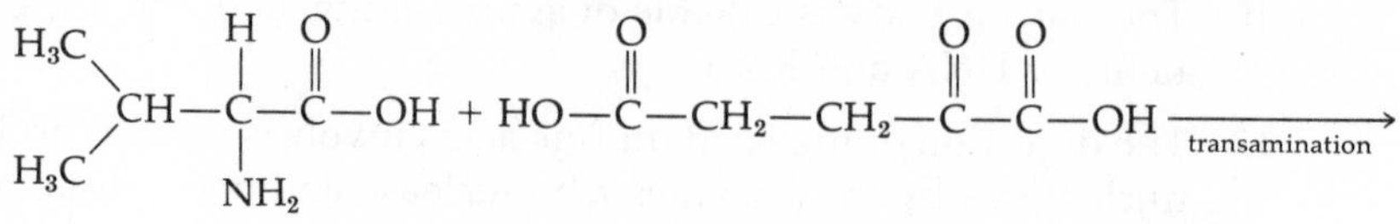

(d) 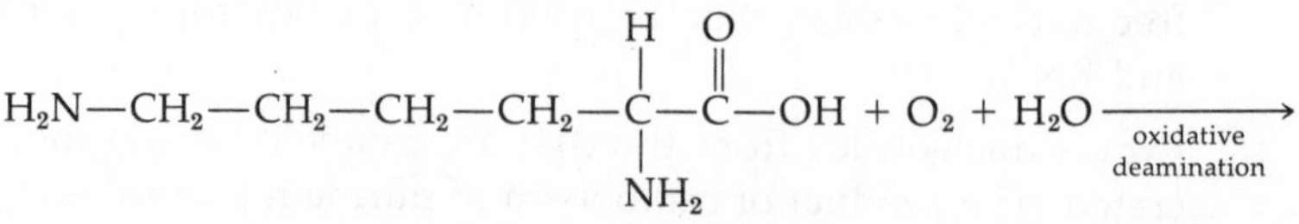 H$_2$N—CH$_2$—CH$_2$—CH$_2$—CH$_2$—C—C—OH + O$_2$ + H$_2$O $\xrightarrow{\text{oxidative deamination}}$
                                    H O
                                    | ||
                                    NH$_2$

6. The transamination of alanine to pyruvic acid is accompanied by the conversion of pyridoxal phosphate to pyridoxamine phosphate. How can pyridoxal phosphate continue to function as a coenzyme when it is continually converted to pyridoxamine phosphate?

SOLUTIONS TO EXERCISES

28-1

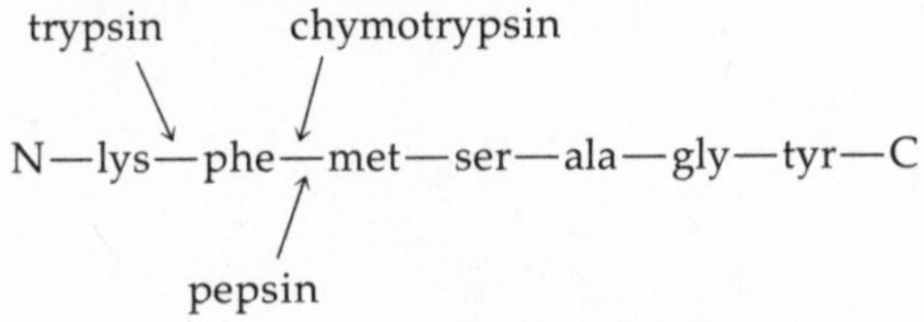

| | trypsin | chymotrypsin |
| | | |

N—lys—phe—met—ser—ala—gly—tyr—C

pepsin

(a) N-lys-phe-C + N-met-ser-ala-gly-tyr-C
(b) lys + N-phe-met-ser-ala-gly-tyr-C
(c) same as (a)
(d) lys + phe + N-met-ser-ala-gly-tyr-C

28-2 A healthy individual on a high-protein diet will be in a state of nitrogen balance. Although more nitrogen is taken in, an equal amount is excreted. A fasting individual, however, will be in a state of negative nitrogen balance because body proteins will be used for energy production.

28-3 (a) 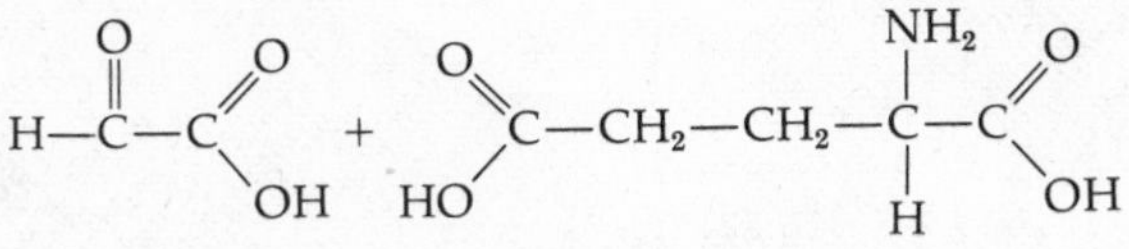 $H-\overset{\overset{O}{\|}}{C}-C\overset{O}{\underset{OH}{}}$ + $HO\overset{O}{\underset{}{}}C-CH_2-CH_2-\overset{\overset{NH_2}{|}}{\underset{H}{C}}-C\overset{O}{\underset{OH}{}}$

(b) 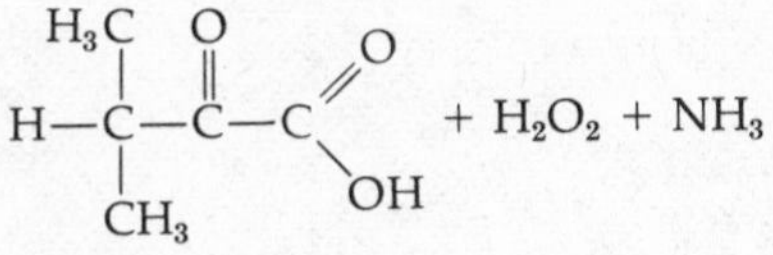 $H-\overset{\overset{H_3C}{|}}{\underset{CH_3}{C}}-\overset{\overset{O}{\|}}{C}-C\overset{O}{\underset{OH}{}}$ + H_2O_2 + NH_3

(c) $H_2O + \frac{1}{2}O_2$

(d) 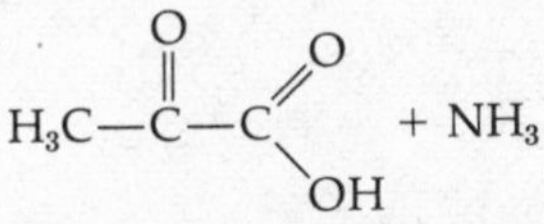 $H_3C-\overset{\overset{O}{\|}}{C}-C\overset{O}{\underset{OH}{}}$ + NH_3

28-4 The glutamic acid produced may then be used to donate an amino group to form urea. The amino group may be removed via reaction 28-7 as ammonia, or donated to oxaloacetic acid in reaction 28-11. The α-ketoglutaric acid is an intermediate in the TCA cycle.

28-5 One is immediately released as inorganic phosphate (P_i). The other is released, as P_i, in the second step of the urea cycle.

28-6 One enters the cycle as free ammonia in step 1 after release from glutamic acid. Alternatively, the amino group may come from an oxidase or deaminase reaction. The second amino group is transferred from glutamic acid to an oxaloacetic acid molecule by aspartic acid transaminase, and it enters the urea cycle as part of aspartic acid.

28-7 No. The synthesis of tyrosine requires a supply of phenylalanine, an essential amino acid.

28-8 **(Optional)** Decarboxylation of an amino acid or amino acid derivative is common to the synthesis of serotonin, epinephrine, and histamine. Hydroxylation is common to the synthesis of serotonin, norepinephrine, and DOPA.

28-9 These enzymes catalyze the hydrolysis of DNA and RNA into component nucleotides.

28-10 No. DNA must be protected against turnover in order to preserve the integrity of the cell's genetic information.

CHAPTER 29

The Integration and Regulation of Metabolism

29-1 INTRODUCTION

On several occasions we have explained biochemical concepts by using analogies between a living cell and a factory, and between a living organism and an industrial society. We are all aware of the interdependence of various industries in society. In this final chapter we shall examine the interdependence of the different metabolic pathways in the cells of the human body. No single metabolic pathway can function without regard to the needs of the whole body. Metabolic pathways are responsive to conditions of supply and demand. For example, when a sufficient supply of ATP is on hand to meet the body's needs, the oxidation of carbohydrates and fats in cells is retarded. We shall see that there are key control points for metabolic pathways which serve to maintain the body in proper working order under normal conditions.

Some biomolecules are common participants in a number of metabolic pathways. For example, acetyl-S-CoA is the starting material for fatty acid and steroid biosyntheses and, among other things, it is also the fuel for the TCA cycle. The fate of any given molecule of acetyl-S-CoA is determined by the demands of the various interconnected pathways.

In order for cellular factories to function smoothly to promote the well-being of the entire body, they must have an adequate supply of raw materials. Part of this chapter will therefore deal with the nutritional requirements of the cells in the human body. Even with proper nutrition, however, there are times when cells do not function properly. Fortunately, most of these disorders can now be detected and treated. The techniques for detection generally involve sampling a patient's blood or other body fluids, and determining what biomolecules are present in abnormally small or large amounts. The identity and quantity of these molecules is an indication of the nature and severity of the metabolic disorder. We have already seen, for example, that the concentration of glucose in the blood is abnormally high in untreated diabetics. Our discussion of diseases and their diagnosis is far from complete. It should serve, however, as a bridge between your study of basic chemical and biochemical concepts in this course and more specific studies of the human body in health and disease.

For centuries people have transformed the earth in order to cultivate crops that are rich sources of proteins, carbohydrates, and the other nutrients required for a balanced diet.

29-2   STUDY OBJECTIVES

After studying the material in this chapter, you should be able to:

1. Describe the interconnections of glycolysis and the TCA cycle, glycolysis and triglyceride metabolism, and the TCA cycle and the urea cycle.

2. Describe the central role of acetyl-S-CoA in metabolism.

3. Understand how the activities of pyruvate carboxylase and acetyl-S-CoA carboxylase are controlled, as well as the necessity for these controls.

4. Predict the effects of a diet that is unusually high in lipids, proteins, or carbohydrates on the operation of the metabolic pathways we have discussed.

5. Define nutrient and essential nutrient, and explain what is meant by the expression "recommended daily allowance" as applied to nutrition.

6. Explain why some nutrients are needed in much larger quantities than other nutrients.

7. List at least three factors that influence the daily nutritional needs of any individual.

8. Outline the major functions of extracellular fluids.

9. Describe what is meant by the terms electrolyte balance, pH balance, acidosis, and alkalosis.

29-3   METABOLIC INTERRELATIONSHIPS

Many different metabolic pathways operate in a typical cell at any given time. Exactly which metabolic pathways are functioning and to what extent depends on several factors, including the supply of nutrients, the need for energy, and the concentrations of the products of various pathways.

In the preceding chapters we have discussed several metabolic pathways, including glycolysis, gluconeogenesis, fatty acid synthesis and catabolism, and the TCA cycle. We also discussed some points of connection between specific pathways, and the involvement of acetyl-S-CoA in a number of pathways. The interrelationships of the major pathways we have studied are summarized schematically in Figure 29-1.

We have also discussed the different types of controls that regulate metabolic activity, including (1) allosteric enzymes, (2) hormones, (3) chemical modification of enzymes, and (4) the concentrations of substrates and products. In addition, you should recognize that different cells in the body are doing different things at any given time. For example, glycogen may be broken down to glucose in the liver, but the glucose is shipped to other cells (including nerve and muscle cells) for glycolysis. We shall now examine how the interconnections and controls of the various metabolic pathways work together for the well-being of the individual cell and the entire body.

Glycolysis and Gluconeogenesis

Recall that glycolysis is the major metabolic pathway for carbohydrate catabolism (Chapter 25). The catabolism (breakdown) of carbohydrates via glycolysis

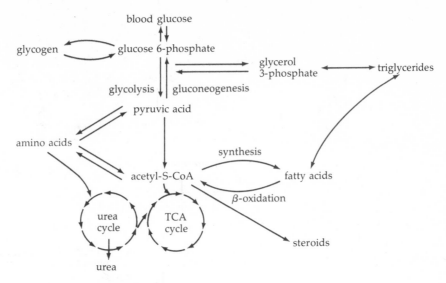

Figure 29-1   The interrelationships of some of the major metabolic pathways in human cells. (The electron transport system and pathways for nucleic acid and protein synthesis are not shown.)

and the β-oxidation of fatty acids are the major sources of metabolic energy in human cells. Glycolysis is interconnected with a number of metabolic pathways, including gluconeogenesis (glucose synthesis), glycogen synthesis and catabolism, the pentose phosphate pathway, the synthesis and catabolism of some amino acids, the synthesis and catabolism of glycerides and phosphoglycerides, and the TCA cycle (see Figure 29-2). All of the reactions that connect these pathways have already been presented in the preceding chapters, as were the three major control points indicated in Figure 29-2. Recall from Chapters 23 and 25 that the hormones epinephrine and glucagon turn off the formation of glycogen and initiate the breakdown of glycogen to glucose 1-phosphate, which is then catabolized via glycolysis in response to an energy demand on the body as a whole (control point 1 in Figure 29-2). The second major control point is responsive to the energy needs of the individual cells in the body. At this point, the rate of glycolysis is increased if the ATP supply is low, and it is decreased if the ATP supply is high. Recall that this occurs by means of the allosteric activation or inhibition of the enzymes fructose 1,6-diphosphatase and phosphofructokinase by ATP and other molecules.

The third control point indicated in Figure 29-2 involves the control of oxaloacetic acid synthesis by acetyl-S-CoA. Acetyl-S-CoA is produced by the catabolism of fatty acids, some amino acids, and glycolysis. When acetyl-S-CoA is produced in excessive amounts, it allosterically activates the enzyme pyruvate carboxylase, which makes oxaloacetic acid (see Figure 29-3). The oxaloacetic acid that is formed can be used in gluconeogenesis or in the synthesis of certain amino acids such as aspartic acid. Oxaloacetic acid is also used in the first step of the TCA cycle. The fate of a particular molecule of oxaloacetic acid will thus depend on other needs and controls. Sometimes more oxaloacetic acid is needed to combine with acetyl-S-CoA in the TCA cycle. At other times an excess of acetyl-S-CoA may indicate that sufficient fuel is already available for the TCA cycle and the oxaloacetic acid may then be used for gluconeogenesis. In the latter case, pyruvic acid is used to make glucose (via oxaloacetic acid) instead of being converted to more acetyl-S-CoA.

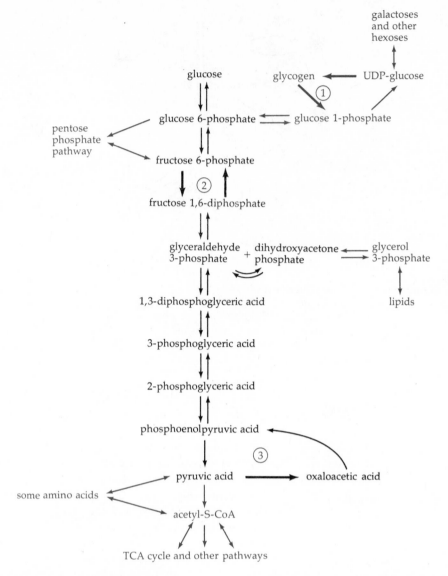

Figure 29-2   Connections between glycolysis and gluconeogenesis (shown in black) and some other metabolic pathways (shown in blue). The heavier arrows represent reactions involved in the three major control points that are discussed in the text.

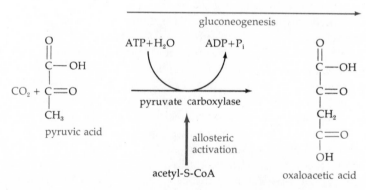

Figure 29-3   Pyruvate carboxylase is allosterically activated by acetyl-S-CoA, thereby increasing the rate of gluconeogenesis.

Lipid Metabolism

Recall that triglycerides are the major energy storage depot in the human body. Other lipids have a variety of functions (Chapter 26). The synthesis and catabolism of all lipids are connected to other metabolic pathways via acetyl-S-CoA (see Figure 29-4). Notice in this figure that metabolism of the glycerol portions of triglycerides is connected to glycolysis, as we explained in Chapter 27.

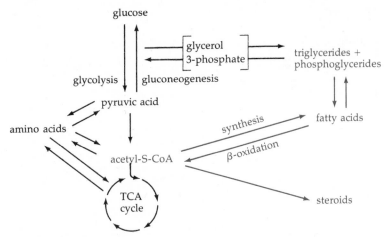

Figure 29-4   Pathways for the metabolism of lipids (shown in blue) are connected to other metabolic pathways by   acetyl-S-CoA and glycerol 3-phosphate.

Fatty acids, you will recall, are components of most lipids. Fatty acids are synthesized from acetyl-S-CoA and can also be broken down via β-oxidation to acetyl-S-CoA. The synthesis and catabolism of fatty acids must be controlled, so that these two pathways do not continually run in a wasteful circle. A major control of fatty acid synthesis appears to be the rate of synthesis of malonyl-S-CoA from acetyl-S-CoA in step 1 of fatty acid synthesis (see Figure 29-5). The activity of this enzyme is allosterically increased by citric acid and is inhibited by palmitoyl-S-CoA. Recall that citric acid is an intermediate in the TCA cycle, and the concentration of citric acid will build up if the concentration of acetyl-S-CoA is large. Palmitoyl-S-CoA, the CoA thioester of the 16-carbon fatty acid palmitic acid, is the major product of fatty acid biosynthesis and acts as an end-product inhibitor of the first step in this pathway.

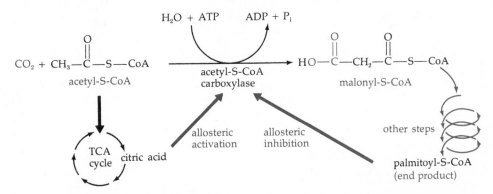

Figure 29-5   The first step in fatty acid biosynthesis, catalyzed by the enzyme acetyl-S-CoA carboxylase, is accel-   erated by excess citric acid and inhibited by excess palmitoyl-S-CoA.

The regulation of fatty acid catabolism is more complex. One major control of fatty acid catabolism is the activity of the lipase enzymes, which cleave triglycerides into glycerol and fatty acids. The hormone insulin slows down fatty acid catabolism by inhibiting lipase action in adipose (fat) cells when sufficient glucose is present to meet the body's energy needs.

The TCA and Urea Cycles

Two major connections to the TCA cycle are glycolysis and the β-oxidation of fatty acids, which provide acetyl-S-CoA—the fuel for this cycle. The interrelationships between the TCA cycle and other metabolic pathways are summarized in Figure 29-6.

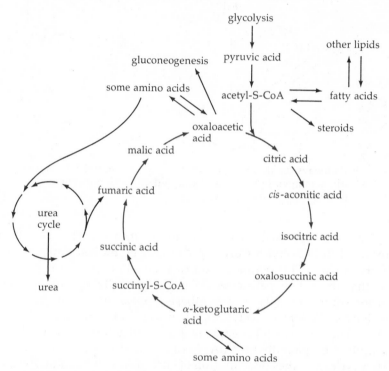

Figure 29-6   Connections between the TCA cycle and other metabolic pathways.

In Chapter 28 we noted that the urea cycle is intimately connected to the TCA cycle. The TCA cycle is also interconnected to the synthesis and catabolism of a number of amino acids. The TCA cycle is additionally connected to gluconeogenesis via oxaloacetic acid, which is a common intermediate in both of these pathways.

We have touched briefly on a few of the major metabolic controls operating in the human body. Much more is known about these metabolic controls. However, there is still a great deal to be learned about human metabolism and its control. Considerable research is being done in this area.

Exercise 29-1
The final product of each round of the TCA cycle is a molecule of oxaloacetic acid, which *may* then combine with another molecule of acetyl-S-CoA to start the TCA cycle again. Explain why this cannot go on indefinitely without supplementing the supply of oxaloacetic acid. (Hint: See Figures 29-2 and 29-3.)

Exercise 29-2
People who consistently eat too much sugar tend to get fat. Under these conditions, explain how fatty acid synthesis is stimulated.

29-4 NUTRITIONAL REQUIREMENTS

In the preceding chapters we generally assumed that cells had sufficient supplies of all the raw materials needed to operate their metabolic pathways. Unfortunately, this is not always the case. In fact, poor nutrition is an ever-increasing problem around the world, not just in underdeveloped nations, but in parts of industrialized nations as well.

Good nutrition involves the regular ingestion of sufficient, but not overwhelming, amounts of all of the necessary nutrients. These **nutrients** include oxygen, water, such energy sources as carbohydrates and lipids, and a wide variety of other substances including proteins, vitamins, and minerals. The amounts of each type of nutrient that are required are shown in Figure 29-7. Notice the large variations in our requirements for different nutrients.

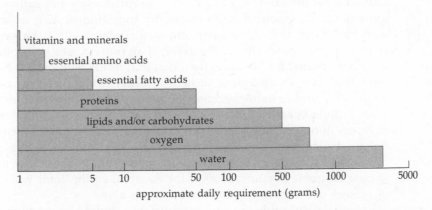

Figure 29-7 Approximate daily requirements (in grams) for the various nutrients required by the human body. Note that a sliding scale is used in this figure so that requirements for water (about 2800 g/day) and essential amino acids (about 2 g/day) can fit on the same graph. In spite of this device, the daily requirements for vitamins and minerals (ranging from 0.2 g/day down to 10^{-6} g/day) cannot be represented accurately on this scale.

Nutrients Required in Large Amounts (more than 100 g/day)

Water is needed in far greater amounts than any other nutrient. On the average, a total of almost 3 kg of water (from liquids and solid foods) is normally taken in daily by an adult, and approximately half of this is excreted in urine. The water in sweat and urine serves as a solvent for the elimination of some by-products of metabolism. The evaporation of water in perspiration also aids in temperature regulation.

The **oxygen** we inhale (about 0.6 to 0.8 kg/day, in about 2500 liters of air) is used almost exclusively as the final electron acceptor in the electron transport system, and is then eliminated in the form of water. Molecular oxygen also serves as a substrate for a few enzyme-catalyzed reactions.

Our daily requirements for **carbohydrates** and **lipids** reflect our daily energy needs. Adults with average energy needs require between 2000 and 3000 kcal* of energy per day. This corresponds to roughly 500 to 600 g of carbohydrates or 250 to 300 g of lipids per day. If a person's diet contains both carbohydrates and lipids, then these amounts are reduced proportionately; for example, 250 to 300 g of carbohydrates *plus* 125 to 150 g of lipids will satisfy the energy needs of the average adult. Note that lipids supply more calories per gram than carbohydrates (Chapter 27). You should keep in mind that these numbers represent *average* requirements. The actual amounts required can be quite different for persons with unusual energy needs. A lumberjack, for example, may require twice these amounts of carbohydrates and/or lipids in order to supply enough energy for the heavy labor required by that occupation.

Nutrients Required in Moderate Amounts (1 to 100 g/day)

Adults typically require about 50 g of **proteins** per day. In the last chapter we saw that the amino acids derived from digestion of proteins enter the body's amino acid pool, where they may be incorporated into newly synthesized proteins, converted (in some cases) to hormones, used for the synthesis of other nitrogen-containing biomolecules, or oxidized for energy production. The figure 50 g of protein per day is a very rough approximation. The actual requirement is for a certain amount of each of the **essential amino acids** (those that cannot be synthesized by human cells). A typical adult may satisfy his or her requirements for essential amino acids by ingestion of 40 g or less of protein daily, provided that the protein ingested contains all of the essential amino acids in proper proportions. However, if an individual's sole source of protein were lima beans, for instance, then that individual would need to ingest much more than 50 g of lima bean protein in order to obtain adequate amounts of methionine and cysteine because the percentage of these two essential amino acids in lima beans is quite low. Because many sources of dietary protein are deficient in one or more of the essential amino acids, nutritionists recommend **balanced diets** that contain several protein sources. For example, a diet containing equal amounts of protein from wheat and beans or rice and beans can supply adequate amounts of all of the essential amino acids with a total of 50 g of protein per day from these combinations alone.

Essential fatty acids are another nutrient required in moderate amounts, as was discussed in Chapter 26. Briefly, the average human requires about 5 g of these polyunsaturated fatty acids per day, an amount easily obtained in a normal diet. This requirement may not be met, however, when hospitalized patients are fed intravenously for extended periods of time.

Nutrients Required in Small Amounts (less than 1 g/day)

The nutrients in this group are the **vitamins** and several **minerals,** including the elements Fe, Zn, F, Mg, Mn, Cu, I, Mo, Se, Cr, Sn, and Co. Minimum daily requirements for vitamins and minerals have been established on the basis of the minimum amount needed to prevent certain deficiency diseases. For example, the lack of an adequate amount of vitamin C can cause a disease called **scurvy,** in which collagen synthesis is impaired. For some vitamins, the daily ingestion of larger amounts may be needed for peak operating efficiency and longevity. For example, it has been suggested that large daily doses of vitamin C may help to prevent colds and heart disease. However, excessive amounts of some vitamins, such as vitamin A, can be harmful. We understand the functions

---

* Recall that the kilocalorie is equivalent to the Calorie unit used by nutritionists.

of some vitamins and minerals, but the precise metabolic functions of others are not clearly understood. The very small amounts of vitamins and elements that are required indicates that they are not used up rapidly by the body. Most, in fact, are used as recyclable coenzymes or as prosthetic groups for proteins.

Variations in Required Daily Allowances

Published lists of recommended daily allowances should be viewed as only rough estimates of human nutritional needs. There are a number of factors that influence the daily nutritional requirements of any given individual, some of which have already been mentioned. These include:

1. The source of the nutrient, as in the case of dietary protein.

2. The health and physical state of the individual—growing children, pregnant and lactating women, and people recovering from an illness may need larger than average amounts of certain nutrients, whereas sedentary individuals need less fat or carbohydrate.

3. The overall diet, which influences the amounts of various components needed (for example, if a person is on a high-protein diet, he or she will need less carbohydrate or fat for energy).

4. A number of other factors, including symbiotic intestinal bacteria (which manufacture some vitamins for us), the composition of ingested foods, and the climate.

Exercise 29-3
On a diet in which all protein is obtained from lima beans, which of the following amino acids is not likely to be oxidized for energy production: (a) methionine, (b) glutamic acid, (c) alanine? Explain your answer.

Exercise 29-4
Figure 29-7 shows some very large differences in recommended daily allowances. Explain why hundreds of grams of carbohydrates or fat are required daily but only 5 g of essential fatty acids.

29-5 BODY FLUIDS

The transport of required nutrients to all of the cells in the body, the transmission of hormonal messages, and several other vital body functions are dependent upon the blood and other extracellular fluids. **Extracellular fluids** include blood plasma, spinal fluid, saliva, and other fluids that occur outside of body cells. About 20% of human body weight is due to extracellular fluids, and three-quarters of this extracellular fluid is found around and between individual cells. This fluid is called **interstitial fluid.** Blood plasma, the noncellular fluid portion of whole blood, accounts for most of the remaining 25% of the extracellular fluid. In previous chapters we have discussed several functions of blood, including the transport of nutrients such as glucose, lipids (bound to albumin or other carrier proteins), and oxygen (bound to hemoglobin in red blood cells), as well as the transport of hormones and metabolic waste products such as urea.

Blood plasma and interstitial fluid also function in the maintenance of the correct pH, ionic, and osmotic environment for body cells. The volumes of these extracellular fluids, and their pH and electrolyte concentrations, are therefore normally kept under precise control. Large variations in extracellular fluid composition cause related changes in body cells that are harmful and often deadly. Likewise, when body cells behave abnormally, there is likely to be a change in the composition of the extracellular fluids. Changes in the composition of body fluids, particularly that of plasma, since it is easy to obtain, are now used to diagnose certain diseases and metabolic disorders (see Table 29-1).

Table 29-1 Some Diagnostic Uses of Blood Plasma

Plasma Component	Normal Concentration	Abnormalities	
		Disorder	Concentrations
H^+ (measured as pH)	7.35–7.45	Acidosis	Less than 7.35
		Alkalosis	Greater than 7.45
CO_2 (measured as P_{CO_2})	35–40 mm Hg	Respiratory acidosis	Above 40 mm Hg
K^+	4.0–4.8 meq/liter	Renal disease	Less than 4 meq/liter
Cholesterol	150–250 mg/100 ml	Obstructive jaundice	250–500 mg/100 ml
		Biliary cirrhosis	Up to 1800 mg/100 ml
Glucose	80–120 mg/100 ml	Diabetes	150 mg/100 ml to 1000 mg/100 ml
Gammaglobulins	0.5–1.6 g/100 ml	Multiple myeloma	Above 2 g/100 ml
Glutamic oxaloacetic transaminase	9–25 Karmen units*	Myocardial infarction (heart attack)	50–400 Karmen units

* The activity of some enzymes is expressed using special units.

Notice that a wide variety of plasma components—from simple ions to complex proteins—are used for the diagnosis of a large number of disorders. During the treatment of many diseases the concentration of plasma components is also monitored regularly to assess the patient's progress and the effectiveness of the treatment.

pH Balance

Three interrelated mechanisms—**kidney function, respiration rate,** and **buffer actions**—are used to maintain blood pH within the normal limits of 7.35 to 7.45. The buffer systems include H_2CO_3/HCO_3^- (the major buffer), $H_2PO_4^-/HPO_4^{2-}$, and the buffering action of many proteins. Recall that carbonic acid, H_2CO_3, is in rapid equilibrium with dissolved CO_2 (reaction 29-1) and also with bicarbonate and H_3O^+ ions (reaction 29-2).

(29-1) $$CO_2(aq) + H_2O \underset{}{\overset{\text{carbonic anhydrase}}{\rightleftharpoons}} H_2CO_3$$

(29-2) $$H_2CO_3 + H_2O \rightleftharpoons HCO_3^- + H_3O^+$$

Normally, on a short-term basis, the positions of these equilibria and hence the blood pH are controlled by the proper breathing rate. If the respiration rate is very low, however, a condition known as **hypoventilation,** not enough $CO_2(g)$ is expelled from the lungs and the partial pressure of $CO_2(g)$ in the lungs increases. As a consequence, the concentration of dissolved CO_2 in the blood plasma increases. Also, $[H_2CO_3]$, $[HCO_3^-]$, and $[H_3O^+]$ rise, as the positions of the equilibria for reactions 29-1 and 29-2 respond to the increase in $[CO_2(aq)]$. This condition, with its abnormally low blood pH, is referred to as **respiratory acidosis** (Table 29-2). Respiratory acidosis can be caused by asthma, pneumonia, drug overdoses, and other dysfunctions where insufficient CO_2 is exhaled.

Table 29-2 Alterations in Blood Plasma During Respiratory Acidosis and Alkalosis

	pH	$[H_3O^+]$	$[HCO_3^-]$	$[H_2CO_3]$
Respiratory acidosis	Below normal	Above normal	Above normal	Above normal
Respiratory alkalosis	Above normal	Below normal	Below normal	Below normal

On the other hand, prolonged **hyperventilation** (overbreathing) during, for example, anxiety, hysteria, crying, or mountain climbing, results in a decrease in the partial pressure of CO_2 in the lungs and a consequent decrease in $[CO_2(aq)]$. In this case $[H_2CO_3]$, $[HCO_3^-]$, and $[H_3O^+]$ decrease as the positions of the equilibria for reactions 29-1 and 29-2 respond to the decrease in $[CO_2(aq)]$. This condition, with its abnormally high blood pH, is referred to as **respiratory alkalosis** (see Table 29-2). The partial pressure of $CO_2(g)$ that can be in equilibrium with blood plasma is used as a measure of the concentration of carbon dioxide dissolved in the blood and also as a measure of $[H_2CO_3]$. Recall that the solubility of a gas is directly proportional to the partial pressure of the gas. The equilibrium constant expression for reaction 29-1 is

$$K_{eq} = \frac{[H_2CO_3]}{[CO_2(aq)]}$$

Therefore, $[H_2CO_3] = K_{eq}[CO_2(aq)]$, and $[H_2CO_3]$ is directly proportional to $[CO_2(aq)]$. In this relationship K_{eq} is actually very small (about 2.5×10^{-3}). Thus when $CO_2(g)$ dissolves in pure water, almost all of the dissolved carbon dioxide exists as $CO_2(aq)$, and very little of it is in the form of H_2CO_3.

The kidneys also help maintain the pH of blood plasma at its proper level. The kidneys can respond to a stress that tends to alter the pH of blood plasma from its normal value with more long-term effectiveness than can be achieved by an alteration of breathing rate.

For example, in an acidosis condition, where the blood plasma pH is below normal, the kidneys produce very acidic urine (pH as low as 4, compared to an average value of about 6). In this process,

1. $CO_2(aq)$ enters from the blood into the kidney cells, where it is used to produce H_3O^+ and HCO_3^-.

2. H_3O^+ ions go into the urine from the kidney cells, and an equal number of positively charged Na^+ ions enter the cells from the urine.

3. Na^+ ions and HCO_3^- ions enter the blood from the kidney cells (see Figure 29-8). The influx of HCO_3^- into the blood and $CO_2(aq)$ out of the blood shifts the equilibrium for reaction 29-2 to the left.

As we mentioned previously in this section, $[H_2CO_3]$ is directly proportional to $[CO_2(aq)]$, and according to equilibrium 29-2, $[H_3O^+]$ in blood plasma is related to the ratio $[H_2CO_3]/[HCO_3^-]$ by the expression

(29-3) $$[H_3O^+] = \frac{[H_2CO_3]}{[HCO_3^-]} \times (8.0 \times 10^{-7} \text{ mole/liter})$$

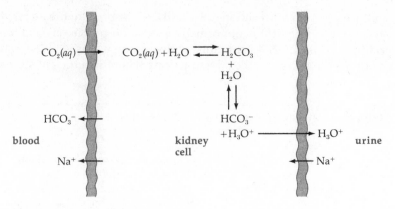

Figure 29-8   The response of kidney cells to a condition of acidosis.

Thus, when $CO_2(aq)$ goes from the blood into the kidney cells, $[H_2CO_3]$ in the blood decreases. This effect and the entry of HCO_3^- into the blood from the kidney cells results in a decrease in the $[H_2CO_3]/[HCO_3^-]$ ratio and a decrease in $[H_3O^+]$ by Eq. 29-3, thus helping to restore the blood pH to its normal level.

Electrolyte Balance

The kidneys also aid in the maintenance of correct concentrations of several other electrolytes in the blood. Figure 29-9 shows the normal concentrations of electrolytes in body fluids. Note that, whereas plasma and interstitial fluid have similar compositions, the composition of the fluid within cells is quite different. Also notice that each fluid is electrically neutral, containing equal amounts of positive and negative equivalents. In this figure the concentration unit milliequivalents per liter is used to put ions with different charges on an equivalent basis.

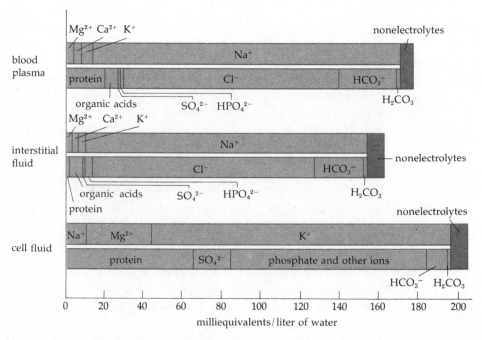

Figure 29-9   The normal concentrations of electrolytes in body fluids.

The term **electrolyte balance** is used to refer to the normal situation in which the concentrations of the electrolytes in body fluids are kept within close tolerances of their normal values in spite of the intake and removal of large volumes of water and solutes. Proper kidney function is primarily responsible for normal electrolyte balance in blood plasma and interstitial fluids. Specific membrane transport proteins are in turn responsible for the maintenance of the correct intracellular fluid composition. For example, the protein Na^+K^+-ATPase pumps K^+ ions into cells and Na^+ ions out of cells into the interstitial fluid.

For electrolyte balance, the overall input of water (in the form of ingested liquids, water in ingested solid foods, and water produced as a by-product of metabolic catabolism) must equal the overall output of water (eliminated in the urine and feces, from the lungs as water vapor, and through the skin as perspiration). There must also be a balance of the water within the body among blood plasma, interstitial fluid, and cell fluid. Recall that there will be a net flow of water across a semipermeable membrane separating two solutions unless the solutions have the same osmotic pressure. Also recall that the osmotic pressure of an aqueous fluid is directly proportional to the concentration of *all* the dissolved or dispersed particles in the fluid.

In arid desert climates, people must be especially careful to maintain sufficient quantities of water in their bodies.

The osmotic pressure of the body fluids (and thus the water flow within the human body) is controlled primarily by the kidneys, which regulate the volume of urine that is eliminated. For example, if you decrease the volume of water you ingest, then both the concentration of electrolytes in the body fluids and the osmotic pressure of these fluids increases. When the osmotic pressure of blood increases, the hormone **vasopressin** is released from the pituitary gland. This hormone triggers the reduction of urine output by the kidneys. The net result is that water is retained by the body, and this stabilizes the electrolyte concentrations and osmotic pressure of the body fluids.

Exercise 29-5

The acidic ketone bodies produced in diabetics often cause a condition called metabolic acidosis, which we shall study shortly. How would you expect kidney cells to respond to this condition?

29-6 METABOLIC DISORDERS

The malfunction of any one of the thousands of proteins in the human body will result in illness and, potentially, death. Other diseases are caused by pathogenic organisms and by external stresses that are placed on the body. An example of the latter case is **emphysema,** a progressive deterioration of the respiratory system caused by, among other things, cigarette smoke.

Let us consider a few common metabolic disorders in order to demonstrate that, because of the intricate interrelationships of metabolic pathways, a defect in one pathway can cause major alterations in the overall metabolic performance of the body. Such is the case in the disease **diabetes mellitus,** which might in the future afflict more than 20% of the population of the United States. Recall that diabetes results from an inability to produce a sufficient amount of active insulin and that the best diagnostic indication of diabetes is the glucose tolerance test (Section 25-12). Also recall that insulin normally functions in promoting glucose uptake by cells, stimulates the formation of glycogen, and inhibits the hydrolysis of triglycerides in adipose tissue. Lack of insulin, therefore, results in

1. Increased blood glucose levels

2. Increased fatty acid oxidation

3. Cessation of fatty acid synthesis

4. Increased gluconeogenesis

The abnormal fatty acid metabolism, items 2 and 3, results in an excessive production of acetyl-S-CoA (see Figure 29-10). Not all of this acetyl-S-CoA can be oxidized in the TCA cycle. The excess is converted to ketone bodies (see Chapter 27).

Recall that two of the ketone bodies are acids, so that an overproduction of ketone bodies causes a higher-than-normal $[H_3O^+]$ in blood and a lower-than-normal blood pH. This metabolic disorder is called **metabolic acidosis.** In an attempt to get rid of the increased amount of ketone bodies, the urine volume is increased. The urine volume can increase to such a point that a diabetic may become severely dehydrated. The normal concentration of ketone bodies in blood is less than 1 mg/100 ml, and the normal rate at which they are excreted in the urine is about 20 mg per day. If a person suffering from acute diabetes remains untreated, the concentration of ketone bodies in the blood may reach 90 mg/100 ml, and the rate at which they are excreted in the urine may go as high as 5000 mg per hour.

Since diabetics do not catabolize much glucose, ingested amino acids are also used for energy production and gluconeogenesis, with a concomitant increase in urea production. Thus, the lack of adequate amounts of a single hormone, insulin, can alter the normal functioning of virtually all of the metabolic pathways. This metabolic disruption results in easily detectable changes in the concentration of several components of extracellular fluids, including increased concentrations of glucose in the blood and urea in the urine, and possibly, a decrease in blood pH. These changes can be monitored during therapy.

Metabolic acidosis is a very serious condition. A decrease in blood pH interferes with the ability of hemoglobin to bind molecular oxygen. Recall that hemoglobin, HHb, and oxyhemoglobin, $HHbO_2$, are both weak acids. The effect of a decrease in blood pH on the binding of oxygen to hemoglobin is apparent when we consider the following equilibrium between hemoglobin, the oxyhemoglobin anion, molecular oxygen, and H^+:

(29-4) $$HHb + O_2 \rightleftharpoons HbO_2^- + H^+$$

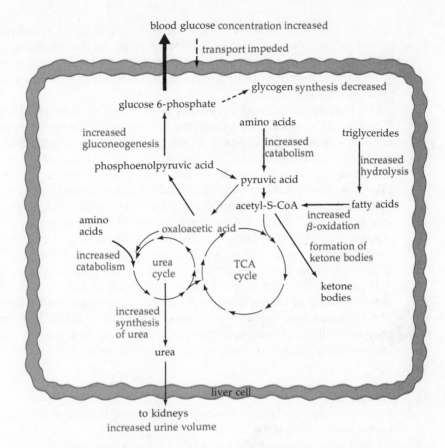

Figure 29-10   A schematic representation of the metabolic malfunctions in diabetes that result from the absence of sufficient active insulin.

A decrease in blood pH, that is, an increase in [H⁺], will shift the position of equilibrium for reaction 29-4 to the left. Thus, a person suffering from metabolic acidosis has difficulty in providing the body cells with sufficient oxygen, and the person's respiration rate increases in an attempt to get more oxygen.

In an attempt to decrease $[H_3O^+]$ and restore the blood pH to its normal value, the position of the H_2CO_3/HCO_3^- buffer equilibrium,

(29-5) $$HCO_3^- + H^+ \rightleftharpoons H_2CO_3$$

is shifted to the right, thus decreasing $[HCO_3^-]$ in blood plasma. Under these circumstances, $[H_2CO_3]$ in blood does *not* increase, as it would appear from reaction 29-5, because the increased breathing rate causes an increase in the amount of $CO_2(g)$ expelled by the lungs. The increased expulsion of $CO_2(g)$ from the lungs more than offsets the shift in the equilibrium for reaction 29-5 to the right, and as a consequence, $[H_2CO_3]$, in blood plasma *decreases*.

Table 29-3 compares blood pH, $[HCO_3^-]$, and $[H_2CO_3]$ (or P_{CO_2}) for metabolic acidosis with normal values. Metabolic acidosis can be caused by impairment of kidney function, starvation, and other disorders in addition to diabetes.

Table 29-3 Alterations in Blood Plasma During Metabolic Acidosis or Alkalosis

	pH	$[H_3O^+]$	$[HCO_3^-]$	$[H_2CO_3]$
Metabolic acidosis	Below normal	Above normal	Below normal	Below normal
Metabolic alkalosis	Above normal	Below normal	Above normal	Above normal

A less common condition than metabolic acidosis is **metabolic alkalosis,** in which the blood plasma pH is above normal. In this disorder, [HCO_3^-] and [H_2CO_3] (or P_{CO_2}) are above normal (see Table 29-3). An overdose of antacids or prolonged vomiting (which reduces the amount of stomach acid) can cause metabolic alkalosis.

Many metabolic disorders are not as complex as diabetes. In **galactosemia** (Chapter 25), for example, an enzyme that catalyzes the conversion of galactose to glucose is missing. Galactose is one-half of the milk disaccharide lactose. The other half is glucose. Clinical symptoms of galactosemia result from a buildup of abnormal amounts of galactose, not from generalized metabolic alterations. Likewise, **phenylketonuria** (PKU) results from a genetic deficiency of the enzyme that converts the essential amino acid phenylalanine to tyrosine. If tyrosine is provided in the diet there will be no difficulty in protein synthesis, but abnormal amounts of phenylalanine may build up unless the patient is placed on a diet that is low in this essential amino acid. The prime clinical symptom of phenylketonuria is an excessive concentration of phenylalanine in the blood.

These are only a few of the diseases that result from metabolic disorders. An understanding of their molecular basis has made effective treatment possible. The chemical and biochemical concepts you have learned in the preceding chapters will enable you to better understand not only normal body functions but also metabolic malfunctions and their treatment.

Exercise 29-6

A person suffering from untreated diabetes mellitus is likely to have which of the following clinical symptoms?

  (a) Below-normal urine pH and below-normal plasma [HCO_3^-]
  (b) Below-normal urine pH and above-normal plasma [HCO_3^-]
  (c) Above-normal urine pH and below-normal plasma [HCO_3^-]
  (d) Above-normal urine pH and above-normal plasma [HCO_3^-]
  (e) Normal urine pH and above-normal plasma [HCO_3^-]

29-7 SUMMARY

1. There are three major controls of carbohydrate metabolism. The first regulates the synthesis of glycogen and its degradation for use in glycolysis. The second controls glycolysis versus gluconeogenesis in response to the cell's energy needs. The third regulates gluconeogenesis in response to variations in the acetyl-S-CoA concentration.

2. Acetyl-S-CoA is a central intermediate connecting glycolysis and the metabolism of fatty acids and amino acids; it also serves as the entry point into the TCA cycle.

3. Fatty acid biosynthesis is stimulated by citric acid, which allosterically activates acetyl-S-CoA carboxylase; it is suppressed by palmitoyl-S-CoA, which allosterically inhibits this enzyme.

4. The TCA cycle and amino acid metabolism, as well as the TCA cycle and the urea cycle, share some common intermediates.

5. Humans require vastly different amounts of various nutrients, ranging from 10^{-6} to 10^3 g per day.

6. The actual nutritional requirements of the human body depend on the age, health, and level of activity of each individual. Recommended daily

allowances are average minimum requirements that have been suggested by nutritionists.

7. The extracellular fluids of the body function to (a) transport nutrients and wastes; (b) maintain the correct pH, ionic, and osmotic environment; and (c) transport hormones between cells.

8. Kidney function, respiration rate, and buffer actions are the mechanisms used to maintain blood pH within normal limits.

9. Hypoventilation can cause respiratory acidosis, whereas hyperventilation can cause respiratory alkalosis.

10. Electrolyte balance and water balance are regulated by the kidneys.

11. Metabolic disorders are diagnosed and monitored by determination of the concentrations of various components in several extracellular fluids (especially blood).

12. Diabetes is a metabolic disorder that involves the malfunction of several metabolic pathways.

13. Metabolic acidosis can be caused by diabetes, impaired kidney function, or starvation, whereas metabolic alkalosis can be caused by an antacid overdose or prolonged vomiting.

PROBLEMS

1. As you walk down a dark alley in the middle of the night, you hear a strange sound. You become frightened and start to run.
   (a) Your fear causes the release of epinephrine from cells in the adrenal gland. What effect will this have on glycogen and triglyceride metabolism?
   (b) In order to run, your muscle cells must use up large amounts of ATP. What effect will this rapid use of ATP have on glycolysis and gluconeogenesis in these muscle cells?
   (c) You continue to run for several miles. (You are really scared!) What is the major product of glucose catabolism in your muscle cells?
   (d) You stop running at last. What happens to the product of glycolysis from part (c)?

2. You are overweight and decide to go on a crash diet. There are several popular fad diets from which you can choose. What are some potential problems you might encounter with the following choices?
   (a) A diet totally devoid of unsaturated fats
   (b) A pure "liquid" protein diet
   (c) A cholesterol-free diet

3. As you climb a mountain, the partial pressure of oxygen in the atmosphere decreases and you may overbreathe to try to increase the partial pressure of oxygen in your lungs.
   (a) What effect, if any, would this overbreathing have on the partial pressure of CO_2 in your lungs? Explain your answer.
   (b) What effect, if any, might this overbreathing have on the pH of your blood plasma?

4. A person suffering from a stomach ulcer ingests an excessive amount of bicarbonate. Indicate by symbols whether the following are likely to have a value that is higher than normal $(+)$, lower than normal $(-)$, or normal (0).
   (a) Plasma pH
   (b) Plasma $[HCO_3^-]$
   (c) Plasma $[H_2CO_3]$
   (d) Respiration rate

5. Would you expect the hormone vasopressin to be released when you engage in prolonged strenuous exercise on a hot day?

6. Consider the two enzymes phosphofructokinase (PFK) and fructose 1,6-diphosphatase (FDP). When a person engages in vigorous exercise, which of the following will occur?
   (a) Both enzymes bind ATP at an allosteric site and the activity of both increases.
   (b) Both enzymes bind ADP at an allosteric site and the activity of PFK increases while the activity of FDP decreases.
   (c) FDP binds ATP at an allosteric site and its activity increases, while PFK binds ADP at an allosteric site and its activity also increases.
   (d) PFK binds ADP at an allosteric site and its activity decreases, while FDP binds ATP at an allosteric site and its activity increases.
   (e) None of the above.

7. Which of the following can result in a blood plasma pH that is above the normal value?
   (a) A woman panting during labor
   (b) Untreated diabetes
   (c) Overstrenuous exercise
   (d) A person nearly drowning
   (e) None of the above

SOLUTIONS TO EXERCISES

29-1 In addition to participating in the TCA cycle, oxaloacetic acid is used in gluconeogenesis and for the synthesis of aspartic acid. Thus, without replenishment, all of the oxaloacetic acid would eventually be drained off from the TCA cycle.

29-2 The large amounts of sugar you eat are catabolized and produce large amounts of acetyl-S-CoA, which allosterically activates acetyl-S-CoA carboxylase—the enzyme required for the first step of fatty acid biosynthesis.

29-3 (a) Methionine, since it is present in only relatively small amounts in lima bean protein.

29-4 Carbohydrate and/or fat must be catabolized to provide energy. Essential fatty acids, however, are generally not catabolized, but used by body cells for the synthesis of comparatively small amounts of more complex lipids.

29-5 Kidney cells will increase the acidity of the urine in an attempt to restore the plasma pH to normal (see Figure 29-8).

29-6 (a) The $[HCO_3^-]$ in blood plasma is below normal (see Table 29-3), and kidney cells will acidify the urine in an effort to raise the plasma pH.

ESSENTIAL SKILLS

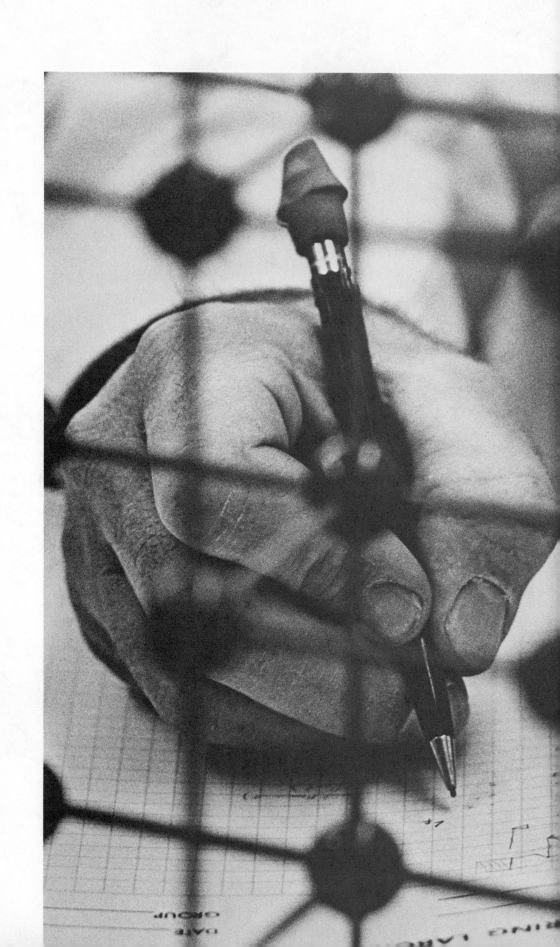

ESSENTIAL SKILLS 8

Naming Organic Compounds

Common and IUPAC names for the various classes of organic compounds are described in detail in Chapters 13 through 18. The basic principles for naming organic compounds are summarized here.

COMMON NAMES

Common names for many organic compounds are formed by combining the name for the functional group with the names for the alkyl groups involved. Thus, CH_3-CH_2-OH is called ethyl alcohol, and $CH_3-CH_2-O-CH_2-CH_3$ is called diethyl ether. The common names for some of the more prevalent alkyl groups are given in Table 1.

Table 1 Common Alkyl Groups

Name	Structural Formula
Methyl	CH_3-
Ethyl	CH_3-CH_2-
n-Propyl	$CH_3-CH_2-CH_2-$
Isopropyl	$CH_3-\overset{\mid}{C}H-CH_3$
n-Butyl	$CH_3-CH_2-CH_2-CH_2-$
Isobutyl	$CH_3-\overset{\overset{\textstyle CH_3}{\mid}}{C}H-CH_2-$
t-Butyl	$CH_3-\overset{\overset{\textstyle CH_3}{\mid}}{\underset{\underset{\textstyle}{\mid}}{C}}-CH_3$

The common names for carboxylic acids, aldehydes, amides, and anhydrides, however, are generally not formed from the names of alkyl groups. Common names for simple compounds belonging to these classes are derived from a natural source of the carboxylic acid with that number of carbon atoms. The structural formulas and common names for a few simple carboxylic acids are given in Table 2. The Greek letters α, β, γ, and so on, are used as prefixes to the common names of carboxylic acids to specify the position of substituents, as in β-hydroxybutyric acid.

Table 2 Common Names for Some Carboxylic Acids

Common Name	Structural Formula
Formic acid	$\overset{\displaystyle O}{\overset{\|}{HC}}-OH$
Acetic acid	$CH_3-\overset{\displaystyle O}{\overset{\|}{C}}-OH$
Propionic acid	$CH_3-CH_2-\overset{\displaystyle O}{\overset{\|}{C}}-OH$
Butyric acid	$CH_3-CH_2-CH_2-\overset{\displaystyle O}{\overset{\|}{C}}-OH$
β-Hydroxybutyric acid	$CH_3-\overset{\displaystyle OH}{\overset{\|}{CH}}-CH_2-\overset{\displaystyle O}{\overset{\|}{C}}-OH$

Common Names for Acids and Bases

Recall that any carboxylic acid solution contains both carboxylic acid molecules and carboxylate ions that exist in equilibrium, and that the proportion of each form that is present depends on the pH. The higher the pH, the greater the proportion that exists as the carboxylate ion. For simplicity, however, we nearly always refer to these solutions as carboxylic acid solutions.

Similarly, bases are in equilibrium with their conjugate acid (the protonated cation). But, for simplicity, we refer to any equilibrium mixture that involves an amine and its conjugate acid by the name for the amine, even though in a sufficiently acidic solution the predominant form is the protonated cation.

Common Names for Complex Compounds

The common names for many other organic compounds, especially some very complex ones, are not formed in any systematic manner. Examples are glucose, toluene, and aspirin.

BASIC PRINCIPLES OF IUPAC NOMENCLATURE

The IUPAC name of an organic compound consists of three parts: a root, an ending, and one or more prefixes:

The **root** of a IUPAC name specifies the longest continuous chain of carbon atoms in a molecule of that compound, that is, the parent alkane molecule (see Table 3).

Table 3 Names for Unbranched Alkanes

Common Name	IUPAC Name	Structural Formula
Methane	Methane	CH_4
Ethane	Ethane	$CH_3—CH_3$
Propane	Propane	$CH_3—CH_2—CH_3$
n-Butane	Butane	$CH_3—CH_2—CH_2—CH_3$
n-Pentane	Pentane	$CH_3—CH_2—CH_2—CH_2—CH_3$
n-Hexane	Hexane	$CH_3—CH_2—CH_2—CH_2—CH_2—CH_3$
n-Heptane	Heptane	$CH_3—CH_2—CH_2—CH_2—CH_2—CH_2—CH_3$
n-Octane	Octane	$CH_3—CH_2—CH_2—CH_2—CH_2—CH_2—CH_2—CH_3$
n-Nonane	Nonane	$CH_3—CH_2—CH_2—CH_2—CH_2—CH_2—CH_2—CH_2—CH_3$
n-Decane	Decane	$CH_3—CH_2—CH_2—CH_2—CH_2—CH_2—CH_2—CH_2—CH_2—CH_3$

The **ending** of a IUPAC name specifies the class of organic compounds to which the molecule belongs or the major functional group of the molecule.

Prefixes are used to specify the position of functional groups and the identity and location of substituents attached to the longest carbon chain. Substituents are identified by name, and by a number that indicates the carbon atom of the longest chain to which they are attached, as follows:

1. The longest continuous chain must be numbered so that the positions of the substituents will have the lowest possible numbers.

2. The prefixes *di-*, *tri-*, and *tetra-* before the name of a substituent indicate two, three, or four of that substituent in the molecule.

3. When a molecule contains more than one substituent, the substituents are arranged alphabetically.

4. IUPAC names are written as a single word with numbers separated from one another by commas and numbers separated from letters by hyphens.

5. No punctuation or space is used between the name of a substituent and the root name.

Naming Alkanes

The common names for the alkanes containing an unbranched chain of one to ten carbon atoms are also given in Table 3. The common and IUPAC names of all alkanes end in *-ane*. The prefix *n-* (for normal) is not needed in the IUPAC name for an alkane. For example, the common name *n-*pentane specifies the compound with the formula $CH_3—CH_2—CH_2—CH_2—CH_3$, which must be

distinguished from its isomer isopentane $CH_3—CH_2—\overset{\displaystyle CH_3}{\underset{\displaystyle H}{C}}—CH_3$. The prefix

n- is not needed in the IUPAC name pentane, since the IUPAC name for

$CH_3—CH_2—\overset{\displaystyle CH_3}{\underset{\displaystyle H}{C}}—CH_3$ is 2-methylbutane, where the root of the IUPAC name,

but-, specifies that the longest carbon chain consists of four carbon atoms, and the prefix 2- specifies that the methyl substituent is bonded to the second carbon in that chain.

Table 4 Common and IUPAC Nomenclature for Some Classes of Organic Compounds

Class	General Formula	Ending IUPAC	Ending Common
Alkenes	$\overset{R_1'}{\underset{R_2'}{>}}C=C\overset{R_3'}{\underset{R_4'}{<}}$	-ene	-ene
Alcohols	R—OH	-ol	alcohol
Ethers	R_1—O—R_2	—	ether
Aldehydes	$R'-\overset{\overset{\displaystyle O}{\|}}{C}-H$	-al	aldehyde
Ketones	$R_1-\overset{\overset{\displaystyle O}{\|}}{C}-R_2$	-one	ketone
Carboxylic acids	$R'-\overset{\overset{\displaystyle O}{\|}}{C}-OH$	-oic acid	-ic acid
Esters	$R_1'-\overset{\overset{\displaystyle O}{\|}}{C}-O-R_2$	-oate	-ate
Anhydrides	$R_1'-\overset{\overset{\displaystyle O}{\|}}{C}-O-\overset{\overset{\displaystyle O}{\|}}{C}-R_2'$	-oic anhydride	-ic anhydride
Amines	$R_3'-\overset{\overset{\displaystyle R_1'}{\|}}{N}-R_2'$	—	amine
Simple amides	$R'-\overset{\overset{\displaystyle O}{\|}}{C}-NH_2$	amide	amide
Thiols	R—SH	-thiol	mercaptan

Table 4 lists some of the major classes of organic compounds and the common and IUPAC endings used in their names. For some classes, common names are almost always used for simple compounds; the IUPAC endings for these compounds are not shown. Notice that the names for aldehydes, anhydrides, and amides are derived from the names for the corresponding carboxylic acids. Also note that esters are named in a manner analogous to carboxylic acid salts.

Drawing the Structural Formula for a Compound When Given Its IUPAC Name

1. Identify the root and ending of the IUPAC name.

2. Draw the longest continuous chain of carbon atoms (specified by the root).

3. Draw the functional group (specified by the ending) at the location on the chain specified by the prefix number for that functional group. (e.g., 2-pentanone has a ketone group at the number 2 carbon atom.)

Structural Formula	Example IUPAC Name	Common Name
$CH_3-CH=CH_2$	Propene	Propylene

$CH_3-CH-CH_3$
     |
    OH
	2-Propanol	Isopropyl alcohol

CH_3-O-CH_3	—	Dimethyl ether
$CH_3-\overset{\displaystyle O}{\overset{\|}{C}}-H$	Ethanal	Acetaldehyde
$CH_3-CH_2-\overset{\displaystyle O}{\overset{\|}{C}}-CH_2-CH_3$	3-Pentanone	Diethyl ketone
$H-\overset{\displaystyle O}{\overset{\|}{C}}-OH$	Methanoic acid	Formic acid
$H_3C-\overset{\displaystyle O}{\overset{\|}{C}}-O-CH_3$	Methyl ethanoate	Methyl acetate
$H_3C-\overset{\displaystyle O}{\overset{\|}{C}}-O-\overset{\displaystyle O}{\overset{\|}{C}}-CH_3$	Ethanoic anhydride	Acetic anhydride
$CH_3-CH_2-CH_2-NH_2$	—	n-Propylamine
$CH_3-\overset{\displaystyle O}{\overset{\|}{C}}-NH_2$	Ethanamide	Acetamide
CH_3-CH_2-SH	Ethanethiol	Ethyl mercaptan

    4. Draw any substituents (specified by prefixes) at the locations specified by their prefix numbers.

Writing the IUPAC Name for a Compound When Given Its Structural Formula

    1. Identify the longest continuous chain of carbon atoms and thus determine the root of the name.

    2. Identify the class of compound and thus obtain the ending for the IUPAC name.

    3. Identify the substituents and number the longest carbon chain so that the substituents have the lowest possible numbers.

    4. Name the substituents as prefixes of the IUPAC name.

    Practice in naming organic compounds can be gained by working the appropriate exercises and problems in Chapters 13 through 18.

ESSENTIAL SKILLS 9

Predicting the Products of Organic Reactions

As you study organic compounds and the chemical reactions they undergo, you will frequently find problems of the following type: "Draw the structural formula(s) for the product(s) of the following reaction, given the structural formula(s) for the reactant(s)." The problem may also state: "If no reaction occurs, write N.R." Some problems may specify the reaction conditions or indicate that an oxidizing agent or a reducing agent is involved in the reaction.

You can avoid some of the difficulties with this type of problem if you develop a systematic approach, such as the one illustrated in the example on page 747. You will find this approach extremely useful if: (1) you have gained the ability to recognize the functional groups in the structural formula for an organic compound; and (2) you know the general types of reactions that the class of compounds specified by a given functional group can undergo. For convenience, the general reactions of organic compounds that are presented in Chapters 13 through 18 are summarized in Table 1.

Table 1 General Types of Organic Reactions

Functional Group Class	General Reaction (In the problems in this text, it is assumed that the appropriate reaction conditions, catalyst, and so on, needed for the reaction to occur are present.)	
$\diagdown C = C \diagup$ Alkenes	Reduction (hydrogenation)	$R'_1-CH=CH-R'_2 + H_2 \longrightarrow R'_1-CH_2-CH_2-R'_2$
	Halogenation	$R'_1-CH=CH-R'_2 + HY \longrightarrow R'_1-CH_2-\overset{\overset{\displaystyle Y}{\displaystyle |}}{C}H-R'_2$ (Y = halogen atom) (Adds according to Markovnikov's rule) $R'_1-CH=CH-R'_2 + Y_2 \longrightarrow R'_1-\underset{\underset{\displaystyle Y}{\displaystyle |}}{C}H-\underset{\underset{\displaystyle Y}{\displaystyle |}}{C}H-R'_2$
	Hydration	$R'_1-CH=CH-R'_2 + H_2O \longrightarrow R'_1-\underset{\underset{\displaystyle H}{\displaystyle |}}{C}H-\underset{\underset{\displaystyle OH}{\displaystyle |}}{C}H-R'_2$ (Adds according to Markovnikov's rule)

Table 1 (Continued)

Functional Group Class		General Reaction

$-\overset{\mid}{\underset{\mid}{C}}-OH$

Alcohols

Dehydration

(1) $R_1'-\overset{\overset{\displaystyle H}{\mid}}{\underset{\underset{\displaystyle H}{\mid}}{C}}-\overset{\overset{\displaystyle OH}{\mid}}{\underset{\underset{\displaystyle H}{\mid}}{C}}-R_2' \longrightarrow H_2O + R_1'-CH\!=\!CH-R_2'$

(2) $R-OH + R-OH \longrightarrow H_2O + R-O-R$
  an ether

Oxidation

(1) $X + R'-CH_2-OH \longrightarrow XH_2 + R'-\overset{\overset{\displaystyle O}{\|}}{C}-H$
  a primary alcohol   an aldehyde

(X is the general symbol for an oxidizing agent, and XH_2 represents a reducing agent.)

(2) $X + \quad R_1-\overset{\overset{\displaystyle OH}{\mid}}{\underset{\underset{\displaystyle H}{\mid}}{C}}-R_2 \longrightarrow XH_2 + R_1-\overset{\overset{\displaystyle O}{\|}}{C}-R_2$
  a secondary alcohol   a ketone

(For the reactions of alcohols with aldehydes, hemiacetals, ketones, and hemiketals, see aldehydes and ketones.)

Ester formation

$R_1-OH + R_2'-\overset{\overset{\displaystyle O}{\|}}{C}-OH \longrightarrow R_2'-\overset{\overset{\displaystyle O}{\|}}{C}-O-R_1 + H_2O$
  an ester

—OH

Phenols

Ionization

—OH + $H_2O \rightleftharpoons$ —$O^- + H_3O^+$

$-\overset{\overset{\displaystyle O}{\|}}{C}-H$
Aldehydes

Reduction

$R'-\overset{\overset{\displaystyle O}{\|}}{C}-H + XH_2 \longrightarrow R'-\overset{\overset{\displaystyle H}{\mid}}{\underset{\underset{\displaystyle H}{\mid}}{C}}-OH + X$
  a primary alcohol

Oxidation

$R'-\overset{\overset{\displaystyle O}{\|}}{C}-H + X + H_2O \longrightarrow R'-\overset{\overset{\displaystyle O}{\|}}{C}-OH + XH_2$
  a carboxylic acid

Hemiacetal formation

$R_1-OH + H-\overset{\overset{\displaystyle O}{\|}}{C}-R_2 \rightleftharpoons R_1-O-\overset{\overset{\displaystyle OH}{\mid}}{\underset{\underset{\displaystyle H}{\mid}}{C}}-R_2$

Acetal formation

$R_1-OH + R_1-O-\overset{\overset{\displaystyle OH}{\mid}}{\underset{\underset{\displaystyle H}{\mid}}{C}}-R_2 \overset{H^+}{\longrightarrow} R_1-O-\overset{\overset{\displaystyle OR_1}{\mid}}{\underset{\underset{\displaystyle H}{\mid}}{C}}-R_2 + H_2O$

Aldol condensation reaction

$2\left[R-\overset{\overset{\displaystyle R'}{\mid}}{\underset{\underset{\displaystyle H}{\mid}}{C}}-\overset{\overset{\displaystyle O}{\|}}{C}-H \right] \overset{OH^-}{\longrightarrow} R-\overset{\overset{\displaystyle R'}{\mid}}{\underset{\underset{\displaystyle H}{\mid}}{C}}-\overset{\overset{\displaystyle OH}{\mid}}{\underset{\underset{\displaystyle H}{\mid}}{C}}-\overset{\overset{\displaystyle R'}{\mid}}{\underset{\underset{\displaystyle R}{\mid}}{C}}-\overset{\overset{\displaystyle O}{\|}}{C}-H$

Table 1 (Continued)

Functional Group Class		General Reaction

Ketones

$$-\overset{\displaystyle |}{\underset{\displaystyle |}{C}}-\overset{\displaystyle O}{\overset{\displaystyle \|}{C}}-\overset{\displaystyle |}{\underset{\displaystyle |}{C}}-$$

Reduction

$$XH_2 + R_1-\overset{\displaystyle O}{\overset{\displaystyle \|}{C}}-R_2 \longrightarrow R_1-\overset{\displaystyle OH}{\underset{\displaystyle H}{\overset{\displaystyle |}{\underset{\displaystyle |}{C}}}}-R_2$$

a secondary alcohol

Hemiketal and ketal formation   As for aldehydes

Aldol condensation   As for aldehydes

Carboxylic acids

$$-\overset{\displaystyle O}{\overset{\displaystyle \|}{C}}-OH$$

Ionization

$$R'-\overset{\displaystyle O}{\overset{\displaystyle \|}{C}}-OH + H_2O \rightleftharpoons R'-\overset{\displaystyle O}{\overset{\displaystyle \|}{C}}-O^- + H_3O^-$$

Salt formation

$$R'-\overset{\displaystyle O}{\overset{\displaystyle \|}{C}}-OH + NaOH \rightleftharpoons R'-\overset{\displaystyle O}{\overset{\displaystyle \|}{C}}-O^-\ Na^+ + H_2O$$

Reduction

$$R'-\overset{\displaystyle O}{\overset{\displaystyle \|}{C}}-OH + XH_2 \longrightarrow R'-\overset{\displaystyle O}{\overset{\displaystyle \|}{C}}-H + H_2O + X$$

Decarboxylation

$$R'-\overset{\displaystyle O}{\overset{\displaystyle \|}{C}}-OH \longrightarrow R'H + CO_2$$

Condensation reactions

Anhydride formation

$$2R'-\overset{\displaystyle O}{\overset{\displaystyle \|}{C}}-OH \longrightarrow R'-\overset{\displaystyle O}{\overset{\displaystyle \|}{C}}-O-\overset{\displaystyle O}{\overset{\displaystyle \|}{C}}-R' + H_2O$$

an anhydride

Ester formation

$$R_1'-\overset{\displaystyle O}{\overset{\displaystyle \|}{C}}-OH + R_2-OH \longrightarrow R_1'-\overset{\displaystyle O}{\overset{\displaystyle \|}{C}}-O-R_2 + H_2O$$

an ester

Amide formation

$$R_1'-\overset{\displaystyle O}{\overset{\displaystyle \|}{C}}-OH + H-\overset{\displaystyle R_2'}{\underset{\displaystyle R_4'}{\overset{\displaystyle |}{\underset{\displaystyle |}{N}}}}-R_3' \longrightarrow R_1'-\overset{\displaystyle O}{\overset{\displaystyle \|}{C}}-\overset{\displaystyle R_2'}{\underset{\displaystyle R_4'}{\overset{\displaystyle |}{\underset{\displaystyle |}{N}}}}-R_3' + H_2O$$

an amide

Thioester formation

$$R_1'-\overset{\displaystyle O}{\overset{\displaystyle \|}{C}}-OH + R_2-SH \longrightarrow R_1'-\overset{\displaystyle O}{\overset{\displaystyle \|}{C}}-S-R_2 + H_2O$$

a thioester

Phosphoester formation

$$R_1'-\overset{\displaystyle O}{\overset{\displaystyle \|}{C}}-OH + HO-\overset{\displaystyle O}{\underset{\displaystyle OH}{\overset{\displaystyle \|}{\underset{\displaystyle |}{P}}}}-OH \longrightarrow$$

$$R_1'-\overset{\displaystyle O}{\overset{\displaystyle \|}{C}}-O-\overset{\displaystyle O}{\underset{\displaystyle OH}{\overset{\displaystyle \|}{\underset{\displaystyle |}{P}}}}-OH + H_2O$$

a phosphoester

Note: The hydrolysis of anhydrides, esters, and any of the other products of condensation reactions involving carboxylic acids can be written as the reverse of the reactions in which they are formed.

Table 1 (Continued)

Functional Group Class	General Reaction	

$\begin{array}{c} R_3' \\ | \\ R_1'\!-\!N\!-\!R_2' \\ \text{Amines} \end{array}$ Ionization $R_1'\!-\!\underset{\underset{}{|}}{\overset{\overset{R_3'}{|}}{N}}\!-\!R_2' + H_2O \rightleftharpoons R_1'\!-\!\underset{\underset{H}{|}}{\overset{\overset{R_3'}{|}}{N^+}}\!-\!R_2' + OH^-$

   Amide formation or hydrolysis   See carboxylic acids

$R\!-\!SH$
Thiols   Disulfide formation $R_1\!-\!SH + R_2\!-\!SH + X \longrightarrow \underset{\text{a disulfide}}{R_1\!-\!S\!-\!S\!-\!R_2} + XH_2$

   Thioester formation and hydrolysis   See carboxylic acids

EXAMPLE Draw the structural formulas for the products of the following reaction. (If no reaction occurs, write N.R.)

$$CH_3\!-\!CH_2\!-\!\overset{\overset{O}{\|}}{C}\!-\!OH + CH_3\!-\!CH_2\!-\!OH \longrightarrow$$

Step 1 Identify the functional groups and determine the class of organic compounds to which each reactant belongs.

This example involves a carboxylic acid and an alcohol.

Step 2 Determine if a reaction can occur that involves a functional group on one reactant and a functional group on the other reactant. Write out the general reaction if one is possible (see Table 1).

In this example, the general reaction is

$$\underset{\text{carboxylic acid}}{R_1'\!-\!\overset{\overset{O}{\|}}{C}\!-\!OH} + \underset{\text{alcohol}}{R_2\!-\!OH} \longrightarrow \underset{\text{ester}}{R_1'\!-\!\overset{\overset{O}{\|}}{C}\!-\!O\!-\!R_2} + H_2O$$

Step 3 Write the structural formulas for the specific products of the reaction in question to conform with: (a) the general reaction in step 2, and (b) the structural formulas for the reactants.

In this example, $R_1' = CH_3\!-\!CH_2\!-$ and $R_2 = CH_3\!-\!CH_2\!-$. Thus,

$$CH_3\!-\!CH_2\!-\!\overset{\overset{O}{\|}}{C}\!-\!OH + CH_3\!-\!CH_2\!-\!OH \longrightarrow$$

$$CH_3\!-\!CH_2\!-\!\overset{\overset{O}{\|}}{C}\!-\!O\!-\!CH_2\!-\!CH_3 + H_2O$$

APPENDIX 3

Vitamins and Coenzymes

Vitamins are organic compounds that people must ingest in small amounts for proper body function. Vitamins are divided into two classes on the basis of

Coenzyme	Vitamin Precursor	Reaction	Group Transferred
Nicotinamide adenine dinucleotide (NAD+)	Niacin (nicotinic acid)	Oxidation-reduction	$H:^-$ (hydride ion)
		$NAD^+ + SH_2 \rightleftharpoons NADH + S + H^+$	
Nicotinamide adenine dinucleotide phosphate (NADP+)	Niacin (nicotinic acid)	Oxidation-reduction	$H:^-$ (hydride ion)
		$NADP^+ + SH_2 \rightleftharpoons NADPH + S + H^+$	
Thiamin pyrophosphate (TPP)	Thiamine (vitamin B$_1$)	Acyl group transfer	$R-\overset{\overset{\displaystyle O}{\|}}{C}-$

their solubility in water. The structural formulas for the fat-soluble (i.e., water-insoluble) vitamins A, D, K, and E are given in Chapter 27, together with what is known about their function.

The functions of the water-soluble vitamins are generally much better understood. All of the water-soluble vitamins are components of coenzymes. The structural formulas of these coenzymes are presented at appropriate places in the text, and are collected in the following table, which also summarizes the types of reactions in which these coenzymes are involved. The vitamin portions of these coenzymes are shown in blue.

Structural Formula

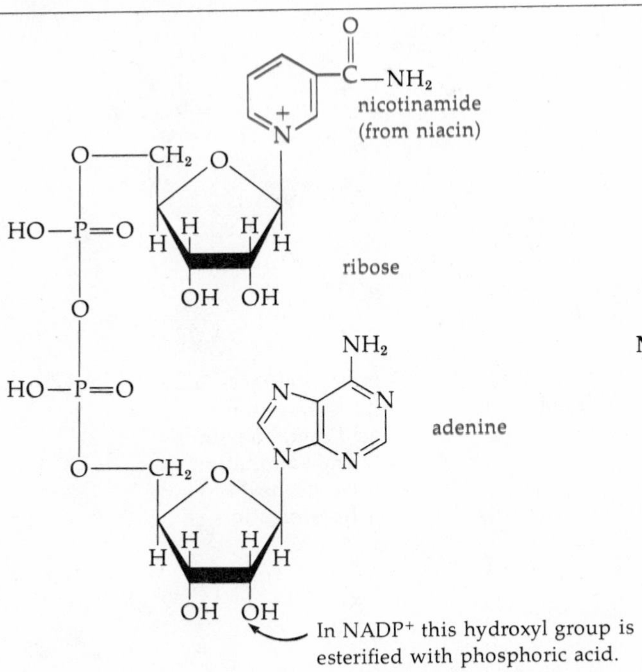

nicotinamide
(from niacin)

ribose

adenine

NAD$^+$

In NADP$^+$ this hydroxyl group is esterified with phosphoric acid.

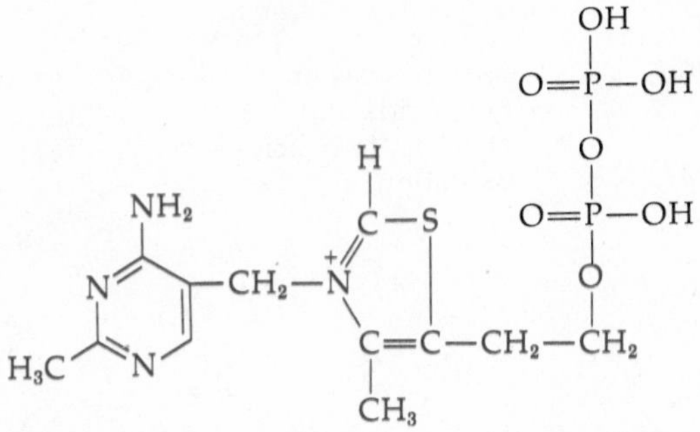

TPP

Coenzyme	Vitamin Precursor	Reaction	Group Transferred
Flavin mono-nucleotide (FMN)	Riboflavin (vitamin B_2)	Oxidation-reduction $$FMN + SH_2 \rightleftharpoons FMNH_2 + S$$	Hydrogen atoms
Flavin adenine dinucleotide (FAD)	Riboflavin	Oxidation-reduction $$FAD + SH_2 \rightleftharpoons FADH_2 + S$$	Hydrogen atoms
Pyridoxal phosphate	Pyridoxine (vitamin B_6)	In several reactions of amino acid metabolism: Transamination Decarboxylation Racemization	NH_3
Coenzyme A (CoA-SH)	Pantothenic acid	Biosynthesis of fatty acids and steroids, fatty acid oxidation	Acyl groups

Structural Formula

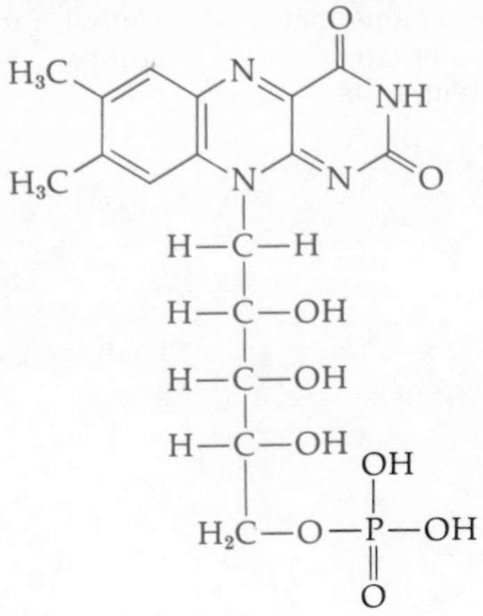

FMN

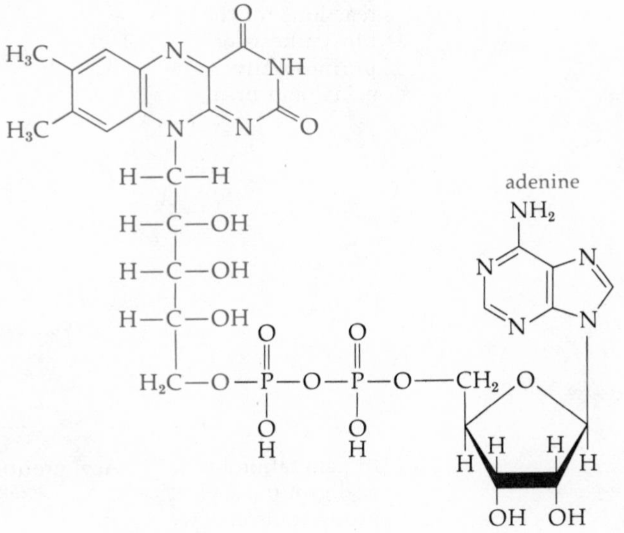

FAD

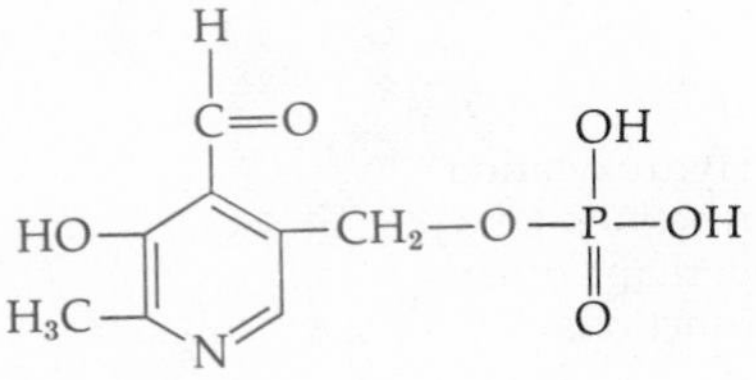

adenine

Pyridoxal phosphate

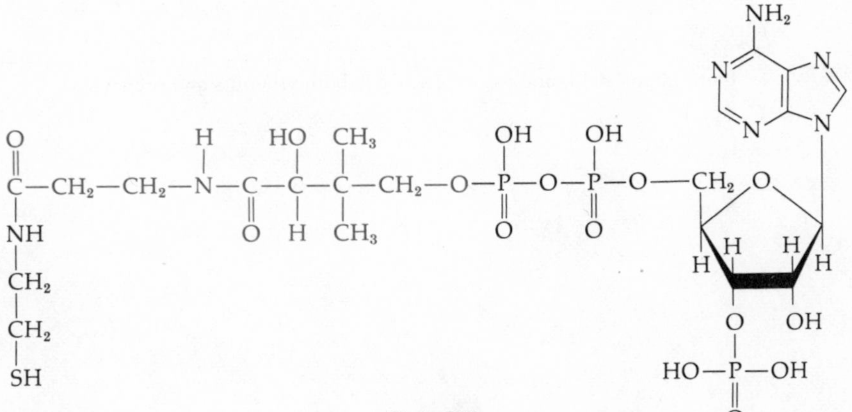

CoA-SH

Coenzyme	Vitamin Precursor	Reaction	Group Transferred
Tetrahydrofolic acid (THFA)	Folic acid	In reactions that transfer single carbon units	Methyl, formyl groups
Coenzyme Q (Ubiquinone)		Electron transport	Hydrogen atoms
Biotin*		In carboxylation reactions for the biosynthesis of purines, fatty acids, and urea	CO_2
Lipoic acid*		In generation of acyl groups, acyl group transfer, and electron transport	Acyl groups
Ascorbic acid* (Vitamin C)		Hydroxylation reactions (and perhaps other functions)	

* Biotin, lipoic acid, and ascorbic acid are both vitamins and coenzymes.

Structural Formula

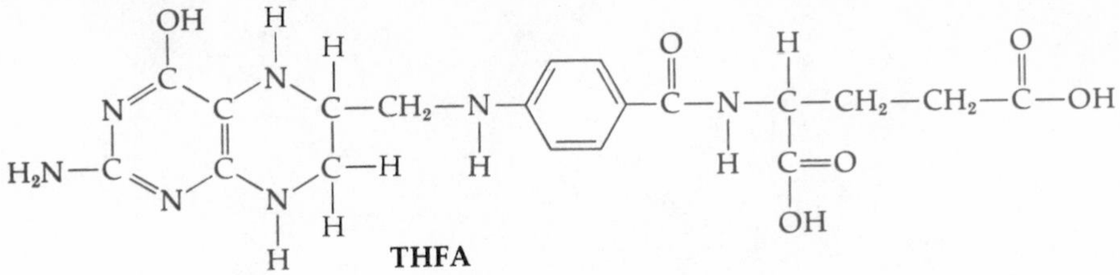

THFA

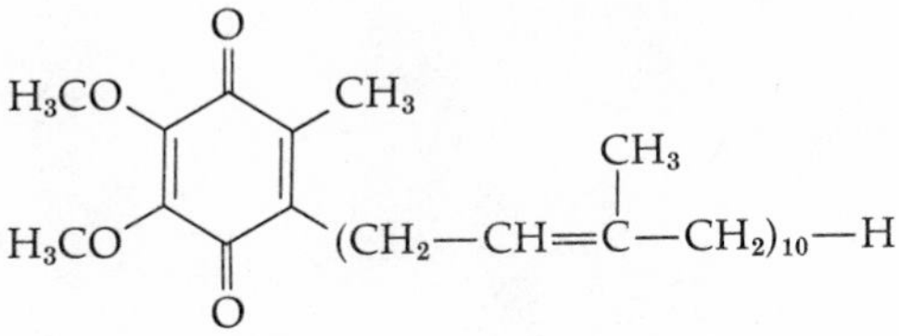

Coenzyme Q

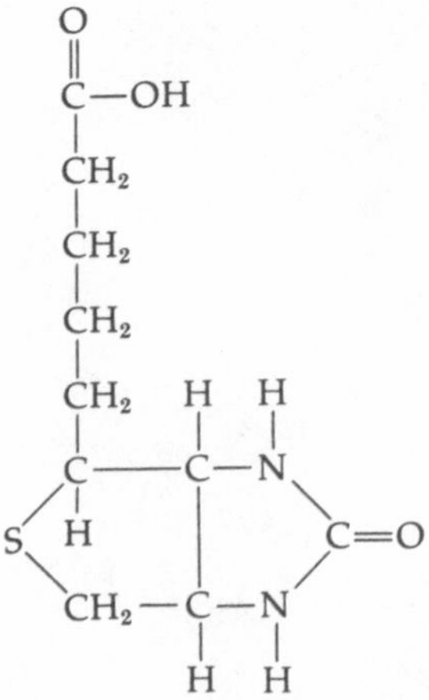

Biotin

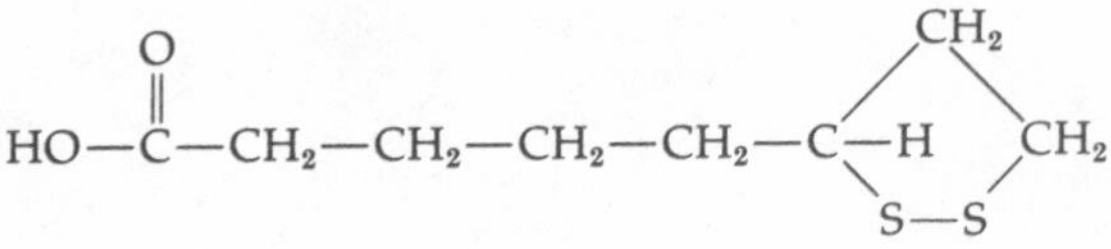

Lipoic acid

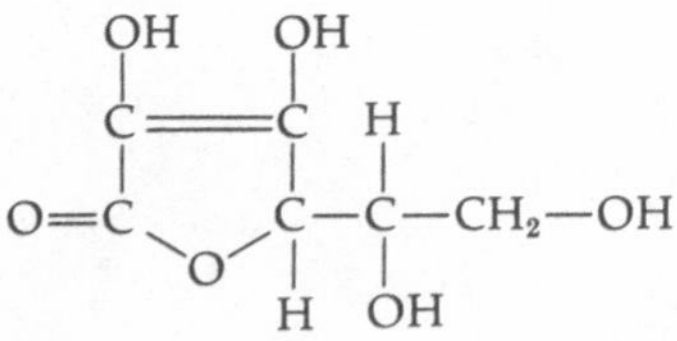

Ascorbic acid

Answers

CHAPTER 13

1. $CH_3-CH_2-CH_2-CH_2-$, n-butyl

$CH_3-CH_2-\overset{|}{C}H-CH_3$, sec-butyl

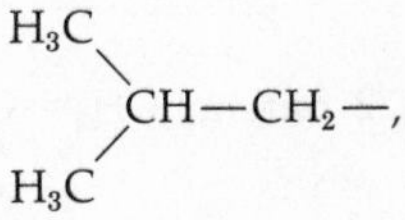

$\underset{H_3C}{\overset{H_3C}{>}}CH-CH_2-$, isobutyl

$H_3C-\overset{\overset{\displaystyle CH_3}{|}}{\underset{\underset{\displaystyle CH_3}{|}}{C}}-$, t-butyl

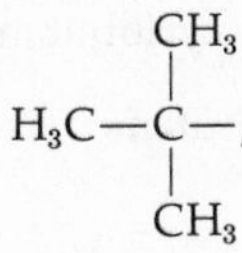

2. (a) 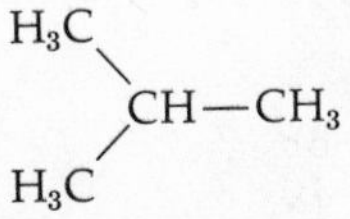 $\underset{H_3C}{\overset{H_3C}{>}}CH-CH_3$

   (b) and (c) $CH_3-CH_2-CH_2-CH_2-CH_2-CH_2-CH_2-CH_3$

   (d) 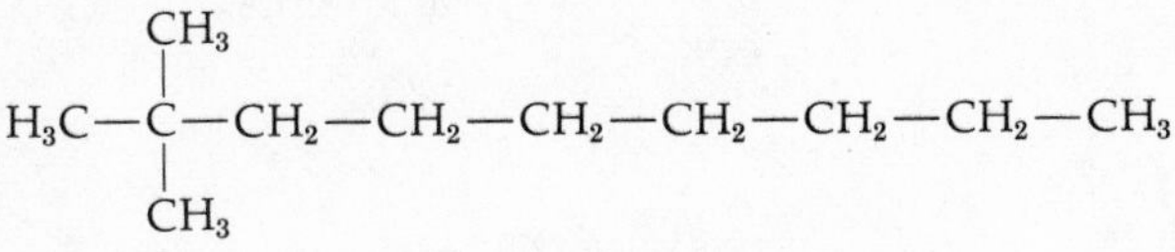 $H_3C-\overset{\overset{\displaystyle CH_3}{|}}{\underset{\underset{\displaystyle CH_3}{|}}{C}}-CH_2-CH_2-CH_2-CH_2-CH_2-CH_2-CH_3$

   (e) 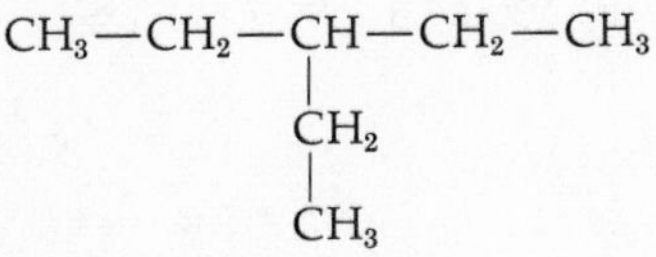 $CH_3-CH_2-\overset{\overset{\displaystyle}{|}}{C}H-CH_2-CH_3$
       $\qquad\qquad\quad\underset{\underset{\displaystyle CH_3}{|}}{CH_2}$

3. (a) and (c) are both pairs of positional isomers

4. (a) 2,2-Dimethylbutane
   (c) Butane
   (e) 2,3-Dimethylpentane

   (b) 2,2,3,4-Tetramethylpentane
   (d) Butane
   (f) 2,3,5-Trimethylhexane

CHAPTER 14

1. (a) Cycloalkanes
   (c) Alkenes
   (e) Alkyl halides

   (b) Cycloalkanes
   (d) Aromatic hydrocarbons

2. (a)

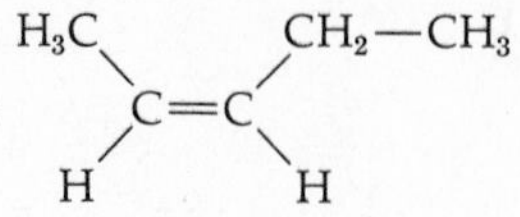

   (b)

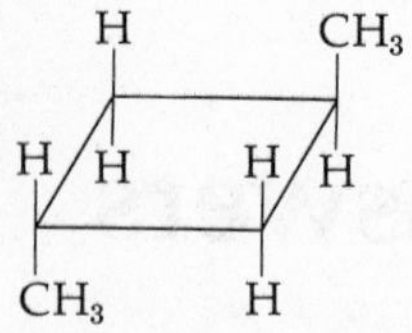

   (c)

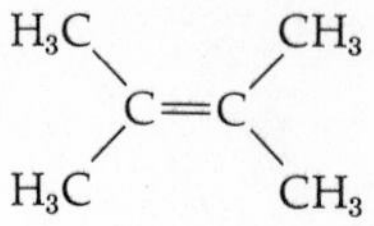

   (d) 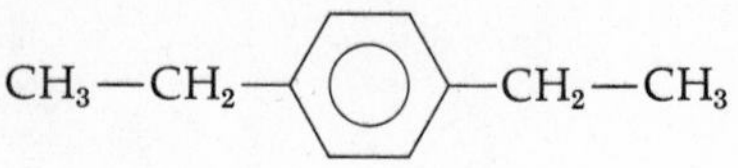 CH_3-CH_2-⟨○⟩$-CH_2-CH_3$

   (e)

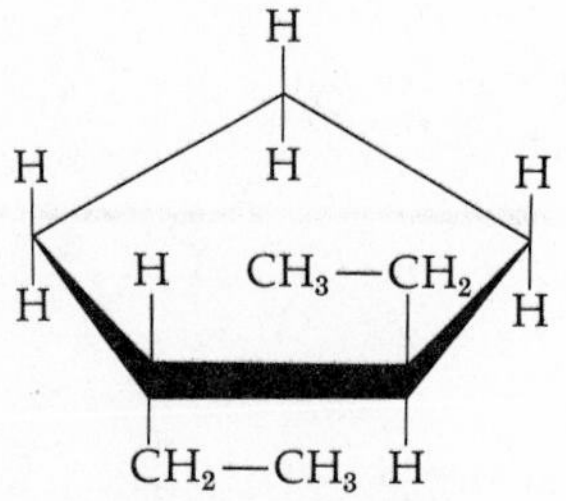

3. (a) 2,3-Dimethyl-2-pentene
   (c) Methylcyclohexane
   (e) *cis*-3,4-Dimethyl-2-pentene

   (b) *cis*-2-Pentene
   (d) 1,3-Diisopropylbenzene

4. $CH_2{=}CH-CH_2-CH_3$, 1-butene; $CH_3-CH{=}CH-CH_3$, 2-butene; $CH_2{=}CH-CH_3$,

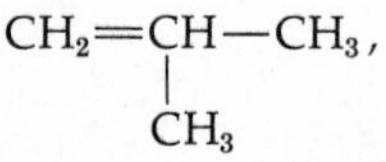

   2-methylpropene; ▷$-CH_3$, methylcyclopropane; and ☐, cyclobutane are all

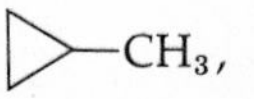

   structural isomers. There are two geometrical isomers for 2-butene: *cis*-2-butene,

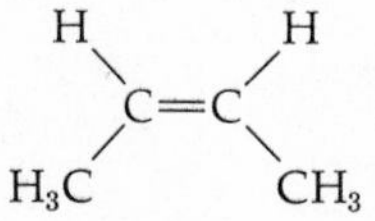

; and *trans*-2-butene, 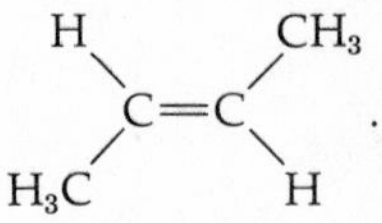 .

5. (a)

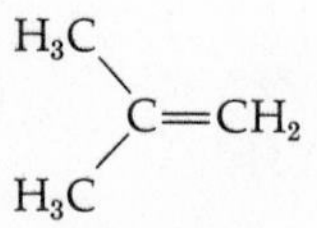

   (b)

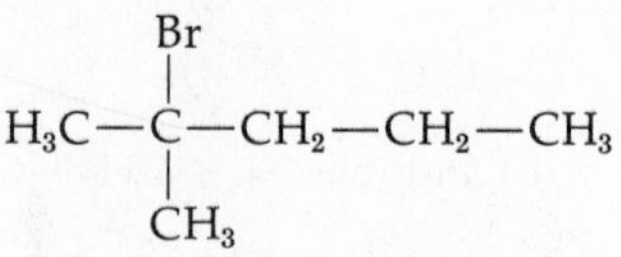

   (c)

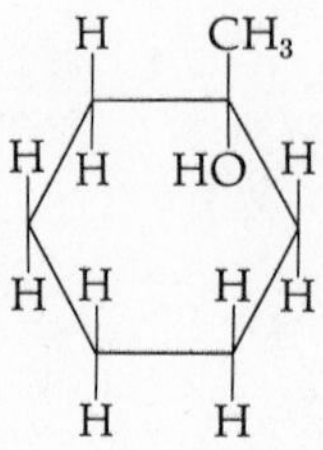

   (d) $CH_3-CH_2-CH_3$

6.

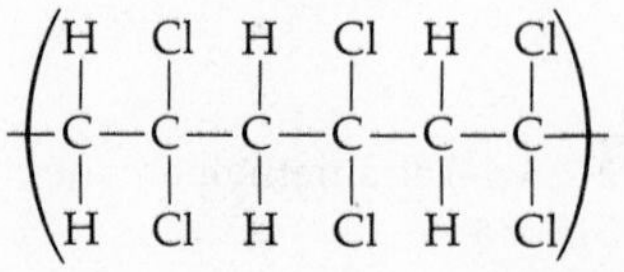

CHAPTER 15

1. (a)

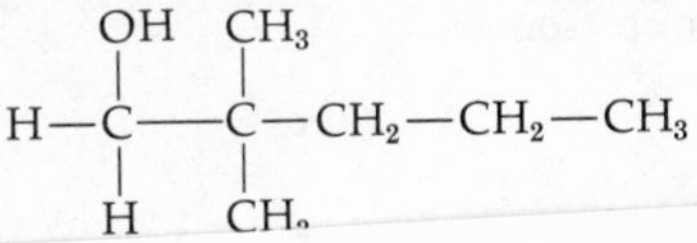

$$H-\underset{\underset{H}{|}}{\overset{\overset{OH}{|}}{C}}-\underset{\underset{CH_3}{|}}{\overset{\overset{CH_3}{|}}{C}}-CH_2-CH_2-CH_3$$

(b) 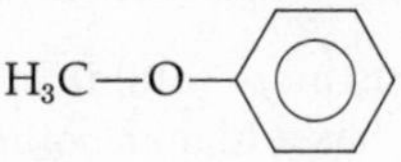 $H_3C-O-\bigcirc$

(handwritten notes overlaid)

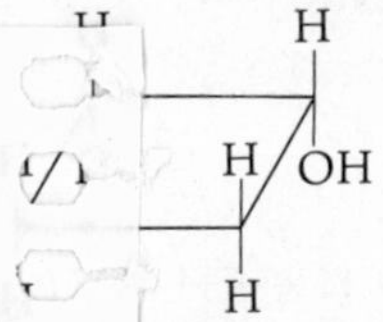

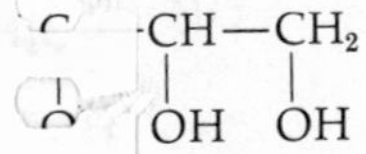

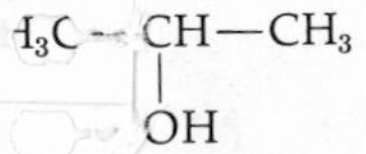

MT — 37.265

Full — 39.360

$$\begin{array}{r} 39.\overset{2}{\cancel{3}}\overset{5}{\cancel{6}}10 \\ 37.265 \\ \hline 2.095 \end{array}$$

product

2.095 yeild

(right margin printed, partially obscured)

H — H, OH, H

·CH—CH₂ with OH OH → $-CH-CH_2$ with OH OH

$H_3C-CH-CH_3$ with OH

...est boiling point.

R_1-O-R_2

N.R.

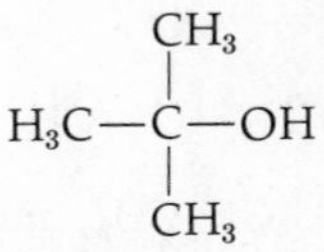

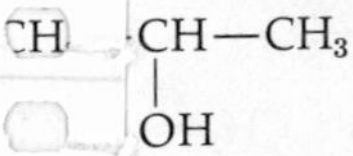

$$H_3C-\underset{\underset{CH_3}{|}}{\overset{\overset{CH_3}{|}}{C}}-OH$$

$CH-CH-CH_3$ with OH

b) 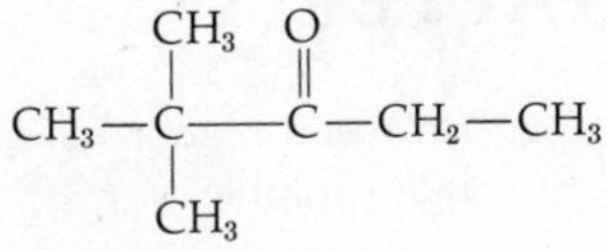 $CH_3-\underset{\underset{CH_3}{|}}{\overset{\overset{CH_3}{|}}{C}}-\overset{\overset{O}{||}}{C}-CH_2-CH_3$

(d) 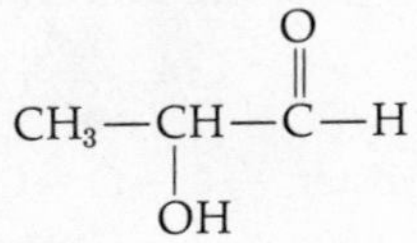 $CH_3-\underset{\underset{OH}{|}}{CH}-\overset{\overset{O}{||}}{C}-H$

2. (a) 2-Methylpropanal
   (c) 3-Hydroxybutanone
   (e) Butanone

   (b) 2,4-Dimethyl-3-pentanone
   (d) Cyclopentanone

3. Compounds (a) and (e) are functional group isomers.

4. (c) is more soluble in water than (a)
   (c) has a higher boiling point

5. (a) N.R.

   (b) 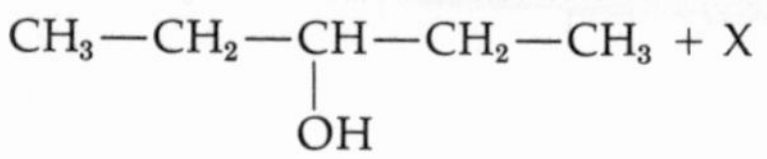 $CH_3-CH_2-\underset{\underset{OH}{|}}{CH}-CH_2-CH_3 + X$

   (c) 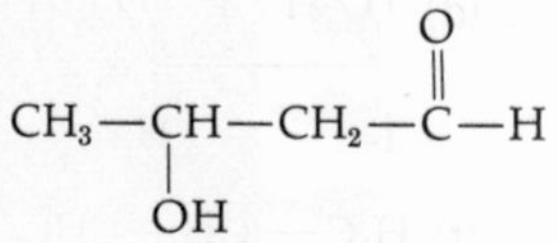 $CH_3-\underset{\underset{OH}{|}}{CH}-CH_2-\overset{\overset{O}{\|}}{C}-H$

   (d) 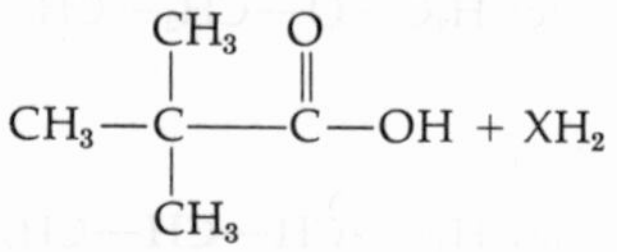 $CH_3-\underset{\underset{CH_3}{|}}{\overset{\overset{CH_3}{|}}{C}}-\overset{\overset{O}{\|}}{C}-OH + XH_2$

   (e) $CH_3-CH_2-CH_2-OH + X$

   (f) 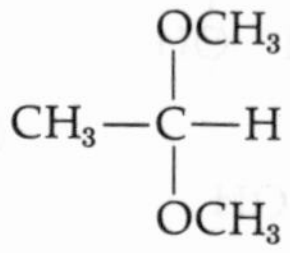 $CH_3-\underset{\underset{OCH_3}{|}}{\overset{\overset{OCH_3}{|}}{C}}-H$

   (g) 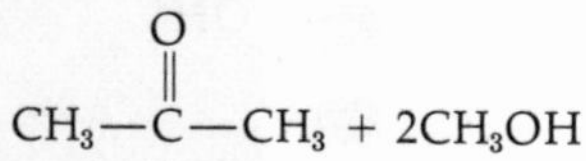 $CH_3-\overset{\overset{O}{\|}}{C}-CH_3 + 2CH_3OH$

   (h) N.R.

   (i) 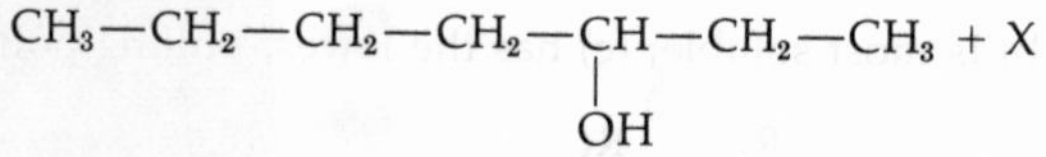 $CH_3-CH_2-CH_2-CH_2-\underset{\underset{OH}{|}}{CH}-CH_2-CH_3 + X$

6. (a) 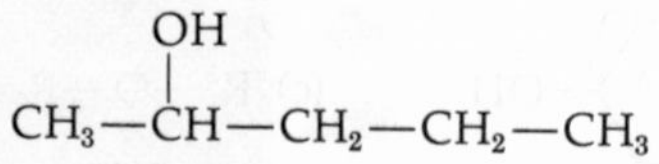 $CH_3-\underset{\underset{OH}{\overset{\overset{OH}{|}}{CH}}}{CH}-CH_2-CH_2-CH_3$

   (b) 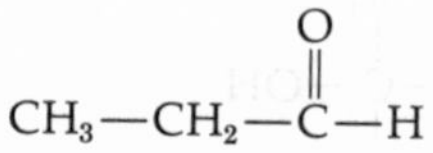 $CH_3-CH_2-\overset{\overset{O}{\|}}{C}-H$

   (c) and (d) 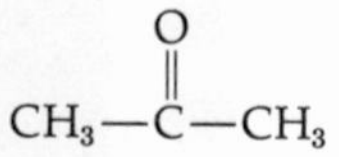 $CH_3-\overset{\overset{O}{\|}}{C}-CH_3$

   (e) and (f) 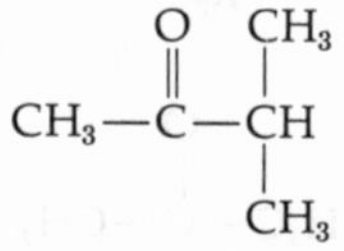 $CH_3-\overset{\overset{O}{\|}}{C}-\underset{\underset{CH_3}{|}}{\overset{\overset{CH_3}{|}}{CH}}$

7. 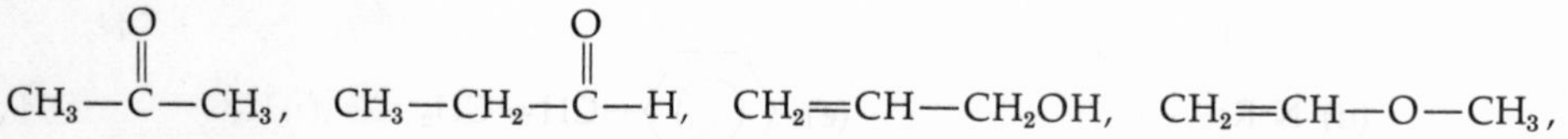 $CH_3-\overset{\overset{O}{\|}}{C}-CH_3$, $CH_3-CH_2-\overset{\overset{O}{\|}}{C}-H$, $CH_2=CH-CH_2OH$, $CH_2=CH-O-CH_3$,

   $\underset{\underset{H_2C-O}{|}}{\overset{\overset{H_2C-CH_2}{| \quad |}}{}}$, and $H_2C\overset{\diagdown \quad \diagup}{\underset{O}{}}CH-CH_3$. Note that a compound with the structural

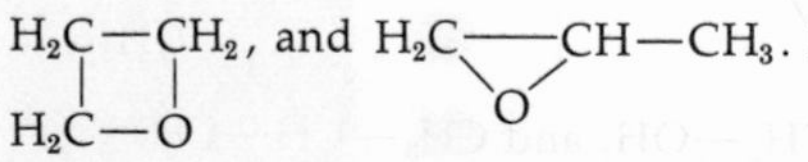

   formula $CH_2=\underset{\underset{OH}{|}}{C}-CH_3$ cannot be isolated.

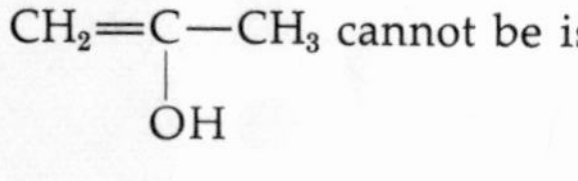

CHAPTER 17

1. (b) has a higher boiling point than (a), and both are soluble in water, although (b) is more soluble.

2. Yes. They are functional group isomers.

   (a) Ester, $-\underset{|}{C}-O-\overset{\overset{O}{\|}}{C}-$

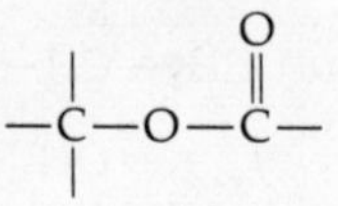

   (b) Carboxylic acid, $-\overset{\overset{O}{\|}}{C}-OH$

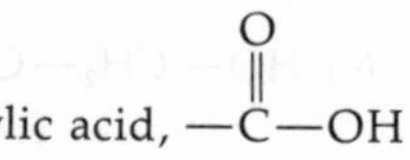

3. (a) 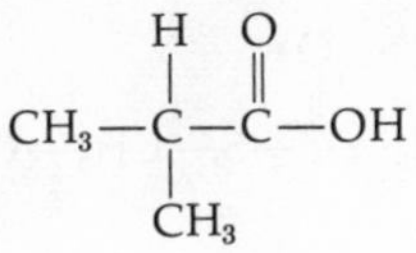 $CH_3-\underset{\underset{CH_3}{|}}{\overset{\overset{H}{|}}{C}}-\overset{\overset{O}{\|}}{C}-OH$

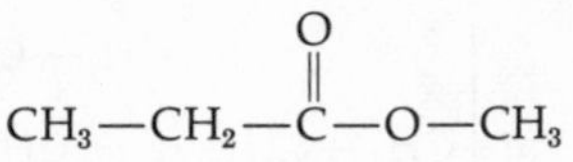

(b) $CH_3-CH_2-\overset{\overset{O}{\|}}{C}-O-CH_3$

(c) $CH_3-\underset{\underset{CH_3}{|}}{\overset{\overset{H}{|}}{C}}-\overset{\overset{O}{\|}}{C}-OH$

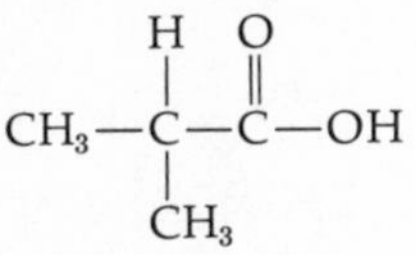

(d) 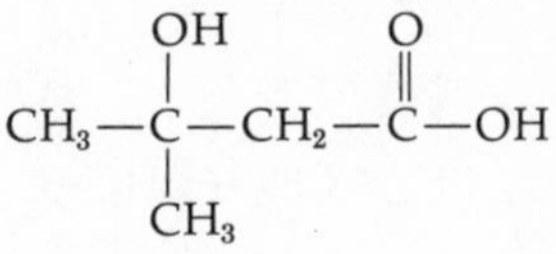 $CH_3-\underset{\underset{CH_3}{|}}{\overset{\overset{OH}{|}}{C}}-CH_2-\overset{\overset{O}{\|}}{C}-OH$

(e) 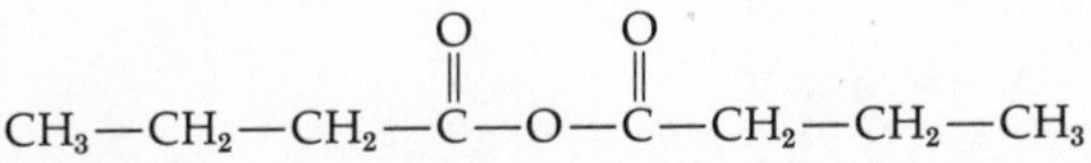 $CH_3-CH_2-CH_2-\overset{\overset{O}{\|}}{C}-O-\overset{\overset{O}{\|}}{C}-CH_2-CH_2-CH_3$

4. (a) α,α-Dimethylpropionic acid
   (b) Isopropyl α,α-dimethylpropionate
   (c) Ethyl propionate
   (d) Methyl benzoate
   (e) Ethyl phosphate

5. (a) 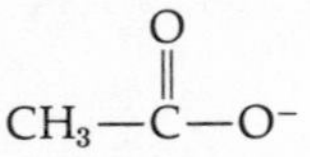 $CH_3-\overset{\overset{O}{\|}}{C}-O^- + CH_3OH$

(b) 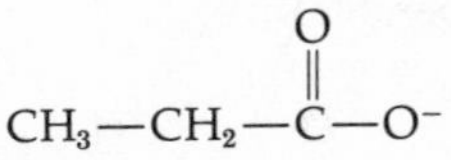 $CH_3-CH_2-\overset{\overset{O}{\|}}{C}-O^- + H_2O$

   (c) and (d) N.R.

(e) CH_3-CH_3

(f) 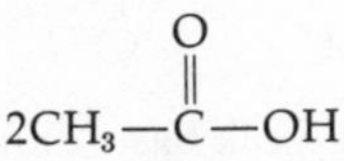 $2CH_3-\overset{\overset{O}{\|}}{C}-OH$

6. (a) is a functional group isomer of the reactant (propionic acid) in (b), (c), and (e).

7. N.R. in cold water. Ethyl acetate, $CH_3-\overset{\overset{O}{\|}}{C}-O-CH_2-CH_3$, is formed with a strong acid at a higher temperature.

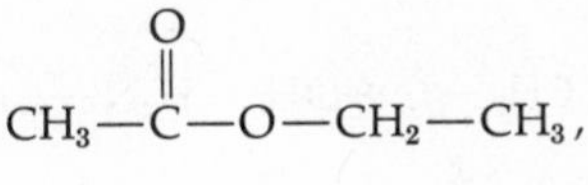

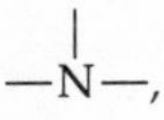

CHAPTER 18

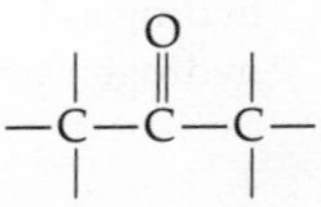

1. (a) Amide, $-\overset{\overset{O}{\|}}{C}-N-$
   (b) Amine, $-N-$, and ketone, $-\overset{|}{\underset{|}{C}}-\overset{\overset{O}{\|}}{C}-\overset{|}{\underset{|}{C}}-$

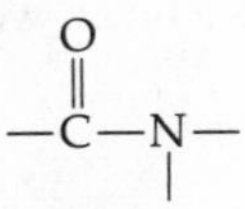

   (c) Amide, $-\overset{\overset{O}{\|}}{C}-N-$
   (d) Disulfide, $-S-S-$
   (e) Thioester, $-\overset{\overset{O}{\|}}{C}-S-\overset{|}{\underset{|}{C}}-$

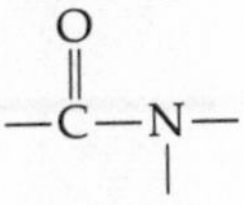

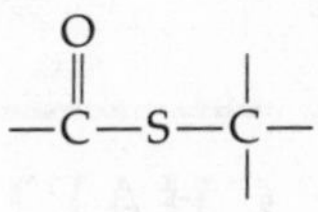

   (f) Ketone, $-\overset{|}{\underset{|}{C}}-\overset{\overset{O}{\|}}{C}-\overset{|}{\underset{|}{C}}-$, and sulfhydryl, $-SH$

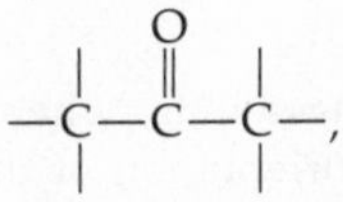

   (g) Tertiary ammonium ion, $-\overset{|}{\underset{|}{C}}-\overset{\overset{H}{|}}{\underset{\underset{-C-}{|}}{N^+}}-\overset{|}{\underset{|}{C}}-$

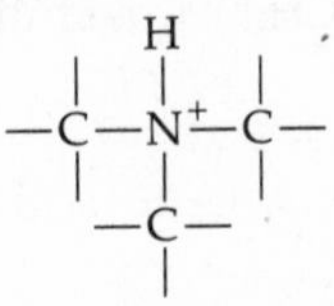

2. (a) N-ethylformamide
   (c) N-methylacetamide
   (d) Methyl disulfide
   (e) Ethyl thioacetate
   (g) Trimethylammonium chloride

3. (a) 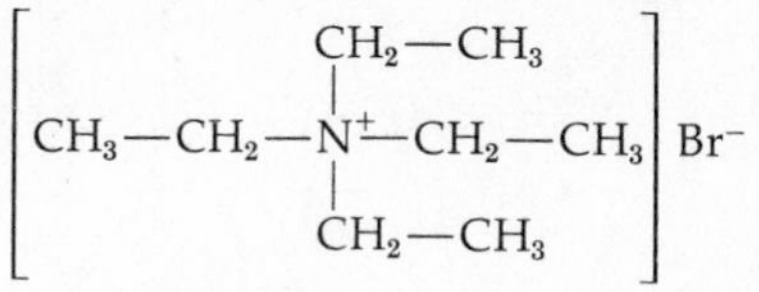 $\left[CH_3-CH_2-\overset{\overset{\displaystyle CH_2-CH_3}{|}}{\underset{\underset{\displaystyle CH_2-CH_3}{|}}{N^+}}-CH_2-CH_3 \right] Br^-$

(b) 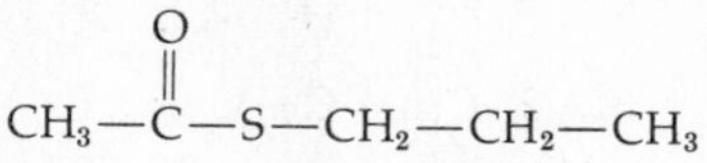 $CH_3-\overset{\overset{\displaystyle O}{||}}{C}-S-CH_2-CH_2-CH_3$

(c) $CH_3-CH_2-CH_2-\overset{\overset{\displaystyle O}{||}}{C}-\underset{\underset{\displaystyle H}{|}}{N}-CH_3$

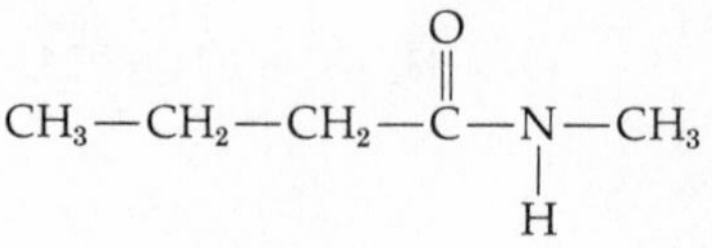

(d) 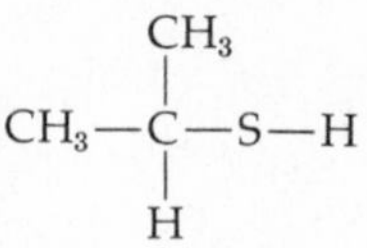 $CH_3-\overset{\overset{\displaystyle CH_3}{|}}{\underset{\underset{\displaystyle H}{|}}{C}}-S-H$

(e) 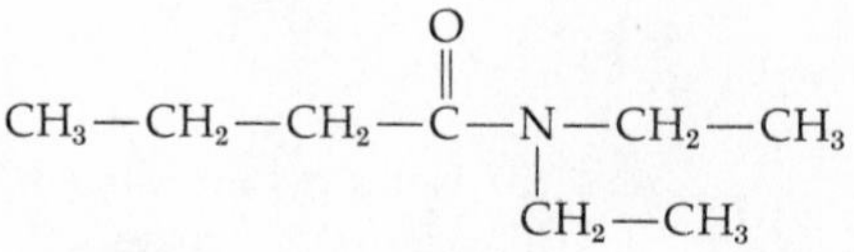 $CH_3-CH_2-CH_2-\overset{\overset{\displaystyle O}{||}}{C}-\underset{\underset{\displaystyle CH_2-CH_3}{|}}{N}-CH_2-CH_3$

4. The ammonium salt should be more soluble in water than an N-substituted amide of comparable size.

5. (a) 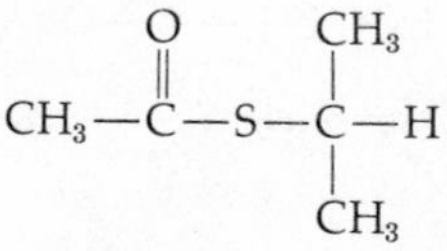 $CH_3-\overset{\overset{\displaystyle O}{||}}{C}-S-\overset{\overset{\displaystyle CH_3}{|}}{\underset{\underset{\displaystyle CH_3}{|}}{C}}-H$

(b) 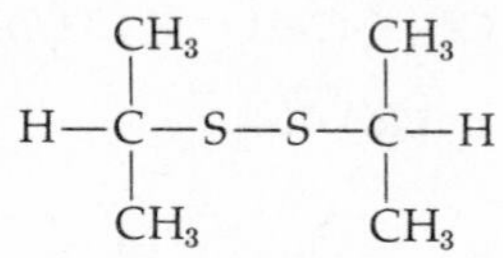 $H-\overset{\overset{\displaystyle CH_3}{|}}{\underset{\underset{\displaystyle CH_3}{|}}{C}}-S-S-\overset{\overset{\displaystyle CH_3}{|}}{\underset{\underset{\displaystyle CH_3}{|}}{C}}-H$

(c) Some $H_3N^+-CH_2-CH_3 + OH^-$, since ethylamine is a weak base, although most of the ethylamine will remain unprotonated.

(d) 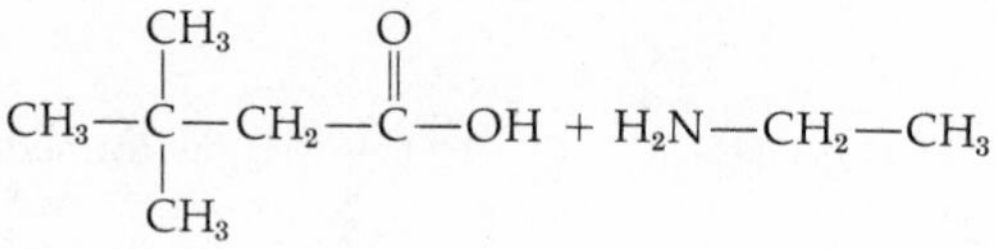 $CH_3-\overset{\overset{\displaystyle CH_3}{|}}{\underset{\underset{\displaystyle CH_3}{|}}{C}}-CH_2-\overset{\overset{\displaystyle O}{||}}{C}-OH + H_2N-CH_2-CH_3$

(e) 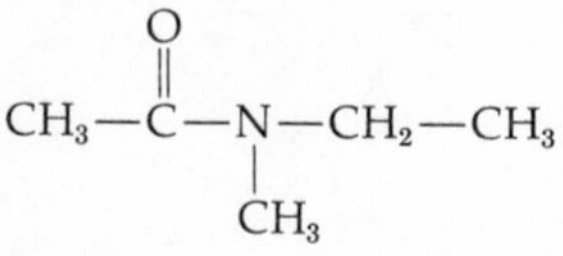 $CH_3-\overset{\overset{\displaystyle O}{||}}{C}-\underset{\underset{\displaystyle CH_3}{|}}{N}-CH_2-CH_3$

(f) CH_3SH

6. Ethyl alcohol has a higher boiling point, because molecules of an alcohol form strong hydrogen bonds with each other, whereas thiols have a much weaker tendency to form hydrogen bonds.

CHAPTER 19

1. Structural isomers have different bonding arrangements, whereas stereoisomers have the same bonding arrangement but a different orientation of the bonds in space. The compounds ethanol, CH_3CH_2OH, and dimethyl ether, CH_3-O-CH_3, are structural

isomers. The compounds D-lactic acid, $H-\overset{\overset{\displaystyle COOH}{|}}{\underset{\underset{\displaystyle CH_3}{|}}{}}-OH$, and L-lactic acid, $HO-\overset{\overset{\displaystyle COOH}{|}}{\underset{\underset{\displaystyle CH_3}{|}}{}}-H$,

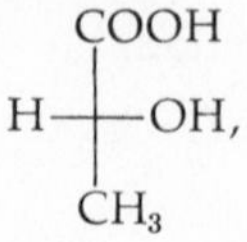

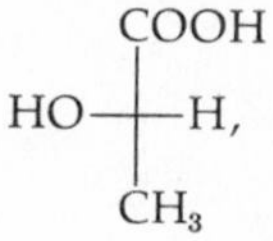

are stereoisomers.

2. (b)

3. Diastereomer pairs: I and III; I and IV; II and III; II and IV
   Enantiomer pairs: I and II; III and IV

4. I and II represent the same meso-type compound.
   I and III, and I and IV, are diastereomers.
   III and IV are enantiomers.

5. (d)

6. (a) and (b)

7. (a)

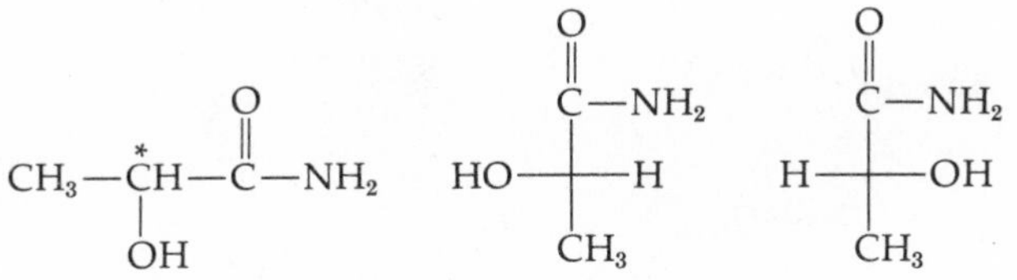

enantiomers

(b) 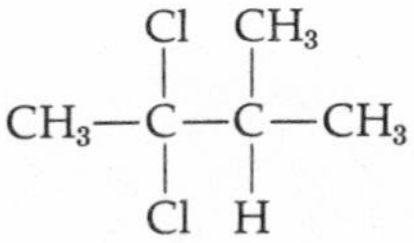 no asymmetric carbon

(c)

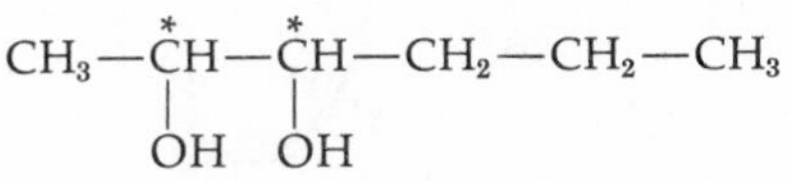

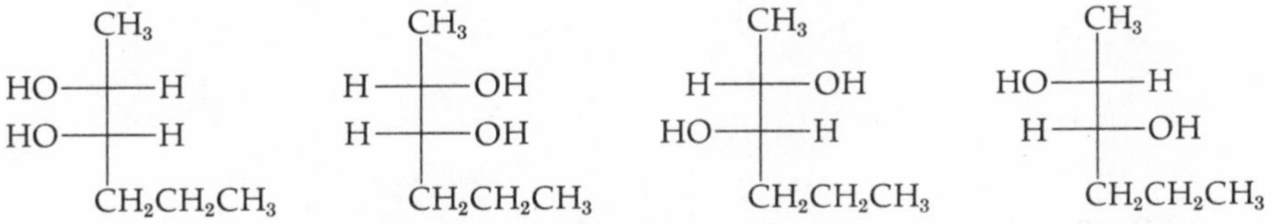

Enantiomer pairs: I and II; III and IV
Diastereomer pairs: I and III; I and IV; II and III; II and IV

(d) 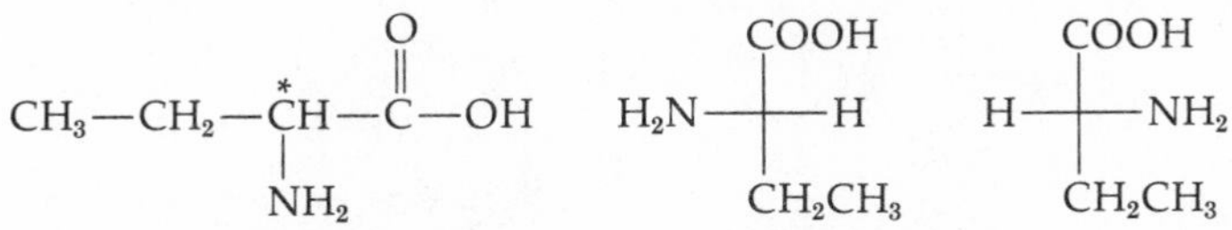 enantiomers

8. (a)

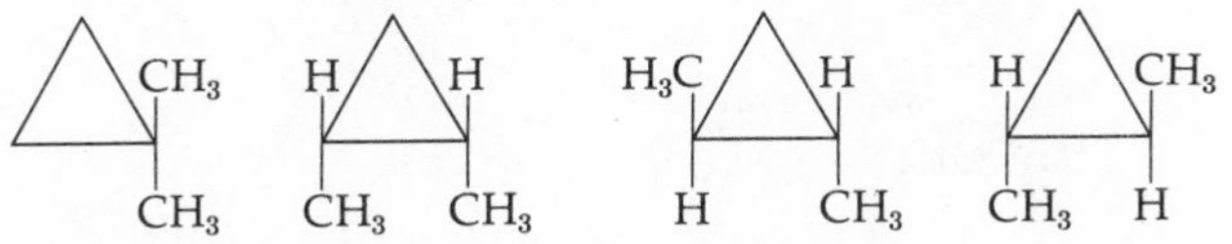

(b) Structural isomers: I and II; I and III; I and IV
    Stereoisomers: II and III; II and IV; III and IV
(c) III and IV are enantiomers.
    II and III, or II and IV, are diastereomers.

9. In linearly polarized light, the electric field is oriented in a single direction. Unpolarized light is a combination of electromagnetic waves with the electric field in all directions in the plane perpendicular to the direction of propagation.

10. You could measure the rotation of a solution of each isomer with a polarimeter. You could then determine which isomer is in which bottle from the results of this experimental test and the fact that D-glyceraldehyde is (+) and L-glyceraldehyde is (−).

CHAPTER 20

1. (a)

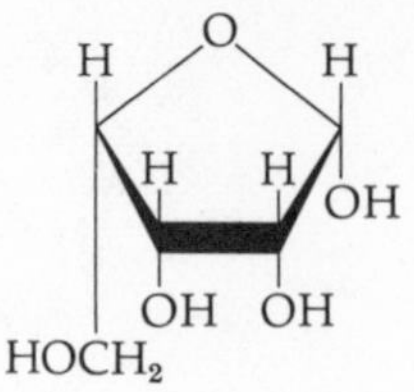

(b)

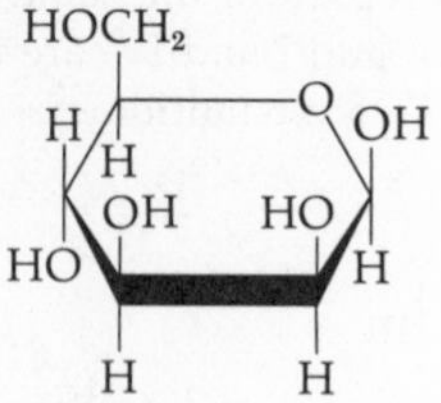

(c)

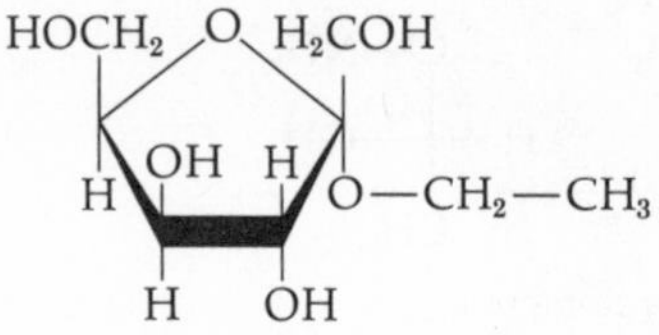

2. (a)

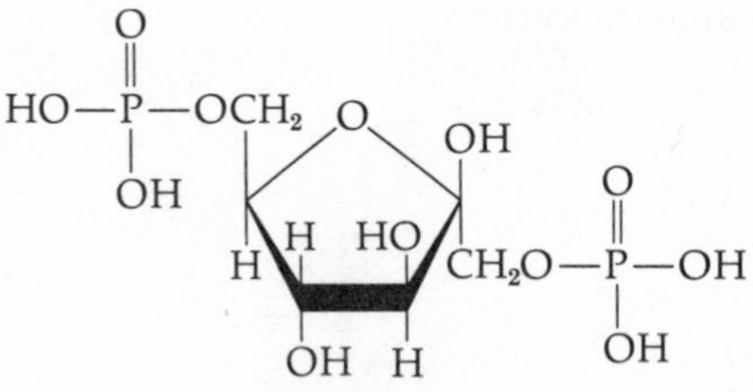

(b)

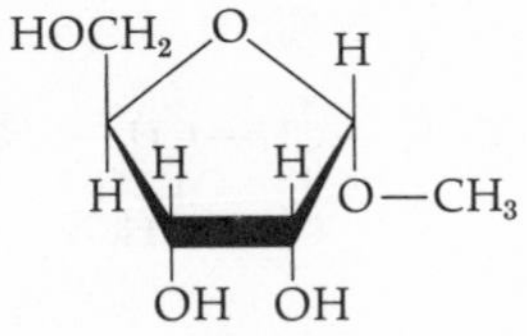

3.

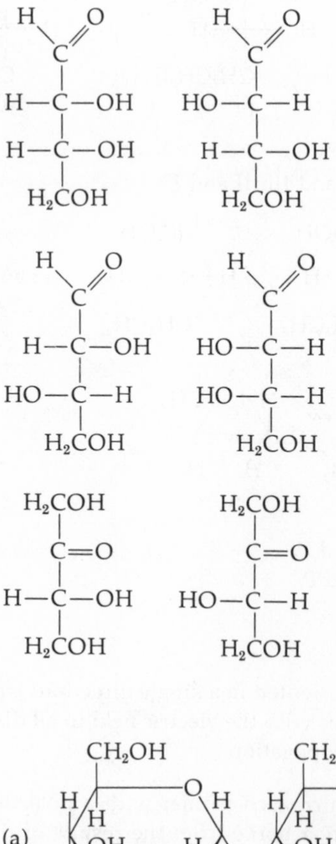

4. (a)

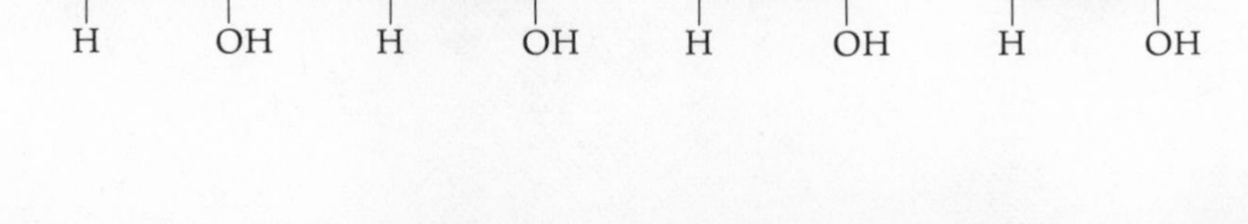

(b)

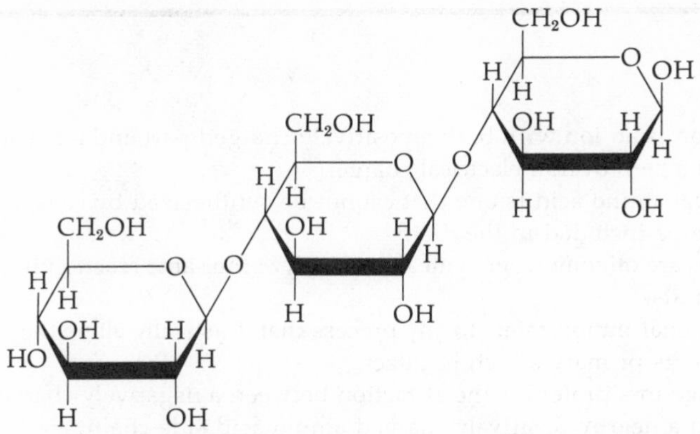

5. The carbon atom in the hemiacetal, hemiketal, acetal, or ketal form of a sugar that is the carbonyl carbon in the open-chain form of the sugar is called an anomeric carbon.

6. maltose:

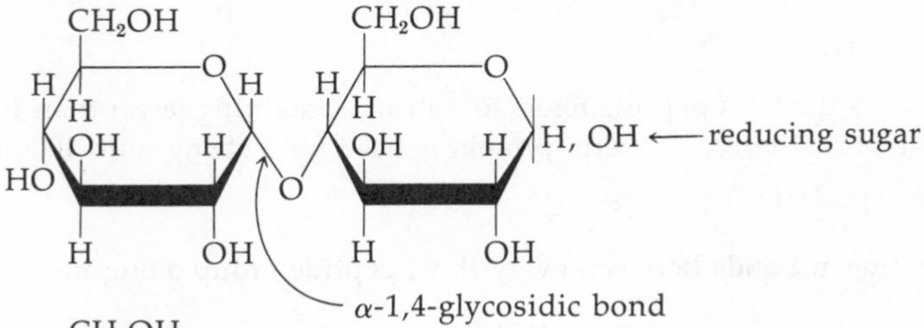

H, OH ←—— reducing sugar

α-1,4-glycosidic bond

sucrose:

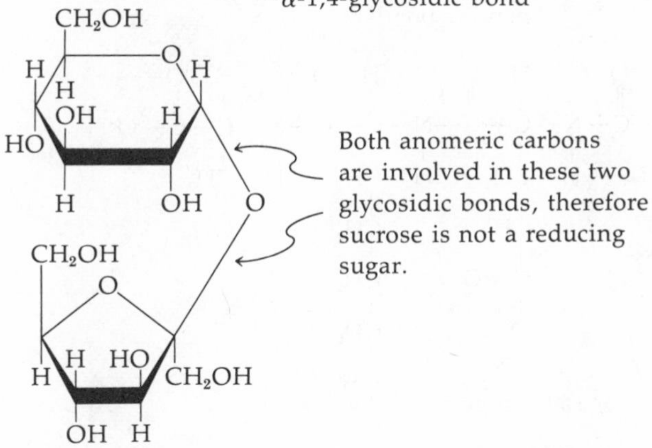

Both anomeric carbons are involved in these two glycosidic bonds, therefore sucrose is not a reducing sugar.

7. (a)

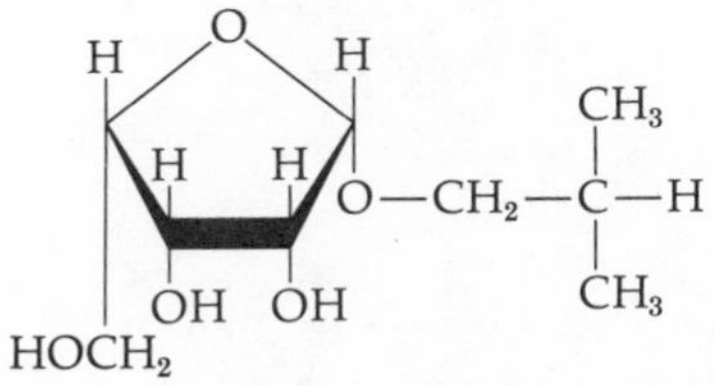

(b) N.R.

8.

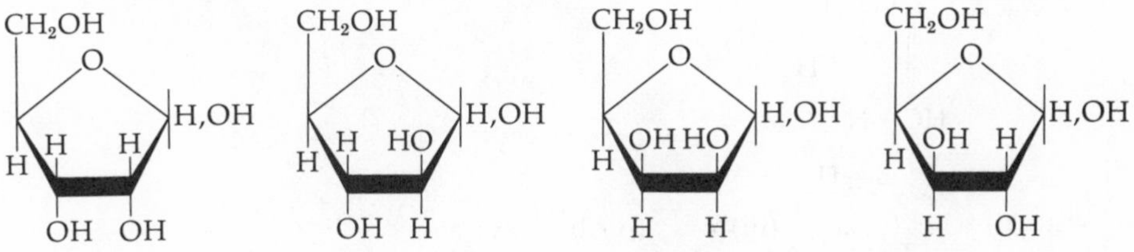

9. In addition to D-glyceraldehyde, they are

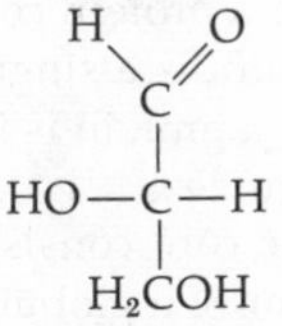

L-glyceraldehyde   and   dihydroxyacetone

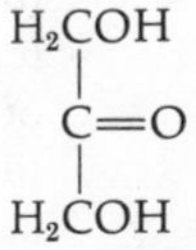

CHAPTER 21

1. (a) A zwitterion is an ion with both a positively charged part and a negatively charged part, but with a zero overall electrical charge.

   (b) An essential amino acid is one that cannot be synthesized by cells in the body and must therefore be included in the diet.

   (c) Isoenzymes are oligomeric enzymes that catalyze the same reaction but have slightly different subunits.

   (d) The term denaturation refers to any process that drastically alters the shape of a protein but leaves its primary structure intact.

   (e) A salt bridge in a protein is the attraction between a negatively charged amino acid side chain and a nearby positively charged amino acid side chain.

2. 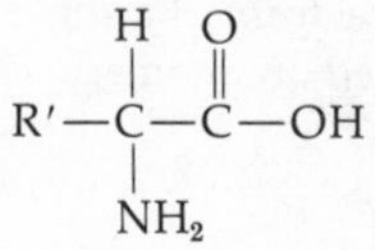
$$R'-\overset{\overset{\displaystyle H}{|}}{\underset{\underset{\displaystyle NH_2}{|}}{C}}-\overset{\overset{\displaystyle O}{\|}}{C}-OH$$

3. Usually the term peptide refers to a string containing fewer than 100 amino acids joined together, whereas the term protein is used for a string with 100 or more amino acids.

4. (a) +1   (b) −1   (c) −2

5. Hydrogen bonds between every third peptide group along the amino acid chain

6.

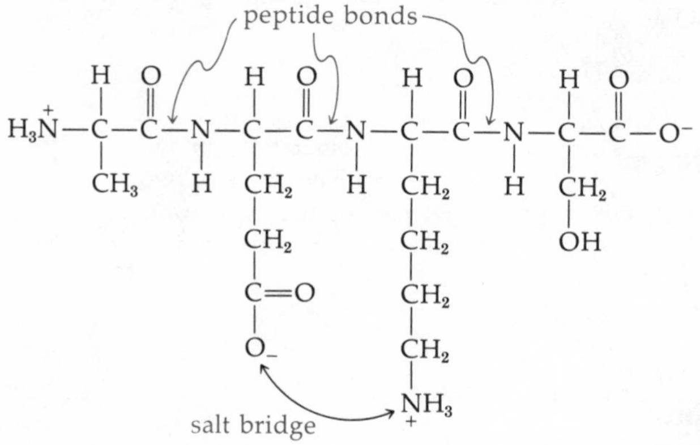

7. Tertiary and quaternary

8. (d) A drastic increase in temperature

9.

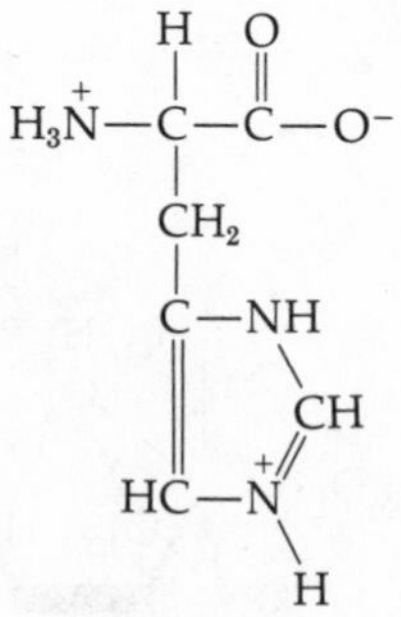

10. (i) a   (ii) a   (iii) c   (iv) b   (v) a

11. (a) An oligomer is a protein composed of subunits.

    (b) A protein subunit is a single polypeptide chain component of an oligomer.

    (c) Coagulation of a protein is its irreversible denaturation that results in the formation of an insoluble complex.

    (d) A hydrophobic core consists of tightly packed side chains of hydrophobic amino acids in the interior of a globular protein.

(e) A β-pleated sheet is a zigzag secondary structural element involving hydrogen bonds between peptide groups of two or more chains of amino acids or two or more portions of the same chain.

12. Disulfide bonds are covalent bonds formed between cysteine side chains, which need not be near each other in the primary structure of a protein. They stabilize tertiary and sometimes quaternary structure.

13.

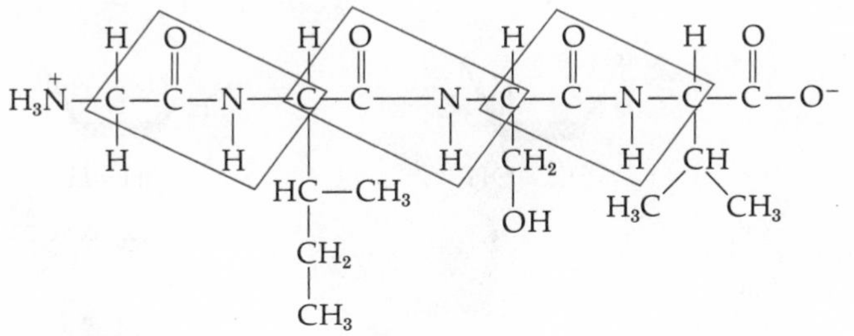

(a) The atoms enclosed by each of the three rectangles lie in a common plane.
(b) +1

14. (d) None of these

CHAPTER 22

1. (c) Active transport

2. (a), (b), (c), and (d)

3. (b) Albumins

4. (d) An enzyme

5. (c) Coenzyme requirement and (e) Water solubility

6. About 150,000; two antigen binding sites and one F_c part.

CHAPTER 23

1. W is an allosteric inhibitor of the enzyme E_1.

2. Molar activities: A = 4.0×10^3 min^{-1}, B = 6.0×10^3 min^{-1}. B is therefore more efficient.

3. 2.4×10^{-2} M/min

4. Hydrolases

5. The substrate, upon binding, induces the appropriate conformational change in the enzyme to allow for catalysis.

6. Lysine, arginine, and/or histidine can form ionic interactions with the negatively charged phosphate groups. Serine, threonine, tyrosine, or other polar amino acids can form hydrogen bonds with the hydroxyl groups on the ribose portion of ATP or the nitrogen atoms of the ring. Phenylalanine, tryptophan, and other amino acids with nonpolar side chains can form hydrophobic interactions with the nitrogen-containing ring of ATP.

CHAPTER 24

1.

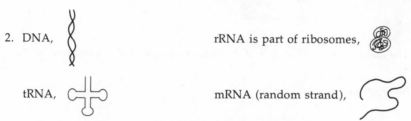

ATP   deoxycytidine

2. DNA,   rRNA is part of ribosomes,

tRNA,   mRNA (random strand),

3. Initiation: Step 1—Binding of met-tRNA$_{met}$ to mRNA and the smaller subunit
                Step 2—Binding of the larger subunit to form the complete complex
   Elongation: Steps 3, 4, 5, 6, etc.
   Termination: Step 8 (not shown in Figure 24-13) involves release of the completed pro-
                tein when a chain-termination codon is encountered.

4. Repression. For drawing see Figure 24-15. The end product in this case is histidine. An
   accumulation of histidine leads to repression.

5. DNA polymerase must recognize and use deoxyribonucleotides involving adenine,
   guanine, cytidine, and thymine, whereas RNA polymerase uses ribonucleotides involving adenine, guanine, cytosine, and uracil.

CHAPTER 25

1. The two steps are step 7, catalyzed by phosphoglycerate kinase, and step 10, catalyzed
   by pyruvate kinase.

2. (a) Products, CO_2 and H_2O; 19 ATP   (b) Product, lactic acid; 8 ATP
   (c) Products, CO_2 and H_2O; 15 ATP   (d) Product, oxaloacetic acid; 5 ATP

3. All six end up in CO_2 molecules. The decarboxylation of two pyruvic acid molecules in
   the reaction catalyzed by pyruvate dehydrogenase accounts for the two carbon atoms
   that do not enter the TCA cycle as part of the two molecules of acetyl-S-CoA.

4. Both pyruvate dehydrogenase and α-ketoglutarate dehydrogenase catalyze the decarboxylation of substrates that are α-ketoacids and produce thioesters. The glyceraldehyde
   3-phosphate dehydrogenase reaction does not involve the decarboxylation of an α-
   ketoacid, but rather the formation of a phosphate ester. All three reactions are coupled to
   the formation of NADH from NAD$^+$.

5. $C_3H_6O_3 + 3O_2 \rightarrow 3CO_2 + 3H_2O$. Eighteen moles of ATP are produced.

6. The hydroxyl group of citric acid is attached to a tertiary carbon atom and cannot be oxidized (see Chapter 15).

7. Cyanide blocks the operation of the ETS. Thus NADH will build up, the supply of NAD^+ will become exhausted, and the TCA cycle will not continue to function.

8. (a) $FADH_2$ transfers its electrons to the ETS after the first site of coupling of the ETS with ATP synthesis.
(b) No. Both form one molecule of water per coenzyme molecule.

9. The two NADH molecules produced in the cytoplasm during glycolysis are shuttled to form two $FADH_2$ molecules in muscle mitochondria, whereas they are shuttled to form two NADH molecules in liver mitochondria. An input of two $FADH_2$ molecules into the ETS yields four ATP molecules, whereas an input of two NADH molecules into the ETS yields six ATP molecules.

10. (a)

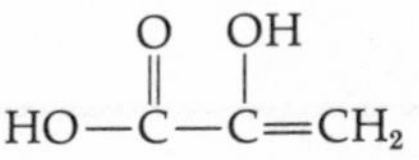

$$\begin{array}{cc} O & OH \\ \parallel & \mid \\ HO-C-C&=CH_2 \end{array}$$

(b)

$$\begin{array}{c} O \\ \parallel \\ C-H \\ \mid \\ P_i + H-C-OH \\ \mid \\ H_2C-O-\textcircled{P} \end{array}$$

(c)

$$\begin{array}{c} O \\ \parallel \\ C-H \\ \mid \\ H-C-O-\textcircled{P} \\ \mid \\ H_2C-OH \end{array}$$

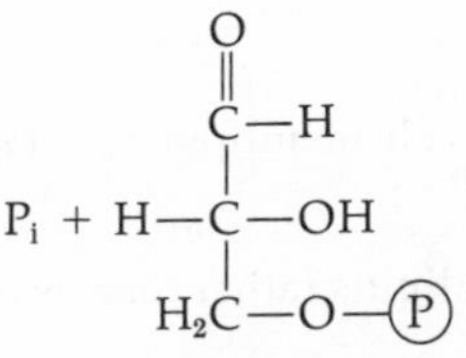

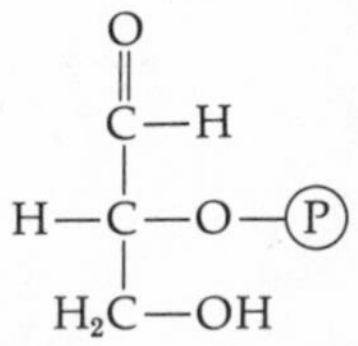

11. 3.8 kcal/g, 1.6 kcal/g in muscle cells

12. **(Optional)** Yes. A component of RNA nucleotides, ribose 5-phosphate, is produced.

13. Oxaloacetic acid.

14. No. The enzyme that galactosemic children cannot produce is also lacking when these children become adults.

15. Lactose catabolism and lactose synthesis involve UDP-galactose. Glycogen synthesis and lactose catabolism involve UDP-glucose.

16. Epinephrine $\xrightarrow{\text{allosteric}}$ adenyl cyclase $\longrightarrow$ cyclic AMP

cyclic AMP $\xrightarrow{\text{allosteric}}$ protein kinase $\xrightarrow[\text{modification}]{\text{chemical}}$ glycogen synthetase and glycogen phosphorylase

17. (a) two (b) seven

18. Many sugarholics produce insulin at elevated levels. They may develop hypoglycemia if their pancreas continues to produce insulin at a high rate even though their sugar intake has decreased.

19. Overproduction of insulin in response to the ingested donuts (and possibly sugar in the coffee) may cause a temporary drop in the blood sugar level.

20. 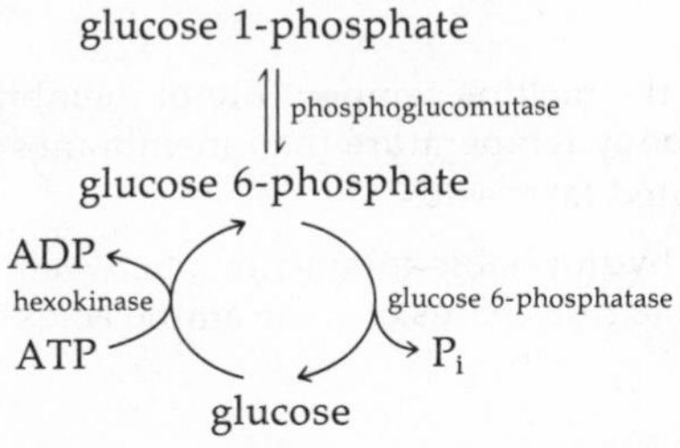   glucose 1-phosphate
       $\updownarrow$ phosphoglucomutase
       glucose 6-phosphate

ADP $\xleftarrow{}$ hexokinase, glucose 6-phosphatase $\xrightarrow{}$ P_i
ATP $\xrightarrow{}$ glucose

21. (a) $CH_3-\overset{\overset{\displaystyle O}{\|}}{C}-S-CoA + NADH + H^+ + CO_2$

(b) $HO-\overset{\overset{\displaystyle O}{\|}}{C}-CH_2-\overset{\overset{\displaystyle O}{\|}}{C}-\overset{\overset{\displaystyle O}{\|}}{C}-OH + ADP + P_i$

(c) $CH_3-\overset{\overset{\displaystyle HO}{|}}{\underset{\underset{\displaystyle H}{|}}{C}}-\overset{\overset{\displaystyle O}{\|}}{C}-OH + NAD^+$

CHAPTER 26

1. Lipids can function as hormones, components of cell membranes, vitamins, and for energy storage and insulation.

2. They exist almost entirely as components of complex lipids rather than as free fatty acids.

3. $\begin{array}{l} H_2C-O-\overset{\overset{\displaystyle O}{\|}}{C}-(CH_2)_{10}-CH_3 \\ \;\;|\;\;\;\;\;\;\;\;\;\overset{\displaystyle O}{\|} \\ H-C-O-\overset{}{C}-(CH_2)_{10}-CH_3 \\ \;\;|\;\;\;\;\;\;\;\;\;\overset{\displaystyle O}{\|} \\ H_2C-O-\overset{}{C}-(CH_2)_{10}-CH_3 \end{array}$

4. $\begin{array}{l} H_2C-OH \\ \;\;| \\ H-C-OH \\ \;\;| \\ H_2C-OH \end{array}$ and $CH_3-(CH_2)_{10}-\overset{\overset{\displaystyle O}{\|}}{C}-O^-Na^+$

5. Micelles that can be washed away are formed with the long hydrophobic alkyl groups of the soap interacting with the hydrocarbon molecules, and the charged carboxylate groups of the soap interacting with water molecules.

6. (a) Abnormal membrane composition and therefore abnormal membrane function
   (b) Inability to manufacture prostaglandins

7.

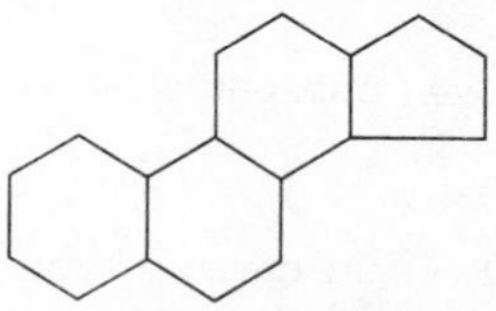

8. They competitively inhibit the formation of proconvertin. These rat poisons cause death as a result of uncontrolled internal bleeding.

9. Siberian children frequently do not receive adequate exposure to sunlight, which catalyzes the formation of vitamin D from precursors in human skin. Sunlamp treatments could be one solution to this problem.

10. Unsaturated fatty acid components lower the melting temperature of membranes, thus giving them more fluid characteristics at body temperature than membranes composed exclusively of higher-melting-point saturated fatty acids.

11. Lipoproteins are held together mainly by hydrophobic interactions between their nonpolar lipid components and the hydrophobic side chains of some amino acids of the protein component.

CHAPTER 27

1.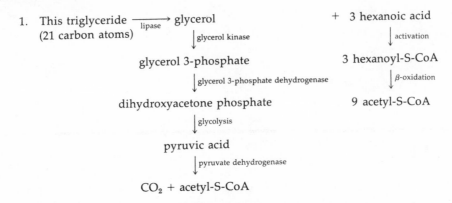

   This triglyceride $\xrightarrow{\text{lipase}}$ glycerol   + 3 hexanoic acid
   (21 carbon atoms)

   glycerol $\downarrow$ glycerol kinase   activation $\downarrow$

   glycerol 3-phosphate   3 hexanoyl-S-CoA

   $\downarrow$ glycerol 3-phosphate dehydrogenase β-oxidation $\downarrow$

   dihydroxyacetone phosphate   9 acetyl-S-CoA

   $\downarrow$ glycolysis

   pyruvic acid

   $\downarrow$ pyruvate dehydrogenase

   CO_2 + acetyl-S-CoA

   Thus, 20 of the carbon atoms become part of acetyl groups, whereas one carbon atom from the catabolism of glycerol becomes part of a CO_2 molecule.

2.

$$
\begin{array}{l}
H_2C-O-\overset{\overset{\text{O}}{\|}}{C}-(CH_2)_4-CH_3 \\
\quad\quad\quad\quad\text{O} \\
H-\overset{\overset{}{|}}{C}-O-\overset{\overset{\text{O}}{\|}}{C}-(CH_2)_4-CH_3 + 3H_2O \xrightarrow[\text{lipase}]{\text{pancreatic}} \\
\quad\quad\quad\quad\text{O} \\
H_2C-O-\overset{\overset{\text{O}}{\|}}{C}-(CH_2)_4-CH_3
\end{array}
$$

$$
\begin{array}{l}
H_2C-OH \\
H-C-OH \\
H_2C-OH \\
+ \\
3\left[HO-\overset{\overset{\text{O}}{\|}}{C}-(CH_2)_4-CH_3 \right]
\end{array}
$$

3. 154 moles of ATP

4. The active "business end" of both coenzyme A and the acyl carrier portion of fatty acid synthetase are identical, ending with a sulfhydryl group, —SH, which combines with a carboxylic acid to form a thioester. The acyl carrier portion of fatty acid synthetase is used exclusively for fatty acid biosynthesis, whereas coenzyme A is involved in a variety of catabolic pathways.

5. They are intermediates that are converted to diglycerides and then to phosphoglycerides.

6. Both are synthesized from nucleoside diphosphate-activated precursors.

7. See Figure 27-8. Fatty acid synthesis from acetyl-S-CoA requires NADPH, which is provided by the pentose phosphate pathway and the pathway involving malic enzyme. Malic enzyme is allosterically activated by the precursor of fatty acids, acetyl-S-CoA. Other controls of fatty acid biosynthesis include allosteric inhibition of isocitrate dehydrogenase by excess ATP, which shunts acetyl-S-CoA to fatty acid synthesis instead of into the TCA cycle; allosteric activation of the enzyme acetyl-S-CoA carboxylase by excess TCA cycle intermediates; and inhibition of both acetyl-S-CoA carboxylase and fatty acid synthetase by excess palmitoyl-S-CoA. The synthesis of triglycerides from fatty acids is stimulated by insulin, whereas the hydrolysis of triglycerides to fatty acids is stimulated by epinephrine and glucagon.

8. (a) CO_2 + $CH_3-\overset{\overset{\text{O}}{\|}}{C}-\overset{\overset{\text{O}}{\|}}{C}-OH$

   (b) $HO-\overset{\overset{\text{O}}{\|}}{C}-CH_2-\overset{\overset{\text{O}}{\|}}{C}-S-CoA$

(c) $CH_3-\overset{\overset{\displaystyle O}{\|}}{C}-CH_2-\overset{\overset{\displaystyle O}{\|}}{C}-S-CoA$

(d) $\begin{matrix} H_2C-OH \\ | \\ H-C-OH \\ | \\ H_2C-OH \end{matrix}$ $+ R_1-\overset{\overset{\displaystyle O}{\|}}{C}-OH + R_2-\overset{\overset{\displaystyle O}{\|}}{C}-OH + R_3-\overset{\overset{\displaystyle O}{\|}}{C}-OH$

CHAPTER 28

1. Protein digestion is the hydrolysis of ingested protein in the digestive tract, whereas protein turnover involves the hydrolysis of body proteins in cells and body fluids.

2. (a) Transamination of pyruvic acid, $CH_3-\overset{\overset{\displaystyle O}{\|}}{C}-\overset{\overset{\displaystyle O}{\|}}{C}-OH$, yields alanine.

   (b) Transamination of oxaloacetic acid, $HO-\overset{\overset{\displaystyle O}{\|}}{C}-CH_2-\overset{\overset{\displaystyle O}{\|}}{C}-\overset{\overset{\displaystyle O}{\|}}{C}-OH$, yields aspartic acid.

   (c) Transamination of α-ketoglutaric acid, $HO-\overset{\overset{\displaystyle O}{\|}}{C}-CH_2-CH_2-\overset{\overset{\displaystyle O}{\|}}{C}-\overset{\overset{\displaystyle O}{\|}}{C}-OH$, yields glutamic acid.

   (d) Asparagine is produced from aspartic acid, $HO-\overset{\overset{\displaystyle O}{\|}}{C}-CH_2-\overset{\overset{\overset{\displaystyle H}{|}}{\underset{\underset{\displaystyle NH_2}{|}}{C}}}{}-\overset{\overset{\displaystyle O}{\|}}{C}-OH$, and ammonia by asparagine synthetase.

3. Aspartic acid provides one of the amino groups that will end up in urea. Aspartic acid is regenerated by the conversion of fumaric acid, produced in the urea cycle, to oxaloacetic acid by TCA cycle enzymes, followed by transamination of this oxaloacetic acid with glutamic acid serving as the amine donor.

4. Both protein and nucleic acid polymers are hydrolyzed to their respective monomeric components.

5. (a) $CH_3-CH_2-\overset{\overset{\displaystyle O}{\|}}{C}-\overset{\overset{\displaystyle O}{\|}}{C}-OH + NH_3$

   (b) $H_2N-CH_2-CH_2-CH_2-CH_2-CH_2-NH_2 + CO_2$

   (c) $\begin{matrix} H_3C \\ \diagdown \\ CH-\overset{\overset{\displaystyle O}{\|}}{C}-\overset{\overset{\displaystyle O}{\|}}{C}-OH \\ \diagup \\ H_3C \end{matrix}$ $+ HO-\overset{\overset{\displaystyle O}{\|}}{C}-CH_2-CH_2-\overset{\overset{\overset{\displaystyle H}{|}}{\underset{\underset{\displaystyle NH_2}{|}}{C}}}{}-\overset{\overset{\displaystyle O}{\|}}{C}-OH$

   (d) $H_2N-CH_2-CH_2-CH_2-CH_2-\overset{\overset{\displaystyle O}{\|}}{C}-\overset{\overset{\displaystyle O}{\|}}{C}-OH + NH_3 + H_2O_2$

6. The amino group of pyridoxamine phosphate is donated to an α-keto acid, which is usually α-ketoglutaric acid, thus regenerating pyridoxal phosphate.

CHAPTER 29

1. (a) Increased hydrolysis of triglycerides by lipases, increased breakdown of glycogen to glucose 1-phosphate, and decreased conversion of glucose to glycogen
   (b) Glycolysis will be accelerated, whereas gluconeogenesis will be inhibited.
   (c) Lactic acid
   (d) Gluconeogenesis in liver cells will convert most of it to glucose.

2. (a) Without a supply of essential fatty acids, you will lack some components of cell membranes and you will be unable to produce prostaglandins.
   (b) You will not be getting required vitamins, minerals, or essential fatty acids.
   (c) None. Humans synthesize cholesterol. It is not needed in the diet.

3. (a) It will decrease as you exhale rapidly.
   (b) It might increase.

4. (a) +
   (b) +
   (c) +
   (d) −

5. Yes

6. (b)

7. (a) and (c)

ILLUSTRATION ACKNOWLEDGMENTS

Chapter 13
p. 344   Van Bucher/Photo Researchers

Chapter 14
p. 362   John Bryson/Photo Researchers
p. 366   Georg Gerster/Photo Researchers
p. 375   Michael Hayman/Photo Researchers

Chapter 15
p. 386   Novosti/Sovfoto

Chapter 16
p. 400   Photoworld/FPG

Chapter 17
p. 414   E.I. du Pont de Nemours & Co.

Chapter 18
p. 428   Bruce Coleman

Chapter 19
p. 448   Jeanloup Sieff/Photo Researchers

Chapter 20
p. 476   Henri Cartier-Bresson/Magnum
p. 487   Georg Gerster/Photo Researchers
p. 492   Henri Cartier-Bresson/Magnum

Chapter 21
p. 498   Carson Baldwin, Jr./FPG
p. 512   Figure 21-10 after Albert L. Lehninger, *Biochemistry*, 2nd ed., Worth Publishers, Inc., New York, 1975
p. 516   Figure 21-17 adapted from Helena Curtis, *Biology*, 2nd ed., Worth Publishers, Inc., New York, 1975
p. 518   Figure 21-21 adapted from Richard E. Dickerson in H. Neurath, ed., *The Proteins*, Academic Press, Inc., New York, 1964

Chapter 22
p. 528   Philip Harrington/Peter Arnold
p. 537   EPA
p. 544   Rockefeller University
p. 547   Figure 22-10 after E.W. Silverton, M.A. Navia, and D.R. Davies, *Proceedings of the National Academy of Sciences,* **74:** 5140, 1977

Chapter 23
p. 552   Bruce Coleman
p. 563   Figure 23-6 after Lehninger, *op. cit.*
p. 566   Figure 23-10 after C.M. Anderson, F.H. Zucker, and T. A. Steitz, *Science* **204:** 375, 1979. Copyright 1979 by American Association for the Advancement of Science

Chapter 24
p. 576   National Foundation of the March of Dimes
p. 582   Figure 24-4 after L. Pauling and R.B. Corey, *Archives of Biochemistry and Biophysics,* **65:** 164, 1956
p. 584   Figure 24-6 (a) from J.D. Watson and F.H.C. Crick, *Nature,* **171:** 737, 1953; 24-6 (b) from *DNA Synthesis* by Arthur Kornberg, W.H. Freeman and Company, copyright © 1974
p. 585   Figure 24-7 (b) reproduced, with permission, from C.G. Kurland, *Annual Review of Biochemistry,* Volume 46, © 1977 by Annual Reviews Inc.; 24-7 (c, left) adapted from S.H. Kim *et al., Science,* **185:** 435, 1974, copyright 1974 by the American Association for the Advancement of Science; 24-7 (c, right) redrawn from S.H. Kim *et al., Proceedings of the National Academy of Sciences,* **71:** 4970, 1974
p. 588   O.H. Miller, Jr., and Barbara R. Beatty, Biology Division, Oak Ridge National Laboratory
p. 592   Figure 24-12 adapted from Kim *et al., op. cit.*
p. 593   Figure 24-13 adapted from Kurland, *op. cit.*
p. 594   Figure 24-14 from O.L. Miller, Jr., Barbara A. Hamkalo, and C.A. Thomas, Jr., *Science,* **169:** 392, 1970. Copyright 1970 by the American Association for the Advancement of Science

Chapter 25
p. 606   Victor Aleman/Photoworld
p. 620   Figure 25-9 adapted from Keith R. Porter in Helena Curtis, *Biology,* 3rd ed., Worth Publishers, Inc., New York, 1979
p. 628   Figure 25-14 adapted from "How Cells Make ATP" by Peter C. Hinkle and Richard E. McCarty. Copyright © 1978 by Scientific American Inc. All rights reserved
p. 649   Becton/Dickinson

Chapter 26
p. 654   Malak/Annan Photo Features/Photo Trends
p. 664   American Heart Association
p. 669   John S. O'Brien
p. 671   Figure 26-11 after S.J. Singer and G.L. Nicolson, *Science,* **175:** 720, 1972. Copyright 1972 by the American Association for the Advancement of Science

Chapter 27
p. 674   Burt Glinn/Magnum

Chapter 28
p. 694   Denis Callewaert
p. 705   George Rodger/Magnum

Chapter 29
p. 716   Burt Glinn/Magnum
p. 729   S. Treval/D.B./Bruce Coleman

Essential Skills
p. 738   Charles Harbutt/Magnum

INDEX

The page on which a term is defined, or on which the structural formula for a compound is given, is indicated in **boldface** type.